I0473600

Introduction to Civil Engineering

This comprehensive new textbook bridges the gap between academic knowledge and professional practice in civil engineering, connecting traditionally separate course units into a cohesive whole that teaches readers to 'think like a civil engineer.'

This book traces civil engineering's evolution through influential historical figures while exploring how technology transforms the field. Across ten chapters, it covers engineering fundamentals, ethics, material science, biomimicry applications, mathematical modeling, design tools, and professional development. Readers gain practical understanding of everything from equation formulation and force analysis to software applications and construction management, with a special emphasis on the iterative nature of design and the transition from theoretical knowledge to real-world applications.

Written primarily for civil engineering students transitioning to industry and early-career professionals seeking to master the profession, this guide also serves as a valuable resource for educators teaching civil engineering courses and experienced professionals requiring a refresher on fundamental principles that unite the various branches of this constantly evolving field.

Introduction to Civil Engineering
A Guide for Students and Professionals

Patrick Ssempeera

CRC Press
Taylor & Francis Group
Boca Raton London New York

CRC Press is an imprint of the
Taylor & Francis Group, an **informa** business

Designed cover image: Shutterstock

MATLAB® and Simulink® are trademarks of The MathWorks, Inc. and are used with permission. The MathWorks does not warrant the accuracy of the text or exercises in this book. This book's use or discussion of MATLAB® or Simulink® software or related products does not constitute endorsement or sponsorship by The MathWorks of a particular pedagogical approach or particular use of the MATLAB® and Simulink® software.

First edition published 2026
by CRC Press
4 Park Square, Milton Park, Abingdon, Oxon, OX14 4RN

and by CRC Press
2385 NW Executive Center Drive, Suite 320, Boca Raton FL 33431

© 2026 Patrick Ssempeera

CRC Press is an imprint of Informa UK Limited

British Library Cataloguing-in-Publication Data
A catalogue record for this book is available from the British Library

Library of Congress Cataloging-in-Publication Data
Names: Ssempeera, P. author
Title: Introduction to civil engineering : a guide for students and professionals / Patrick Ssempeera.
Description: First edition. | Abingdon, Oxon ; Boca Raton, FL : CRC Press, 2026. | Includes bibliographical references and index.
Identifiers: LCCN 2025038908 (print) | LCCN 2025038909 (ebook) | ISBN 9781032911519 hbk | ISBN 9781032911502 pbk | ISBN 9781003561620 ebk
Subjects: LCSH: Civil engineering—Textbooks
Classification: LCC TA147 .S74 2026 (print) | LCC TA147 (ebook)
LC record available at https://lccn.loc.gov/2025038908
LC ebook record available at https://lccn.loc.gov/2025038909

ISBN: 9781032911519 (hbk)
ISBN: 9781032911502 (pbk)
ISBN: 9781003561620 (ebk)

DOI: 10.1201/9781003561620

Typeset in Times
by codeMantra

To my dear mother, Betty Serina Nakimera, whose unwavering wisdom and intelligence have been a constant source of inspiration and guidance.

Contents

Foreword   xix
Preface   xx
Acknowledgments   xxii
Author Biography   xxiii

1 What Is Civil Engineering?   **1**
1.1   Introduction   1
1.2   History of Civil Engineering   1
    1.2.1   Ancient Civil Engineering Structures   2
1.3   STEM Courses for an Aspiring Civil Engineer   3
    1.3.1   Primary Education   3
    1.3.2   Secondary Education   3
    1.3.3   Tertiary Education   4
1.4   How to Become a Professional Civil Engineer   5
1.5   Inculcating Lifelong Interdisciplinary Learning Skills   5
    1.5.1   Civil Engineering Students   7
    1.5.2   Professional Engineers   7
1.6   Branches of Civil Engineering   8
    1.6.1   Transportation Engineering   8
    1.6.2   Water Resources Engineering   9
    1.6.3   Environmental Engineering   9
    1.6.4   Geotechnical Engineering   10
    1.6.5   Structural Engineering   11
    1.6.6   Municipal Engineering   11
    1.6.7   Construction Engineering   13
    1.6.8   Civil Engineering Surveying   13
1.7   The Institution of Civil Engineers, United Kingdom   13
    1.7.1   What Does ICE Mean to the World of Civil Engineering?   14
1.8   American Society of Civil Engineers (ASCE)   14
1.9   Notable Civil Engineers in History   14
    1.9.1   John Smeaton (1724–1792)   14
    1.9.2   Thomas Tredgold (1788–1829)   15
    1.9.3   Thomas Telford (1757–1834)   15
    1.9.4   Isambard Kingdom Brunel (1806–1859)   16
    1.9.5   Sir Joseph Bazalgette (1819–1891)   16
    1.9.6   Benjamin Wright (1770–1842)   17
    1.9.7   Archie A. McCorkindale (1923–2018)   17
    1.9.8   Terzaghi Karl (1883–1963)   17
    1.9.9   Elsie Eaves (1898–1983)   18
    1.9.10   Dr.Tokujiro Yoshida (1888–1950)   18
    1.9.11   Sir Mokshagundam Visvesvaraya (1861–1962)   19
    1.9.12   William Hunter Dammond (1873–1956)   19
    1.9.13   Mao Yi-sheng (1896–1989)   19
    1.9.14   Dr. Albert Isimbwa Rugumayo (1958–2018)   20
1.10   Conclusion   21
References   21

2 Philosophy of Civil Engineering   **23**
2.1   Introduction   23
2.2   Primary Civil Engineer's Attributes   23
    2.2.1   Stewardship   23
    2.2.2   Service to Humanity   24
    2.2.3   Health, Safety, and Risk Management   24

		2.2.4	Sustainability	25
		2.2.5	Harnessing New and Advancing Technology	25
		2.2.6	Innovation and Progress	26
		2.2.7	Collaboration and Teamwork	26
		2.2.8	Ethical Practice	26
		2.2.9	Continuous Learning	27
		2.2.10	Support Others to Grow Professionally	27
		2.2.11	Effective Communication and Interpersonal Networking	27

2.3 The Role of Philosopher, Engineer, Technologist, and Technician — 27
  2.3.1 Overview — 27
  2.3.2 Who Is a Philosopher? — 27
  2.3.3 Who Is an Engineer? — 28
  2.3.4 Who Is a Technologist? — 30
  2.3.5 Who Is a Technician? — 30
2.4 Engineering Ethics — 31
  2.4.1 Advance of engineering ethics — 32
  2.4.2 Ethical code of conduct — 32
  2.4.3 Typical moral dilemmas faced by civil engineers in history — 32
  2.4.4 Trolley problem — 35
2.5 Prominent Figures in Civil Engineering — 36
  2.5.1 Leonardo da Vinci (1452–1519) — 36
  2.5.2 Karl von Terzaghi (1883–1963) — 42
  2.5.3 Henry Gantt (1861–1919) — 42
  2.5.4 Henri Fayol (1841–1925) — 43
2.6 Civil Engineering Management — 44
  2.6.1 Key Attributes of Civil Engineering Management — 45
  2.6.2 Principles of Management — 46
  2.6.3 Ms of Management — 49
2.7 Civil Engineering Impact on the Society and Environment — 49
  2.7.1 Improved Infrastructure — 49
  2.7.2 Fostering Economic Growth and Development — 49
  2.7.3 Saving Lives — 49
  2.7.4 Disaster Mitigation — 50
  2.7.5 Resilient and Sustainable Communities — 50
2.8 Mind Mapping Question — 51
2.9 Conclusion — 51
Notes — 51
References — 51

3 Civil Engineering Technology — **53**
3.1 Introduction — 53
  3.1.1 Staying Ahead of the Curve: Why Civil Engineers must Embrace Emerging Technologies — 54
  3.1.2 Engineering versus Technology — 54
3.2 Technology, Codes, and Standards of Practice — 57
  3.2.1 Evolution of Codes and Standards — 59
  3.2.2 Role of Standards and Codes — 59
  3.2.3 Selected Civil Engineering Standards — 59
  3.2.4 Selected Civil Engineering Codes — 64
  3.2.5 How Does Technology Shape Codes and Standards and Vice Versa? The Engineer's Role — 65
  3.2.6 AI and ML Advancing Civil Engineering Standards and Codes — 66
3.3 Scientific Laws That Have Advanced Civil Engineering Technology — 67
  3.3.1 Sir Isaac Newton (1643–1727) — 68
  3.3.2 Robert Hooke (1635–1702) — 69
  3.3.3 The Law of Conservation of Energy — 70
  3.3.4 Typical Civil Engineering Technologies at the Peak of Scientific Laws — 77
3.4 Systems of Units of Measurement — 83
  3.4.1 Evolution of Systems of Units of Measurement — 83
  3.4.2 Imperial Units — 84
  3.4.3 Metric System — 84
  3.4.4 Conversion Factors — 85
  3.4.5 Measurement of Angles — 85

3.5   Civil Engineering Software Technologies   86
   3.5.1   Structural Analysis and Design   86
   3.5.2   Building Information Modeling (BIM)   86
   3.5.3   Computer-Aided Design (CAD)   86
   3.5.4   Geotechnical and Foundation Engineering   86
   3.5.5   Transportation Engineering   86
   3.5.6   Hydrology and Hydraulic Engineering   86
   3.5.7   Construction Management and Estimating   86
   3.5.8   Surveying and Mapping   86
   3.5.9   AI Tools for Civil Engineering   86
3.6   Mind Mapping Questions   86
3.7   Conclusion   87
Notes   87
References   87
Further reading   88

4 Forces in Civil Engineering   **89**
4.1   Introduction   89
4.2   Quick Insights   89
4.3   The Need to Understand and Address Forces   89
4.4   Loads versus Forces   91
4.5   Outline of Forces That Civil Engineers Encounter   91
   4.5.1   Gravity   91
   4.5.2   Friction   92
   4.5.3   Tension   92
   4.5.4   Compression   93
   4.5.5   Shear   93
   4.5.6   Bending   94
   4.5.7   Torque (Torsion)   94
   4.5.8   Seismic Forces   94
   4.5.9   Hydrodynamic Forces   95
   4.5.10 Hydrostatic Forces   95
   4.5.11 Geotechnical Forces   95
   4.5.12 Thermal Forces   95
   4.5.13 Aerodynamic Forces   95
   4.5.14 Combined Loading   96
4.6   What Is Stress in Civil Engineering?   96
   4.6.1   Axial Stress   96
   4.6.2   Tensile Stress   96
   4.6.3   Compressive Stress   96
   4.6.4   Shearing Stress   97
4.7   Understanding Loads in Civil Engineering   97
4.8   Mind Mapping Questions   97
4.9   Conclusion   97
References   98

5 The Science of Materials   **99**
5.1   Introduction   99
5.2   Why Study the Science of Materials?   101
5.3   Temperature Effects on Materials   102
   5.3.1   Thermal Expansion   102
   5.3.2   Viscosity–Temperature Relationship of Binder Materials Such as Bitumen   109
   5.3.3   Pipeline Design   110
   5.3.4   Conclusion   110
5.4   Pressure Effect on Materials   110
   5.4.1   Key Differences between Pressure and Stress   110
5.5   Strength of Materials   111
   5.5.1   Example [5.3] for Illustrative Purposes: A 1.0-in-Diameter Aluminum Rod Loaded in Pure Tension   111
   5.5.2   Common Structural Members   112
   5.5.3   Why Civil Engineers Study Strength of Materials?   112
   5.5.4   Materials and Structures   118

	5.5.5	Stability of Structures	119
	5.5.6	Roadmap for Specifying Materials for a Civil/Structural Project	119
5.6	The Need to Select Appropriate Materials for Civil Engineering Projects		121
	5.6.1	Strength and Durability	121
	5.6.2	Reliability	121
	5.6.3	Corrosion Resistance	121
	5.6.4	Cost-Effectiveness	122
	5.6.5	Low-Carbon Footprint	122
	5.6.6	Safety and Health Concerns	122
	5.6.7	Aesthetics	122
	5.6.8	Standards and Regulations Compliance	122
	5.6.9	Up-to-Date Materials	122
5.7	Main Categories of Construction Materials		122
	5.7.1	Composites	122
	5.7.2	Alloys	123
	5.7.3	Admixtures	123
	5.7.4	Additives	123
	5.7.5	Binders	123
	5.7.6	Concrete and Masonry	123
	5.7.7	Polymers and Plastics	123
	5.7.8	Other Materials	123
5.8	Common Civil Engineering Materials		123
	5.8.1	Reinforced Concrete	123
	5.8.2	Steel Reinforcement Bars	124
	5.8.3	Structural Steel	124
	5.8.4	Composite Steel and Concrete	124
	5.8.5	Structural Glass	125
	5.8.6	Aluminum	126
	5.8.7	Timber	126
	5.8.8	Wood	126
	5.8.9	Other Common Materials	126
5.9	Mind Mapping Questions		126
5.10	Conclusion		127
	Notes		127
	References		127
	Further reading		127
6	**Mimicking the Natural World**		**128**
6.1	Introduction		128
	6.1.1	Blood Vessels and Pipelines	129
	6.1.2	Tree and Columns	130
	6.1.3	Appreciating Nature in General	131
6.2	Mimicking Anatomy, Physiology, and the Natural World		131
	6.2.1	Anatomy, Physiology, and Civil Engineering	132
	6.2.2	The Natural Environment and Civil Engineering	144
6.3	The Gym and the Civil Engineer		145
	6.3.1	Why Do People Go to the Gym?	146
	6.3.2	Lessons for a Civil Engineer	148
	6.3.3	Relationship between the Gym and Civil Engineering	149
6.4	Mind Mapping Questions		160
6.5	Conclusion		160
	Note		160
	References		160
	Further reading		161
7	**Equations and Their Formulation**		**162**
7.1	Introduction		162
7.2	Mathematics: A Primary Tool for Civil Engineers		163
	7.2.1	Communication	163
	7.2.2	Modeling	163

	7.2.3 Prediction	163
	7.2.4 Design Optimization	164
	7.2.5 Interpreting Data	164
	7.2.6 Driving Innovation	164
	7.2.7 Breaking Down Complex Problems	164
7.3	Primary Mathematical Techniques for Engineers	164
	7.3.1 Analytical Methods	164
	7.3.2 Empirical Methods	166
	7.3.3 Numerical Methods	167
	7.3.4 Example [7.5]: Numerical Method Example	167
7.4	Formulae and Equations	167
	7.4.1 Equation and Its Characteristics	167
	7.4.2 Formula and Its Characteristics	168
	7.4.3 Key Differences between Formula and Equation	168
	7.4.4 Theoretical Equations	168
	7.4.5 Empirical Equations	169
	7.4.6 Numerical Methods in Civil Engineering	176
	7.4.7 On Dimensional Consistency	176
	7.4.8 Grouping Equations	178
7.5	Building Equations in Civil Engineering	181
	7.5.1 How Newton's Discoveries Inspired Engineers and Scientists	181
	7.5.2 Fundamental Equations That Have Advanced Civil Engineering	183
	7.5.3 Questioning Mathematical Operators in Equations and Formulas	185
	7.5.4 Building Equations through R&D	188
7.6	Role of Coefficients in Equations	188
	7.6.1 Scaling	188
	7.6.2 Proportionality	188
	7.6.3 Weighting	189
	7.6.4 Parameterization	189
	7.6.5 Simplification	189
	7.6.6 Physical Significance	189
	7.6.7 Conversion Factors	189
	7.6.8 Correction Factors	189
	7.6.9 Typical Examples of How Coefficients Are Used in Civil Engineering	189
7.7	Formative Research and Development: The Civil Engineer's Effective Design Guide	194
	7.7.1 Role of Formative Research to a Civil Engineer	194
	7.7.2 Key Components of Formative Research	195
	7.7.3 Formative Research Methods	196
	7.7.4 The Role of Dependent and Independent Variables	196
7.8	Logarithms and Its Applications in Engineering Science	197
	7.8.1 Simplify Complex Calculations	197
	7.8.2 Logarithmic Scales	199
	7.8.3 Log–Log Plots	199
	7.8.4 Logarithms Help Deal with Large Numbers	200
	7.8.5 Common Logarithm	200
	7.8.6 Binary Logarithm	200
	7.8.7 Natural Logarithm	201
7.9	Versatile Mathematical Fields for Civil Engineers	205
	7.9.1 Calculus	205
	7.9.2 Probability and Statistics	215
	7.9.3 Other Versatile Mathematical Fields for Civil Engineers	219
7.10	Mathematical and Computational Tools for Civil Engineers	220
	7.10.1 MATLAB	220
	7.10.2 Simulink	220
	7.10.3 MAPLE	220
7.11	Logic and the Engineer	220
	7.11.1 Logic Example [7.21]	220
	7.11.2 Logic Example [7.22]	221
	7.11.3 Logic Example [7.23]	221
	7.11.4 Logic Example [7.24]	221

	7.11.5 Logic Example [7.25]	221
	7.11.6 Analysis	221
7.12	Mind Mapping Questions	222
7.13	Conclusion	222
Notes		222
References		223
Further reading		223

8 Common Civil Engineering Tools **224**

8.1	Introduction	224
8.2	Structural Engineering—A Gateway to Other Branches	224
8.3	Introduction to Support and Connection Types	225
8.4	The Three Connection or Support Types	225
	8.4.1 Roller Supports	225
	8.4.2 Pinned Supports	227
	8.4.3 Fixed Supports	227
	8.4.4 Key Takeaways	227
	8.4.5 Reaction Forces and Statistical Determinacy and Indeterminacy	228
8.5	Primary Civil Structural Analysis/Graphical Tools—Fundamentals	229
	8.5.1 Free-Body Diagram	229
	8.5.2 Steps/Procedure for Drawing FBDs	231
	8.5.3 2D FBDs	231
	8.5.4 Shear Force Diagram	233
	8.5.5 Bending Moment Diagram	234
8.6	Essential Drawings for Civil Engineers	235
	8.6.1 Design Drawings	236
	8.6.2 Architectural Impressions	236
	8.6.3 Site Layout (Plan) Drawings	236
	8.6.4 Architectural Drawings	239
	8.6.5 Structural Drawings	239
	8.6.6 Section Drawings	239
	8.6.7 Plan Drawings	239
	8.6.8 Elevation Drawings	239
	8.6.9 Isometric Drawings	239
	8.6.10 Orthographic Drawings	242
	8.6.11 Detail Drawings	242
	8.6.12 Mechanical Drawings	242
	8.6.13 Electrical Drawings	243
	8.6.14 Plumbing Drawings	243
	8.6.15 Shop Drawings	243
	8.6.16 Construction Drawings	243
	8.6.17 As-Built Drawings	243
	8.6.18 Excavation Drawings	243
	8.6.19 Site Drawings	243
	8.6.20 Flow Charts and Diagrams	243
	8.6.21 Survey Drawings	243
	8.6.22 Geotechnical Drawings	243
	8.6.23 Hydraulic and Hydrologic Drawings	243
	8.6.24 Bar Bending Schedule	243
8.7	Special Drawings and Charts	244
	8.7.1 Structural/Geotechnical Engineering: Mohr's Circle	244
	8.7.2 Water Resources/Hydraulics: Nomograms and Moody Diagram	244
	8.7.3 Hydrology: IDF Curves	245
8.8	Pedagogical Tools in Civil Engineering	245
	8.8.1 Software	245
	8.8.2 Artificial Intelligence	245
	8.8.3 Drawings	246
	8.8.4 FBDs	246
	8.8.5 Graphs and Charts	246

	8.8.6	Roller, Hinge, and Fixed Support Idealization	246
	8.8.7	Computers and Associated Hardware/Software	247
	8.8.8	Scale Models	247
	8.8.9	The Gym Equipment	247
	8.8.10	Visual Aids	247
	8.8.11	Conclusion	247
8.9	Software Tools for Civil Engineers		247
	8.9.1	Types of Civil Engineering Software	247
	8.9.2	Key Features of Civil Engineering Software	248
	8.9.3	Development Process for Software Tools for Civil Engineers	248
	8.9.4	Software Customization and Adaptation Techniques	250
8.10	Mind Mapping Questions		250
8.11	Conclusion		251
References			251
Further Reading			251

9 Think like a Civil Engineer   **252**
9.1	Introduction		252
9.2	Civil Engineers Piece Together		252
	9.2.1	Understand the Context in Design	253
	9.2.2	Logic and Design Intuition	255
	9.2.3	Intuition Informs Logic	257
	9.2.4	Logic Refines Intuition	257
	9.2.5	Failure: An Option in Design, not in Construction	258
	9.2.6	Convergent Thinking versus Divergent Thinking	258
	9.2.7	The Power of Playing with Sketches	259
	9.2.8	The Power of Developing Scale Models	259
	9.2.9	Physical Prototypes	259
	9.2.10	The Jigsaw	259
	9.2.11	The Engineer's Thought Pattern	260
9.3	Transitioning from Academia to Industry		261
9.4	The Design Process		262
	9.4.1	The DAE Cycle	262
	9.4.2	Iterative Process	264
9.5	Construction Techniques		266
	9.5.1	Traditional Construction Techniques	266
	9.5.2	Modern Construction Techniques	267
9.6	Need for Appropriate Tools to Piece a Project through Design and Construction		269
	9.6.1	Quality Control	269
	9.6.2	Smooth Project Workflow	270
	9.6.3	Timely Completion	270
	9.6.4	Minimizing Errors	270
9.7	The Importance of Managing Procedures in Construction		270
	9.7.1	Improved Safety	271
	9.7.2	Minimizes Reworks	273
	9.7.3	Quality Control	273
	9.7.4	Cost Control	274
	9.7.5	Builds Reliability in the Project	274
	9.7.6	Minimizes Risks	274
9.8	1D, 2D, and 3D Models in Civil Engineering		274
	9.8.1	1D Analysis	274
	9.8.2	2D Analysis	275
	9.8.3	3D Analysis	275
9.9	Case Study [9.3]: Water Reservoir Tank Planning and Design, Kisoro UG		275
9.10	Conclusion		281
Note			281
References			281
Appendix 9.1			282

10 Must-Know Terminologies in Civil Engineering   **283**

   10.1 Introduction   283

   10.2 A   283

       10.2.1   Abutment   283

       10.2.2   Analogizing Nature   283

       10.2.3   Active Forces   283

       10.2.4   Aquifer   283

       10.2.5   Asphalt   283

       10.2.6   Analytical Model   283

       10.2.7   ASTM   283

       10.2.8   ASCE   284

       10.2.9   Ambient Temperature   284

       10.2.10   Anchor   284

       10.2.11   Angle of Repose   284

       10.2.12   Aggregate   284

       10.2.13   Artificial Intelligence   284

       10.2.14   Architect   284

   10.3 B   284

       10.3.1   BSI   284

       10.3.2   BS   284

       10.3.3   Brittleness   284

       10.3.4   Blend   284

       10.3.5   Beam   284

       10.3.6   Bearing Capacity   284

       10.3.7   Bending Moment   284

       10.3.8   Bitumen   285

       10.3.9   Bridge   285

       10.3.10   Building Code   285

   10.4 C   285

       10.4.1   CAD   285

       10.4.2   Closed-Form Solution   285

       10.4.3   Caisson   285

       10.4.4   Creep   285

       10.4.5   Colloid   285

       10.4.6   Camber   285

       10.4.7   Cement   285

       10.4.8   Column   285

       10.4.9   Concrete   285

       10.4.10   Construction Management   285

       10.4.11   Contract   286

       10.4.12   Culvert   286

       10.4.13   Cantilever   286

       10.4.14   Cofferdam   286

       10.4.15   Compaction   286

       10.4.16   Compression   286

       10.4.17   Concrete Mix Design   286

       10.4.18   Cracking   286

       10.4.19   Cantilever   286

       10.4.20   Compression Member   286

       10.4.21   Consolidation   286

       10.4.22   Construction Survey   286

       10.4.23   Curb   286

   10.5 D   287

       10.5.1   DCP Test   287

       10.5.2   Degrees of Freedom   287

       10.5.3   Detailing a Drawing   287

       10.5.4   Ductility   287

       10.5.5   Dam   287

       10.5.6   Dead Load   287

       10.5.7   Density   287

       10.5.8   Drainage System   287

10.5.9	Drawdown	287
10.5.10	Deflection	287
10.5.11	Dewatering	287
10.5.12	Dowel	288
10.6	E	288
10.6.1	Energy	288
10.6.2	Elasticity	288
10.6.3	Earthquake	288
10.6.4	Elasticity	288
10.6.5	Elevation	288
10.6.6	Embankment	288
10.6.7	Erosion	288
10.6.8	Excavation	288
10.6.9	Expansion Joint	288
10.6.10	Emulsion	288
10.6.11	Earthquake Engineering	288
10.6.12	Embankment	288
10.6.13	Excavation	289
10.6.14	Earthwork	289
10.6.15	Embankment Dam	289
10.6.16	Excavator	289
10.7	F	289
10.7.1	Factor of Safety	289
10.7.2	Force	289
10.7.3	Fatigue	289
10.7.4	Flexure	289
10.7.5	Footing	289
10.7.6	Foundation	289
10.7.7	FIDIC	289
10.8	G	289
10.8.1	Gabions	289
10.8.2	Geosynthetics	290
10.8.3	Granular Materials	290
10.8.4	Grading	290
10.8.5	Geotechnical Engineering	290
10.8.6	Girder	290
10.8.7	Grade	290
10.8.8	Grade Beam	290
10.8.9	Groundwater	290
10.8.10	Grout	290
10.9	H	290
10.9.1	Hardness	290
10.9.2	Health	291
10.9.3	Hydrostatic Pressure	291
10.9.4	Hydrology	291
10.10	I	291
10.10.1	ICE	291
10.10.2	I-beam	291
10.10.3	Infill	291
10.11	J	291
10.11.1	JSCE	291
10.11.2	Joist	291
10.12	K	291
10.12.1	K—Kerb	291
10.13	L	291
10.13.1	Landfill	291
10.13.2	Landslide	291
10.13.3	Lateral Load	292
10.13.4	Lateral Force	292
10.13.5	Live Load	292
10.13.6	Load-Bearing Wall	292

10.13.7	Load Factor	292
10.14 M		292
10.14.1	Mass of Material	292
10.14.2	Moment of Inertia	292
10.14.3	Monolithic Connections	292
10.14.4	Malleability	292
10.14.5	Machinability	292
10.14.6	Masonry Construction	292
10.14.7	Micro-Pile	292
10.14.8	Moisture Barrier	293
10.14.9	Masonry	293
10.14.10	Modulus of Elasticity	293
10.14.11	Moment of Inertia	293
10.15 N		293
10.15.1	NEC	293
10.15.2	Numerical Models	293
10.15.3	Negative Pressure	293
10.16 O		293
10.16.1	Offset	293
10.17 P		293
10.17.1	Power	293
10.17.2	Poisson Ratio, v	293
10.17.3	PUNDIT	293
10.17.4	Plasticity	293
10.17.5	Pavement	294
10.17.6	Pile	294
10.17.7	Pipe	294
10.17.8	Pipeline	294
10.17.9	Prestressed Concrete	294
10.17.10	Pavement	294
10.17.11	Permeability	294
10.17.12	Pile Foundation	294
10.17.13	Plate Load Test	294
10.17.14	Point Load	294
10.17.15	Post-tensioning	294
10.17.16	Purlin	294
10.18 Q		294
10.18.1	Quantity Surveyor	294
10.19 R		295
10.19.1	Reactive Forces	295
10.19.2	River Training	295
10.19.3	Resilience of Material	295
10.19.4	RCC	295
10.19.5	Riprap	295
10.19.6	Rockfill	295
10.19.7	R-Value	295
10.19.8	Raft Foundation	295
10.19.9	Rebar	295
10.19.10	Reinforced Concrete	295
10.19.11	Retaining Wall	295
10.19.12	Rigid Frame	295
10.20 S		296
10.20.1	Slenderness Ratio	296
10.20.2	Statically Determinate Structure	296
10.20.3	Statically Indeterminate Structure	296
10.20.4	Strength of a Material	296
10.20.5	Stiffness	296
10.20.6	Suspension	296
10.20.7	Safety	296
10.20.8	Smart Systems	296

10.20.9	Serviceability Limit State	296
10.20.10	Schedule	296
10.20.11	Settlement	297
10.20.12	Shear Force [3]	297
10.20.13	Soil Mechanics	297
10.20.14	Soil Stabilization	297
10.20.15	Steel Reinforcement	297
10.20.16	Structural Analysis	297
10.20.17	Substructure	297
10.20.18	Superstructure	297
10.20.19	Sedimentation	297
10.20.20	Shear Wall	297
10.20.21	Shoring	297
10.20.22	Site Investigation	297
10.20.23	Slurry Wall	297
10.20.24	Soil Nailing	297
10.20.25	Soil Stabilization	298
10.20.26	Spillway	298
10.20.27	Spread Footing	298
10.20.28	Stability Analysis	298
10.20.29	Steel Reinforcement	298
10.20.30	Stormwater Management	298
10.20.31	Stress Concentration Point	298
10.20.32	Structural Integrity	298
10.20.33	Structural System	298
10.20.34	Subsurface Drainage	298
10.20.35	Subsidence	298
10.20.36	Superimposed Load	298
10.20.37	Survey Monument	298
10.20.38	Suspended Floor	298
10.20.39	Surveying	298
10.20.40	Shear Force	298
10.20.41	Shear Strength	298
10.20.42	Slab	298
10.20.43	Slope Stability	299
10.20.44	Soil Mechanics	299
10.20.45	Surveying	299
10.20.46	Span	299
10.20.47	Span-to-Depth Ratio	299
10.20.48	Specific Gravity	299
10.20.49	Splice	299
10.20.50	Spread Footing	299
10.20.51	Strain	299
10.20.52	Stress	299
10.20.53	Structural Engineering	299
10.20.54	Subgrade	300
10.21 T		300
10.21.1	Toughness	300
10.21.2	Tunnel	300
10.21.3	Tensile Strength	300
10.21.4	Tension	300
10.21.5	Timber	300
10.21.6	Topography	300
10.21.7	Topographical Survey	300
10.21.8	Traffic Engineering	300
10.21.9	Truss	300
10.21.10	Tailings	300
10.21.11	Tensile Strength	300
10.21.12	Tieback	300
10.21.13	Topographic Survey	300

10.21.14	Torsion	300
10.21.15	Traffic Count	300
10.21.16	Transverse Slope	300
10.21.17	Trenching	301
10.21.18	Tunnel Boring Machine	301
10.21.19	Turnkey Project	301
10.21.20	Tension	301
10.21.21	Topographic Map	301
10.21.22	Torsion	301
10.21.23	Truss	301
10.22 U		301
10.22.1	UPV Test	301
10.22.2	Underpinning	301
10.22.3	Uplift	301
10.22.4	UIPE	301
10.22.5	Ultimate Limit State	301
10.22.6	User-Centered Design (UCD)	302
10.23 V		302
10.23.1	Void Ratio	302
10.24 W		302
10.24.1	Work	302
10.24.2	Welfare	302
10.24.3	Wall	302
10.24.4	Water Supply System	302
10.24.5	Water Table	302
10.24.6	Weld	302
10.24.7	Wind Load	302
10.24.8	Welding	302
10.24.9	Work Breakdown Structure	302
10.24.10	Wall Footing	302
10.24.11	Water Main	302
10.24.12	Weathering	302
10.24.13	Weep Hole	303
10.24.14	Working Stress	303
10.25 X		303
10.25.1	X-bracing	303
10.26 Y		303
10.26.1	Young's Modulus, E	303
10.26.2	Yield Strength	303
10.27 Z		303
10.27.1	Zero Lot Line	303
10.27.2	Zoning	303
10.27.3	Zoning Ordinance	303
10.28 Conclusion		303
References		303
Index		305

Foreword

Civil engineering is one of the oldest and most impactful professions, shaping societies and the built environment across history. From ancient aqueducts and monumental structures to today's sustainable infrastructure and smart cities, the field has continually evolved by integrating science, technology, ethics, and creativity. It not only builds but also safeguards human well-being and the sustainability of our planet.

This book, *Introduction to Civil Engineering: A Guide for Students and Professionals*, is a timely and valuable contribution that presents civil engineering as both an academic discipline and a professional calling. It introduces the foundations of the profession, its history and global institutions, and the role of ethics and philosophy in practice. The chapters address scientific principles, forces, materials, design tools, and emerging approaches such as biomimicry, while also offering insights into mathematics as the language of engineering. By bridging theory with practice, this book prepares students and early-career engineers to think critically, design iteratively, and transition with confidence from academia to industry.

The strength of this work lies in connecting fundamentals with broader philosophy and real-world applications. It not only explains 'what civil engineering is,' but also inspires readers to think like civil engineers, applying knowledge and creativity to address the challenges of our time. I commend the author for this achievement and strongly recommend this book as a valuable guide for students, educators, and professionals alike.

Professor Sanjay Kumar Shukla,
PhD, MTech, BSc Eng, F.ASCE, FIEAust,
FIEIndia, FIGS, CPEng NER, APEC Engineer, IntPE(Aus)

Founding Editor-in-Chief, International Journal of Geosynthetics and Ground Engineering, Switzerland and Founding Geotechnical and Geoenvironmental Engineering Research Group Leader, School of Engineering, Edith Cowan University, Joondalup, Perth, Australia

Preface

This book is intended primarily to bridge the gap between academia and the industry, promoting knowledge and understanding of civil engineering. Civil engineering course units are taught separately, and a graduate is left to think on their own to grasp civil engineering. This book shares how the units jigsaw to nurture a professional. It supports the graduate to transition from academia to industry, grasping the basic techniques as much as possible. It is intended to simplify and make everything look whole to inspire a graduate to think like a civil engineer but not as a specialist falling under civil engineering. It emphasizes that to think as a civil engineer requires the graduate to be well grounded in the bachelor's degree.

This book inspires those interested in civil engineering to master the art and techniques of the profession and shapes constantly evolving civil engineers to meet industry demands. It takes the reader from how civil engineering has evolved through the years and how prominent scientists changed the face of civil engineering and advanced civil engineering technology.

This book illustrates the application of technology in engineering practice, how new and advancing civil engineering technology advances scientific research, and how scientific research advances civil engineering technology. This combination of scientific facts reciprocally influencing new and advancing civil engineering technology must be at the fingertips of anyone wishing to master the civil engineering profession, especially a continuously evolving civil engineer.

This book hints at the art of formulation of techniques in the design and construction of civil engineering structures and how these techniques are borrowed from scientific fields to solve an engineering problem. The world has advanced dramatically over the past two or three decades, and information technology is rapidly shaping the face of civil engineering, especially in the design of civil engineering structures and management systems. This book will prove very helpful in this area.

This book illustrates the formulation and grouping of dimensionally consistent equations as a primary technique engineers use to solve engineering problems. In the same way, it includes probability and statistical distributions that are harnessed to model complex engineering situations. The engineer has to learn how to analyze and formulate equations and when to use them relying on the role of coefficients as much as applicable. This analysis of techniques and tools to use to solve an engineering problem is essential to achieve accurate results as it involves how to customize techniques and tools to suit a situation. It also illustrates the metric and English units. This book provides an account of the derivation of theoretical equations from scientific laws and outlines how empirical equations are built to ease an engineer's work.

Chapter 1 (What Is Civil Engineering?) defines civil engineering, how the field has evolved, and the STEM courses needed to become a civil engineer. It further outlines the steps to become a professional civil engineer and the need to inculcate lifelong learning skills as a student of civil engineering and adopting an interdisciplinary approach because civil engineering constantly evolves. It also deals with the branches of civil engineering. The founding of the Institution of Civil Engineers (ICE), United Kingdom, as the oldest institution in the world, and her counterpart, the American Society of Civil Engineers (ASCE), are also featured in the book for the reader to appreciate how the profession has evolved through the years and what the ICE means to the world of civil engineering. This chapter outlines the most influential civil engineers throughout history who have shaped the built environment from a global perspective.

Chapter 2 (Philosophy of Civil Engineering) shares engineering ethics for civil engineers, outlining the role of the philosopher, the engineer, and the technician. This chapter shares the prominent people who have shaped the discipline from diverse fields of science, technology, and engineering. In particular, the works of Leonardo Da Vinci and Karl von Terzaghi and others are of great significance. It features the 5 Ms of management. This chapter considers how civil engineering work affects society, environmental and social issues civil engineers deal with, and how civil engineers have more work to do than ever because the sustainability of Earth is primarily championed by civil engineers.

Chapter 3 (Civil Engineering Technology) provides in-depth knowledge about the scientific laws that have changed the face of civil engineering and advanced civil engineering technology. The works of Sir Isaac Newton and Robert Hooke are of paramount importance in the history of civil engineering. In addition, the law of conservation of energy influenced the profession and advanced civil engineering technology. The civil engineer appreciates the flexibility in using energy equations in different branches of civil engineering. This chapter outlines how technology has influenced civil engineering throughout history and why a constantly evolving civil engineer must appreciate the role of new and advancing technology in their practice, especially in the development of codes and standards of practice. It shares several civil engineering software technologies shaping the world. The evolution of metric and English units is provided in this chapter to support the engineer in learning the necessary conversions.

Chapter 4 (Forces in Civil Engineering) outlines the forces that civil engineers deal with in their day-to-day planning, design, construction, operation, and maintenance of civil engineering structures.

Chapter 5 (The Science of Materials) approaches material science by defining the most important attributes, that is, temperature and pressure which influence or modify the physical and chemical properties of materials in the civil engineering context. This chapter outlines civil engineering materials, such as metals, concrete, bricks, masonry, ceramics, composites, polymers, bituminous materials, timbre, wood, sand, aggregates, and glass. It provides the importance of learning the strength of materials and the chemical and physical properties of materials. It also illustrates the differences between materials and structures, providing an understanding of the stability, strength, and stiffness of materials/structures. It concentrates on the reasons for adopting suitable materials for civil engineering purposes.

Chapter 6 (Mimicking the Natural World) shares how the civil engineering industry mimics the natural world and provides how the human anatomy and physiology compares with civil engineering structures (the gym and the civil engineer) specifically, with structural forces and moments, how arteries and veins provide insights for understanding fluid mechanics, and how the heart pumps blood around the body, giving insights into pumps pumping water through pipelines. This chapter also describes the digestive system compared with mechanical parts in an engineering system, that is, how water, drinks, and sauce act as lubricants aiding food movement through the throat down to the digestive system and how smart structures mimic a human's senses, intelligence, and adaptability. It also outlines how sustainable drainage systems (SuDS) mimic the natural world. This will be very helpful to civil engineering course instructors.

Chapter 7 (Equations and Their Formulation) shares how mathematics is a primary tool for engineers and the importance of logarithms in engineering science. It further deals with how standard equations are built from scientific theories and how empirical equations are formulated to guide the work of a civil engineer. The reader benefits from the understanding of how equations are used to communicate engineering information. This chapter shares the difference between a formula and an equation, the use and application of calculus in engineering, the relationship between dependent and independent variables, and their use in industry and formative research and development. The grouping of equations with dimensional consistency taken into account is properly outlined in this chapter. The reader is taken through the role of coefficients in equations and how management decisions are taken based on the coefficients obtained from the data analysis or laboratories while testing different materials for use in civil engineering design, analysis, and evaluation of structures. In the same way, this chapter outlines the role of factors of safety (FOS) in the design of civil engineering structures. The application of probability and statistical distributions is well illustrated, making the reader understand its role in quality control and management of the design and construction of civil engineering structures. This chapter outlines how logic is harnessed in the design and management of engineering functions, emphasizing that logic is

the output of engineering education. It provides how complex graphs are created to communicate engineering information.

Chapter 8 (Common Civil Engineering Tools) illustrates the application of free-body diagrams, shear force diagrams, and bending moment diagrams. It provides an understanding of the pedagogical tools used in engineering education, shedding light on tools that remain abstract in academia. This chapter outlines the different types of drawings a civil engineer must know. In this chapter, the reader learns concepts such as rollers, fixed support, and hinged support and gains a deep understanding. It also outlines the development process for computer software tools for civil engineers. This can be very helpful for readers interested in civil engineering software customization and adaptation techniques.

Chapter 9 (Think Like a Civil Engineer) takes the reader through the approach to designing and constructing civil engineering structures, emphasizing that design is an iterative process. This chapter also shares the process of transitioning from academia to a professional engineer, outlining the ethical code of conduct for civil engineers, professionalism, costs, and safety considerations as key issues in engineering practice. It promotes the idea that logic is the outcome of engineering education which guides the design, analysis, and evaluation. It tackles construction techniques and the importance of managing procedures in construction. It also provides an understanding of the application of 1D (e.g., water pipelines design), 2D (e.g., beam analysis and design), and 3D (e.g., structures) models in civil engineering design, analysis, and evaluation. It outlines how the iterative process is done during design, the software applications in design and management, and how engineering technology is creatively harnessed to advance new solutions to civil engineering problems. It emphasizes the need for the engineer to have appropriate technological tools to piece together a project through the design or construction. It chapter inspires the graduate to think like a civil engineer.

Chapter 10 (A Must-Know Terminologies in Civil Engineering) provides must-know civil engineering terminologies for engineers who wish to master the profession.

This book provides an in-depth understanding of what civil engineering is all about and what a graduate needs to learn to think like a civil engineer. Civil engineers need to have skills such as management skills, design competencies, and commercial awareness skills. This book serves as a potential guide for engineers in the making and training to learn the importance of having these skills. In overall terms, it sheds light on grasping several techniques and how communication of engineering information can be simplified and compressed into equations, coefficients, and drawings.

Acknowledgments

This book would not have been possible without the extraordinary support of Hon. Berunado Ssebbugga-Kimeze, MP, a professional civil engineer. His unwavering moral courage and unshakeable support, especially during the time when all other professionals shied away from the truth, are truly immeasurable. I'm deeply grateful for the opportunity to have collaborated with him on several projects. His guidance, expertise, and leadership have not only elevated our work but also contributed significantly to this book. I extend my sincere gratitude to the following for their support:

- Michael Daka, Dickson Berabose, Richard Li Chenghze, Prof. Sanjay Kumar Shukla, Owek. John Baptist Walusimbi, Mahad Uzairu Magala, Robert Mukiibi, Moses Gava Kagimu, Joanita Ndagire, Patrick Dino, Robert K. Kakiiza, Rogers Sabiiti, Allan Amayo, Christine Salanza Nafuna, Robert Kasule, John Kibuuka Ssemikisa, Fred Byiringiro, Bashir Mubiru, Alex Kasanya, Yasin Naku Ziraba (PhD), Peter O. Lating (PhD), Allan Kiberu, Raymond Mutyaba, James Othieno, Kizito K. Charles, Andrew Grace Naimanye (PhD), Prof. Jackson Mwakali, Jane Frances Nnantamu (PhD), Benjamin Olobo, Isaac Mutenyo (PhD), Martin Kayemba (RIP), and Joseph Mubiru.
- Senior Publisher, Taylor & Francis Group—Joseph Clements.
- Editorial Team, Taylor & Francis Group—Ghosh Khaustav, Sanger Khyati, Andrew Stow, Nanda Anahita, and Gupta Radhika. Particularly, Sanger Khyati and Andrew Stow for their crucial role in the critical stage of guiding the idea through the Taylor & Francis Group's proposal vetting process.
- Reviewers, Taylor & Francis Group.
- Gym Team, Michael Kabuusu, Easter Nassaka, and Joseph Lubega.
- Chetan Kanojia and SHAKTI Pumps (I) Ltd, UG Team.
- Art, design, and drafts team, Ivan Tebugambe, Ronald Mayanja, Festo Titus Malobo, and Eric Kayongo.
- My children, Mark Ssempeera Jr and Patricia Nampeera, for their assistance with key physics illustrations that underpin the fundamental principles of civil engineering.
- Maria Galea, Customer Services, ICE Publishing, Thomas Telford Ltd, One Great George Street, London, UK SW1P 3AA for exceptional customer care.
- The Institution of Civil Engineers, United Kingdom, for exceptional CPD programs.
- Weihai International Economic & Technical Cooperative Co., Ltd (WIETC).
- China Railway Construction Engineering Group Co., Ltd (CRCEG).
- Associated Design and Build Engineers Limited, Kampala UG.
- Ministry of Education and Sports officials, UG, for saying 'No' after being prequalified and passing both written and oral interviews for the World Bank's UG Secondary Education Expansion Project (USEEP), 2022/2023, individual consultant's recruitment process. Your chaos inspired me to author this book!
- Makerere university CEDAT professors for their Markets of Wandegeya Chronicles!

TO THE GLOBAL COMMUNITY

Special thanks to all individuals and organizations whose work, whether in the public domain or copyrighted, has contributed significantly to building a sustainable world and provided invaluable foundations for this book.

In line with UN SDGs, it is thus essential to note that

As long as we continue watering bad seeds, sustainable development will not happen. A person who fails to understand that they write because people wrote and that they owe a duty to future generations to write is a bad seed.

Patrick Ssempeera CEng MICE
Team lead, Unlocking Africa's sustainable
development project.

Author Biography

Patrick Ssempeera is a chartered civil engineer and a member of the Institution of Civil Engineers, United Kingdom. He has previously authored the books, *Integrated Drainage Systems Planning and Design for Municipal Engineers* and Unlocking Africa's Sustainable Development. He holds a BSc Civil Engineering and master's degree in Engineering (Civil) and has practiced widely. He is affiliated with CRBNetwork, an organization that integrates initiatives and leads the delivery of sustainable infrastructure projects that exceed expectations.

What Is Civil Engineering?

1

1.1 INTRODUCTION

Civil engineering deals with the planning, designing, construction, operation, and maintenance of the physical and built environment, which includes roads, power plants, railroads, bridges, canals, aqueducts, tunnels, buildings, airports, dams, water supply and distribution systems, sewage systems, airfields, schools, offices, and hospitals. It deals with everything built around us by applying physical and scientific principles to the planning, designing, constructing, operating, and maintaining of artificially created and natural environments: the things we take for granted but would find life very hard to live without. So, a civil engineer has saved life more than anyone else!

Today's 'civil engineer' performs a range of jobs and can work in different industries. Civil engineers originate from different backgrounds, but all spend years training, learning, and getting qualifications.

The term **civil engineer** was established by John Smeaton in 1750 to contrast engineers working on civil projects with military engineers who worked on armaments and defenses. The word *civil* relates to ordinary citizens and their concerns, distinct from military or ecclesiastical matters. Smeaton must have noted the need to have civilians work on the public infrastructure and not leave everything for the military.

Therefore, civil work projects represent any construction done by civilians for civilian purposes intended for collective or public use. Civil works may also require a license from a governmental agency to be undertaken.

The term 'civil engineer' first appeared in the Minutes of the Society of Civil Engineers, formed in **1771**. By using this title, founder members of the society recognized a new profession in Great Britain which was distinct from the much earlier profession of military engineer [1].

John Smeaton was among the founder members of the Society of Civil Engineers. The Society, which still exists, was later renamed as the Smeatonian Society of Civil Engineers after its principal founder, John Smeaton, and was the precursor of, but distinct from, the Institution of Civil Engineers (ICE), which was formed in 1818, with Thomas Telford as its first President.

The Smeatonian Society of Civil Engineers is the oldest society of engineers in the world and embraces engineers of all disciplines. It predates the ICE founded in 1818. Its members are known as 'The Smeatonians.'

Unlike the ICE, the Society of Civil Engineers was not intended to be a learned Society but rather a dining club, and members of the Society still meet today at the ICE headquarters [2].

The ICE was granted a Royal Charter on 3 June 1828 (updated by Queen Elizabeth II in 1975) and has the status as the leading institution for civil engineering in the United Kingdom. Thus, the title 'chartered civil engineer' is protected by law.

It is from this background that the entire world that uses or adopts the term 'civil engineer' commemorates the roots of the profession.

The ICE defines civil engineering as follows:

> The general advancement of mechanical science, and more particularly for promoting the acquisition of that species of knowledge which constitutes the profession of a civil engineer being the art of directing the great sources of power in nature for the use and convenience of man.
>
> *—Thomas Tredgold, 1828*

Since the term 'civil engineer' was established by John Smeaton, who was probably the first person to describe himself as a 'civil engineer' [2], the world has grown to have many civil engineers shaping it and saving humanity. Identifying with Smeaton's creativity as a 'civil engineer,' you derive a sense of fulfillment, having saved humanity and made the world a better place to live.

In 2007, the ICE Council updated its definition of civil engineering as follows:

> Civil Engineering is a vital art, working with the great sources of power in nature for the wealth and well-being of the whole of society. Its essential feature is the exercise of imagination to engineer the products and processes, and develop the people needed to create and maintain a sustainable natural and built environment. It requires a broad understanding of scientific principles, a knowledge of materials and the art of analysis and synthesis. It also requires research, team working, leadership and business skills. A civil engineer is one who practices all or part of this art.
>
> *—ICE Council, 2007 [3]*

1.2 HISTORY OF CIVIL ENGINEERING

Civil engineering has been integral to human life since ancient times. It is a mirror of the history of human beings on Earth. Humans used the old shelter caves to protect themselves from weather and harsh environment. They also used tree trunks to cross rivers, which demonstrates ancient civil engineering. Early evidence of civil engineering can be seen in the

DOI: 10.1201/9781003561620-1

developments of the ancient Egypt and Mesopotamia between 4000 and 2000 BC, where humans built shelters and improved transportation systems, including the wheel and sailing.

Until modern times, there was no clear distinction between civil engineering and architecture, and the terms engineer and architect were mainly geographical variations referring to the same occupation and often used interchangeably. Engineers kept developing techniques for constructing primarily building structures, including cathedrals and castles, during the medieval and Renaissance periods. In the 18th century, the term civil engineering was coined to denote all things that are civilian, as opposed to military engineering (see Section 1.1). One of the notable first institutions for teaching civil engineering is the École Nationale des Ponts et Chaussées (renamed École des ponts ParisTech), which was established in France in 1747 by Daniel-Charles Trudaine and is one of the oldest and most prestigious French Grandes Écoles to date. So the 18th century witnessed better civil engineering emerging across Europe, and it remains on record that the term 'civil engineer' was established by John Smeaton in 1750 in the United Kingdom.

The first private college to teach civil engineering in the United States was Norwich University, founded in 1819 by Captain Alden Partridge. The first degree in civil engineering in the United States was awarded by Rensselaer Polytechnic Institute in 1835. The first such degree to be awarded to a woman was granted by Cornell University to Nora Stanton Blatch in 1905.

Beginning from the earliest human settlements, civil engineering has played a vital role. In ancient Egypt and Mesopotamia, around 4000–2000 BC, humans developed construction skills and transportation technologies like the wheel and sailing to support their growing communities.

Around 312 BCE to 226 CE, the Romans constructed aqueducts throughout their Republic and later Empire to bring water from external sources into cities and towns. This period saw the development of advanced water supply systems, with some aqueducts still partly in use today. They were made from a series of pipes, tunnels, canals, and bridges. They also supported mining operations, milling, farms, and gardens. Aqueducts facilitated movement of water through gravity alone, along a slight overall downward gradient within conduits of stone, brick, or concrete; the steeper the gradient, the faster the flow. The most recognizable feature of Roman aqueducts is the bridges constructed using rounded stone arches.

1.2.1 Ancient Civil Engineering Structures

Some of the ancient historic civil engineering structures still standing today are as follows: Pyramids in ancient Egypt, Ziggurats of Mesopotamia, Acropolis and Parthenon in Greece, Roman aqueducts, Via Appia, and Colosseum. Other ancient structures such as Hanging Gardens of Babylon and the Pharos of Alexandria are no longer standing.

1.2.1.1 Pyramids in ancient Egypt

The Egyptian pyramids are ancient pyramid-shaped masonry structures located in Egypt. The Pyramids of Giza are a complex of ancient Egyptian royal monuments located on the Giza Plateau near Cairo. Most of them were built as tombs for the pharaohs and their consorts during the Old and Middle Kingdom periods. Limestone and sandstone were the main building stones used in ancient Egypt. The Pyramids of Giza are among the most recognizable symbols of ancient Egyptian civilization.

A view of the pyramids at Giza from the plateau to the south of the complex is shown in Figure 1.1 [4]. From left to right of the figure, the three largest are as follows: the Pyramid of Menkaure, the Pyramid of Khafre, and the Great Pyramid of Khufu. These pyramids were built as tombs for pharaohs and their consorts during the Fourth Dynasty of the Old Kingdom, around 2600–2500 BC.

1.2.1.2 Acropolis and Parthenon in Greece

Perched atop a rugged hill overlooking Athens, the Acropolis is an ancient fortress housing several historically significant

FIGURE 1.1 Pyramids at Giza, Egypt [5].

structures, with the iconic Parthenon standing out. Under the artistic direction of Phidias, who oversaw the sculptural embellishments, architects Ictinos and Callicrates led the construction process between **447 BC and 432 BC**. Although the main structure was completed by 432 BC, decorative work persisted until at least 431 BC. Skilled marble workers, adept at precision-cutting blocks to exact specifications, built the Parthenon. These craftsmen were highly skilled, and any blocks of lesser quality were rejected by the architects. The Parthenon was built primarily by men who knew how to work on marble. These quarrymen had exceptional skills and were able to cut the blocks of marble to very specific measurements. If the marble blocks were not up to the standard, the architects would reject them.

1.2.1.3 Ziggurats of Mesopotamia

Ziggurats were significant architectural and religious structures in ancient Mesopotamia, reflecting the region's cultural and spiritual practices. In ancient Mesopotamia and western Iran, ziggurats were towering, which were stepped structures that resembled pyramids but with receding levels. Built using mud-bricks, these structures likely served as temples dedicated to the region's gods. The tradition of creating a ziggurat was started by the Sumerians, but other civilizations of Mesopotamia, such as the Babylonians, the Akkadians, and the Assyrians, also built ziggurats for local religions.

Characterized by square or rectangular bases and sloping walls, ziggurats were constructed to honor the primary deity of a city. They began as a platform (usually oval, rectangular, or square) and was a mastaba-like structure with a flat top. The sunbaked bricks formed the core of the construction with facings of fired bricks on the outside. Each step was slightly smaller than the level below it.

Key features of the ziggurats

- The number of floors of the Ziggurats ranged from two to seven.
- Part of larger temple complexes with multiple buildings.
- Constructed using sun-dried bricks with fired brick facings.
- Featured stepped structures with flat tops, typically ranging from two to seven levels.
- Built by various Mesopotamian civilizations, including the Sumerians, the Akkadians, the Babylonians, and the Assyrians.
- Ziggurat bases were square or rectangular.
- Had sloping walls.
- Had a mastaba-like structure with a flat top.

1.3 STEM COURSES FOR AN ASPIRING CIVIL ENGINEER

STEM stands for science, technology, engineering, and mathematics and refers to any subjects that fall under these four disciplines.

The STEM subjects generally include physics, chemistry, biology, mathematics, biochemistry, computer science, psychology, and statistics. The list is far more exhaustive, but this gives you an idea of the subjects included under STEM.

The acronym STEM originated from the discussions about the lack of qualified graduates to work in high-tech jobs in the United States. Since its creation, governments and universities worldwide have attracted students to STEM courses to address this shortfall.

1.3.1 Primary Education

The formative years are a great source of inspiration. A child inspired by civil engineering should take primary education in a conducive environment to attain proper social, intellectual, physical, and emotional growth.

Between 0 and 8 years of a child's life is where they learn more quickly than at any other time. These are the years a child experiences rapid cognitive (intellectual), social, emotional, and physical development. So, secure attachments are necessary to make a child fit for professionalism later in the years.

A child aspiring to become a civil engineer needs a proper environment to nurture their logic. At the epitome of the civil engineering education, logic is the outcome.

Primary education is essential. It deals with the child's cognitive (intellectual), social, emotional, and physical development. Nurturing communication skills, logic, interpersonal skills, culture, and sustainability enthusiasm in a child is the core of primary education. Today, a civil engineer needs a mix of these skills to lead the sustainability of our planet.

Everything taught in primary is necessary. From mathematics to languages to science and social studies, the child learns a multitude of subjects that trigger their intellectual growth, making their mind resonate with the past, present, and future while practicing engineering later in years. The world of engineering needs no more people thinking in the present only—the main reason for today's global challenges! As a point of reference, environmental engineering, a branch of civil engineering, has taught the world the need to shape engineers who can resonate with yesterday, today, and tomorrow.

Creativity, a key attribute engineers possess, has a strong bearing on nature and nurturing in their formative years. Making an archeological dig into the nurturing of highly creative people depicts an association with creativeness-triggering stimuli early in the formative years. Sometimes, it is either nature or nurture or the both. So, primary education is essential.

1.3.2 Secondary Education

This typically involves ordinary (O-levels) and advanced (A-levels) secondary education. The naming differs in several countries worldwide. However, the English-speaking and Commonwealth countries commonly use the titles O- and A-levels: Ordinary level (11–15 years old) and A-levels (16–18 years old). However, due to changing legislations, the naming constantly varies.

Therefore, it is better to consider the age ranges, that is, primary education (up to 11 years old), O-level (11–15 years), A-levels (16–18 years), and university (above 18 years).

In the United Kingdom, O-level is the General Certificate of Secondary Education (GCSE) O-level, and A-level is the General Certificate of Education (GCE) A-level. So, GCSE exams are taken at the age of 15/16 in the United Kingdom, while A-levels are taken at the age of 17/18. GCE is used for university admissions.

In English-speaking countries and Commonwealth countries like Uganda, the same applies; the O-level is Uganda Certificate of Education (UCE), and the A-level is Uganda Advanced Certificate of Education (UACE). These are the usual routes to secure a place at university and progress to professional engineer status after graduation.

The essential core STEM subjects are mathematics and physical sciences—primarily physics and chemistry. Learning languages such as English is also essential because it proves helpful in communicating fluently. And today, as English being the most spoken language in the world, you do yourself a disservice if you fail to master it. The desirable subjects are geography, art, technical drawing, design, and technology.

Stay on at school or college and study the required subjects. As a piece of advice, having gaps in your education may not help if your goal is to quickly become a professional engineer after attaining the qualifications and training necessary.

1.3.2.1 Ordinary level

The O-level (11–15 years old) qualifications, that is, the GCSE in the United Kingdom, require one to focus on math, English, and physical sciences. Similarly, most of the English-speaking countries, particularly the Commonwealth nations and those that were once members, put a major focus on those subjects at O-levels. Other subjects are also desirable, such as art, geography, information and communications technology (ICT), design, and technology.

Having achieved the GCSE (or equivalent) in the United Kingdom, you can carry on studying in the sixth form college, start an apprenticeship, or join a vocational course.

1.3.2.2 Advanced level

The A-level qualifications are the usual route to secure a place at the university and progress to becoming a chartered engineer in the United Kingdom or professional engineer in similar English and Commonwealth countries after graduation. Usually, mathematics and physics are the essential subjects required to gain access to higher education civil engineering courses. The desirable courses include information and communications technology (ICT), geography and geology, languages, art, and design. Note that environmental engineering, a branch of civil engineering, also draws on such disciplines as chemistry and biology.

The General Certificate of Education Advanced Level, commonly known as the GCE A-Level in the United Kingdom, is the usual route to gain access to universities. The GCE A-Level is a school-leaving qualification offered by secondary schools, sixth form colleges, and further education colleges in the United Kingdom, some Commonwealth countries, and many international schools around the world.

1.3.3 Tertiary Education

This can be termed 'further and higher education.' Once the A-levels (usual route) are completed, one may choose to enroll on an apprenticeship, degree, or vocational course. These qualifications can give you the academic base to become professionally qualified as a chartered engineer or engineering technician in the United Kingdom.

1.3.3.1 Apprenticeship

An apprenticeship is where you work and study part-time, which again can lead to gaining qualifications such as an engineering technician or incorporated engineer status from the Engineering Council of the United Kingdom.

1.3.3.2 Vocational course

Vocational qualifications provide practical and work-related skills. For example, the UK Business and Technology Education Council (BTEC) qualifications are available at different levels and stages of academic and professional careers. BTECs are specialist work-related qualifications available in a range of sectors.

BTEC Level 3 Diploma or Extended Diploma in Construction and the Built Environment—Civil Engineering may lead you to professional qualification as an Engineering Technician or Incorporated Engineer with the ICE, United Kingdom.

From O-levels, you may choose to enroll on BTEC Level 3 Diploma or Extended Diploma in Construction and the Built Environment—Civil Engineering. Note: BTEC Level 3 is equivalent to A-levels.

From A-levels, you may choose to enroll on BTEC Level 4–5 Diploma or Extended Diploma in Construction and the Built Environment—Civil Engineering. Note: BTEC Level 4–5 is equivalent to first- and second-year undergraduate studies. These qualifications may lead to professional qualification as an Engineering Technician or Incorporated Engineer with the ICE, United Kingdom.

1.3.3.3 Degree

Completion of the A-levels is the usual route to secure a place at a university. As outlined in Section 1.3.3.2, BTEC Level 3 is equivalent to A-levels, and BTEC Level 4–5 is equivalent to first- and second-year undergraduate studies. Either way, you gain a place to study at university. So, it is your choice to choose which route to take after the GSCEs O-levels.

Taking a civil engineering degree at an accredited university is essential if your goal is to become accredited by professional civil engineering institutions in future and gain professional recognition, for example, as a chartered engineer.

You must plan all the way to the end because some universities are not accredited, and some courses may be outdated, which could potentially harm your career. So, looking out for institutions that are accredited and enrolling on civil engineering courses that are up to date as required is not a bad idea.

Progressing to professional registration, such as chartered engineering, requires you to have taken accredited courses in universities licensed by the government to teach those courses.

Some professional institutions would require a master's degree to register as an engineer. For example, an integrated Master of Engineering (MEng) degree offered by several accredited universities in the United Kingdom combines undergraduate and postgraduate studies, fulfilling the educational requirements for the chartered engineer status. Once you complete it at an accredited university, you may automatically qualify for the academic base required for a chartered engineer in the United Kingdom.

1.4 HOW TO BECOME A PROFESSIONAL CIVIL ENGINEER

Section 1.3 outlined the STEM courses necessary for an aspiring civil engineer. However, to become a professional civil engineer, that is not enough. To become a professional civil engineer, you must have the right attitude, academic base, skills, experience, and ability.

The right attitude is an essential component for someone aspiring to become a professional civil engineer. More than ever, the world faces challenges that require people of the right thinking, behaviors, and emotionally competent to contribute to sustainable promotion of the built environment. Those are people who keep an ethical code of conduct and see saving humanity as the highest good in the hierarchy of doing good, while delivering their expertise connecting the built environment. Having the right attitude means staying optimistic and positive in all circumstances while delivering professional expertise.

As a professional, you meet several challenges, and you have to overcome them ethically because the overarching goal of civil engineers is to save humanity. Civil engineers have saved humanity more than any individuals of other professions throughout history—from constructing earthquake-resistant structures to developing sewerage systems to save communities from cholera outbreaks, from building bridges to connect communities and transport goods to developing massive irrigation systems that facilitate the production of large amounts of food to save communities from hunger, and much more. The modern world cannot function without civil engineers!

Civil engineers understand that we have no other place to go when our planet becomes unsafe and strive to save it through their actions by combating climate change, protecting interest groups in every decision, and promoting dignity, human values, equality, equity, and justice. That means a civil engineer's work is at the heart of the society, influencing the built environment at the core of every decision. In that process, civil engineers play a pivotal role in listening to the views of stakeholders and differing interest groups.

The right academic courses must be taken to gain the right knowledge and understanding. These are outlined in Section 1.3. Professional institutions demand candidates to have the right knowledge and understanding of civil engineering and to have received the training necessary for the grade of membership they aim for. Specifically, the ICE—the oldest institution in the world—has several grades, that is, engineering technician, engineer, fellow, and associate member, that call for different academic credentials for one to join the fraternity. Anyone may join as a student or graduate to smoothly attain an appropriate grade.

Taking courses at accredited universities is important. The first step an institution would do, such as the ICE, is to assess your academic base. The academic base would direct you to the grade of membership you can apply for.

Training to gain sufficient skills after attaining academic qualifications is necessary. You train and gain hands-on skills and apply scientific knowledge to solve problems under the guidance and supervision of professionals. In pursuit of professional registration, you learn, relearn, and unlearn relentlessly. At some point, you have to stand on your own, take the lead, and make independent decisions on civil engineering projects as this is necessary to demonstrate that you have become the person fit to practice on your own. In that regard, you creatively harness the great sources of power in nature to innovate safe practical solutions to civil engineering problems.

Skills: the skills you need to grow as an engineer include interpersonal skills, communication skills, and commercial skills. These skills are gained through work experience and training. But also one may gain them through continuous learning, learning on the job, and a relentless effort with passion to improve the built environment.

Experience and ability: you must have taken responsibility and provided accountability on projects of a civil engineering nature and demonstrated management and leadership as an engineer.

You must develop the skills, techniques, and knowledge that professional engineers need and learn the underpinning science and mathematics. Engineering is a creative and analytical subject.

Membership in outstanding professional institutions is stage-managed, and you must pass a series of examinations and/or interviews. However, the most important element is to show evidence of gaining sufficient training and readiness to practice on your own as a civil engineer.

1.5 INCULCATING LIFELONG INTERDISCIPLINARY LEARNING SKILLS

The need to develop an interdisciplinary approach is key for today's civil engineers because they require an understanding of diverse disciplines to solve complex engineering problems, especially in leading and managing multicultural teams. Through an interdisciplinary approach, civil engineers are able to combine insights and methods from diverse fields, such as science, art, humanities, and social sciences.

The exponential growth of technological innovations and global challenges has created an unprecedented need for lifelong interdisciplinary learning skills (LILS), essential for driving innovation, adaptability, and collaborative problem-solving in today's complex world. You don't simply need to acquire lifelong learning skills (LLS) but LILS to remain competitive. This is because civil engineering is constantly evolving, and in the 21st century, it relates to becoming more interdisciplinary.

Simply put, LILS refers to the abilities, knowledge, and attitudes necessary for individuals to continuously learn and adapt across multiple disciplines and contexts, integrate knowledge from diverse fields to tackle complex challenges, and navigate dynamic, interconnected environments.

Therefore, inculcating LILS as a student of civil engineering or a constantly evolving civil engineer is essential because of the rapid evolution of technology and materials, increasing complexity of infrastructure projects, growing focus on sustainability and environmental considerations, need for interdisciplinary collaboration, and constant updates in codes, regulations, and standards. All these need an engineer who remains vigilant, adaptable, and engages in continuous learning in multiple disciplines to remain relevant and competent, spreading horizontally not just vertically.

Therefore, to stay ahead in the field, you must cultivate a combination of qualities: vigilance in anticipating challenges, adaptability in navigating change, and a commitment to continuous learning across diverse disciplines—that is, adopt a dynamic mindset. This holistic approach enables you to broaden your expertise horizontally, complementing the depth of knowledge with the breadth of understanding.

Just imagine being hired, good at highway design, but not conversant with the latest technologies that you must use and that require customization to solve a problem in a specific region. How will you go over it? Or imagine you are hired to present a project proposal embracing brevity as much as possible in the short time you have. You must understand the appropriate current techniques, tools, and technologies to shoot this task amid several multicultural audiences—whether online or in-person.

The shifting societal needs and expectations, increasing complexity of projects and systems, and globalization and interdisciplinary collaboration all need an all-embracing professional.

Therefore, up-to-date engineers need to move with the trend to deliver effective and efficient engineering solutions. And, of course, this trend is not specific, but rather multicultural. If you can navigate multicultural dynamics effectively in the world of civil engineering, you're well positioned to succeed and go places, delivering exceptional professional expertise. So, embracing LILS is essential for all engineers at all career stages, from entry-level to senior professionals.

Specifically, interdisciplinary learning empowers students to combine frameworks and concepts from multiple disciplines to examine a theme or solve a problem from different perspectives. This means, in practice, students will gain the skillset to tackle complexity and change effectively throughout their careers; thus, they become more adaptable.

For a constantly evolving civil engineer whose overarching goal is saving the planet, for example, by reducing carbon footprint while delivering professional expertise, embracing LILS is an essential practice and a must-go-to approach to deliver sound engineering solutions for the 21st century.

The benefits of LILS for civil engineers are numerous and impactful. These benefits include staying up to date with industry advancements, such as embracing Building Information Modeling (BIM) and Geographic Information Systems (GIS); enhancing problem-solving and critical thinking skills to tackle complex projects such as sustainable infrastructure development and disaster resilience; and expanding knowledge of new materials, techniques, and software. Learning programing languages such as C++, Auto LISP, VBA, and C# support civil engineers in various ways, which is a step in the right direction.

Additionally, LILS enables civil engineers to adapt to changing project requirements and client needs, such as incorporating green building materials and energy-efficient designs from environmental engineering and architecture. It also expands their knowledge of new materials, techniques, and software, including advanced construction management tools and cutting-edge technologies such as Artificial Intelligence (AI) and Machine Learning (ML) from computer science and data analytics.

Furthermore, LILS paves the way for advancing to leadership roles or specialized fields, such as structural engineering (e.g., seismic design), water resources (e.g., hydrology), transportation planning (e.g., smart traffic management), construction management, geotechnical engineering, and environmental engineering.

Professional engineering organizations now emphasize lifelong learning. These include the American Society of Civil Engineers (ASCE), ICE, United Kingdom, National Society of Professional Engineers (NSPE), Institution of Engineering and Technology (IET), and most professional engineering bodies around the world. These organizations provide resources, training, and certifications to support lifelong learning and professional development.

The ASCE launched its first multidisciplinary, open-access journal, in December 2022 [5]. It is ASCE's first fully open-access journal and its first fully multidisciplinary journal. ASCE OPEN's first editor-in-chief, **Ertugrul Taciroglu**, clearly spelt out that the vision of the journal is to publish interdisciplinary research papers that converge civil engineering subdisciplines with expertise from social sciences, economics, and other fields. This convergence research addresses complex sustainability and resilience challenges, filling a gap in existing ASCE journals.

ASCE OPEN access is likely the first platform to routinely disseminate such innovative, multidisciplinary work.

> There are many problems in civil engineering—especially in the areas of sustainability and resilience—that require knowledge from multiple disciplines. ASCE doesn't have a journal that can routinely disseminate those kinds of papers. The ASCE OPEN will be the first one. I'm pretty excited about it,
>
> **Ertugrul Taciroglu**

This shows that the relevance of interdisciplinary learning is gaining momentum in the 21st century. In summary, LILS are essential for all engineers, regardless of discipline or career stage, to stay up to date, adapt to changing environments, and maintain expertise especially today when sustainability is the hot topic of the 21st century.

1.5.1 Civil Engineering Students

Civil engineering students need to embrace LILS to become promising in their journey to becoming professionals later in their careers. To begin with, thriving in the rapidly evolving industry impacted by climate change, students must prioritize resilience and adaptability as essential skills for success.

Learn to manage time and be adaptable. Adaptability as a skill is very important. Sometimes, people reared in environments of surplus fail in adaptability compared to those reared in crises or shortage. Every environment has its advantages—those with surplus environments will have access to whatever learning they need, but remember a crisis is also an opportunity to grow, on the other hand. Those grown on urban balconies with plenty of everything may not quickly understand what it means to have nothing! Now, that sustainability has knocked on our doors, smartness in decisions we do, and the innovations and creatives we make as engineers, make us unique. Be unique—turn whatever situation you encounter in your career and personal life into an opportunity to learn. This will help you in the long run.

Students need to continuously learn technical skills such as BIM, GIS, and CAD. Learn these skills to be competitive during your early career stages, because at this stage, you barely can't make many sound decisions but rather support teams with hands-on skills. To stand out, continuously learn software tools necessary for becoming a promising multiculturally sound student.

Students need to acquire communication and collaboration skills. For example, boosting your vocabulary enables effective communication, leveraging brevity to convey ideas with clarity and precision. Enhanced vocabulary helps convey ideas, thoughts, and emotions more effectively. It helps you develop better comprehension, and understanding complex texts, articles, and conversations becomes easier.

As a civil engineering student, you need skills that support environmental sustainability and foster sustainable growth. Consequently, joining and contributing to sustainability and environmental awareness campaigns is crucial. Engaging with advocacy groups is a valuable strategy to nurture skills in sustainability and environmental awareness. Advocacy groups expose you to multicultural teams that will support your interdisciplinary learning.

Developing leadership and management skills is crucial in the constantly evolving civil engineering industry. Once mastered, these skills unlock one of the essential attributes necessary for civil engineers to deliver successful projects. This is because complex projects require effective coordination and communication. The evolving technologies demand adaptability and strategic thinking. Also, stakeholder expectations necessitate strong relationships and negotiation. Project timelines and budgets demand efficient resource allocation. What about team dynamics that constantly require emotional intelligence and conflict resolution? Don't you want to develop emotional competence and learn conflict resolution? All these need sound leadership and management skills.

As a civil engineering student continuously engage in data analysis and interpretation of sorts and develop critical thinking and problem-solving abilities to stand out as an engineer. Critical thinking sets engineers apart from the rest, in most cases.

Chapter 2 outlines ethics and professional development. Students need to be exposed to ethics at an early stage in their career. As a student, you need to know that ethics are the industry's Holy Writ, providing moral guidance and standards for professionals. Professional societies issue the ethical codes of conduct to guide the work of engineers. Through interdisciplinary learning, you can boost your understanding of ethics from time to time with a relentless effort to be of value to the industry.

More than ever, networking and relationship-building have never been important as they are today because of the growing power of the interconnected world and globalization. Staying connected with colleagues, making new friends, peers, and mentors, is crucial. Continuously develop networking skills such as gym-going, partying, or Rotaract if possible; join writers' clubs, develop social media skills, and attend networking events, among others. All these provide a platform to continuously learn from multiple fields.

Specifically, committing to a gym routine requires self-discipline that involves setting goals, planning workouts, and sticking to a schedule. This discipline often translates into other areas of life, demonstrating that gym-going is a remarkably resourceful venture in which to invest time. You might find yourself more focused on tasks, better at time management, and more diligent in achieving personal or professional goals. Section 6.3 outlines the connection between the gym and the civil engineer and what a civil engineer can potentially learn from the gym.

Other avenues to explore include reading or listening to podcasts and making podcasts if possible. Set goals and start new things, whether they are work- or hobby-related while keeping a running list of things that interest you. Seek and embrace change and take courses, from free and quick to a degree- or certificate-focused.

When you do all or most of these, be assured to meet the demands of the ever-evolving civil engineering industry because you will meet what the society expects of you—the civil engineer in the making or training!

1.5.2 Professional Engineers

Professional engineers need to stay current with emerging trends such as sustainability and climate resilience, digitalization and industry 4.0, AI and ML, and Internet of Things (IoT) and connected systems.

Therefore, lifelong learning strategies for professional engineers include conducting continuing education courses and workshops. This can be by attending in-person workshops delivered through a lecture-style or hands-on training. Engineers can also attend conferences, seminars, and webinars. These may be live or pre-recorded sessions. Attending online tutorials and certifications as appropriate can help develop your skills further.

Video courses, that is, self-paced, online modules, are a vital resource for conducting LLS. Also, mentorship and coaching, helping young engineers to grow professionally, helps the mentor in furthering their skills.

Another rich area that requires exploring is publishing. As a professional engineer, you can get involved in industry publications and research journals to boost your skills and remain up to date with emerging trends, hence boosting your reputation.

Participation in professional organizations is crucial to remain up to date. Do not only limit yourself to engineering bodies. As mentioned in Section 1.5, the need for interdisciplinary learning to meet the demands of a sustainable world is growing daily. So, engage with several professional societies to bring new knowledge from multiple fields to advance engineering research across the profession and related disciplines that will help inform, develop, and inspire sustainable action, overall.

Project-based learning and knowledge sharing are important for a professional engineer. Here, engineers work on real-world projects, integrating subject matter, critical thinking, and problem-solving skills. The tools for project-based learning and knowledge sharing include project management tools (e.g., Asana and Trello), collaboration platforms (e.g., Slack and Microsoft Teams), and learning management systems (e.g., Canvas and Moodle).

1.6 BRANCHES OF CIVIL ENGINEERING

Civil engineering has several branches, and sometimes, it may not be easy to delineate the standalone subdisciplines because of the nature of crosscutting knowledge from one sub-discipline to another. Some civil engineering disciplines overlap with others. That means the naming of a sub-discipline may differ in some books. Nevertheless, the main branches of civil engineering are outlined below.

1.6.1 Transportation Engineering

Transportation engineering or transport engineering is a branch of civil engineering that deals with the planning, construction, and management of transportation facilities. It specifically deals with the application of technology and scientific principles to the planning, design, operation, and management of facilities for any mode of transportation in order to provide safe, efficient, rapid, comfortable, convenient, economical, and environmentally compatible movement of people and goods (transport). See Figure 1.2 for an interchange.

There are six divisions related to transportation engineering, that is, highway, air transportation, waterway, aerospace, coastal and ocean, and urban transportation. However, the major specialties of transportation engineering include the following.

1.6.1.1 Traffic engineering

This is a sub-discipline of transportation engineering that focuses on the infrastructure necessary for transportation.

1.6.1.2 Highway engineering

This is a sub-discipline of transportation engineering that focuses on the major roadways and transportation systems involving automobiles. Highway engineering usually involves the construction and design of highways.

1.6.1.3 Railway systems engineering

This is a sub-discipline of transportation engineering that focuses on the design, construction, and operation of all types

FIGURE 1.2 Interchange.

of rail transport systems. A rail systems engineer is primarily responsible for providing insights and technical engineering expertise on railway projects and systems such as traction power, train and traffic signal controls, fare collection, and rail vehicles.

1.6.2 Water Resources Engineering

Water resource engineering is a branch of civil engineering that primarily deals with the design and construction of hydraulic structures. These structures include dams, breakwaters, canals, water distribution systems, and sewage conduits; the management of waterways, such as erosion protection and flood protection; and environmental management, such as prediction of the mixing and transport of pollutants in surface water. Therefore, the water resources engineer is responsible for the design of the hydraulic structures as well as their implementation and safety precautions that must be closely adhered to when dealing with them. Figure 1.3 shows a hydroelectric power-generating dam.

The most familiar applications of resource engineering include hydroelectric power development, water supply, irrigation, and navigation, which involve utilizing water for beneficial purposes. The concern for preserving the environment has increased the importance of water resource engineering.

The quantitative study of the hydrologic cycle is conducted through water resource engineering, that is, the distribution and circulation of water linking the Earth's atmosphere, land, and oceans. The field develops water-related systems and resources, applying engineering principles and techniques that address water-related challenges such as water supply, water quality, flood control, and water resource sustainability.

1.6.3 Environmental Engineering

Environmental engineering is a profession that applies mathematics and science to utilize the properties of matter and sources of energy in the solution of problems of environmental sanitation. These include the provision of safe, palatable, and ample public water supplies; the proper disposal of or recycle of wastewater and solid wastes; the adequate drainage of urban and rural areas for proper sanitation; and the control of water, soil, and atmospheric pollution, and the social and environmental impact of these solutions. Furthermore, it is concerned with engineering problems in the field of public health, such as control of arthropod-borne diseases; the elimination of industrial health hazards; the provision of adequate sanitation in urban, rural, and recreational areas; and the effect of technological advances on the environment [6,7].

The discipline is largely defined by problems rather than by technical or scientific methods. The typical problems include remediation of a contaminated site (fixing the past), treatment of a dirty effluent (dealing with the present), and pollution avoidance (planning for the future). It deals with the development of processes and infrastructure for the supply of water, the disposal of waste, and the control of pollution of all kinds.

FIGURE 1.3 Hydroelectric power dam.

FIGURE 1.4 Sewage treatment plant.

These endeavors protect public health by preventing disease transmission, and they preserve the quality of the environment by preventing the contamination and degradation of air, water, and land resources. Figure 1.4 shows a sewage treatment plant system, with sludge scrapers used to remove sludge from sedimentation tanks.

Environmental engineering draws on such disciplines as chemistry, geology, ecology, hydraulics, hydrology, microbiology, economics, and mathematics.

Environmental engineering was traditionally a specialized field within civil engineering and was called sanitary engineering until the mid-1960s when the more accurate name *'environmental engineering'* was adopted. However, today, the discipline is slowly transforming into its own field. This is because environmental engineering requires more chemistry and biology than other STEM subjects.

1.6.4 Geotechnical Engineering

Geotechnical engineering is a branch of civil engineering that systematically applies the techniques which allow construction on, in, or with geomaterials, i.e., soil and rock. It uses the principles of soil and rock mechanics to solve engineering problems. It also relies on the knowledge of geology, hydrology, geophysics, and other related sciences. Geotechnical engineering deals with the engineering behavior of earth materials.

Every civil engineering structure is related to soil in some way. Its design depends on the properties of the soil or rock. Figure 1.5 shows a geotechnical drilling rig collecting soil samples to provide engineers and geologists with essential data for geotechnical applications.

Geotechnical operations are vital concerning soil sampling, investigating geomaterial properties, controlling groundwater level and flow, and environmental and hydrological interactions. Foundation engineering, excavations and supporting ground structures, underground structures, dams, natural or artificial fills, roads and airports, subgrades and ground structures, and slope stability assessments are examples of geotechnical engineering applications in practice. Geotechnical engineering also has several applications in military engineering, mining engineering, petroleum engineering, coastal engineering, and offshore construction. The fields of geotechnical engineering and engineering geology have overlapping knowledge areas.

Until the last century, geotechnical engineering was largely empirical and based on observation and careful reflection. Remarkable scientific advancement in geotechnical engineering was achieved in the post-World War II era and continues today with use of high-performance computers, sensors, data visualization, and advanced soil testing.

Geotechnical engineering was defined in a Memorandum of Understanding (1999) on the proposed unification of the ICE Ground Board (United Kingdom) and the British Geotechnical Association (BGA). Today, the BGA performs the role of the Ground Board for the ICE. Appendix A of that memorandum [8], establishing the British Geotechnical Association, sets out the following definition:

Geotechnical engineering is the application of the sciences of soil mechanics and rock mechanics, engineering geology and other related disciplines to civil engineering construction, the extractive industries and the preservation and enhancement of the environment.

FIGURE 1.5 Geotechnical drilling rig at work.

1.6.5 Structural Engineering

Structural engineering is a branch of civil engineering that deals with the structural analysis and design of structures. It specifically concerns the application of the laws of physics, mathematics, and empirical knowledge to safely design the 'bones and joints' that create the form and shape of human-made structures.

The load-bearing elements create the form and shape of human-made structures. These active structural elements hold the weight of the elements above them by transferring their weight to a foundational structure below them.

Structural engineers are responsible for making sure that the structures we use in our daily lives, like bridges and tall buildings (storied structures and skyscrapers), are safe and stable and do not collapse under applied loads. Figures 1.6 and 1.7 show a super high-rise building. Structural engineers calculate the stability, strength, rigidity, and earthquake-susceptibility of built structures.

Structural engineers analyze and design structures that safely bear or resist stresses, forces, and loads. The design must satisfy the project specifications while meeting all safety regulations. The structure must endure massive loads as well as natural disasters and climate changes.

The structural designs produced by structural engineers are integrated with those of other designers such as architects and building services engineers and often supervise the construction of projects by contractors on site.

Structural engineers can also be involved in the design of machinery, medical equipment, and vehicles where structural integrity affects the functioning and safety.

Structural Engineering is often considered the 'King' of all civil engineering branches because it offers the foundation of infrastructure projects, including buildings, bridges, and dams. It is important to note that it is one of the oldest branches of civil engineering. And because of its complexity, for example, dealing with complex loads, stresses, and materials, requiring advanced mathematical modeling and analysis, it is exceptional. In terms of safety, structural integrity is critical to all civil engineering projects to ensure public safety, making structural engineers responsible for designing and analyzing structures that can withstand various loads and hazards. In other words, due to the fact that all branches of civil engineering cater for loads, structural engineering serves as the backbone providing safety guidelines to avoid hazards to people, goods, and environment as a whole. It offers an interdisciplinary approach—structural engineering overlaps with other branches, such as the following: geotechnical engineering (soil–structure interaction), transportation engineering (bridge design), water resources engineering (dam design), and materials science (material properties). That is why structural engineers are in high demand worldwide, with opportunities in various industries, including construction, oil and gas, and renewable energy. It is often considered among the most prestigious branches, with iconic projects like skyscrapers, long-span bridges, and historic landmarks.

1.6.6 Municipal Engineering

Municipal engineering is a branch of civil engineering that deals with the design, construction, operation, and maintenance

FIGURE 1.6   International Commerce Centre, 1 Austin Road West, Kowloon, Hong Kong. Image credits: Richard Li Chengzhe.

FIGURE 1.7   CITIC Centre, Central Business Division, 4 Jianguomen Outer St, Chao Yang Qu, Bei Jing Shi, China, 100020. Image credits: Richard Li Chengzhe.

FIGURE 1.8 Urban sewer pipeline development.

of several types of infrastructure, facilities, and utility systems in an urban environment. Specifically, municipal engineering is concerned with municipal infrastructure, and this involves specifying, designing, constructing, and maintaining streets, pavements, sidewalks, water supply networks, sewer systems, street lighting, municipal solid waste management and disposal, storage depots for bulk materials used for maintenance and public works, public parks, and cycling infrastructure.

Sometimes, municipal engineering is referred to as urban engineering. So, it deals with problems peculiar to urban life. Figure 1.8 shows an urban sewer pipeline under development.

1.6.7 Construction Engineering

Construction engineering as a branch of civil engineering is a specialized field within civil engineering that focuses on the planning, design, construction, and management of infrastructure projects, that includes buildings, bridges, and roads. It bridges the gap between design and construction, emphasizing practical application, project management, and ensuring projects are completed safely, on time, and within the allocated budget. Construction engineering covers different civil engineering materials—suitability, merits and demerits, and advancing smart materials. It also looks at quality control during project execution, maintenance of the structures, use of sensor technologies, robotics technology in construction, construction management, retrofitting and rehabilitation, and sustainable development.

1.6.8 Civil Engineering Surveying

Civil engineering surveying is a branch of civil engineering that focuses on determining the relative positions of points on, above, or below the Earth's surface. It involves measuring and recording data about land features, structures, and other elements to create maps, plans, and other representations for use in civil engineering projects. It involves remote sensing, GIS, GPS, electronic equipment, and application of drone technology in civil engineering and construction. In essence, civil engineering surveying provides the essential data needed for planning, designing, and construction of various civil engineering projects, such as roads, buildings, bridges, dams, and pipelines.

1.7 THE INSTITUTION OF CIVIL ENGINEERS, UNITED KINGDOM

In 1818, a small group of brilliant young engineers founded the iconic ICE, the world's first professional engineering body.

Since its founding, the institution's purpose has been:

To improve lives by ensuring the world has the engineering capacity and infrastructure systems it needs to enable our planet and our people to thrive.

Headquartered in London, the ICE is an international engineering body whose current membership is over 97,000 members practicing on projects around the world. Since its founding, the ICE has attracted some of the world's most famous engineers.

The ICE seeks to keep growing its status as an international engineering body but not as a competitor of local engineering institutions, and anyone with appropriate knowledge, skills, experience, and ability may join from anywhere around the world.

Commemorating the founding of this icon civil engineering institution is essential because it is a learned society which holds in custody the portrait for John Smeaton, the first person to describe himself as a civil engineer, painted by George Romney (1734–1802).

1.7.1 What Does ICE Mean to the World of Civil Engineering?

Associating with the ICE or becoming a member of the ICE is an achievement to celebrate. It means gaining a reputation for identifying with the right minds that have shaped the built environment and sustainability that the world deserves for over 200 years.

Identifying with the ICE as a member means gaining a reputation in the public for your expertise and competence to solve problems of a civil engineering nature—a professional to rely on to advise governments, agencies, and organizations about civil engineering and infrastructure development, which is key to developing economies around the world. Identifying with the ICE puts you in a class apart from the rest of the pack as a civil engineer. The ICE is well known for its diversity sensitivity, which makes it stand out worldwide.

As a member of the ICE, you get the privilege to learn and grow continuously over the years with tailor-made lectures, courses, and webinars from the institution. The rich history that the institution has makes it a home for all the information necessary for a civil engineer to develop and learn, relearn, and unlearn.

Joining the institution makes you gain a deep insight into how the profession has diverse fields—the employability of civil engineers is just amazing! It is not about steel and concrete only! In today's changing world, the flexibility of civil engineers has been growing daily.

Just as one may be a professional football referee and coach or simply a professional coach or/and footballer is the same way the ICE provides several grades of membership to suit all categories of members. With the right attitude, academic credentials, and experience, one may easily find their membership grade and join the institution from anywhere around the globe.

The ICE is a fully established organization delivering sustainable development through knowledge, skills, and professional expertise for over 200 years. Once you join it, you automatically become a sustainable development enthusiast; you get interested in shaping zero to save the planet. It seeks to continue developing its status globally as an international body, but not as a competitor with local organizations. Anyone with appropriate skills, knowledge, experience, and ability may join the institution from anywhere at any time around the globe.

A society established for the general advancement of Mechanical Science, and more particularly for promoting the acquisition of that species of knowledge which constitutes the profession of a civil engineer, being the art of directing the great sources of power in nature for the use and convenience of man.

Thomas Tredgold, 1828—about the ICE

1.8 AMERICAN SOCIETY OF CIVIL ENGINEERS (ASCE)

Founded in 1852, 34 years after the ICE was founded, the **ASCE** is a tax-exempt professional body and the oldest national engineering society in the United States. Its constitution was based on the older **Boston Society of Civil Engineers** from 1848. Headquartered in Reston, Virginia, the ASCE equally plays a vital role in shaping the civil engineering profession, promoting excellence, and addressing global challenges.

The ASCE sets and publishes standards, guidelines, and codes for civil engineering projects, ensuring safety, quality, and best practices. It also engages in professional development by providing training, education, and networking opportunities for civil engineers, helping them stay updated on industry developments. The ASCE also influences public policy and advocates for infrastructure investment, sustainability, and community resilience. It disseminates knowledge through publications, conferences, and research, fostering innovation and collaboration in the civil engineering community.

1.9 NOTABLE CIVIL ENGINEERS IN HISTORY

This list of notable engineers provides their contributions to civil engineering and the routes they took to become civil engineers. Some took apprenticeships, while others went through vocational institutes and universities, but they all spent years training, learning, and getting necessary qualifications.

The full list is far more exhaustive, but the list below gives you an idea of the notable civil engineers in history who have influenced the civil engineering profession.

This list of notable civil engineers includes those members of the field after the title 'civil engineer' was first used officially in 1771. Therefore, the list of notable engineers is provided indicatively, not exhaustively.

1.9.1 John Smeaton (1724–1792)

John Smeaton, a British national (Figure 1.9), was the first engineer probably to describe himself as a 'civil engineer' around 1760. He was the founding member of the Society of Civil Engineers in 1771 [1].

John Smeaton is a highly regarded figure in the history of civil engineering for his role in breaking up from military

FIGURE 1.9 John Smeaton.

engineering to found a new profession (civil engineering) in Great Britain, which would later spread all over the world.

The name Smeatonian Society of Civil Engineers was derived from Smeaton, the leading engineer of the mid-eighteenth century who coined the name of the civil engineering profession around 1760, and led the formation of the Society [9].

1.9.2 Thomas Tredgold (1788–1829)

Thomas Tredgold (Figure 1.10) was an English engineer and author, well known for his work on railroad construction. He was the 47th member of the ICE and the author of the definition of civil engineering in its Royal Charter. He defined civil engineering, forming the basis of the charter of the ICE in 1828.

The following is the definition Thomas Tredgold proposed on which the charter of the ICE based itself in 1828:

> A Society for the general advancement of Mechanical Science, and more particularly for promoting the acquisition of that species of knowledge which constitutes the profession of a Civil Engineer; being the art of directing the great sources of power in Nature for the use and convenience of man, as the means of production and of traffic in states, both for external and internal trade, as applied in the construction of roads, bridges, aqueducts, canals, river navigation, and docks, for internal intercourse and exchange; and in the construction of ports harbours, moles, breakwaters, and lighthouses, and in the art of navigation by artificial power, for the purposes of commerce; and in the construction and adaptation of machinery, and in the drainage of cities and towns.
>
> *[10]*

Tredgold's definition of civil engineering was updated by the ICE Council in 2007 [3]. Several scholars and engineers have since shared recommendations on the same.

1.9.3 Thomas Telford (1757–1834)

Thomas Telford (Figure 1.11), a Scottish civil engineer, was one of the founders and the first president of the ICE, United Kingdom, supporting the institution to get the ICE's Royal Charter in 1828, a post he held for 14 years until his death.

This charter (updated by Queen Elizabeth II in 1975) gives ICE the status as the leading institution for the civil engineering profession in the United Kingdom and in the Commonwealth countries. Telford designed numerous infrastructure projects, including harbors and tunnels, in his country, Scotland. During his lifetime, he was involved in many projects such as canals, bridges, roads, and harbors. He was widely consulted from different parts of the world including the Russian and Swedish governments.

FIGURE 1.10 Thomas Tredgold.

FIGURE 1.11 Thomas Telford.

FIGURE 1.12 Isambard Kingdom Brunel.

FIGURE 1.13 Sir Joseph William Bazalgette.

At the age of 14, he was apprenticed to a stonemason, and some of his earliest work can still be seen on the bridge across the River Esk in Langholm in Dumfries and Galloway. Telford went to London in February 1782, where he met architects Robert Adam and Sir William Chambers, and worked as a stonemason on the building of Somerset House [11].

1.9.4 Isambard Kingdom Brunel (1806–1859)

Isambard Kingdom Brunel (Figure 1.12) was a British civil engineer and mechanical engineer. Brunel took an apprenticeship going through the footsteps of his father, Marc Isambard Brunel—a civil engineer behind Thames Tunnel and also known for making ship blocks.

Isambard Kingdom Brunel is one of the most celebrated civil engineers of all time. Taking on a vast variety of projects throughout his lifetime, and revolutionizing them along the way, he designed tunnels, bridges, railway lines, ships and much more [12].

Isambard Kingdom Brunel's most significant project is the Royal Albert Bridge, a railway bridge that spans the River Tamar in England between Plymouth, Devon and Saltash, Cornwall.

Brunel built dockyards, the Great Western Railway (GWR) as the chief engineer, a series of steamships including the first purpose-built transatlantic steamship, and numerous important bridges and tunnels. His designs revolutionized public transport and modern engineering in the United Kingdom. Brunel is considered one of the most ingenious and prolific figures in engineering history and a notable figure in the Industrial Revolution.

According to BBC history, Brunel was one of the most versatile and audacious engineers of the 19th century. In 2002, Brunel was placed second in a BBC public poll to determine the '100 Greatest Britons.'

1.9.5 Sir Joseph Bazalgette (1819–1891)

Sir Joseph William Bazalgette (Figure 1.13) was an English civil engineer in the 19th century who built London's first sewer network that helped wipe out cholera in the capital. This sewer network is still in use today.

He was called the savior of the Great Stink for saving Londoners from the cholera epidemics. Bazalgette's sewage system was developed in response to the Great Stink of 1858. He also designed the Albert, Victoria, and Chelsea embankments, which housed the sewers in central London [13].

Although numerous individuals contributed to the construction and engineering of London's inaugural sewerage system, Bazalgette was intimately involved throughout the process. He scrutinized hundreds of proposals to identify the optimal design. While no solution is completely impervious to future challenges, Bazalgette's emphasis on utilizing more extensive tunnels and resilient Portland cement enabled the Victorian sewerage system to accommodate London's expanding population.

> Bazalgette probably did more good, and saved more lives, than any single Victorian official.
>
> —*John Doxat, historian and author*

Sir Joseph William Bazalgette is believed to have spent his early career articled to Sir John Macneill (1793–1880), an Irish civil engineer, working on railway projects and obtained sufficient experience (partly in China and Ireland) in land drainage and reclamation. This experience enabled him to set up his own London consulting firm in 1842. Sir John Macneill, Bazalgette's trainer, was a close associate of Thomas Telford—the first president of the ICE.

Sir Joseph Bazalgette was the 24th president of the ICE, United Kingdom, in 1884.

FIGURE 1.14 Benjamin Wright.

1.9.6 Benjamin Wright (1770–1842)

In 1969, the ASCE declared Benjamin Wright (Figure 1.14) the 'Father of American Civil Engineering' in the United States. Benjamin Wright was a surveyor by training and education and also a judge and civic leader.

Benjamin Wright received much of his training while living with an uncle, during which he learned law and the basics of surveying.

During his career, Wright became involved in railroads and conducted route location surveys for tracks in New York, Virginia, and even Cuba. During his career, Benjamin Wright played a key role in the development of the transportation infrastructure in the young nation—setting the stage for westward expansion and America's 'manifest destiny.'

Wright was the chief engineer of the Chesapeake and Ohio Canal from 1828 to 1831 and was a consultant for the St. Lawrence Canal project in 1833 [14].

1.9.7 Archie A. McCorkindale (1923–2018)

Archie A. McCorkindale, a Scottish civil engineer, was the first civil engineer to emerge in both East and Central Africa in the 20th century. After graduating in Civil Engineering from the University of Glasgow in 1944, McCorkindale worked for 2 years with the Ministry of Aircraft Production in the United Kingdom on aircraft design. He then joined a civil engineering contractor in the United Kingdom and came with them to East Africa in 1948.

He was the first Chairman of the Kenya's Association of Consulting Engineers in 1968. McCorkindale was the representative of the ICE, United Kingdom, in East Africa, Somalia, and Mauritius from 1969 to 1989. He was the president of the Institution of Engineers Kenya from 1983 to 1985 [15].

He was involved in the design and construction of over 2,000 km of roads in Kenya as well as roads in Tanzania, Uganda, Yemen, and many other countries. He worked on airport, water supply, road, and other civil engineering projects in East Africa and Mauritius between 1953 and 1962, when he was appointed Chief Engineer of Sir Alexander Gibb & Partners. In 1964, he became Resident Partner of the firm responsible for the firm's work in East Africa, Rhodesia, and Mauritius.

McCorkindale played a major role in civil engineering in Kenya and throughout East Africa until his retirement in 1988. In the same year, Queen Elizabeth II honored McCorkindale by making him an Officer of the Most Excellent Order of the British Empire (OBE) for his lifelong engineering service in East Africa. This made McCorkindale the foremost engineer in East Africa.

1.9.8 Terzaghi Karl (1883–1963)

Terzaghi Karl (Figure 1.15), an Austrian, is considered the father of soil mechanics and remains the only engineer to receive the Norman medal four times, which is the highest award of the ASCE [16]. Terzaghi graduated in 1904 with a 'Diplom-Ingenieur' (MSc) in mechanical engineering from Technical University in Graz, Germany.

Karl Terzaghi had a passion for geology, and his experience working on dams and hydroelectric power plants made him identify knowledge gaps in geotechnical engineering. He published Erdbaumechanik auf Bodenphysikalischer Grundlage (Earthwork Mechanics based on the Physics of Soils) in 1925, launching the beginning of modern geotechnical engineering. The book contains the theory of effective stress to explain the behavior of soils under loads. The success of the book led to positions at MIT and the Technical University of Vienna, and he left Austria to permanently settle in the United States in 1938.

FIGURE 1.15 Terzaghi Karl.

His formulation of the effective stress principle and its influence on settlement analysis, strength, permeability, and erosion of soils was the most extraordinary contribution to civil engineering [16].

Terzaghi's ideas were met with skepticism in several civil engineering circles, but he continued to write, lecture, and demonstrate the validity of his concepts through their practical application. Terzaghi's consulting work extended to earth dams, the stabilization of landslides, and foundations for buildings, waterfronts, highways, and airports. During his lifetime, he wrote several books and published over 100 papers on topics ranging from the stability of slopes and conditions for the failure of soils to vibration problems and drainage mechanics.

1.9.9 Elsie Eaves (1898–1983)

Elsie Eaves (Figure 1.16) graduated from the University of Texas at Austin with a degree in civil engineering in 1926. Eaves was the first woman professional engineer in the state of New York in 1930. She went on to become the first female ASCE life member, first female member of Chi Epsilon, and the first honorary female member of the American Association of Cost Engineers [17]. Elsie Eaves was a pioneering structural engineer who made significant contributions to the design and construction of large dams and other major infrastructure projects [18].

Elsie Eaves once served as the President of the Society of Women Engineers. Eaves' legacy as a pioneering female engineer and advocate for women in STEM serves as a powerful reminder of what can be accomplished, inspiring and empowering the next generation to do even more. Eaves was a champion for women's rights and worked to promote the inclusion of women in engineering and other male-dominated fields.

She was one of the first female engineers to work on the design of the Hoover Dam, which remains one of the largest and most impressive engineering projects in history [18]. Eaves was also instrumental in the design and construction of other major dams, including the Parker Dam and the All-American Canal. Her work helped improve the safety and efficiency of these hydraulic structures and paved the way for modern dam design and construction techniques.

1.9.10 Dr. Tokujiro Yoshida (1888–1950)

Dr. Tokujiro Yoshida, shown with his hammer (Figure 1.17), was a Japanese civil engineer and the 37th president of the Japanese Society of Civil Engineers (JSCE). According to the JSCE, the Yoshida Award was founded to honor the outstanding professional achievements and contributions of Dr. Tokujiro Yoshida to the establishment of concrete technology in Japan.

Dr. Tokujiro Yoshida's groundbreaking research included the 1921 research, titled 'Studies on the cooling of fresh concrete in freezing weather,' on advanced concrete technology in Japan. His contribution to the discovery of segregation in fresh concrete, innovation and fabrication of high-strength concrete (unbreakable for almost ten years!), and reinforced concrete design standards is remarkable.

It is upon this special contribution to concrete technology that the Yoshida Award was established by the JSCE. The Yoshida Award is conferred on those who have made remarkable accomplishments and outstanding contributions to the advancement of concrete engineering, either in research, planning, design, or construction, with these accomplishments and contributions being made in the form of a paper and other written presentations or through practice.

Dr. Tokujiro Yoshida, the 37th president of the JSCE, said in his presidential address in 1950 that 'Japanese civil engineering technology is at least 20 to 30 years behind UK or USA' [19]. Thirty years later, in 1979, Japan grew to become a magnificent nation of technology, to the extent that American sociologist Ezra Vogel wrote the book *Japan as Number One*.

FIGURE 1.16 Elsie Eaves.

FIGURE 1.17 Dr. Tokujiro Yoshida.

FIGURE 1.18 Sir Mokshagundam Visvesvaraya.

1.9.11 Sir Mokshagundam Visvesvaraya (1861–1962)

Sir Mokshagundam Visvesvaraya (Figure 1.18) was an Indian civil engineer and statesman regarded in India as one of the foremost civil engineers, whose birthday, 15 September, is celebrated every year as Engineer's Day in India, Sri Lanka, and Tanzania.

He was responsible for significant engineering projects in India, including the Krishna Raja Sagara Dam in Mandya, which helped convert nearby wasteland into fertile land and became the primary source of drinking water for many neighboring cities [20]. The dam continues to draw thousands of visitors every year.

Beyond his astonishing achievements in engineering, Sir M. Visvesvaraya is credited for his extraordinary work in accelerating industrialization and education in India.

He is well known for designing flood protection systems. He was also the chief engineer of the Laxmi Talav Dam near Kolhapur, in the Indian state of Maharashtra. In 1899, Visvesvaraya joined the Indian Irrigation Commission where he implemented an intricate system of irrigation in the Deccan Plateau and designed and patented a system of automatic weir water floodgates that were first installed in 1903 at Khadakvasla Dam near Pune.

He received an honorary membership from the ICE, United Kingdom, and a fellowship from the Indian Institute of Science, Bangalore, and several honorary degrees from eight universities in India.

Visvesvaraya was a civil engineer for the Government of British India and later as Prime Minister of the Kingdom of Mysore. In 1906/1907, the Government of British India sent Visvesvaraya to the British Colony of Aden (present-day Yemen) to study water supply and drainage systems. The project he prepared was successfully implemented in Aden. After India attained independence, Visvesvaraya received the Bharat Ratna, India's highest civilian honor, in 1955.

Visvesvaraya received recognition in many fields. The most notable ones were in education and engineering. Some of them are listed below:

- Visvesvaraya Technological University in Belagavi (to which most engineering colleges in Karnataka are affiliated) was named in Visvesvaraya's honor.
- University Visvesvaraya College of Engineering, Bangalore, established in 1917. In order to develop a skilled force among the boys from Mysore, Bharat Ratna Sir M Visvesvaraya envisioned the college in 1917 [21].
- Sir M. Visvesvaraya Institute of Technology, Bangalore.
- Visvesvaraya National Institute of Technology, Nagpur.
- Visvesvaraya Industrial and Technological Museum, Bangalore.

1.9.12 William Hunter Dammond (1873–1956)

William Hunter Dammond (Figure 1.19) was an African–American civil engineer and the inventor of a pioneering train signaling system that drastically enhanced rail safety. He was the first Black graduate of the University of Pittsburgh in 1893.

In 1903, he received a patent for a 'Signaling System,' an alternating-track, circuit-based technology designed to replace human hand signals used to direct trains.

He published pieces on train crashes and rail safety throughout his career, resulting in him being celebrated as both a leader and a pioneer in the field of locomotive safety [22].

1.9.13 Mao Yi-sheng (1896–1989)

Mao Yi-sheng (Figure 1.20) born in Zhenjiang, Jiangsu Province, China, was a well-known civil engineer, bridge

FIGURE 1.19 William Hunter Dammond.

FIGURE 1.20   Mao Yi-sheng.

engineer, and engineering educator. After graduating from the Tangshan Engineering Institute in 1916, he went to study in the United States. He received an MCE in civil engineering from Cornell University in 1917 and a PhD in civil engineering from Carnegie Institute of Technology (now Carnegie Mellon University) in 1919. This was the first PhD granted by that university. His dissertation on secondary stresses in trusses was a major contribution to bridge theory at that time.

He was elected as the academician of the Chinese Academy of Sciences in 1955 and foreign academician of the American National Academy of Engineering in 1982. He was also elected as a foreign associate of the National Academy of Engineering in 1982. Dr. Mao was recognized for his distinguished leadership in the development of China's transportation system. His significant accomplishments as a bridge designer and his guiding role in engineering education were remarkable.

Dr. Mao was the President of the Institute of Railway Technology, the China Academy of Railway Sciences, the China Engineers' Association, and the Chinese Civil Engineering Society. He was also Vice President and then honorary president of the China Association for Science and Technology. Under his leadership for 30 years, the China Academy of Railway Sciences developed into a research institute, providing research support for railway transportation and construction and training for a large number of engineers and scientists [23].

1.9.14 Dr. Albert Isimbwa Rugumayo (1958–2018)

A chartered civil engineer with the Engineering Council of the United Kingdom and a fellow of the ICE, United Kingdom, Dr. Albert Isimbwa Rugumayo (Figure 1.21), gave birth to the idea of starting engineering as a course in several institutions of higher learning in Uganda, East Africa.

FIGURE 1.21   Dr. Albert Isimbwa Rugumayo.

A civil engineering graduate of the University of Zambia, Dr. Rugumayo went on to obtain an MSc in engineering hydrology from the University College Galway, Ireland, in 1988.

He mentored many engineers throughout his career. He was highly dedicated to teaching engineering at several universities in East Africa. Highly involved in the ICE activities, Rugumayo was an editor of the ICE municipal engineer journal [24]. In his research publications, Rugumayo contributed to hydrology and water resources courses, which he taught in several universities and developed several hydrologic techniques helpful in software customizations applicable to the East African region. Through consultancy work, he was involved in several dam projects in East Africa.

Dr. Rugumayo is highly remembered for mentoring and shaping several outstanding engineers in Uganda, and his thoughtfulness, popularity, influence, and connections helped many of his students to gain professional qualifications and better their lives. While some lecturers were hesitant to support students in furthering their careers and growing professionally, Dr. Rugumayo was all optimistic. His belief in professional commitment as the benchmark to evaluate anyone who bargains for professional accreditation will keep his legacy live on.

With tremendous charisma, Dr. Rugumayo inspired thousands of his students and described him as a 'lecturer who was human.' His exemplary leadership taught the students and associates the need to have soft skills and 'follow your heart but take your brain with you.' Leonardo da Vinci, a historical icon in science and civil engineering, was a profound source of inspiration for him. He was the **Uganda Institution** of **Professional Engineers (UIPE)** President in 2000/01. This man deserves not only special acknowledgment, but the UIPE should also establish an award in his honor for his contributions to the engineering fraternity.

1.10 CONCLUSION

This chapter shares the history of civil engineering and what you need to learn to become a civil engineer. The notable civil engineers in history took different routes to become professionally qualified, but the most important thing is that they all spent years learning, training, and getting qualifications.

The list of notable civil engineers highlights the routes taken to become professional civil engineers. Some took apprenticeships, learning on the job, while others went for university and vocational training.

The ICE defines civil engineering as follows:

> The general advancement of mechanical science, and more particularly for promoting the acquisition of that species of knowledge which constitutes the profession of a civil engineer being the art of directing the great sources of power in nature for the use and convenience of man.

Thomas Tredgold (1788–1829)

In the 18th and 19th centuries, civil engineers had not taken much from nature. Today, civil engineers give back to nature what they've taken out by designing and constructing sustainable structures. Thus, civil engineers direct the great sources of power in nature for the use and convenience of man in a sustainable manner.

The term 'civil engineer' originated in the United Kingdom and spread worldwide. This makes civil engineers appreciate the history of the roots of the profession and to benchmark on the ICE learned society to further civil engineering education. The profession has saved humanity more than any other in history. Unlike doctors who bury their mistakes in the grave, civil engineers don't. The public quickly learns about their flaws in designs and construction.

Civil engineers build bridges, canals, airports, sewer networks, hospitals, schools, dams, and much more. It is obvious that the modern world cannot function without civil engineers!

STEM courses required to become a civil engineer are listed, and routes to becoming professionally qualified with engineering bodies such as the ICE, United Kingdom, are properly elaborated in the chapter. Mathematics and physics are primarily the essential subjects to take through secondary school if your goal is to become a civil engineer. You can then enroll on an apprenticeship, degree, or vocational course. These qualifications can give you the academic base to become professionally qualified. There is no doubt that becoming a civil engineer is one of the most fulfilling professions. As a civil engineer, you see how your work solves societal problems and makes the world a better place to live.

REFERENCES

1. Civil engineers' commemorative plaques. Biographical notes on the civil engineers whose names are commemorated on the façade of the Civil Engineering Building. Civil Engineering Department. Imperial College. London 2000. Retrieved from: https://www.imperial.ac.uk/media/imperial-college/faculty-of-engineering/civil/public/about/History_Plaques_Booklet.pdf. Accessed 2/Jan/2024.
2. Institution of Civil Engineers. Library and Information Services. Guide to the ICE Archives 2010. Retrieved from: https://www.ice.org.uk/media/fkzj0hua/guide-to-ice-historical-archives.pdf. Accessed 2/Jan/2024.
3. Governance Handbook 2022–23. Retrieved from: governance-handbook-2022-23.pdf (ice.org.uk). https://www.ice.org.uk/media/n3cb4ako/governance-handbook-2022-23.pdf. Accessed 3/Jan/2024.
4. By KennyOMG—Own work, CC BY-SA 4.0, Retrieved from: https://commons.wikimedia.org/w/index.php?curid=40998161
5. ASCE launches its first multidisciplinary, open-access journal. ASCE launches its first multidisciplinary, open-access journal I ASCE. Retrieved from: https://www.asce.org/publications-and-news/civil-engineering-source/article/2022/12/13/asce-launches-first-multidisciplinary-open-access-journal
6. ASCE (1973). *Official Record*. American Society of Civil Engineers, New York.
7. ASCE (1977). *Official Record, Environmental Engineering Division, Statement of Purpose*. American Society of Civil Engineers, New York.
8. Memorandum of Understanding (1999). The proposed unification of the Institution of Civil Engineers Ground Board (UK) and the British Geotechnical Association (BGA).
9. The Smeatonian Society of Civil Engineers. History of the society. Retrieved from: https://www.smeatonians.org/history. Accessed 5/Jan/2024.
10. Institution of Civil Engineers (Great Britain) (1870). *Minutes of Proceedings of the Institution of Civil Engineers*. The Institution. p. 215, note 1.
11. Thomas Telford I Institution of Civil Engineers (ICE). Retrieved from: https://www.ice.org.uk/what-is-civil-engineering/who-are-civil-engineers/thomas-telford. Accessed 3/Jan/2024.
12. Isambard Kingdom Brunel - Chief engineer for the Great Western Railway. Why you might have heard of Isambard Kingdom Brunel? Retrieved from: https://www.ice.org.uk/what-is-civil-engineering/who-are-civil-engineers/isambard-kingdom-brunel. Accessed 5/Jan/2024.
13. Sir Joseph Bazalgette. Civil engineer, 1819–1891. Why you might have heard of Joseph Bazalgette? Retrieved from: https://www.ice.org.uk/what-is-civil-engineering/who-are-civil-engineers/sir-joseph-bazalgette. Accessed 5/Jan/2024
14. American Society of Civil Engineers (ASCE). Benjamin Wright. Retrieved from: https://www.asce.org/about-civil-engineering/history-and-heritage/notable-civil-engineers/benjamin-wright. Accessed 5/Jan/2024.
15. Past Presidents of IEK. . Retrieved from: https://www.iekenya.org/honour-board. Accessed 26/Oct/2025.
16. American Society of Civil Engineers (ASCE). Retrieved from: https://www.asce.org/about-civil-engineering/history-and-heritage/notable-civil-engineers/karl-terzaghi. Accessed 3/Jan/2024.
17. Celebrating the first women members of ASCE. Retrieved from: https://www.asce.org/publications-and-news/civil-engineering-source/article/2022/03/14/celebrating-the-first-women-members-of-asce. Accessed 4/Jan/2024.

18. 20 Historic Female Engineers Who Shaped Our World: Honoring Their Achievements on International Women's Day. Retrieved from: https://www.borntoengineer.com/historic-female-engineers-shaping-our-world-international-womens-day. Accessed 2/Jan/2024.

19. Japanese Crises and the Future of Civil Engineering - from the Perspective of Our Origins and Culture – (1/3). Prof. Ieda delivered a Presidential Speech at the 2020 JSCE Annual Meeting. The IAC News introduces the speech from this issue to No.100. Retrieved from: https://www.jsce-int.org/system/files/IAC_News_No.98%28En%29rev4_%EF%BC%88%E8%8B%B1%E8%AA%9E%E7%89%88%29.pdf. Accessed 3/Jan/2024.

20. Sir Mokshagundam Visvesvaraya. Retrieved from: https://www.ice.org.uk/what-is-civil-engineering/who-are-civil-engineers/sir-mokshagundam-visvesvaraya. Accessed 3/Jan/2024.

21. University of Visvesvaraya College of Engineering. Retrieved from: https://uvce.ac.in/. Accessed 3/Jan/2024.

22. 5 pioneering Black civil engineers who paved the way for future generations | Institution of Civil Engineers (ICE). Retrieved from: https://www.ice.org.uk/news-and-insight/ice-community-blog/october-2021/5-pioneering-black-civil-engineers. Accessed 4/Jan/2024.

23. Yi-Sheng Mao 1896–1989. Retrieved from: https://www.nae.edu/188912/YISHENG-MAO-18961989 https://www.nae.edu/File.aspx?id=188913. Accessed 3/Jan/2024.

24. Editorial | Proceedings of the Institution of Civil Engineers - Municipal Engineer (icevirtuallibrary.com). Retrieved from: https://www.icevirtuallibrary.com/doi/10.1680/jmuen.2017.170.4.187. Accessed 4/Jan/2024. 6:00 pm.

Philosophy of Civil Engineering

2

2.1 INTRODUCTION

The philosophy of civil engineering is rooted in the pursuit of creating and developing safe, sustainable, reliable, and innovative infrastructure solutions to support and sustain humanity. This approach integrates a range of moral, ethical, social, economic, and environmental considerations that ultimately guide the profession. Civil engineers recognize that saving humanity is the highest good in the hierarchy of doing the good. In that regard, civil engineers have saved humanity more than any other professional throughout history.

The key aspects of the philosophy of civil engineering profession include stewardship, service to humanity, health, safety, welfare and risk management, sustainability, innovation and creativity, harnessing technology, teamworking and collaboration, effective communication and interpersonal networking, ethical practice, continuous learning, and supporting others to grow professionally. Therefore, mastering these primary attributes is what defines a competent civil engineer.

Throughout history, prominent people from diverse backgrounds have contributed to the development of the profession, thus refining the way civil engineers approach problem-solving in the modern-day scenario: from harnessing the power of creativity to innovating civil engineering marvels and to learning and understanding nature and how forces of nature can be manipulated to improve lives.

2.2 PRIMARY CIVIL ENGINEER'S ATTRIBUTES

What occupies a civil engineer's thinking and actions—what really defines their attitude in delivering civil engineering expertise? [1].

2.2.1 Stewardship

Civil engineers recognize their responsibility to manage resources wisely and minimize environmental impacts. This is a practice committed to high ethical value that embodies the responsible planning and management of resources. By virtue of their practice, civil engineers deal with materials, a great deal of which comes from our planet, Earth. Therefore, minimizing environmental damage is a core aspect of the profession. That means managing resources responsibly is a primary prerequisite for civil engineers—because, if not properly managed, climate consequences and material depletion and water and air pollution are potential dangers that can arise out of mismanagement as civil engineers play their roles.

Civil engineers, therefore, prepare and control sound budgets using knowledge of statutory and commercial frameworks. In this regard, civil engineers require an appreciation of several commercial arrangements, having proper judgment on statutory, contractual, and commercial issues throughout the intended jurisdiction. Because without being able to prepare and control budgets and to understand contractual and commercial issues in a given jurisdiction, they risk failing to successfully manage cost-effective projects and running into cost overruns. Under-costed projects impair the project's reliability and delay projects unnecessarily, and contractual dilemmas may stall projects, overall.

Civil engineering requires sound management of quality processes, that is, constantly engaging in quality improvement and continuously looking for better solutions to improve existing systems and to plan, direct, and control tasks, people, and resources. In that regard, leading large teams and developing staff/human resources to meet changing technical and managerial needs is the thing civil engineers perform on a daily basis and cherish. Therefore, continuously improving systems and products through quality management is the hallmark of great civil engineering practice. This is because technology keeps evolving. Yet, application of technology in civil engineering practice cannot be escaped.

It is well known that technology becomes obsolete. So, for a civil engineer to be up to date, they must engage in managing quality processes all the time, refine their knowledge and understanding of advancing technologies, and properly align available resources with goals. Engineering is like 'management'; if there were only one agreed definition for all time, there would not be so many books on it. The requirements are not static; they shift with the changing demands of our profession [2].

Renowned engineers from the early 20th century, who achieved remarkable feats in their time, would face significant challenges adapting to modern civil engineering practices if they were to return today. The profound technological advancements and paradigm shifts that have transformed the field over the past century would render their expertise largely obsolete. Thus, there is the need to continuously evolve and stay ahead of the curves, be up to date with emerging technologies, and record continuing professional development records as a good practice.

DOI: 10.1201/9781003561620-2

2.2.2 Service to Humanity

Civil engineers strive to create infrastructure that benefits the society, promotes public health, and enhances the quality of life. It is important to note that civil engineers have saved humanity more than any other profession in history.

Saving humanity is the highest good done by civil engineers. Service to humanity is about making a positive impact, leaving a lasting legacy, and contributing to the greater good.

They promote environmental sustainability, social justice, and human rights in all their endeavors and abstain from doing things that contravene environmental justice.

Your primary goal for joining the civil engineering profession should be to save humanity and not to leverage it for commercial benefits. Other benefits are secondary. See Chapter 1 for the notable civil engineers in history—their work was never about money-making! They served humanity with their whole heart. We learn from them to advance their legacies forward for the greater good.

In serving humanity, you must unlearn, relearn, and learn new ways, approaches, and knowledge that advance the civil engineering profession.

There is no doubt that pursuing a lucrative career as a professional civil engineer is a commendable goal. However, to achieve success, focus on adding value to the profession through innovation and creativity. Develop practical solutions, advance existing technologies, or pioneer new approaches that benefit the industry. By doing so, financial rewards will inevitably follow.

Let us recognize the bitter truth: pursuing financial rewards as the primary goal just reduces one's worth. There is no way that you can innovate great stuff and without financial rewards. Never. Not so many people can do what you can! Just remember that, and do the needful, going an extra mile to create, recreate, and innovate cool stuff to serve humanity, as shown in Sections 2.2.1–2.2.11. As civil engineers play an increasingly vital role in addressing the planet's most pressing challenges, our profession has never been more critical to shaping a sustainable future.

Civil engineers are built to serve humanity and put things in order, thus saving our planet as we devise sustainable habitats and infrastructures, while keeping the ecosystem intact. Therefore, civil engineers' primary role is to ensure humanity is served with one heart and soul; thus, commercial benefits will definitely pour. Be of value, as echoed by Albert Einstein (1879–1955), sometime in history!—one of the tenets civil engineers borrow from Einstein.

2.2.3 Health, Safety, and Risk Management

Ensuring the safety of the public and mitigating risks are paramount in civil engineering decisions. In all their endeavors, civil engineers ensure that their work contributes to the safety of the public. Civil engineers are deemed to have an appreciation of and be able to identify and manage risks to all those engaged and affected by their works.

Whether dealing with minor hazards or reducing the potential for major accidents, managing health and safety risks effectively is an essential requirement for technicians, engineers, and managers throughout their careers. The civil engineering structures they develop must be safe and reliable. That means, they should be able to perform the intended function over a designated period. In doing so, they must not harm people, animals, goods, and the environment. It does no good, for example, to construct a bridge that is unsafe, risking the lives of road users after its opening or the lives of construction workers during its construction. So, civil engineers do not just add the dimension of safety to their work, but they incorporate it throughout the project life cycle!

The society expects a lot from civil engineering as it deals with infrastructure development and environmental conservation. So, managing health and safety risks is crucial. There are several social and legal reasons for managing health and safety risks. International and national laws govern civil engineering business conduct, thus setting standards for health and safety. The consequences of non-compliance with the law related to health and safety include enforcement action, prosecution, fines, and imprisonment. The societal expectations have become increasingly high in the modern world as they drive legal standards, and they just get increasingly high over time, leading to more stringent laws that keep evolving. Therefore, health and safety laws are amended to meet rising societal expectations. The employer bears the primary obligation for ensuring workplace health and safety. The key responsibilities of employers include the following:

Key health and safety responsibilities of employers

These include providing safe systems of work, safe place of work, safe plant and equipment, training, supervision, and competency of staff.

Safe systems of work

Employers must ensure that there are established procedures for safely conducting all work activities, covering all possible scenarios, routine and non-routine tasks, occasional or on–off activities, and potential emergencies. Procedures should consider all possible scenarios, including different weather conditions, unforeseen events, and various operating situations, for example, the operation of drilling equipment in different types of weather.

Safe place of work

Employers are responsible for providing a reasonably safe workplace, considering the type of work, and ensuring safe access to and from the workplace. This includes creating a work environment that is free from health risks and hazards.

Safe plant and equipment

Employers must ensure that all machinery, tools, and equipment are reasonably safe and without health risks, with the level of care and maintenance varying depending on the type of work and level of risk involved. The greater the risk involved with the tool/equipment, the greater the care that must be taken. Machinery maintenance varies by industry: high-risk industries, e.g., steel-making, would need regular inspections, servicing, repairs, and replacements. Low-risk industries, e.g., office work, would probably need a simple inspection regime.

Training, supervision, and competency of staff

Employers must provide workers with the following:

- **Training:** On hazards, risks, safe systems, and emergency procedures. Workers must be aware of the hazards and risks involved in their work, the safe systems of work, and the emergency procedures. Workers must be able to carry out the necessary procedures.

- **Information and instruction:** To reinforce training.
- **Supervision:** Supervise workers to ensure minimal risk, without requiring constant monitoring.
- **Competence:** Ensuring all workers, supervisors, and managers have necessary skills and knowledge.

In this context, '**competent**' means that each person has sufficient training, knowledge, experience, and other abilities or skills to be able to carry out their work safely and without risk to health.

On the other hand, civil engineers and those involved in civil engineering business such as technicians and technologists play a pivotal role in ensuring the employer meets these standards. Civil engineers can be involved with construction, design, operation, and maintenance of structures. They can work as principal designers, designers, supervising engineers, contractors, and operations engineers. There are several roles that civil engineers can play on projects, and they have a direct responsibility to ensure health and safety laws are complied with.

The construction (design and management) (CDM) regulations, for example, give principal designers, designers,[1] and contractors roles to play to ensure that safe working environments are achieved through the pre-construction and/or construction phases. For example, according to CDM Regulation 9 (2&3) (2015), designers must eliminate risk [3].

Elimination of risk (Designers)

(2) When preparing or modifying a design, the designer must take into account the general principles of prevention and any pre-construction information to eliminate, so far as is reasonably practicable, foreseeable risks to the health or safety of any person:

- carrying out or liable to be affected by construction work;
- maintaining or cleaning a structure; or
- using a structure designed as a workplace.

(3) If it is not possible to eliminate these risks, the designer must, so far as is reasonably practicable, do the following:

- take steps to reduce or, if that is not possible, control the risks through the subsequent design process;
- provide information about those risks to the principal designer; and
- ensure appropriate information is included in the health and safety file.

Designers should use existing checks in the process to allow the health and safety implications of the design to be assessed. Designers could utilize the knowledge and experience of Contractors to assist in the process.

Civil engineers recognize working in safe and controlled environments to avoid harm to people, staffs, and workers. Thus, they must possess a sound knowledge of legislation, hazards, and safe systems of work. As mentioned earlier, the highest good is saving humanity. So health and safety always come first!

As part of the philosophy of civil engineering, managing risks, health, safety, and welfare is essential and cannot be avoided because without risk management, projects can translate into length litigation processes, thus harming the people directly engaged in construction or the general public and stalling the projects. Enough care must be exercised both during pre-construction and construction phases. Employers, engineers, staff, workers, contractors, and everyone involved in the construction business have a role to play to ensure safety is achieved.

A civil engineer, therefore, must possess the skill and knowledge to lead continuous improvement in health, safety, and welfare from time to time. Having these skills enables civil engineering to keep constantly evolving and interesting.

2.2.4 Sustainability

In 1987, the United Nations' Brundtland Commission defined sustainability as "meeting the needs of the present without compromising the ability of future generations to meet their own needs." With future generations in mind, civil engineers aim to design and build infrastructure that is resilient, adaptable, and environmentally conscious. They are at the forefront of saving the planet more than any other professional because their work greatly impacts the planet Earth, that is, the environment; thus, they consider the environmental and social impacts in their planning, design, construction, operation, and maintenance of civil engineering structures. Therefore, civil engineers maintain a sound knowledge of sustainable development best practices. For example, civil engineers manage engineering activities that contribute to sustainable development and then leading continuous improvement in sustainable development pillars, that is, social, economic, and environmental.

To lead continuous improvement in sustainable development, civil engineers set clear sustainability goals and metrics and regularly track progress. Civil engineers aim at offsetting embedded and operational carbon in project development. Civil engineers encourage employee engagement and empowerment in the sustainability campaign, foster a culture of innovation, and implement circular economy principles. They collaborate with stakeholders, stay updated on best practices, embed sustainability into core business operations, develop sustainable supply chain management practices, invest in renewable energy, and support education and training. They promote transparency and accountability, develop sustainable water management practices, and support biodiversity conservation. They also foster partnerships and collaborations, invest in climate change adaptation and resilience-building efforts, and continuously adapt and evolve to drive collective impact and sustainable growth.

2.2.5 Harnessing New and Advancing Technology

Civil engineers harness technology to advance and solve problems that challenge humanity. As a primary role, they maintain and extend a sound theoretical approach in enabling the introduction and exploitation of new and advancing technology. Section 3.2 outlines how civil engineers utilize technology to advance the profession borrowing scientific principles as they evolve from time to time to create new solutions to civil engineering problems. In that regard, constantly evolving civil engineers engage in the creative and innovative development of engineering technology and continuous improvement systems.

As an attribute, civil engineers must possess the knowledge and understanding of engineering principles and, thus, be able to maintain and extend a sound theoretical approach in enabling the introduction and exploitation of a new and advancing technology through an evidence-based approach. Additionally, civil engineers conduct appropriate research, relative to design or construction, and appreciate its relevance within their area of responsibility. Thus, they undertake the design and development of engineering solutions and evaluate their effectiveness following an evidence-based approach. They ensure the practicality of the design and its cost-effectiveness.

2.2.6 Innovation and Progress

Civil engineers embrace innovation and technological advancements to drive progress and improvement in the field. They, thus, engage in the creative and innovative development of engineering technology and continuous improvement systems. One of the attributes that top-notch engineers develop is the ability to invent and innovate practical solutions. Engineers can become inventors at some point—that means creating novel solutions or products that did not exist before. However, this often requires a deep understanding of the technical principles and the ability to think creatively. Engineers can also become innovators, applying existing knowledge and technologies in new and impactful ways. Innovation can be just as valuable as invention as it often involves finding practical solutions to real-world problems. It is important to note that invention and innovation are not necessarily hierarchical. Never are they complementary concepts. Invention focuses on creating something new, while innovation focuses on applying and improving existing ideas—invention can lead to innovation, that is, new inventions can be the foundation for innovative products or services. On the other hand, innovation can drive invention, whereby the process of innovating can lead to new ideas and inventions. So, as an engineer, to get to the climax of the profession and stand out, your engineering impact must be felt within the society.

At the epitome of civil engineering, civil engineers identify the limits of personal knowledge and skills. They exercise sound independent engineering judgment and take responsibility. They identify the limits of a team's skill and knowledge and exercise sound holistic independent judgment and take responsibility. Civil engineers embed the concept of holism in their work. The concept of holism, where the whole is more than the sum of its parts, is highly relevant to civil engineering design and construction. In civil engineering, a design consists of various interconnected components, such as structural elements (beams, columns, and foundations), materials (concrete, steel, and asphalt, among others), systems (drainage, mechanical, electrical, and plumbing), and environmental safeguards (for wind, seismic activity, and water flow). Civil engineers view things as a whole not an isolated component. This enables them to take holistic independent judgment and thus drive innovation and progress.

Some specific applications of holism in civil engineering design include systems thinking for infrastructure planning and management, integrated building design (e.g., combining structural, mechanical, and electrical systems), sustainable design and green infrastructure, and resilience-based design for natural hazard mitigation, among others.

In civil engineering, it is good to know that you don't know how to do some things—just like any other learned fraternity—because it is the genesis of wisdom. It is okay not to know! It is extremely practical to judge your knowledge and expertise about something before you make lasting decisions. You must be able to sense that the limits of your knowledge can have disastrous outcomes if not exercised properly in a given field. In that regard, as a civil engineer, you outsource or work with competent people or take precautions, relying or benchmarking on the knowledge of others.

As a civil engineer, knowing the limits of your knowledge is the first rule to develop better things or infrastructure—all the time. Being complacent about your knowledge is terrible. 'The only true wisdom is in knowing you know nothing,' says ancient Greek philosopher Socrates. Socrates also said, 'I know one thing, that I know nothing.' This creates room for constant improvement.

2.2.7 Collaboration and Teamwork

Civil engineers recognize the importance of interdisciplinary collaboration and teamwork to achieve common goals. In today's fast-paced world, civil engineers are now finding it necessary to nurture interdisciplinary learning to catch up with the demands of the ever-evolving profession.

Interdisciplinary collaboration refers to the practice of working together and sharing knowledge, expertise, and methods across different disciplines, fields, or sectors to achieve a common goal or solve complex problems. It involves integrating diverse perspectives and approaches, sharing knowledge and expertise, combining methodologies and tools, fostering open communication and active listening, embracing diverse backgrounds and experiences, encouraging creativity and innovation, and building trust and respect among team members.

2.2.8 Ethical Practice

Civil engineers adhere to ethical principles, such as integrity, honesty, and transparency, in their professional endeavors. A strong moral fabric is required to reach the epitome of civil engineering practice.

Civil engineering requires professional commitment. It is through professional commitment that a civil engineer understands and complies with the ethical code of conduct, as outlined by the institution they subscribe to. They also keep proactively engaged with the institution's activities to continuously promote its sustainability.

They stay committed to the profession all their lives, by understanding and complying with the codes of conduct, engaging with activities that boost the profession and professional bodies they subscribe to.

They exercise responsibilities in an ethical manner, demonstrating appropriate professional standards and recognizing obligations to the society, the profession, and the environment.

2.2.9 Continuous Learning

Civil engineers commit to the ongoing learning and professional development to stay abreast of evolving technologies and challenges. They engage in continuous learning throughout their lives—continuous professional development (CPD) to maintain their knowledge in a constantly evolving profession. So, they plan, carry out, and record continuing professional development and encourage others to do the same—that is what creates a civil engineer role model, per se.

2.2.10 Support Others to Grow Professionally

By embracing this philosophy, civil engineers play a vital role in shaping the built environment and improving the lives of individuals and communities worldwide. They are a pivot of economies around the world, based on whom many infrastructures and associated products are realized. Of late, they have primarily championed sustainability.

Never fail to support others to grow professionally. You do a great disservice to the profession once you fail to support others to grow and become civil engineers. Civil engineers serve humanity.

2.2.11 Effective Communication and Interpersonal Networking

Effective communication cannot be separated from a civil engineer. You must develop effective interpersonal skills and communication to succeed as a civil engineer.

You must be able to communicate effectively with others at all levels through a language legally acceptable in the jurisdiction you practice. Effective communication means being able to communicate orally and in writing. That means being able to discuss complex ideas and plans competently and with confidence, having effective personal and social skills, ability to manage diversity issues, and finally be able to communicate new concepts and ideas to technical and nontechnical teams.

Failing to develop effective communication and interpersonal networking skills deters you from sharing your ideas widely, convincing technical and nontechnical teams that your plans, ideas, or concept are feasible. In some cases, engineers perform a variety of roles that demand a high degree of soft skills such as marketing, promotion, proposal writing, defending clients in courts of law, expert witnessing/testimony, writing columns in newspapers, some go along to become journal editors, and so on. So effective communication and networking abilities are prerequisites for a civil engineer without compromise.

2.3 THE ROLE OF PHILOSOPHER, ENGINEER, TECHNOLOGIST, AND TECHNICIAN

2.3.1 Overview

Philosophy is the mother of all disciplines. It is the foundation for many other fields of study such as science, mathematics, logic, and ethics. Philosophy's emphasis on critical thinking, logical reasoning, and conceptual analysis has made it a foundational discipline for many areas of study. It can be clearly seen that philosophy links all trades involved in engineering. Philosophical inquiry and critical thinking shape the methodologies and approaches used in various disciplines.

Philosophy asks fundamental or foundational questions about existence, knowledge, morality, and reality, which form the underlying principles that other disciplines build upon and investigate further; essentially, it provides the critical-thinking framework and core concepts that underpin various academic areas such as science, ethics, politics, and art.

Philosophy explores relationships between different fields, fostering a deeper understanding of their interconnections and shared concerns. Philosophical concepts, such as ethics, logic, and metaphysics, provide essential frameworks for understanding and analyzing complex issues across disciplines. Philosophy has influenced the development of many disciplines, from science and mathematics to art and literature.

2.3.2 Who Is a Philosopher?

A philosopher is an individual who engages in the study, research, and critical thinking about fundamental questions and concepts related to existence, knowledge, values, reason, and reality. The term 'philosopher' comes from the ancient Greek 'philosophos,' meaning 'lover of wisdom.'

The term 'philosophy' is believed to have been coined in ancient Greece by the philosopher and mathematician, Pythagoras. Pythagoras (c. 570–490 BCE) needed a term for a certain kind of individual, one who prized truth and knowledge above all things. It is believed that he combined the ancient Greek terms for love, philein and wisdom, sophia to produce philosophos, one who loves wisdom. Philosophy, then, is the love of wisdom. It's not as though Pythagoras invented philosophy, of course. What he was describing, our love of knowledge, of wisdom, has been with us perhaps since we first began to think. But he did invent a term that has proven useful in referring to this most basic and urgent of human abilities, the pursuit of knowledge: philosophy.

Philosophy is a way of thinking about certain subjects such as ethics, thought, existence, time, meaning, and value. That 'way of thinking' involves 4 Rs: responsiveness, reflection, reason, and re-evaluation. The aim is to deepen understanding. The hope is that by practicing philosophy, we learn to think better, act more wisely, and thereby help improve the quality of all our lives [4].

Philosophers explore and examine various aspects of life, including the nature of reality and knowledge (metaphysics and epistemology), ethics and morality, logic and reasoning, aesthetics and beauty, politics, and society.

Engineering is a practical application of scientific knowledge, and philosophers contribute to the ethical, logical, and societal aspects of engineering. In fact, there is a growing field called 'Engineering Philosophy' or 'Philosophy of Engineering' that explores the fundamental principles, methods, and values underlying engineering practice.

As STEM fields continue to push the boundaries of human knowledge and innovation, the need for philosophical examination and reflection grows exponentially—there is more work to do for philosophers than ever.

Philosophers are essential to navigating the complex ethical, metaphysical, and epistemological implications of scientific advancements, ensuring that we not only harness the power of technology but also wield it wisely and responsibly.

As STEM fields advance, philosophers often raise fundamental questions about ethics, morality, and human values. Philosophers can help address these questions, ensuring that technological innovations are developed and used responsibly.

The artificial intelligence (AI) question must be answered for the good of humanity. The development of AI, in particular, requires philosophical examination to ensure it benefits humanity—navigating the consequences of scientific advancements.

2.3.2.1 Roles of a philosopher in relation to civil engineering

A philosopher can play a crucial role in civil engineering by providing a framework for ethical considerations, promoting critical thinking, and fostering a deeper understanding of the profession's impact on the society. They can help engineers navigate ethical dilemmas, consider the broader societal implications of engineering projects, and refine their design philosophies.

The philosopher searches into nature and discovers her laws and promulgates the principles and adapts them to our circumstances. Philosophy addresses fundamental questions about existence, knowledge, reality, and values, which underlie all other fields of study.

It is worth to note that philosophy has not paid sufficient attention to engineering. Nevertheless, engineers should not use this as an excuse to ignore philosophy. The argument here is that philosophy is important to engineering for at least three reasons [5]. First, philosophy is necessary so that engineers may understand and defend themselves against philosophical criticisms. In fact, there is a tradition of engineering philosophy that is largely overlooked, even by engineers. Second, philosophy, especially ethics, is necessary to help engineers deal with professional ethical problems. Third, because of the inherently philosophical characteristic of engineering, philosophy may actually function as a means to greater engineering self-understanding.

Philosophy examines the nature of knowledge, truth, and reality, providing a foundation for understanding how we know and understand the world. Thus, the roles of a philosopher include questioning assumptions by challenging prevailing beliefs and ideas to uncover new insights. It also includes critical thinking, analyzing arguments, evaluating evidence, and developing logical reasoning. A philosopher inspires critical thinking and encourages others to think deeply and question assumptions. In detail, the work of a philosopher influences art, literature, and creativity, fostering critical-thinking and problem-solving skills.

A philosopher engages in idea generation, proposing new concepts, theories, and perspectives, and in dialogue and debate. That means engaging with others to refine ideas, challenge assumptions, and foster understanding.

Another role of a philosopher is reflection and introspection, which means that the philosopher examines personal beliefs and values to gain a deeper understanding. They also engage in writing and teaching, sharing philosophical ideas and insights through written works and educational settings.

All in all, philosophers play a vital role in advancing knowledge and understanding, shaping cultural and intellectual discourse. Their work helps inform ethics and moral principles in several disciplines, including engineering.

Throughout history, philosophers have made significant contributions to various fields, from science and politics to art, religion, and spirituality. Their ideas and insights continue to impact contemporary thought and society. With modern-day advances in science, technology, and engineering, there is more work to do for philosophers!

2.3.3 Who Is an Engineer?

Engineers distill complex technical and philosophical concepts into clear, concise language, avoiding jargon and technical terms that might confuse the project's stakeholders and/or laypeople. An engineer is primarily tasked with effective communication—thus juggling between a philosopher and technician.

> From the early years of the profession, until now, an engineer has been known as mediator between the philosopher and the working mechanic; and like an interpreter between two foreigners must understand the language of both. The philosopher searches into nature and discovers her laws, and promulgates the principles and adapts them to our circumstances. The working mechanic, governed by the superintendence of the engineer, brings his ideas into reality. Hence, the absolute necessity of possessing both practical and theoretical knowledge.

The key aspect that differentiates an engineer from a technician or technologist is the possession of sound theoretical knowledge! In many industries and countries, the terms 'engineer' and 'technologist' are used interchangeably, or the distinction between the two roles may be indistinct. However, having advanced theoretical understanding of engineering principles is what sets engineers apart. Therefore, in essence, an engineer is a professional who applies scientific and mathematical principles to design, build, maintain, and improve structures, machines, systems, and processes. Engineers use their knowledge and skills to develop innovative solutions to real-world problems, often with the goal of making things faster, stronger, more efficient, and sustainable.

2.3.3.1 Key differences between an engineer, technologist, and technician

- **Engineer:** Typically holds a bachelor's or advanced degree in engineering (e.g., mechanical engineering, electrical engineering, or civil engineering). Engineers design, develop, and test complex systems, often focusing on the theoretical and conceptual aspects.
- **Technologist:** May hold a bachelor's degree in a field such as engineering technology, applied science, or a related field. Technologists often focus on the practical application and implementation of technical systems, processes, and products. Note the statement 'may hold a bachelor's degree.'
- **Technician:** Typically requires a post-secondary certificate, diploma, or associate's degree in a specific technical field (e.g., electrical, civil, mechanical, or IT). Technicians specialize in the installation, maintenance, repair, and operation of technical equipment, systems, and processes.

2.3.3.2 Roles of an engineer

Engineers play a variety of roles. They design, build, and maintain systems, products, and structures. They also ensure that these are safe and effective. In essence, engineers design, create, and innovate complex applications, structures, and systems.

The engineering sector is broad, with many fields that encompass additional subsets in which you can specialize. Each subset of engineering may contribute to different outcomes, but professionals in this field ultimately apply their knowledge of engineering systems, design, and integration to construct both consumer and commercial products. While the projects can differ between engineering specialties, the day-to-day responsibilities engineers take on include several tasks.

2.3.3.2.1 Planning
Engineers are vital in planning engineering systems, processes, and products. This encompasses the tasks such as planning, designing, and testing prototypes for various applications.

2.3.3.2.2 Design and analysis
Engineers take pride in having the appropriate knowledge and skill to design processes, systems, and products. They are good at creating and developing plans, blueprints, and models for projects and products. The possession of design skills primarily sets engineers apart. Civil engineers, in particular, have the knowledge and skill to design the infrastructure such as roads and highways, water supply schemes, airports, high-rise buildings, and dams, among other infrastructures. Engineers are responsible for the overall design, analysis, and development of new technologies and systems.

2.3.3.2.3 Research and development
Research and development (R&D) is a key role for a civil engineer. This is the ability to conduct appropriate research and disseminate findings to solve engineering problems in an innovative approach. They create potential designs, research existing technologies, and work with other engineers.

Civil engineers continuously explore new materials, methods, and technologies to improve sustainability and efficiency, contributing to advancements in construction practices.

Civil engineers, therefore, conduct experiments, gather data, and analyze results to advance engineering knowledge and improve processes. Thus, they create potential product designs and redesign existing products.

2.3.3.2.4 Implementation
An engineer is tasked with implementing plans and designs to translate into tangible outputs. In this case, the engineer oversees the installation, deployment, and integration of systems and products. In terms of civil engineering, the engineer particularly manages and leads construction projects—or oversees construction projects.

2.3.3.2.5 Quality control and assurance
Quality assurance and quality control are essential aspects of an engineer's responsibilities. Quality assurance is a proactive, process-oriented approach focused on preventing defects by improving processes and systems, while quality control is a reactive, product-oriented approach that focuses on identifying and correcting defects in the final product.

Engineers are responsible for running tests on products to ensure quality and durability. They are tasked to ensure that products produced are of the desired quality. Thus, they ensure that quality is controlled throughout the manufacturing or construction process, and they undertake procedures to ensure that all standards, codes, and regulations are strictly followed to achieve the desired quality.

2.3.3.2.6 Safety
Engineers ensure that they design, develop, and build safe systems. They conduct tests on products they develop to ensure that they are safe to perform the intended purpose. Engineers ensure that safety protocols are observed throughout the project lifecycle, and this includes identifying potential hazards, developing safety plans, implementing safety measures, and monitoring compliance throughout the project lifecycle. Thus, safety protocols protect people, the environment, and infrastructure from potential risks and harm.

2.3.3.2.7 Consulting
Engineers provide expert advice and guidance to clients and stakeholders. These are termed 'Consulting engineers.' A consulting engineer provides expertise and leadership in the planning, design, modification, or rehabilitation of public and private infrastructures.

A consulting engineer is a professionally qualified engineer in private practice who provides expert advice and technical services, often on a project-by-project basis, to clients in both the public and private sectors. They act as technical advisors, helping clients make informed decisions about the design, construction, and operation of infrastructure and facilities. Consulting engineers may specialize in various disciplines, including civil, structural, mechanical, electrical, and environmental engineering.

2.3.4 Who Is a Technologist?

He/she may hold a bachelor's degree in a field like engineering technology, applied science, or a related field. Technologists often focus on the practical application and implementation of technical systems, processes, and products. Note the statement 'may hold a bachelor's degree.'

2.3.4.1 Roles of a technologist

Technologists emphasize **practical application**, implementation, and troubleshooting of existing technologies:

- They focus on the practical application and implementation of engineering designs.
- They bridge the gap between the theoretical design and the practical implementation of technology.
- They are skilled at troubleshooting and maintaining complex systems and equipment.
- They often have strong hands-on skills and practical experience.

2.3.5 Who Is a Technician?

A technician is a skilled professional who primarily focuses on the practical application of engineering principles, often performing hands-on tasks such as building prototypes, setting up equipment, conducting tests, and maintaining systems, typically working under the direction of engineers to translate complex designs into tangible results. They usually have a lower level of formal education compared to engineers but possess a strong technical skillset and knowledge of specific engineering disciplines. In brief, technicians primarily focus on the hands-on operation, maintenance, and repair of equipment. The role of an engineering technician is intermediate to a skilled craft worker and a technologist. They build or set up equipment, conduct experiments, and collect data and calculate results.

2.3.5.1 Roles of a technician

A technician, more particularly, a civil engineering technician, applies engineering, math, and science principles to perform technical tasks in civil engineering. They work on a variety of projects, including construction, design, and maintenance. Their primary roles include testing construction materials, preparing sketches and tabulations, assisting with cost estimates, performing technical tasks in civil engineering research, and assisting in the design, construction, operation, maintenance, and repair of structures.

Civil engineering technicians range from highway technicians, building inspectors, bridge inspectors, construction inspectors, civil engineering survey technicians, mapping technicians, to geotechnical technicians. Sections 2.3.5.1.1–2.3.5.1.8 share the primary roles for civil engineering technicians.

2.3.5.1.1 Testing materials
The technician is responsible for performing and monitoring tests in the lab or field. In civil engineering applications, technicians usually conduct tests for civil engineering structures.

Tests range from geotechnical tests, hydrologic tests, hydrogeological tests, materials testing, structural tests, to environmental tests.

Civil engineering technicians conduct tests to ensure that construction projects comply with required standards and codes of practice for safety, desired quality, reliability, and environmental regulations. Civil engineering technicians usually carry out tests in the laboratory or *in situ*. Table 2.1 lists some of the civil engineering tests technicians usually conduct.

2.3.5.1.2 Support design and construction processes
Technicians assist in the design, construction, operation, maintenance, and repair of structures. In this regard, civil engineering technicians are usually proud of the required skills to enable them support design and construction when piecing together a project. Such skills include the knowledge and art of using engineering software such as AutoCAD, Civil 3D, or Revit; CAD skills; computing and math skills; and accounting techniques. These particularly apply in design processes. During construction, the civil technician typically possesses the following construction skills:

- For the case of a building structure such as high-rise building, they should possess the ability to interpret site plans and buildings layout, and ensure accurate placement of construction elements and structures.

TABLE 2.1 Select civil engineering tests

TYPE OF TESTS	EXAMPLES
Structural tests	• Rebar testing for tensile strength and bend test. • Concrete cylinder testing; compressive strength test. • Structural integrity testing; non-destructive testing (NDT) methods like ultrasonic, radiography, and impact-echo testing.
Geotechnical tests	• **Dynamic cone penetration test (DCPT)**: a geotechnical test that measures the strength of soil and other materials. It's used to evaluate the mechanical properties of soil and to help with engineering design and construction. It measures the soil density and strength. • **Plate load test**: measures the soil bearing capacity.
Materials testing	• **Fresh and hard concrete testing**: slump test, compressive strength test, and concrete density test. • **Soil testing**: density test, moisture content test, and California Bearing Ratio (CBR) test. • **Aggregate testing**: sieve analysis, specific gravity test, and abrasion test. • **Asphalt testing**: density test, Marshall stability test, and asphalt content test.
Environmental tests	• **Water quality testing**: measures pH, turbidity, and bacterial contamination. • **Soil contamination testing**: measures heavy metal and chemical contamination. • **Air quality testing**: measures particulate matter, NOx, and SOx.

- During construction, material testing and inspection are essential. So, the knowledge of material testing procedures for materials such as concrete, soil, and asphalt is essential.
- Familiarity with various construction methods and techniques, including foundation work, framing, and finishing, is essential.
- Knowledge of safety protocols and procedures (procedural management; see also Section 9.7). Understanding of safety regulations, including Occupational Safety and Health Administration (OSHA) guidelines, and ability to identify potential hazards are a must without compromise.

2.3.5.1.3 Installation, operation, and maintenance

Civil engineering technicians are responsible for installing, maintaining, and repairing structures, equipment, machinery, and systems to ensure that they function efficiently and effectively.

They specifically handle the installation, operation, maintenance, and piecing together a civil engineering project, overall.

They are the chief operators and maintainers of civil engineering projects handling day-to-day activities to ensure that structures perform as expected—building reliability in civil engineering projects throughout lifecycle phases of the projects.

A water treatment plant operator is a typical example of a civil engineering technician whose role is to oversee the operation and maintenance of a water treatment plant, ensuring that safe and clean drinking water is supplied to people. On the other hand, a construction inspector monitors construction sites to ensure compliance with building codes, regulations, and project specifications.

2.3.5.1.4 Documentation and reporting

Technicians are tasked with maintaining accurate records, writing reports, and documenting maintenance and repair work. These include the following:

- Test reports, that is, records of testing and calibration activities, including the results and any necessary adjustments.
- Maintenance logs, that is, records of routine maintenance activities, such as inspections, cleaning, and lubrication.
- Field logs entered in a field log book for civil engineering technicians, which is a record of activities and observations made while working on a construction site. It can include details of surveys, sketches, and notes.

2.3.5.1.5 Troubleshooting

This is the process taken to trace and correct faults in a mechanical, structural, or electronic system. As you know, civil engineering projects are an integral of several components for the engineering structure to be operable and maintainable, serving the intended function through the installation, operation, and maintenance phases. Therefore, civil technicians usually identify and resolve technical issues, often working with other professionals to diagnose and fix complex problems.

2.3.5.1.6 Inspection and analysis

Civil engineering technicians conduct regular inspections, analyze data, and interpret results to recommend improvements or repairs. These include building inspectors, bridge inspectors, and construction inspectors.

2.3.5.1.7 Collaboration and communication

Civil engineering technicians work with other teams to communicate technical information and provide guidance and support to junior technicians.

2.3.5.1.8 Continuous learning

Staying updated with the latest technologies, attending workshops, and pursuing certifications to enhance their skills and knowledge are a must for constantly evolving civil engineering technicians. The need to keep the pace as a constantly evolving civil engineering technician cannot be overemphasized—it speaks for itself.

The world is changing fast, more than ever, due to technological advancements and scientific breakthroughs. So, keeping pace with the demands of this fast-paced world is vital for anyone committed to excellence in their profession. This does not need any votes!

Overall, technicians play a vital role in ensuring the smooth operation of equipment, machinery, and systems across various industries.

2.4 ENGINEERING ETHICS

Engineering ethics is the field of system of moral principles that apply to the practice of engineering. Engineering ethics is the set of moral principles and values that guide engineers in their work, ensuring they act with integrity, responsibility, and respect for the public and the environment. Thus, through engineering ethics, engineers commit to sustainability and environmental stewardship.

It involves considering the social, environmental, and economic impacts of their designs, decisions, and actions. Engineers can make a positive impact on the society when they embrace engineering ethics to earn public trust and contribute to a better future. Engineering ethics is essential for engineers because it protects public safety and well-being.

Engineering ethics is essential for engineers because it promotes sustainable and environmentally responsible practices. Trust and credibility in the profession are maintained through upholding engineering ethics. Engineering ethics is essential for engineers because it encourages responsible innovation and technological advancement, guiding decision-making in complex, uncertain situations.

Engineers have a professional and moral obligation to uphold ethical standards, including honesty and transparency. Respect for colleagues, clients, and stakeholders is essential, as well as providing accountability for their work and decisions. They engage in continuous learning and professional development.

Engineering ethics examines and sets the obligations by engineers to the society, to their clients, and to the profession. As a scholarly discipline, it is closely related to subjects such as the philosophy of science, the philosophy of engineering, and the ethics of technology. Ethics is more important in the engineering field than you may think. There is often a lot of pressure to create innovate designs or adhere to contracts.

2.4.1 Advance of engineering ethics

Engineering ethics began to take on supreme importance as people learned that lives depended on a well-crafted design—not compromising safety considerations. Modern engineering ethics now emphasize the importance of protecting public safety and welfare. Engineers are expected to report any unsafe practices and ensure their work adheres to the highest safety standards. These advancements helped ensure that engineering practices not only meet technical standards but also uphold ethical principles, ultimately benefiting the society as a whole.

The field of engineering saw significant advancements in ethics, driven by the increasing complexity and impact of engineering projects on the society and the environment.

Professional organizations—like the National Society of Professional Engineers (NSPE), United States; the American Society of Civil Engineers (ASCE); and the Institution of Civil Engineers, United Kingdom—established comprehensive codes of ethics. These codes guide engineers in making decisions that prioritize public safety, honesty, and integrity. These organizations place a strong emphasis on licensure, ethics, and professional development.

Incorporation into education. Engineering ethics was later incorporated into curricula. Today, engineering ethics are a fundamental part of engineering curricula in many engineering institutions around the world. Universities and colleges started incorporating ethical training to prepare future engineers to handle ethical dilemmas they might face in their careers. Ethical guidelines are now often illustrated through case studies that reflect real-world scenarios. This approach helps engineers practice ethical reasoning and apply ethical principles to their work.

Today, there is growing emphasis on sustainable engineering practices. Engineers are encouraged to consider the environmental impact of their projects and strive for solutions that promote sustainability. With the globalization of engineering projects, there is a push toward creating universal ethical standards that can be applied across different countries and cultures.

2.4.2 Ethical code of conduct

An **'ethical code of conduct'** is a set of principles and guidelines that outline expected behavior for individuals or organizations within a specific field, aiming to promote high ethical standards and ensure responsible actions in all situations, often encompassing values such as honesty, integrity, fairness, and respect.

2.4.2.1 What constitutes an ethical code?

The code of conduct typically includes the following:

- **Ethics:** this refers to upholding honesty, integrity, and transparency. Engineers are required to discharge professional duties with integrity and behave with integrity in relation to all conduct bearing upon the standing, reputation, and dignity of the institution and of the profession of civil engineering—that means undertaking work that you are competent to do, thus maintaining honesty and fairness in all professional dealings.
- **Objectivity:** this refers to striving for impartiality and objectivity when dealing with others in the civil engineering business.
- **Confidentiality:** that is, protecting sensitive information. Engineers hold a duty to keep information about projects confidential as far as contracts with clients dictate.
- **Safety:** prioritizing public safety and well-being. Having full regard for the public interest, particularly in relation to matters of health and safety, and in relation to the well-being of future generations.
- **Responsibility and due care:** taking ownership of one's work and decisions, performing duties with diligence and care.
- **Sustainability:** considering environmental and social impacts, that is, showing due regard for the environment and for the sustainable management of natural resources.
- **Professionalism:** maintaining competence, objectivity, and respect for colleagues and clients. Maintaining and improving technical knowledge and skills. Keep developing professional knowledge, skills, and competence on a continuing basis and give all reasonable assistance to further the education, training, and continuing professional development of others. Benchmarking on the knowledge of others: designers could utilize the knowledge and experience of contractors to assist in the process.
- **Professional behavior:** this involves acting in a manner that reflects positively on the profession.

2.4.3 Typical moral dilemmas faced by civil engineers in history

Civil engineers have faced terrible moral dilemmas throughout history. These examples highlight the importance of ethical awareness and moral courage, professional responsibility and codes of conduct, whistleblowing and speaking out against injustice, and considering the broader social implications of one's work. It serves as a reminder of the need for engineers to prioritize human values, dignity, and safety above all. Four typical examples of moral dilemmas faced by civil engineers in history are outlined. These examples illustrate the complex moral dilemmas civil engineers have faced throughout history, often involving trade-offs between safety, cost, and political or social pressures.

2.4.3.1 The Tacoma Narrows bridge collapse (1940)

Engineers designed a bridge with a flawed aerodynamic profile, leading to its collapse. They faced a moral dilemma between saving costs and ensuring safety. The Tacoma Narrows

Suspension Bridge, which cost US$ 6.4 m, that collapsed in 1940, merely 4 months after its completion, presents an exciting legendary example of how successful project management is premised on sound knowledge and understanding of the elements of the project. In that case, it helps mitigate risks and have proper project planning and scope management. The project manager must understand why they are doing whatever they are doing to lead a project that can maneuver through its lifecycle reliably, operably, and sustainably.

The bridge was primarily funded by the federal government's Public Works Administration. It was intended to connect Seattle and Tacoma with the Puget Sound-Navy Yard at Bremerton, Washington. The bridge collapsed due to poor planning and unforeseen technological effects. It is reported that the bridge experienced severe wind-induced vibrations. Interestingly, the Deer Isle Bridge along the coast of Maine was of similar construction and smaller in size. It had opened 1 year before the Tacoma Bridge and is still in service today.

The evaluation process conducted during feasibility studies for civil engineering structures takes on a holistic perspective to assess potential risks that would be encountered. The engineer of the Deer Isle Bridge had the foresight and sound judgment to add wind fairings along the bridge's length to give it better aerodynamic properties. He also provided diagonal cable bracings to provide greater stiffness. The engineer of Deer Isle Bridge had good knowledge and understanding of the project at hand. However, the engineer of the Tacoma Narrows Suspension Bridge did not correctly account for aerodynamic forces. Many motorists crossing the Tacoma Bridge complained of acute seasickness brought about by the bridge's rise and falling, which was nicknamed 'Galloping Gertie' [6].

The lesson to learn from the Tacoma Bridge is that effective project management would be premised on a thorough knowledge and understanding and accounting for aerodynamic forces by both the chief engineer in charge and the local construction engineer. Mr Charles Andrews, who was the chief engineer in charge of construction, reported that the local construction engineer substituted open girders in the construction of the bridge's sides for flat solid girders that deflected the wind rather than allowing it to pass. Therefore, the local construction engineer did not understand the implications of the decisions he undertook. The one who takes the decision to piece together a construction project must have technical competence beyond project control and management.

2.4.3.2 Quebec Bridge disaster (1907)

The Quebec Bridge was to be one of the engineering wonders of the world. When completed, it would be the largest structure of its kind and the longest bridge in the world, outstripping the famous Firth of Forth Bridge in Scotland. American engineer Theodore Cooper was appointed to design it.

High above the St. Lawrence River, on a hot August day in 1907, a worker named Beauvais was driving rivets into the great southern span of the Quebec Bridge, when he noticed that a rivet that he had driven no more than an hour before had snapped clean into two. Just as he called out to his foreman to report the disquieting news, the scream of twisting metal pierced the air. The giant cantilever dropped out from under them, crashing into the river with such force that people in the city of Quebec, 10 km away, believed that an earthquake had struck [7].

Of the 86 workers on the bridge that 29 August 1907, 75 died, many of them local Caughnawaga, famous for their high steel work. Some of the dead had been crushed by the twisted steel; others by the fall. Still others drowned before the rescue boats could reach them.

The Quebec Bridge collapse on 29 August 1907 was a tragic event in Canadian history. During its construction, the bridge's south section failed, causing the structure to collapse into the St. Lawrence River. This disaster resulted in the deaths of 75 workers, many of whom were Mohawk ironworkers from the Kahnawake community.

The collapse was attributed to design flaws and inadequate supervision. The bridge was intended to be the longest cantilever bridge in the world, but the ambitious design led to excessive stress on the structure. Despite this setback, the bridge was eventually completed in 1917 and remains a significant engineering landmark.

The ill-starred bridge suffered a second disaster on 11 September 1916 when a new center span being hoisted into position fell into the river, killing 13 men. The bridge was finally completed in 1917, and the Prince of Wales (later Edward VIII) officially opened it on 22 August 1919.

The Royal Commission of Inquiry investigating the calamity excoriated John Deans (chief engineer) for his poor judgment in allowing work to continue when it was obvious that the bridge was in danger. The brunt of the blame, however, was placed on the shoulders of Theodore Cooper, who had committed grave errors in design and his calculation of loads. There was criticism of the bridge company for placing profit above safety and for engineers who neglected their professional and moral duties [7].

Cooper chose the cantilever structure as the 'best and cheapest plan' to span the broad St. Lawrence. That word 'cheapest' would come back to haunt him. In order to reduce the cost of building the piers farther out in the river, Cooper increased the bridge span from 490 to 550 m.

Cooper refused to supervise the construction on site, claiming ill health, and trusted Peter Szlapka, who was little more than a desk engineer. By the summer of 1907, the consequences of Cooper's design and of the lack of leadership on the site began to show up on the structure itself, especially in the 'compression members'—the lower outside horizontal pieces running along the length of the bridge.

2.4.3.3 Holocaust gas chambers, World War II

The design and construction of gas chambers for the Holocaust during World War II is a particularly disturbing example of moral dilemma faced by engineers. The engineers involved in this project were confronted with an unbearable choice, either to:

- Participate in designing and building gas chambers (Figure 2.1), which would be used to murder millions of innocent people, including Jews, Romani people, homosexuals, disabled individuals, and others deemed undesirable by the Nazi regime.
- Refuse to participate, which would likely result in severe consequences, including imprisonment, torture, or even death.

FIGURE 2.1 Gas chamber.

Many engineers and architects involved in this project were coerced or manipulated into contributing to the Nazi regime's atrocities. However, some did resist or tried to sabotage the efforts.

2.4.3.4 *The Johnstown Flood (1889)*

The Johnstown Flood of 1889 was a catastrophic disaster caused by the failure of the South Fork Dam, which was owned by the South Fork Fishing and Hunting Club [8]. The causes of the disaster included poor dam design and construction—the dam was rebuilt in 1880 with inadequate engineering oversight, using poor materials and techniques. The main spillway had insufficient capacity and was unable to handle the heavy rainfall on 30–31 May 1889, causing the lake to overtop the dam. Another cause was negligence and complacency as the club's wealthy and influential members who prioritized their interests over public safety, ignoring warnings and concerns about the dam's integrity.

The dam's failure led to a massive flood that killed over 2,209 people and destroyed the town of Johnstown, Pennsylvania. Engineers failed to properly inspect and maintain a dam, leading to a catastrophic failure. They faced a moral dilemma between saving resources and prioritizing public safety. The scale of the Johnstown Flood of 1889 is difficult to visualize. Summarizing the flood's impact in statistics and facts is a quick way to convey the enormity of the event. The list of some of the most descriptive facts/statistics about the great Johnstown Flood disaster are as follows:

- 2,209 people died.
- 99 entire families died, including 396 children.
- 124 women and 198 men were left widowed.

- More than 750 victims were never identified and rest in the Plot of the Unknown in Grandview Cemetery.
- Bodies were found as far away as Cincinnati, and as late as 1911.
- 1,600 homes were destroyed.

The articles on the South Fork Dam failure by the engineering reporters reflected where civil engineering stood in 1889. Some subjects currently integral to the profession, such as surveying and hydraulics, were fairly well understood then. Others, such as geotechnical engineering, had yet to be established beyond scattered principles. It was noted that the dam suffered poor workmanship and flimsy sheet piles used to clumsily barricade the old drainage outlet, and the sag in the dam, which the remnants still indicated—and no engineering advice or supervision was taken.

Civil engineers, especially licensed professional engineers (PEs), can learn critical lessons from the South Fork Dam failure and Johnstown Flood of 1889. The entirely preventable tragedy underscores that those in the profession must navigate project challenges primarily using technical expertise, not business or managerial acumen [8]. This principle holds true throughout the whole lifecycle of a project, from design to construction to operations and maintenance to even, if need be, failure analyses. The lesson is especially applicable for geotechnical engineers, who deal with highly variable subsurface materials.

The disaster also highlights how essential it is for civil engineers to meet the contemporary standard of care in their work. The flood reminds those in the profession that the current standard memorializes victims of either unforeseen circumstances or violations of previous standards. The catastrophe also reflects the impact of the technical work civil engineers perform, ultimately, on human and how the consequences of successes in the field might only be exceeded by those of failures.

2.4.4 Trolley problem

The trolley problem is a classic thought experiment in ethics and moral philosophy. The trolley problem was created by British philosopher Philippa Foot (1920–2010) in 1967. It is a series of thought experiments that a civil engineer ought to know in ethics involving stylized ethical dilemmas of whether to sacrifice one person to save a larger number.

It raises questions about consequentialism: is it morally justifiable to sacrifice one person to save others? It again raises questions about deontology: is it morally wrong to actively cause harm to one person even, if it would save others?

The trolley problem's core question, 'Do you sacrifice one for the greater good?', is at the heart of several civil engineering dilemmas.

The trolley problem explores moral absolutism versus moral relativism. It also explores utilitarianism versus Kantianism and the doctrine of double effect. It also explores personal versus impersonal moral obligations.

These experiments teach engineers the moral gravity of harming an innocent person and seek to minimize harm whenever possible. This, again, teaches engineers the need to be responsible and take holistic independent decisions that incorporate safety measures to the maximum, avoiding tortious liabilities that can accrue out of negligence.

To avoid falling into such a paradox, that is, 'sacrifice one person to save a larger number,' civil engineers always owe a duty to the public and are guided by the fact every human life has equal value and should be treated with equal respect and dignity.

The application of the trolley problem in civil engineering becomes even more intriguing when you, as the engineer in charge, are the 'one person to sacrifice.' This realization drives home the importance of being responsible and vigilant in ensuring safety within your area of expertise. Any complacency or compromise can have severe consequences, ultimately leading to being 'sacrificed for the greater good,' as the engineer. In this case, being sacrificed for the greater good is bearing the consequences for noncompliances and nonconformities. As an engineer tasked to take responsibility, do the appropriate thing all the time to avoid being sacrificed for the greater good!

2.4.4.1 The site safety dilemma: a typical illustrative example

You're the site manager at a construction project, overseeing the deep excavation of a high-rise building. As you're inspecting the site, you notice a worker stepping into the path of an approaching heavy dump truck, unaware of the danger. The truck's brakes have failed, and it's headed straight for the worker.

You have two options:

1. Divert the truck, that is, you can divert the truck into a nearby excavation pit, but this will put the lives of several workers at the bottom of the pit at risk.
2. Do nothing, that is, if you don't intervene, the truck will hit and kill the single worker in its path.

Question: What do you do?

Discussion

The principle "Bearing in mind that every human life has equal value and should be treated with equal respect and dignity" is rooted in various ethical theories including utilitarianism, Kantianism, and human rights. However, from a purely utilitarian perspective, diverting the truck into the excavation pit might be considered the 'lesser evil.' This option would potentially save one life but put others at risk.

This dilemma is meant to spark discussion, but not to provide a clear-cut solution. Possible solutions are rooted in different philosophical schools of thought! It highlights the complexity of real-world decision-making, where there are no perfect answers. Ultimately, the decision would depend on various factors, including the following:

- The site's safety protocols and emergency procedures.
- The number of workers in the excavation pit and their potential risk.
- The possibility of finding an alternative solution, such as warning the worker or stopping the truck using other means.

  Diverting the truck into the excavation pit is a decision that would prioritize the immediate life in danger, that is, the worker in the truck's path. While this choice puts others at risk, it's essential to consider the following:

  - **Immediacy:** The worker in the truck's path is in imminent danger, and every second counts.
  - **Certainty:** Diverting the truck would likely save the worker's life, whereas doing nothing would almost certainly result in their death.
  - **Potential risk versus certain harm:** While diverting the truck puts others at risk, the harm to the worker in the truck's path is certain and immediate.

    This decision is not taken lightly, and it's crucial to acknowledge the potential consequences. However, in this extreme situation, prioritizing the immediate life in danger seems to be the most humane choice.

    This is a classic ethical dilemma, known as the 'Trolley Problem.' There's no one 'right' answer as it depends on an individual's moral principles and values. Consider some perspectives below:

- **Utilitarian approach**: sacrifice the worker in the path to save the lives of the people in the excavation pit. This prioritizes the greater good and minimizes overall harm.
- **Deontological approach**: save the worker stepping into the path as it's morally wrong to actively cause harm to an innocent person, even if it means putting others at risk.

  - **Note:** A utilitarian approach judges the morality of an action based on its consequences, aiming to maximize the greatest good for the greatest number of people, while a deontological approach focuses on whether an action adheres to set moral rules or duties, regardless of the outcome, meaning the 'ends do not justify the means' in a deontological perspective; essentially, utilitarianism is concerned with the results of an action, while deontology is concerned with the inherent rightness or wrongness of the action itself.

- **Virtue ethics approach**: consider the character and moral virtues of the driver. Would a compassionate and empathetic person prioritize the pedestrian's life or the lives of their passengers?
- **Legal perspective**: in most jurisdictions, drivers have a duty to exercise reasonable care and avoid harming pedestrians, in this case the worker stepping into the path. Knocking the worker stepping into the path might be considered negligent or reckless.
- **Emotional response**: many people's initial reaction would be to save the pedestrian as it is a more personal and direct harm.

Ultimately, as far as the above perspectives are concerned, the decision depends on individual values, moral principles, and the specific circumstances. There is no easy answer, and it's a thought-provoking dilemma that challenges our moral intuitions. However, in conclusion, note that ramming into the worker stepping into the path should be considered only as a last resort, when all other options have been exhausted. This approach prioritizes avoiding harm to the pedestrian, if possible, while also considering the safety of the driver and workers at the bottom of the pit. In an ideal scenario, the driver would consider three steps:

1. Attempt to stop the vehicle safely.
2. Swerve to avoid the pedestrian (if possible, without putting others at risk).
3. Use emergency braking or other safety features.

Only if these options (1–3) are unavailable, and the driver is faced with an unavoidable choice, would ramming into the pedestrian be considered. This approach acknowledges the moral gravity of harming an innocent person and seeks to minimize harm whenever possible. Please note that ramming into an innocent person to maximize the greater good has never been a line of defense—and it won't garner any pardon for wrongful criminal actions. This case is not straightforward, and the culprit may or may not get waivers as circumstances may dictate after careful examinations. However, in many jurisdictions, things may rotate around 'exercising reasonable care or failure to' and laws interpreted accordingly. Because of this, civil engineers exercise the highest level of care so that they don't sacrifice themselves by lowering the standards of practice to please clients or some sections of the public. In a reciprocal version, civil engineers act as the one person scarified for the greater good in the trolley problem, whenever standards are compromised, putting safety and lives of the public at risk—lowering standards of practice as an engineer means you risk sacrificing yourself for the greater good whenever things go wrong. The fines and compensations and even potential sentencing, to remedy the wrongful actions, aim to recover the greater good, and they come your way whenever you fail to exercise utmost care.

2.5 PROMINENT FIGURES IN CIVIL ENGINEERING

The civil engineering industry has evolved over years being supported by several influential figures, not necessarily civil engineers. These people have contributed to the knowledge and understating of how civil engineering structures function safely and reliably. From scientists, to artists, management theorists, scholars, and engineers themselves, the civil engineering field has been enriched.

This section describes prominent people not necessarily civil engineers, who changed the face of civil engineering by sharing insightful ideas, concepts, and management theories. Note that this list is not exhaustive but provides some of the most famous figures in the history of civil engineering.

2.5.1 Leonardo da Vinci (1452–1519)

Leonardo da Vinci, a polymath of the High Renaissance, is always celebrated for his influential contributions to the arts field, with iconic masterpieces like the 'Mona Lisa' and 'The Last Supper' remaining unparalleled. However, Leonardo's A-class talent transcended artistic expression, penetrating the fields of science and engineering, where his innovative spirit left an enduring legacy. Born in Anchiano, Vinci, Italy, on 15 April 1452, Leonardo da Vinci was a true Renaissance man, an Italian polymath who made lasting impacts in various fields, including art (Mona Lisa and The Last Supper), engineering (designs for machines and bridges), anatomy (detailed human body studies), mathematics (geometric calculations), and science (observations on light, motion). His curiosity and innovative spirit continue to inspire masses worldwide.

High renaissance is an era spanning from the late 15th to early 16th century—an era of great cultural and intellectual change in Europe characterized by significant developments in art, literature, and science. Leonardo da Vinci, a famous figure in the Renaissance era, made many contributions to the civil engineering field, including designing canals, bridges, and machines. His work helped pioneer structural mechanics and influenced later bridge builders. His designs for machines such as the helicopter, parachute, and armored vehicle, though not realized in his lifetime, have significantly influenced modern engineering [9]. He was an artist, a scientist, an engineer, and an inventor.

Sections 2.1.1–2.1.11 outline the primary attributes a civil engineer must possess. Da Vinci's collection makes the world admire his genius, as it leaves unparalleled evidence of his exceptional skills, which encompass nearly all of these attributes. You can imagine at that time when engineering and science were in their infancy! Notably, Da Vinci excelled in effective written communication and idea generation through his ingenious sketches and paintings.

His expertise in anatomy helped him create better portraits and sculptures and also helped him make sense of mechanics and engineering. He himself defined innovation as connecting the unconnected. He was among the first to conceptually mimic the natural world through his art and engineering concepts. He explored mechanics, anatomy, and nature, fusing art and science. See Chapter 6 for how a sustainable world requires mimicking nature. He was a rare breed of individuals who lived centuries ahead of their time, leaving an indelible mark on history.

The mere mention of creativity to an engineer conjures up the visionary spirit of Leonardo da Vinci, whose groundbreaking work continues to inspire engineers worldwide. His fusion of art and science left an indelible mark on engineering, reminding all engineers that creativity is the spark that ignites innovation.

Leonardo da Vinci's work on art, perspective, and geometry indirectly influenced the development of puzzles like the 9-dot puzzle, which require creative thinking and problem-solving skills. This famous 9-dot puzzle teaches engineers the need to think out of the box and around the box but not limiting to the box—they need to be creative. Section 1.2.6 showed how creativity is a key attribute that engineers possess.

2.5.1.1 The Vitruvian Man

Named after the Roman architect, writer, and engineer Marcus Vitruvius Pollio who lived during the 1st century BC, the Vitruvian Man is an iconic drawing that has inspired many engineers and scientists.

The Vitruvian Man (Figure 2.2) is a famous masterpiece by Leonardo da Vinci that represents the perfect union of art, science, and mathematics. Created around 1490, it depicts a nude male figure inscribed within a circle and square, demonstrating the ideal proportions of the human body described earlier by Vitruvius.

The *Vitruvian Man*, created by Leonardo da Vinci, was initially published in Luca Pacioli's book *Divina Proportione*. The architectural principles of Vitruvius, a prominent ancient Roman architect and engineer, influenced Da Vinci's *Vitruvian Man* design.

Vitruvius is best known for his treatise 'De Architectura' (on architecture), which is one of the most influential architectural texts of all time.

In 'De Architectura,' Vitruvius described the ideal proportions of the human body, which he believed should be reflected in the design of buildings and other structures. He wrote that the human body inscribed within a circle and square should demonstrate perfect proportions, with the navel serving as the center of the circle and the arms and legs extending outward in harmony.

Leonardo da Vinci's famous drawing, the Vitruvian Man, was inspired by Vitruvius' writings on human proportions and the relationship between the human body and geometric shapes. Referred to as 'sacred geometry,' the Vitruvian Man embodies the perfect proportions and harmonies found in nature and the universe.

Sacred geometry is one of the most profound phenomena found all over the world. This universal concept explains the divine connection between all things, pointing to a design or creation by intelligence, and not chaos. Da Vinci's work was

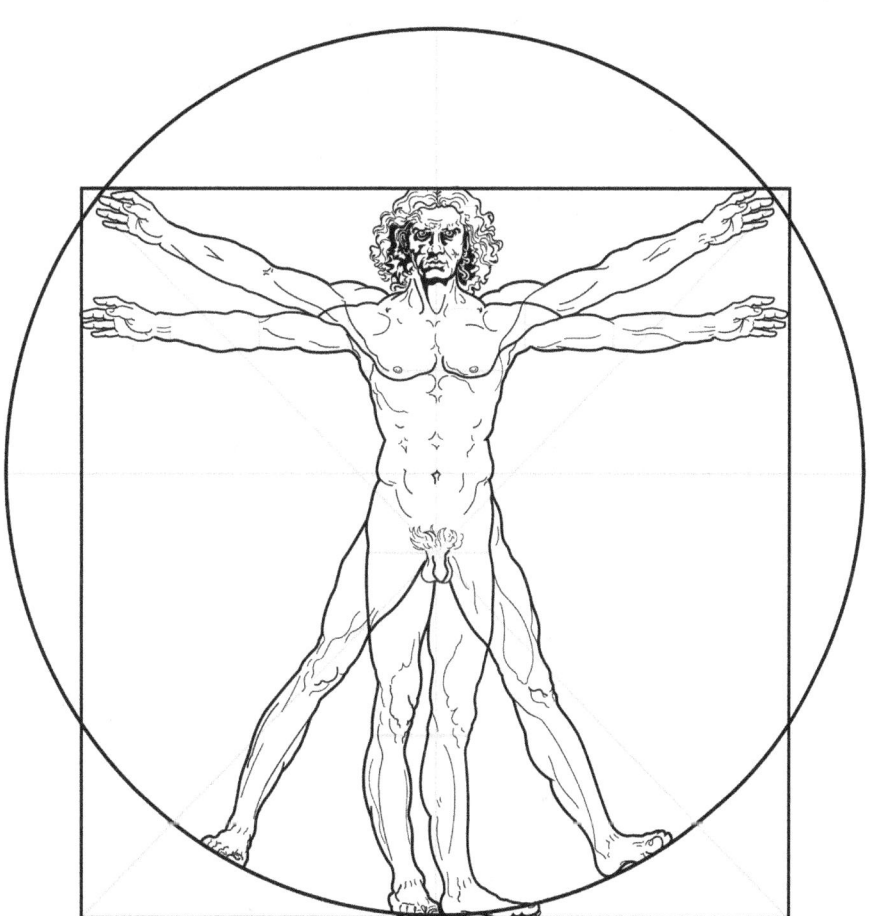

FIGURE 2.2 Vitruvian Man. (Image credits: Shutterstock.)

highly philosophical, sharing insights about the greatness of the Supreme Being.

'The length of a man's outspread arms is equal to his height,' according to Leonardo da Vinci.

2.5.1.1.1 Key lessons from the Vitruvian Man

The Vitruvian Man, a study of the ideal proportions of the human form, remains an iconic symbol of the Renaissance and continues to inspire artists, scientists, and thinkers to this day. This famous drawing has been so widely respected in history.

The Vitruvian Man is a powerful symbol of innovation and creativity, embodying the principles of harmony, proportion, balance, and interconnectedness. These lessons about proportion, harmony, balance, and the interconnectedness of art and science remain relevant and influential even in modern times.

Leonardo da Vinci's Vitruvian Man represents the 'perfect man,' based on the ancient knowledge of ratios and proportions present in human anatomy. The illustration depicts the naked, ideal, healthy form of a man—a study of human physiology—a perfectly proportionate rendering of the human form, as determined by the application of geometry and mathematics. The Vitruvian Man demonstrates the perfect ratios and proportions found in human anatomy.

2.5.1.1.1.1 Innovation and creativity   Overall, Leonardo da Vinci's work on the Vitruvian Man showcases his innovative and creative approach to art, science, and engineering, inspiring future generations to think outside the box.

Thinking outside the box is the cornerstone of innovation and creativity. The drawing embodies the principles of harmony, proportion, balance, and interconnectedness.

The Vitruvian Man was drawn with precise mathematical proportions, reflecting the underlying order of the universe. This attention to detail inspires innovative thinking.

The Vitruvian Man represents the human body as a microcosm of the universe, highlighting the interconnectedness of all things—microcosm–macrocosm analogy. This perspective encourages innovative thinking and creativity.

2.5.1.1.1.2 Proportion, geometry, and harmony   The Vitruvian Man illustrates the mathematical principles that underlie the natural world, demonstrating how geometric shapes and proportions can be found in human anatomy.

The drawing illustrates the mathematical proportions of the human body, as described by Vitruvius. It shows how the body's dimensions are related to each other and to the geometric shapes that surround it. Leonardo is said to have hidden in the drawing a solution to an old mathematical problem, the squaring of the circle.

The drawing emphasizes the precise measurements and ratios of the human body, showcasing how the human form can be inscribed within geometric shapes like a circle and square, demonstrating the ideal proportions described by the Roman architect Vitruvius.

2.5.1.1.1.3 Interconnectedness of art and science   The Vitruvian Man embodies the Renaissance ideal of combining artistic and scientific knowledge. Da Vinci's work demonstrates how art and science can inform and enrich each other.

The Vitruvian Man is important in art, history, and anatomy. It displays the human body's ideal natural proportions as they have been depicted for centuries.

Leonardo da Vinci combined his deep understanding of anatomy with artistic expression, demonstrating how art and science could be integrated to study and depict the human form. The Vitruvian Man embodies a holistic approach, considering the human form as an integral part of the natural world.

2.5.1.1.1.4 Humanism   The **Vitruvian Man** embodies the Renaissance humanist ideal, where humans were seen as the center of the universe, with their bodies reflecting divine perfection and order. He, thus, supported the growth of humanistic theory indirectly.

The humanistic theory is a *psychological theory* that emphasizes self-actualization, free will, and the potential for growth in healthy people. It's based on the idea that people are inherently good and motivated to improve themselves.

Da Vinci emphasized the importance of physical and mental well-being in his writings and art. He believed that a strong, healthy body was essential for a sharp and creative mind.

In his 1478 Codex Atlanticus notebook, Da Vinci wrote, 'The noblest pleasure is the joy of understanding' [10]. The quote means that learning and understanding new things can be a great source of satisfaction and fulfillment. It can be exciting to gain new knowledge and perspectives on the world.

His work reflected a *holistic approach* to human well-being, which integrated physical, mental, and spiritual aspects.

Da Vinci's drawing combines a circle and a square in one representation. The drawing is brought out into a spiritual context. For example, the circle is a symbol of femininity and the square a symbol of masculinity.

He emphasized that 'a healthy mind lives in a healthy body.' His ideas and writings conveyed a message about the interconnectedness of physical and mental health.

2.5.1.1.1.5 Impact on architecture   The Vitruvian Man's principles of proportion also apply to architecture, influencing the design of buildings to achieve harmony and balance. See Chapter 6, mimicking the natural world.

Leonardo da Vinci's Vitruvian Man shares timeless insights that embody a fundamental concept in architectural philosophy. Through depicting the human form inscribed within a circle and square, the drawing demonstrates how human proportions conform to mathematical ratios. This suggests that buildings should be designed with the same precision, balance, and attention to detail that characterize the human body.

2.5.1.2 Connecting the dots

You can't connect the dots looking forward; you can only connect them looking backwards. So, you have to trust that the dots will somehow connect in your future. You have to trust in something—your gut, destiny, life, karma, whatever. This approach has never let me down, and it has made all the difference in my life.

Steve Jobs—Stanford commencement speech (2005)—a fundamental quote about innovation and creativity [11]

According to Steve Jobs, when moving forward in life, you can't always see how your experiences, decisions, and actions will connect and lead to future outcomes or successes. It's only when you look back that you can see how the dots (your experiences) connected to form a larger picture. What you have become is a culmination of many experiences, how the dots unfolded and connected, and here you are today, a civil engineer in making, training, or professionally qualified. At some point, you felt unsure of the decisisons you were making. Somehow, you felt losing trust, maybe. It so happens that Leornardo Da Vinci was among the first to experiment with Steve Job's concept of 'connecting the dots.' Da Vinci's creativity and innovativeness showcased his trust in the process—his dots are constantly connecting even in the modern world!

The lesson is that you can't predict the future completely or see how events will unfold while they're happening (looking forward). Only in hindsight (looking backward) can you see how experiences, decisions, and events are connected and led to where you are today. That's why, politicians, especially those opposing the government, always have more to say because they criticize the government with a hindsight bias—after many events have unfolded! They connect the dots moving backward.

Steve's quote is about trusting the process, being patient, and having faith that seemingly unrelated experiences and decisions (the 'dots') will eventually connect and make sense in the future. You must trust that the choices you make and that the path you take will lead to a meaningful and coherent future, even if you can't see it now. Steve Jobs was also noted saying, 'never lose hope,' especially when innovating cool stuff. Keep moving and trust the process. This trust can be in your intuition (gut), a higher power (destiny), the natural course of life (life), and the universe's balance (karma principle).

Steve Jobs shared his personal philosophy that worked for him—trusting in the unknown and having faith that the 'dots' (experiences and decisions) would connect in a meaningful way. In essence, the quote encourages us to embrace uncertainty as civil engineers (risk taking), trust in the process, have faith in ourselves and the universe, and to make decisions with confidence, even when the outcome is unclear.

Overall, nurturing this kind of mindset allows you to move forward with purpose and confidence, knowing that the connections between events will become clear in time. That is what civil engineers do—they trust the process, whether designing infrastructure or innovating cool stuff. They persist, believe in themselves, and never lose hope. They think out of the box and around it.

2.5.1.3 The '9-dot puzzle'

To put things into perspective, let's consider the famous 9-dot puzzle shown in Figure 2.3. Without lifting your pencil, can you connect all of the dots below with four straight lines? [12]

We can only connect the 9-dots by moving outside the box. See Figure 2.4. There is no option of connecting dots keeping inside or within the boundaries of the box. This illustrates being able to think out of the box and around it. The 9-dot puzzle requires a paradigm shift to solve. Initially, you may try to connect the dots within the boundaries of the square formed by the dots. However, the solution requires drawing lines outside the square, connecting the dots in a creative and

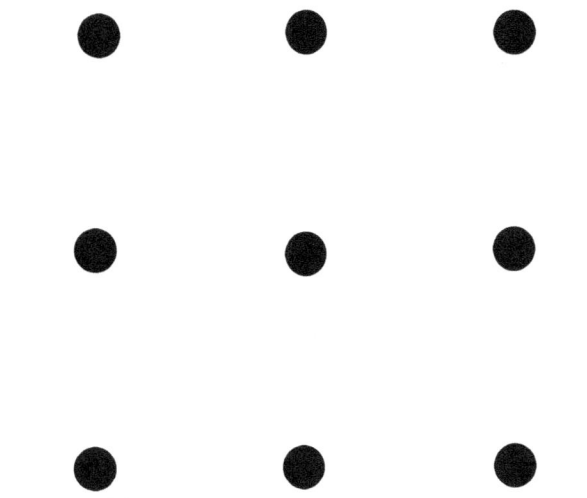

FIGURE 2.3 9-dot puzzle.

Solution

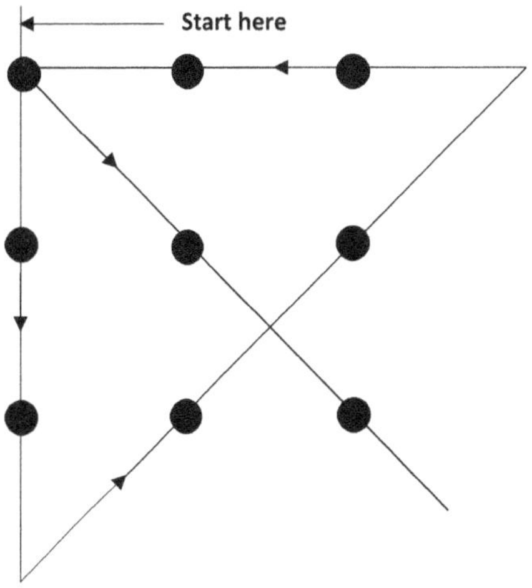

Start here

FIGURE 2.4 9-dot puzzle connected.

unconventional way. This unconventional approach is the gateway to solving problems innovatively.

Civil engineers primarily engage in creative and innovative development of engineering technology and continuous improvement systems. In this process, they do things in an unconventional approach to be able to discover new technologies and approaches. Without thinking out of the box, you can do little to innovate and develop practical solutions for the industry. In that case, you must trust the process. You must believe and trust the real tenets of innovation—visionary thinking! These include risk taking, collaboration, open mindedness, embracing failure, being adaptable, being client-centric, and being able to experiment with and refine ideas.

2.5.1.4 Cross-cutting lesson from Steve Job's quote and 9-dot puzzle

Both the 9-dot puzzle and Steve Jobs' quote encourage thinking beyond conventional boundaries and trusting unconventional connections. In the 9-dot puzzle, you need to think outside the grid to find the solution. Similarly, Steve Jobs' quote encourages you to trust that the seemingly unrelated 'dots' in your life will connect in meaningful ways, even if you can't see it at the moment. In essence, both the puzzle and the quote promote the following:

- **Lateral thinking:** exploring unconventional solutions and connections. That's the only way engineers manage to breakthrough what seems impossible to create the possible.
- **Trust:** having faith that things will work out, even when the path ahead is unclear; trust the process.
- **Creativity:** embracing new perspectives and possibilities—think out of the box and around it.

Finally, both Steve Jobs' quote and the 9-dot puzzle encourage an unconventional approach, trusting the process, and thinking out of the box and around it as the gateway to creativity and innovation. Initially, you may try to connect the dots within the boundaries of the square formed by the dots. The same thing happens when you try to innovate cool stuff; you may try to rely on what you think works. However, you may have to move out of the box to see that the innovation comes true by trusting the process!

2.5.1.5 Leonardo da Vinci's contribution to civil engineering

Leonardo da Vinci was a visionary genius whose groundbreaking contributions revolutionized the field of civil engineering. A true polymath of the High Renaissance, he epitomized a rare breed of individuals who lived centuries ahead of their time, leaving an indelible mark on history. His work in mechanical, physical, and hydraulic disciplines showcased his innovative spirit and problem-solving skills, influencing future engineers and inventors. Da Vinci's designs often preceded modern technologies, demonstrating his visionary approach to science and engineering.

As an engineer, Leonardo conceived ideas vastly ahead of his own time, conceptually inventing the parachute, the helicopter, an armored fighting vehicle, the use of concentrated solar power, the car and a gun, a rudimentary theory of plate tectonics, and the double hull, among others. Da Vinci left behind a collection of engineering sketches and notes.

Even after looking at bridges, canals, engines, gears, and human anatomy, only a small sample of Leonardo's genius is understood. One of the most astonishing facts is that some historians estimate 75% of the material from his technical notebooks has been lost over the centuries [13]. A prolific inventor and engineer, too far ahead of his time to be fully appreciated, Leonardo has missed a great deal of the recognition he deserves for his discoveries. In many ways, Da Vinci represents the ideal engineer because his visionary inventions dealt with theories that had not been addressed during his lifetime. The innovations he drew in his notebooks were simply problem-solving based on observation and rumination—the epitome of engineering.

Da Vinci traveled a lot, and after returning to Florence, he was charged in 1503 to study various hydraulic engineering projects, often for military purposes. While developing his knowledge accumulated in Lombardy, Leonardo drafted a project for joining Florence with the sea by means of a navigable canal that would pass Prato and Pistoia on to Pisa.

While in France, Leonardo's latest studies regarding waterways were undertaken during the French years. In 1518, particularly, he drafted a project for an irrigation canal between Tours, Blois, and the River Saone for construction of a royal palace at Romorantin.

2.5.1.5.1 Hydraulic engineering [14]
Da Vinci made significant contributions to hydraulic engineering. He showed early interest in the field, which was documented in his letter to Ludovico Il Moro, in which he stated that he knew how to conduct water between locations. During his first time in Florence, Italy (Florentine period), Leonardo explored water's potential uses, including the following: channels for communication and water-powered machinery. His work highlighted his innovative thinking in harnessing water's power.

2.5.1.5.1.1 Vertically sliding sluice gate   He envisioned a sluice gate that would regulate the opening of the lock by means of two successive movements being raised by the rotation of the bar about itself and, once having reached its highest point, it would be lifted by a winch. Using this method, it would be possible to allow even the tall watercraft with sails to pass through the lock. Da Vinci's study of a sluice gate examined the possibility of constructing locks for overcoming altitude differences along watercourses with irregular flow.

2.5.1.5.1.2 Canal bridge   This canal bridge, drawn by Leonardo da Vinci in the years after his return to Florence, in central Italy, would allow a watercraft to pass over another watercourse by means of a lock. The lock would be constituted of a basin closed upstream and downstream by two pairs of flap gates, allowing watercraft to pass over the difference in height of the water.

He designed it in such a way that once the craft entered into the lock, the gates upstream would close, and in the downstream gate, a small, secondary door would open, allowing discharge of the water until the water level in the lock reaches the same height as the downstream canal.

After this, the gates would open to allow the watercraft to continue along its course. In this manner, the opening of the gates would be facilitated since they did not have to overcome the resistance of the water pressure in order to open. Da Vinci designed a gated canal system to connect Florence, Italy, to the sea. His design for a sluice gate was similar to the modern locks used in the Panama Canal.

2.5.1.5.1.3 Observation of the Naviglio   Leonardo da Vinci is said to have dedicated much attention particularly to the Naviglio Grande (Grand Canal), executing its drawings on various folios, and highlighting its economic importance not only in terms of transportation but also for irrigation. Besides the canals, Leonardo studied other works of hydraulic engineering in Lombardy, northern Italy. He described an ingenious work of land reclamation for marshlands effected by means of a water stairway during his stay in Vigevano (Italy) in 1493.

2.5.1.5.1.4 Lock with flap gates   Leonardo da Vinci proposed the model of a lock of the flap gate type where the two parts of the gate close together to form an obtuse angle with

one another, pressing against the groove of a level difference, allowing them better to resist the hydrostatic pressure exerted by the water.

2.5.1.5.1.5 Doors of the locks

The doors of the locks of the Martesana canal were described by Leonardo da Vinci. They tell a fundamental episode of the history and identity of the city of Milan. After 50 years in storage, the Museum initiated a comprehensive study, investigation, and restoration project in 2016. This effort aims to preserve and showcase the artifact in the Museum's galleries, making it accessible to the public once again.

2.5.1.5.1.6 Maritime dredge

Leonardo da Vinci's innovative dredging watercraft features a dual-hulled design connected by a central platform. A long slot in the platform accommodates a double cable attached to wheels at the hulls' extremities, supporting an excavating bucket. The watercraft is stabilized by six anchors on each side, mounted at the hulls' ends. The double cable allows easy drawing and lifting of the bucket, even when the forces of traction could be significant. The numerous anchors counterbalance the stresses on the dredge, ensuring stable and effective maritime port excavation.

2.5.1.5.1.7 Mud-digging dredge

Leonardo da Vinci's innovative dredge design, conceptualized during his later years in Milan, features a cylindrical drum with four arms terminating in dredging blades, positioned upon two boats. It is positioned upon two boats and formed of a cylindrical drum on which four arms are mounted terminating with dredging blades. The device's mechanism allows for efficient excavation and collection of mud or stones, which fall into a raft moored between the boats. The depth of excavation is regulated by vertically adjusting the drum, ensuring precise control. The design also incorporates a self-propulsion system, where a cord connected to the bank wraps around the drum's axle as the wheel turns, causing the dredge to advance along the excavation zone. This ingenious maritime excavation device showcases Da Vinci's remarkable engineering prowess.

2.5.1.5.1.8 Studies and proposals for improvements

As he observed and surveyed the locks in use in the Milanese waterway system, Leonardo described and perhaps projected certain improvements, although it is difficult to establish with certainty whether his drawings represent already existing construction works or original proposals.

2.5.1.5.1.9 Waterways in Venice, Tuscany, Lazio, and France

In 1500, Leonardo da Vinci worked for the Republic of Venice, after leaving Milan. In order to defend against the advancing Turkish army, he was sent to Friuli. While there, he designed a defensive plan involving a blockade of watercourses to flood part of the Isonzo valley, specifically between Gorizia and Gradisca, as a strategic measure to protect the region.

2.5.1.5.2 Bridge Design

In 1502, the Sultan of the Ottoman Empire came to Rome to hire a team of civil engineers to design a bridge to stretch across the Golden Horn at Istanbul. Bayezid II, the Sultan (ruler) of the Ottoman Empire, wanted to connect the cities of Constantinople and Galata, which were separated by a river estuary called the Golden Horn. Leonardo drafted his proposal in response to Sultan Bayezid II's 1502 appeal for bridge designs [15].

In 1502, Leonardo da Vinci designed a single-span bridge for Sultan Bayezid II of the Ottoman Empire. The bridge was intended to span the Golden Horn, an inlet in Istanbul, and would have been the longest bridge of its time at 280 m (about 919 feet) long. However, Leonardo's pitch was radically different than any presented previously. According to a press release, he proposed building a single flattened arch tall enough to allow sailboats to pass below and stabilizing against lateral motion—an issue linked with the region's many earthquakes—by adding splayed abutments, or load-carrying supports, to either side of the bridge.

Da Vinci offered his services to the Sultan, modeling a structure in his notebook that represented a beautiful synergy of creative artistry and civil engineering [13].

Bayezid was unimpressed by Leonardo's complicated blueprints, and the bridge the artist envisioned was never built—at least until MIT engineer John Ochsendorf stepped in to test the 500-year-old design's feasibility.

The bridge would have spanned over 280 m of water, and the arch would have been high enough for a ship with sails to pass under. Although the Sultan turned down Da Vinci's proposal, believing the architectural endeavor to be impossible, a modern Swiss scientist, D.F. Stussi, concluded that the plans were 'technically feasible' [13].

Although it was never built, a modern version was constructed in Norway in 2001, proving the soundness of his design. In 2001, inspired by Da Vinci's design, an artist in Norway named Vebjorn Sand decided to construct his bridge. Today, Da Vinci's bridge design stands over a highway in Norway as a monument to his genius. Research conducted by engineers at MIT in 2019 suggests that one of the Renaissance giant's unbuilt designs—a bridge poised to be the world's longest—would have worked if the artist had actually followed through on his plans [16].

2.5.1.5.3 Engines and gears

Although James Watt is credited with inventing the modern steam engine, Da Vinci had designed a much simpler form of Watt's engine that operated by flywheel and crank [13]. Once again, the answer to a more modern dilemma was contained within Da Vinci's drawings—he had designed what is now called a flywheel, or a heavy wheel with high angular momentum.

In conjunction with a crank-and-rod system, the flywheel resists changes in rotational motion caused by irregular strokes of the piston and thus steadies the rotation of the shaft. However, Watt never saw Da Vinci's design and was reluctant to incorporate a flywheel system into his steam engine. As Watt struggled with inventing a working steam engine in the mid-18th century, he worked with complicated transmission systems because engineers feared that a simple crank-and-rod motion would not work with the irregular stroke of the steam piston.

2.5.1.5.4 Structural engineering

Da Vinci's contributions to structural mechanics are evident in his designs and observations. Da Vinci was a prolific designer of machines. He designed various machines, which included flying machines (ornithopters and gliders), armored vehicles (tank-like designs), and submarines (hand-powered submersibles). His work in mechanical, physical, and hydraulic disciplines has greatly impacted the future of the industry. Thus, he contributed to machines and inventions. He lived centuries ahead of their time, leaving an indelible mark on history. Da Vinci explored the mechanics of human movement, and he considered the idea of creating a robot.

2.5.1.5.5 Topography and cartography

Da Vinci's work in topography and cartography led to the development of modern iconographic mapping. His detailed landscapes and topographical measurements were used to alter borders and features in Florence, Imola, and Milan—thus contributing to topography and cartography. He created detailed maps and landscapes, often incorporating precise topographical measurements.

His maps of cities like Florence, Imola, and Milan in Italy showcased his skill in representing geographical features accurately. Da Vinci's cartographic work laid groundwork for future mapping techniques, and his approach to combining art and science in map-making contributed to the evolution of cartography.

2.5.1.5.6 Human anatomy

Da Vinci explored the mechanics of human movement; he considered the idea of creating a robot. Within his analytical mind, he realized that the joints and muscles of the human body reflect the simple gears and pulleys that make up machines.

Da Vinci studied human movement and mechanics, envisioning a robot-like device. He drew parallels between human joints and muscles and machine components, recognizing similarities between the two. Within his analytical mind, he realized that the joints and muscles of the human body reflect the simple gears and pulleys that make up machines. By analyzing joints and muscles, he saw analogies with mechanical systems, laying groundwork for future innovations.

For example, he found that the system of muscles within the human body that support the head and neck in an upright position were analogous to the framework of ropes that support a ship's mast [17].

He designed a robot dressed in knight's armor that could move by the operation of simple machines within its armor. Although he never built the robot, it would have been able to wave its arms, move and rotate its head, and open and close its jaw. Several engineering groups have attempted to build robots based on Da Vinci's design. A technology named after Da Vinci, called the Da Vinci Surgical System, is a human-controlled surgical robot. It is a robotic surgical platform that enables surgeons to perform minimally invasive procedures with enhanced precision and control. The creators of the system named their technology after Da Vinci to acknowledge the fact that he was the inventor of the first robot and also because of his unprecedented anatomical research. The system consists of a surgeon console and an integrated patient-side cart with four robotic arms.

The console displays a magnified three-dimensional image to the doctor while he operates the controls, which scale his movements and translate them to movement in the four robotic arms. Designed primarily for heart and prostate surgeries, the system is shown to increase surgical precision of the doctor, minimize incision size, lessen the risk of transfusion, and shorten recovery time. Even 500 years in the future, Leonardo da Vinci's ingenuity still inspires engineers today to better the quality of medicine and the understanding of the human body.

2.5.2 Karl von Terzaghi (1883–1963)

Terzaghi's formulation of the effective stress principle and its influence on settlement analysis, strength, permeability, and erosion of soils was the most extraordinary contribution to civil engineering.

Terzaghi's ideas were met with skepticism in several civil engineering circles, but he continued to write, lecture, and demonstrate the validity of his concepts through their practical application.

Terzaghi's consulting work extended to earth dams, the stabilization of landslides, and foundations for buildings, waterfronts, highways, and airports.

During his lifetime, he wrote several books and published over 100 papers on topics ranging from the stability of slopes and conditions for the failure of soils to vibration problems and drainage mechanics.

2.5.3 Henry Gantt (1861–1919)

Gantt was an American mechanical engineer and management consultant. He is best known for developing the Gantt chart, a type of bar chart used for project management and scheduling. Gantt's work has had a significant impact on modern management practices, and his chart remains a widely used tool in project management today. He is also known for his groundwork in the development of scientific management.

Named after him, the Gantt chart is a famous invention made in 1910, which helps visualize project timelines, tasks, and dependencies. Gantt applied engineering principles to industrial management, focusing on efficiency, productivity, and quality control. Gantt is also recognized as an early proponent of the social responsibility of businesses.

A Gantt chart is a bar chart that illustrates a project schedule. It was designed and popularized by Henry Gantt around the years 1910–1915. A Gantt chart is a project management tool that visualizes work progress over time against a planned schedule. It is a visualization that helps in scheduling, managing, and monitoring specific tasks and resources in a project. It consists of a list of tasks and bars depicting each task's progress. It typically consists of two main sections: the left side lists all tasks or activities, while the right side features a horizontal timeline, with bars representing each task's duration (which adds up to the entire project duration). These schedule bars clearly show the start and end dates of tasks, making it easy to see what's in progress, what's upcoming, and what's completed.

Gantt charts are popular because they offer a clear, visual way to manage project schedules, helping teams track milestones, deadlines, and task dependencies—how each task connects to and impacts others. With this bird's-eye view, you can manage complex projects more effectively, ensuring everyone is aligned and on track to meet shared goals.

A Gantt chart is a popular project management tool that aids in planning and scheduling projects of all sizes. It provides a visual representation of the project's timeline and assists project managers, stakeholders, and team members in understanding the following: a project's roadmap, task progress, key milestones, work dependencies, resources required, and risks involved. The modern Gantt charts also show the dependency relationships between activities and the current schedule status. It is widely used in project management.

2.5.4 Henri Fayol (1841–1925)

Henri Fayol is widely acknowledged as a founder of modern management methods. A French mining engineer, mining executive, author, and director of mines, Henri Fayol developed a general theory of business administration that is often called Fayolism, around 1900.

Fayolism is a management theory that emphasizes organizational structure and human behavior to increase productivity. Well renowned as the 'Father of Modern Management Theory,' Henri Fayol worked at the French mining company, **Commentry-Fourchambault and Decazeville**. Fayol worked at the French mining company, where he started as an engineer but worked his way up to become the general manager and then the organization's director from 1888 to 1918. When Fayol took on the managerial role at the mining company, he chose to rely not on his technical skills but also on his ability as an organizer and his skills at handling people.

He and his colleagues developed this theory independently of scientific management but roughly contemporaneously. Like his contemporary Frederick Winslow Taylor, he is widely acknowledged as a founder of modern management methods.

Henri Fayol introduced theories that could be applied to all levels of management and for any department. Organizations and managers still practice Fayol's principles of management to ensure an efficient and successful business.

He developed the 14 principles of management based on his experience while working for a large mining company in France. He published the 14 principles of management in the book *Administration Industruelle et Generale*. He lived when the mechanical model of management was more prevalent. Widely influential in the early 20th century, **Henri Fayol** wrote the book on management theories and work organization, *Administration Industrielle et Générale*.

Henri Fayol developed the 14 principles of management toward the tail end of the Industrial Revolution. It was the perfect timing, too; the world had been subjected to massive changes, and new and improved working styles were needed. Therefore, Fayol's Principles of Management influence the present management theory significantly. Fayol's management principles are universally accepted and applicable in modern organizations today; they are not absolute in terms of application.

Henri Fayol and F.W. Taylor greatly contributed to classical theories—concerned with the structure and activities of formal, or official, organization. Issues such as division of work, the establishment of a hierarchy of authority, and the span of control were seen to be of utmost importance in the achievement of an effective organization. Henry Fayol (1841–1925) and F.W. Taylor (1856–1915) were its two greatest exponents.

2.5.4.1 Principles of Fayolism

Henry Fayol developed the famous 14 principles of management, which is his most remarkable contribution to management that civil engineers borrow in their day-to-day practice. He looked at an organization from a top–down approach to support managers get the best from employees and easily run the business.

From Henry Fayol's original work, the 14 principles of management are outlined from Section 2.5.4.1.1 to 2.5.4.1.14. Fayol's original work outlined 14 principles of management, which were first published in his book *Administration Industrielle et Générale* in 1916.

Later, management theorists and practitioners condensed and adapted Fayol's 14 principles of management into various frameworks, including the five principles (functions) of management, that is, planning, organizing, coordinating, leading, and controlling.

We can say that Henri Fayol originally identified five functions of management, but later, management theorists and practitioners condensed them into four main functions. These were planning, organizing, commanding (or leading), coordinating, and controlling. The coordinating function is now often considered a part of the organizing function as it involves synchronizing and integrating different activities and resources.

So, while Fayol's original work identified five functions, the four functions—that is, planning, organizing, commanding (or leading), and controlling—are the most commonly accepted and widely used today.

2.5.4.1.1 Division of work or labor
Work should be divided among individuals and groups so that they can focus on their portion of the task, build up skills, and become more productive (specialization). The more the people specialize, the more efficiently they can perform their work.

2.5.4.1.2 Authority and responsibility
Fayol defined it as the right to give orders and the power to exert obedience. With authority comes responsibility and accountability. Managers have the right to give orders so that they can get things done. Wherever authority is exercised, responsibility arises.

2.5.4.1.3 Discipline
For the best interest of an organization, there should be complete obedience, diligence, effort, and outward marks of respect, which are equally applicable to everybody regardless of the rank. To establish and maintain discipline, there must be clearly defined roles, rules, and regulations for individuals and groups, e.g., code of conduct and ethics. Therefore, members of an organization need to respect the rules and agreements that govern the organization. It is essential for the smooth running of business, without discipline no enterprise could prosper.

2.5.4.1.4 Unity of command
An employee should only receive commands from one superior. In other words, each employee must receive his or her

instructions about a particular operation from only one person. Violation of this principle creates confusion in reporting lines. Fayol believed that if an employee was responsible to more than one boss, conflicting instructions or confusion would result.

2.5.4.1.5 Unity of direction

An organization should be moving toward a common objective. This brings about harmony of effort toward the mission and vision. The efforts of employees should be coordinated and directed by only one manager in order to void different policies and procedures.

2.5.4.1.6 Subordination of individual interests to the general interests

The interests of one employee should not be allowed to become more important than those of the organization. In other words, the interest of employees should not take precedence over the interests of the organization as a whole. An organization should come up with approaches that ensure personal interests and organizational interests are aligned as closely as possible.

2.5.4.1.7 Remuneration

Payment should be fairly proper and satisfactory to the employee. Compensation for work done should be fair to both the employee and the employer. This in turn creates a harmonious relationship and a pleasing atmosphere for work. It should include financial and nonfinancial compensation.

2.5.4.1.8 Centralization

Decision-making is made at the top management level, while decentralization of decision-making is distributed downward among all levels of an organization. Fayol recommended an appropriate balance of the two depending on the size, nature of work, situation, and weight of decision.

Decreasing the role of subordinates in decision-making is centralization; in contrast, decentralization increases their roles. Fayol believed that managers should retain the final responsibility but that they also need to give their subordinates enough authority to do their jobs properly. The problem is to find the proper amount of centralization in each case.

The question of centralization or decentralization is a simple question of proportion. The share of initiative to be left to (subordinates) depends on the character of the manager, the reliability of the subordinates, and the condition of the business. The degree of centralization must vary according to different cases.

2.5.4.1.9 Scalar chain

Every organization has a hierarchy, and employees should be aware of where they stand in the organization's hierarchy or chain of command, e.g., subordinates report to superiors. The line of authority in an organization, often represented by the neat boxes and lines of the organization chart, runs in order by rank from top management to the lowest level of the company. Fayol pointed out that if a hasty decision was needed, it was appropriate for people at the same level of the chain to communicate directly, as long as their immediate superiors approved. It provides for the usual exercise of some measure of initiative at all levels of authority.

2.5.4.1.10 Order

It is defined as systematic, orderly, equal management and distribution of people, places, and materials. Everything should have its place. Materials and people should be in the right place at the right time. In particular, people should be in the jobs or positions best suited for them.

2.5.4.1.11 Equity

The management should treat all its employees in a fair and just manner at all levels. In other words, managers should be both friendly and fair to all subordinates.

2.5.4.1.12 Stability of tenure of personnel

The principle states that an organization cannot run smoothly if it suffers from constant employee attrition, and organizations should make an effort to retain employees. A high attrition rate will cost an organization time, resources, and perhaps some crucial organizational memory. A high employee turnover rate is not good for the efficient functioning of an organization. Instability of tenure is at one and the same time the cause and effect of bad management.

2.5.4.1.13 Initiative

The management should provide the opportunity or freedom to employees to suggest new ideas, experiences, and more convenient methods of work to ensure effectiveness in the organization. Subordinates should be given the freedom to formulate and carry out their plans. It is essential to encourage and develop this capacity to the full. The manager must be able to sacrifice some personal vanity in order to grant this satisfaction to subordinates. A manager able to do so is infinitely superior to one who cannot.

2.5.4.1.14 Espirit De Corps (teamwork)

Organizations should harness harmony, team spirit, and unity among personnel. Promoting team spirit will give the organization a sense of unity. Harmony, union among the personnel of a concern, is a great strength. Effort, then, should be made to establish it.

2.6 CIVIL ENGINEERING MANAGEMENT

Civil engineering management involves the planning, organization, leading, and control of civil engineering projects and teams. Thus, it combines technical engineering knowledge with business and management skills to ensure projects are completed efficiently, effectively, safely, and within budget.

Once mastered as a skill, civil engineering management, construction management, project management, and other basic site-related management bring out one of the essential attributes civil engineers need to deliver successful projects—stewardship; see Section 2.2.1.

Effective civil engineering management ensures projects are completed on time, within budget (without much cost overruns), and to the required quality standards, while also ensuring safety, sustainability, and environmental responsibility.

It also involves strong leadership and communication skills, technical knowledge and expertise, strategic planning and problem-solving abilities, attention to detail and quality, ability to adapt to changing circumstances (adaptability), and strong analytical and decision-making skills.

Civil engineering management benchmarks on the four principles of management, well known as the 'Four Functions of Management,' introduced by Henri Fayol, a French management theorist, in the early 20th century. These are planning, organizing, leading, and controlling. These represent the core principles of managing an organization or a project, where a manager first plans, then organizes resources according to the plan, leads the team to execute it, and finally controls the process to ensure goals are met.

The lifecycle phases of a construction project are planning, design, construction, operation, and maintenance. Throughout these stages (lifecycle phases), the four principles of management can be employed to plan, organize, lead, and control people and resources to achieve the set goals and objectives. These four principles of management are interconnected and interdependent, and they provide a foundation for effective management practices in civil engineering.

To deliver a successful project, civil engineering management philosophy also benchmarks on the 5 Ms. The 5 Ms represent the key elements or factors that managers need to focus on to achieve organizational goals. They encompass the essential resources, processes, and factors that impact an organization's performance. The 5 Ms are often used as a framework to identify key areas for improvement, allocate resources effectively, develop strategic plans, and monitor and evaluate organizational performance.

Civil engineering managers work in various settings which include construction companies, government agencies, consulting firms, private practices, and research institutions. Their goal is to deliver successful projects that meet technical, financial, and social requirements while ensuring safety, quality, and sustainability. They use various tools and techniques, such as project management software (e.g., Primavera and MS Project), cost estimation and scheduling tools, risk assessment, and management methodologies.

Civil engineering managers work on various projects, including infrastructure development (roads, bridges, and airports), building construction (commercial, residential, and industrial), water resource management (dams and water treatment plants), transportation systems (highways, railways, and ports), and urban planning and development.

2.6.1 Key Attributes of Civil Engineering Management

Civil engineering management requires a combination of technical skills, project management skills, and leadership abilities. Sections 2.6.1.1–2.6.1.6 outline the key aspects of civil engineering management.

2.6.1.1 Project management

Project management is the application of processes, methods, skills, knowledge, and experience to achieve specific project objectives according to the project acceptance criteria within agreed parameters. Project management has final deliverables that are constrained to a finite **timescale and budget**. This involves coordinating and overseeing projects from conception to completion.

2.6.1.2 Technical skills

Civil engineering management requires the ability to apply engineering knowledge to design, construct, and maintain infrastructure with proper **attention to detail**. Thus, precision is essential to ensure the safety of structure. This requires technical expertise to manage from a well-informed perspective because you simply cannot manage what you don't know and understand!

2.6.1.3 Team management

This involves leading, controlling, and directing teams of engineers, technicians, laborers, and contractors. A typical construction site, for example, requires civil engineering managers to lead and direct people at different levels to successfully deliver the project within budget and on time.

2.6.1.4 Risk management

Civil engineering management involved identifying and mitigating potential risks and hazards by actively pinpointing possible dangers or threats within a project and then take proactive steps to reduce their likelihood or impact, essentially minimizing the potential for negative consequences. This involves a process of hazard identification, risk assessment, and implementing control measures to manage those risks effectively.

2.6.1.5 Communication

Civil engineering management requires the ability to communicate effectively—a good command of the language of communication for a project to be successful, in writing and orally in order to be able to coordinate stakeholders, clients, and team members. Good project communication means encouraging technical workers to ask questions instead of assuming things, which can cause problems avoiding communication gaps and feedback and response is prompt. Managers should prioritize sharing important information promptly to prevent delays and ensure project teams have accurate details for their tasks. Proactive communication practices are key to keeping projects on track and avoiding misunderstandings.

Better communication between managers and workers is essential for project success. Good communication is essential for project managers overseeing diverse engineering teams, for example, in multiple locations. Using tools like Microsoft Teams helps keep everyone informed about project updates and design changes. The project manager's main task is to gather and share information efficiently among these teams. ensuring effective communications is simply to mandate, through rules and procedures, the manner in which project team members must communicate with each other and the types of information they must routinely share—this is key to project success.

2.6.1.6 Other attributes

Civil engineering management involves several other functions including the following:

- Cost control (budgeting and cost management)—managing project finances and ensuring cost-effectiveness and controlling expenses.
- Sustainability and environmental management involving minimizing the environmental impact of projects.
- Time management which includes meeting deadlines and milestones.
- Quality control, ensuring compliance with standards, regulations, and specifications.
- Planning and scheduling, the ability to plan and schedule construction projects—developing and managing project timelines.
- Resource allocation, that is, managing resources such as materials, equipment, and personnel.
- Safety management, ensuring a safe working environment and compliance with safety regulations.
- Contract administration by overseeing contracts and ensuring compliance.

2.6.2 Principles of Management

The four core principles of management are Planning, Organizing, Leading, and Controlling, often referred to as the 'P-O-L-C' framework, which outlines the key functions a manager should perform in their role. Originally identified by Henri Fayol as five elements, with the 5th as coordinating, there are now four commonly accepted functions of management that encompass the necessary skills: planning, organizing, leading, and controlling.

Management functions are the sets of activities inherent in most of the managerial jobs. These activities are grouped into four general functions: Planning and Decision-Making, Organizing, Leading, and Controlling. The relationships among these functions are shown in Figure 2.5. Consider what each of these functions entails, as well as how each may look in action. More than just specialized knowledge, the management requires an ability to navigate numerous procedural, structural, and interpersonal challenges in the process of guiding one's team to the completion of various goals.

2.6.2.1 Planning and decision-making

This involves determining the organization's goals and deciding how best to achieve them. The purpose of planning is to provide managers with a blue print of what they should be doing in the future. The manager looks at plans to determine what course has been charted for the organization.

The planning process in general consists of three steps. First, goals and objectives are established. This is usually done by the top management. Next, strategic plans are developed. Strategic plans serve as the broad, general guidelines that chart the organizations' future. Strategic planning is also performed by the top management. Finally tactical plans are developed,

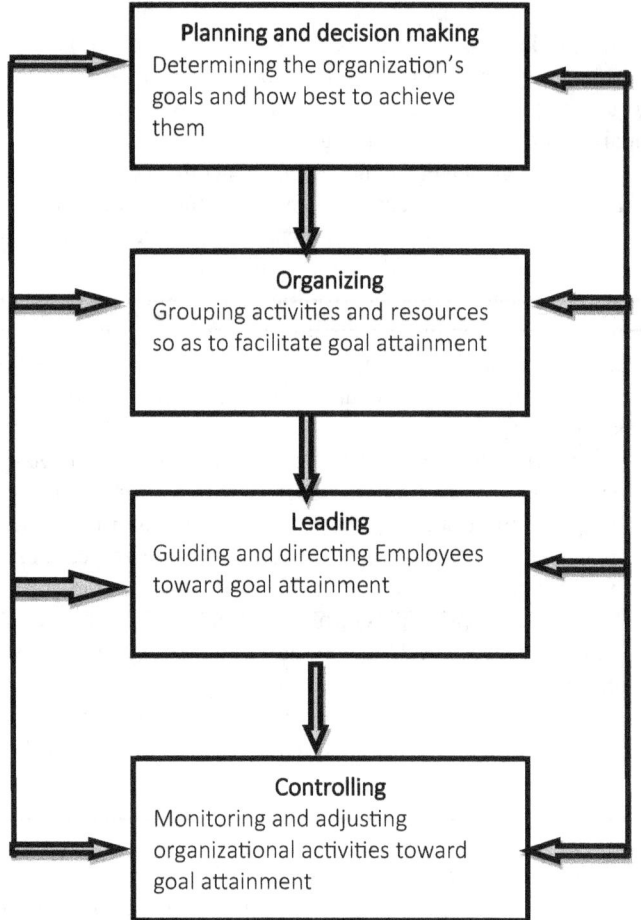

FIGURE 2.5 Basic management functions.

often by middle managers. Decision-making pervades each of these activities.

Planning civil engineering projects involves defining goals, objectives, and strategies for achieving them. It typically involves identifying the need for the project goals and defining the scope of the project. This includes defining the organizations or project's mission and vision. Planning also involves setting specific, measurable, achievable, relevant, and time-bound (SMART) goals; developing strategies and action plans; and allocating resources.

It is a systematic process of defining the scope, goals, timeline, budget, resources, and execution strategy for a civil/construction project, including detailed design, site analysis, permitting, environmental considerations, and risk management, to ensure its successful completion within specified parameters and quality standards.

Construction planning is the process of identifying the steps required to build a structure. It involves defining what actions need to be completed, creating an ordered timeline of events, staffing the project, and determining the necessary materials and equipment. A well-crafted construction plan is essential to keep the project on schedule and within budget. It can also help ensure the overall quality of the project meets your client's standards. Finally, having a construction plan can increase your team's productivity and efficiency by streamlining the communication.

As given in Sections 2.6.2.1.1–2.6.2.1.11, the planning process is all tailored to the role of a civil engineering manager.

What does a civil engineering manager typically do when planning a project?

2.6.2.1.1 Project initiation

This means identifying the need for the project, establishing project goals, and defining the project scope. For example, planning to develop a specialized hospital, end-user assessments are typical activities conducted to ascertain the need for the project. These can help define the project goals and objectives and what the project scope would look like.

2.6.2.1.2 Feasibility study

This includes evaluating the technical and economic viability of the project, including potential risks and constraints. It is an initial assessment of a project's viability, focusing on technical feasibility (can it be done?), financial feasibility (is it cost-effective?), operational feasibility (will it work in practice?), and social feasibility (is it acceptable to stakeholders?)

2.6.2.1.3 Project appraisal

Project appraisal, as part of the planning process, is a more detailed evaluation of a project's potential, usually conducted after the feasibility study. It involves cost–benefit analysis (CBA), risk assessment, sensitivity analysis, and economic and financial evaluation.

2.6.2.1.4 Design process

The design of civil engineering projects usually goes through a series of steps from sketches, to conceptual designs, and detailed designs. It usually involves conducting site investigations and analysis. The surveys are typically conducted to gather data on topography, soil conditions, utilities, and environmental factors.

Conceptual designs are preliminary design concepts, considering functional requirements and aesthetics, while detailed designs are comprehensive plans and specifications for the project, including structural calculations, material selection, and construction details.

2.6.2.1.5 Permits

Permits and approval are parts of the planning process. This means obtaining necessary permits and licenses from regulatory agencies, including zoning, environmental, and construction permits.

2.6.2.1.6 Scheduling

This is the process of creating a project timeline with milestones and critical path analysis to manage project duration.

2.6.2.1.7 Cost estimation

This is the process of calculating the expected project budget, including labor, materials, equipment, and contingencies.

2.6.2.1.8 Resource allocation

Identifying and assigning necessary personnel, equipment, and materials to project tasks is part of the planning process. Making decisions on how much resources are needed to successfully implement a project requires considerable planning.

2.6.2.1.9 Risk management

Risk management involves identifying potential risks and developing mitigation strategies and contingency plans.

2.6.2.1.10 Environmental impact assessment

Analyzing potential environmental impacts of the project and developing mitigation measures contributes to project sustainability in the long run. Addressing environmental impacts in planning is important to ensure sustainable development, protecting natural resources, and promoting human well-being.

2.6.2.1.11 Quality control planning

Establishing quality standards and inspection procedures to ensure project quality is achieved, and quality assurance and control are properly planned and sequenced.

2.6.2.1.12 Stakeholder engagement

This involves communicating project plans and updates to stakeholders, including clients, community members, and regulatory agencies. Stakeholder participation in the planning process builds confidence in the project and enables seamless coordination of activities, leading to sustainable project outputs and outcomes.

2.6.2.2 Organizing

Organizing is the process of grouping activities and resources in a logical and appropriate fashion. In the basic sense, it is creating the organizational chart for a firm. Organizing involves arranging and structuring resources to achieve the planned goals. It includes the following:

- Defining roles and responsibilities
- Establishing organizational structures and hierarchies
- Allocating tasks and resources
- Coordinating and integrating activities.

Managers create a formal structure when they decide how to divide and coordinate work of the organization. These decisions, in part, reflect their interpretation of contingencies affecting the business and in part the views of the stakeholders. Together with many informal arrangements, these decisions create a structure that balances mechanistic and organic forms that affect organizational performance. Figure 2.6 illustrates the structural issues that managers in all organizations need to resolve.

Organizational structure describes the way work is divided, supervised, and coordinated. When people join a department or take a job within the structure, this gives a fairly clear signal about what they should do. The director of engineering, for example, is expected to deal with engineering, not administration. Various operating policies reinforce the signal from the basic structure. These cover matters such as recruitment, selection, and appraisal and reward which manager design to influence employees to act in ways that support wider objectives. The word structure implies organization. People who work in an organization are grouped so that their efforts can be channeled for maximum efficiency. See functional organizational structure in Figure 2.7. The structure of an organization is the sum total of ways in which it divides its labor into distinct tasks and then achieves coordination among them.

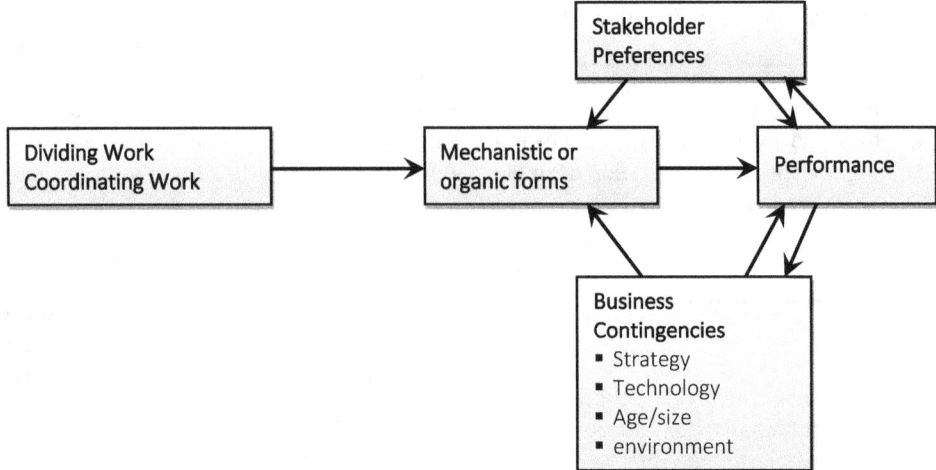

FIGURE 2.6 Alternative structures and performance.

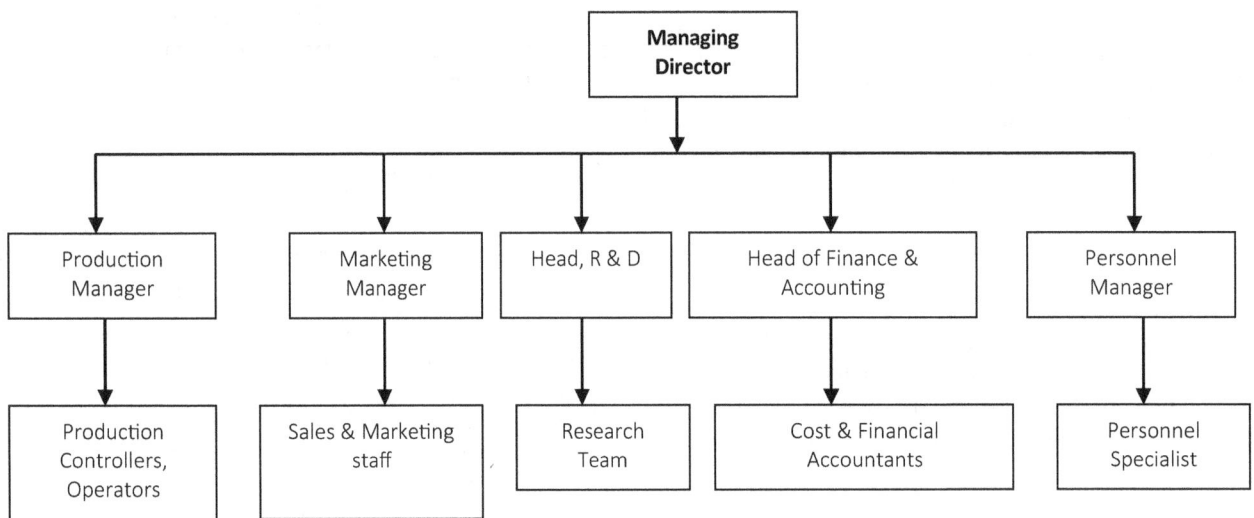

FIGURE 2.7 Functional organizational structure.

2.6.2.3 Leading

Leading is the set of processes associated with guiding and directing employees toward goal attainment. Key parts of leading are motivating employees, managing group processes, and dealing with conflict and change. Note that each of these relates to behavioral concepts and processes. Leading involves motivating, influencing, and directing people to achieve the planned goals. It includes the following:

- Communicating effectively with team members
- Building and maintaining relationships
- Motivating and inspiring team members
- Providing guidance and direction.

The success or failure of managers depends on their leadership qualities. They can be successful leaders by helping subordinates find solutions to their problems. Managers are involved with bringing together resources, developing strategies, and organizing and controlling activities in order to achieve objectives. At the same time, managers, as leaders, have to select the goals and objectives of an organization, decide what is to be done, and motivate people to do it. Thus, leadership is that function of management which is largely involved with establishing goals and motivating people to help achieve them. Leaders set goals and help subordinates find the right path to achieve these goals.

A person may be an effective manager—a good planner and an organized administrator—but lack the motivational skills of a leader. Another may be an effective leader—skilled at inspiring enthusiasm and devotion—but lack the managerial skills to channel the energy he/she arouses in others. Given the challenges of dynamic engagement in today's business world, most organizations today are putting a premium on managers who also possess leadership skills.

2.6.2.4 Controlling

Controlling is the process of monitoring and adjusting organizational activities toward goal attainment. Controlling refers to the process of monitoring progress toward goal attainment and making necessary adjustments. Controlling involves monitoring, evaluating, and correcting performance to ensure that goals are met. It includes the following:

- Establishing performance standards and metrics
- Monitoring and evaluating performance
- Identifying and correcting deviations
- Taking corrective action to get back on track

2.6.3 Ms of Management

The **5 Ms** of management represent a framework for effective management with roots in earlier management theories. This framework can be used to analyze and manage various aspects of an organization or project.

These 5 Ms provide valuable insights for managers and organizations engaged in civil engineering and construction. These 5 Ms particularly form the benchmark for **sound holistic independent judgment** for a project's success. To determine whether the proposed or ongoing project will be a success, evaluation of the 5 Ms is essential.

The 5 Ms provide a comprehensive framework for managers to focus on key areas and ensure successful organizational performance. Therefore, by balancing and integrating these elements, managers can achieve their goals and objectives.

2.6.3.1 Money

These are the financial resources required to fund operations and growth of the organization or project. The project needs money to be successful. This enables the procurement of materials and goods used in production or construction possible.

In construction, particularly, projects require enough funds to be successful. This must be adequate ensuring that all items of the project are well covered. In essence, the project's success requires proper management of finances, budgets, and resources to achieve the mission. How much money is available for the project? That is the guiding question while appraising or running a project. How can we source for more funds or minimize the cost?

2.6.3.2 Manpower

This involves recruiting, training, and managing employees to ensure they are productive and satisfied. Recruit, train, and develop the right team members to successfully manage a project or organization. How much manpower is needed—skilled, semi-skilled, and/or unskilled? That is the guiding question while appraising or running a project.

2.6.3.3 Materials

This refers to the raw materials and resources needed for production or construction. Acquire and manage the necessary resources, equipment, and supplies in order to achieve the project goals. What kind of materials are needed to carry the project through?—that is, the quality of materials, type, etc. This is the guiding question while appraising or running a project.

2.6.3.4 Machinery

This refers to technology required for the task—generally, the tools and equipment used in the production or construction process. Utilize and maintain the necessary tools, equipment, and systems for the success of the project. What kind of technology and/or machinery is needed for the task? That is the guiding question while appraising or running a project.

2.6.3.5 Methods

These are the processes and procedures used to achieve organizational goals. This means developing and implementing efficient processes and procedures in order to achieve project goals and objectives. What approach or method should be taken to accomplish a task/project? That is the guiding question while appraising or running a project.

2.7 CIVIL ENGINEERING IMPACT ON THE SOCIETY AND ENVIRONMENT

Civil engineering greatly impacts both the society and the environment. Concerned primarily with infrastructure development, it has both positive and negative impacts on the society and the environment. Therefore, working toward creating more sustainable, resilient, and environmentally conscious infrastructure that benefits both people and the planet is the primary goal of civil engineering.

2.7.1 Improved Infrastructure

Civil engineers design and develop critical infrastructure, such as roads, bridges, airports, and public transportation systems, which facilitate connectivity, commerce, and economic growth. They advise governments, organizations, and communities about infrastructure development, which is key to developing economies around the world.

The impact of civil engineering on the society and environment is huge. From developing roads and highways to transport goods and people to connecting communities, to building power plants to supply electricity to building water supply and sewerage systems that enormously support communities, the impact is felt.

2.7.2 Fostering Economic Growth and Development

Civil engineering projects, such as transportation systems and public buildings, create jobs, stimulate economic growth, and contribute to the overall development of a region. Civil engineering greatly impacts the society, boosting economies all around the world. Driving innovation and growth is what civil engineering is known for throughout history—supporting increased agricultural production, facilitating the movement of goods and people, technological advancements, and industrial growth. Civil engineering supports almost all sectors of an economy because it deals with tangible infrastructure assets that facilitate trade and investment.

2.7.3 Saving Lives

The civil engineering industry has saved lives more than any other profession in history. No doubt about that! From building bridges to connecting communities to building hospitals and sewer systems to wipe out outbreaks. Imagine how people failed to move from one point to another, for example, to cross rivers, valleys, and hills. Imagine how people died of outbreaks such as cholera and COVID-19. Civil engineers improvise and develop bridges to facilitate the movement of people and goods, touching many people's lives and sometimes relieve them from danger and pain. They build sewer networks and hospitals, sometimes in emergency situations, to support communities overcome outbreaks.

Sir Joseph Bazalgette, a 19th-century civil engineer who designed London's first sewer system, helping eradicate cholera in the city is a typical example of civil engineers who have saved lives and was the mastermind behind the Metropolitan Board of Works.

In 1853, the cholera endemic had claimed the lives of more than 10,000 Londoners in the United Kingdom. Bazalgette's groundbreaking solution to London's sanitation conundrum not only resolved a pressing issue but also exemplifies the profound impact of infrastructure on public health, demonstrating its potential to be a literal lifesaver—civil engineers at the heart of society touching people' lives. This monumental achievement in engineering featured 82 miles of brick-lined intercepting sewers and 1,100 miles of street sewers, all precisely integrated within multiple embankments, including the iconic Victoria Embankment. His emphasis on utilizing more extensive tunnels and resilient Portland cement enabled the Victorian sewerage system to accommodate London's expanding population.

Emergency services are critical to public health and well-being. Civil engineers help build hospitals and emergency units around the world. In the wake of the COVID-19 pandemic in 2019/2020, many civil engineers around the world provided a rapid response, helping build hospitals, for example the Military Assessment Team for 170 (Infrastructure Support) Engineer Group built Nightingale Hospital in just a few weeks [18]. The result was time saving, which is a priority in any emergency. The Nightingale Hospital Birmingham was set up in just 18 days.[2]

In summary, civil engineering enhances public health because it plays a crucial role in designing and implementing water treatment and sanitation systems and sewerage systems that significantly improve public health and reduce the risk of waterborne diseases.

2.7.4 Disaster Mitigation

Civil engineers design and construct buildings, dams, and other structures to withstand natural disasters, such as earthquakes, hurricanes, and floods, which helps protect human life and property. Typical examples of how civil engineers' work is felt in the society in terms of disaster management are outlined henceforth.

- Civil engineers deploy disaster-resistant measures. For example, they specify fire-resistant materials to protect structures from wildfires and also design structures to resist landslide forces and stabilize slopes.
- Civil engineers design and develop earthquake-resistant structures using seismic design principles

to create structures that can absorb and dissipate earthquake forces.
- Civil engineers also design reinforced foundations that can transfer loads to the ground without collapsing.
- Civil engineers design and develop flood-resistant structures on elevated foundations or pilings to protect against floodwaters.
- Civil engineers also use waterproofing and flood-proofing materials to prevent water intrusion into buildings.
- Civil engineers design and develop hurricane-resistant structures withstanding high winds and flying debris.
- Civil engineers specify storm shutters and impact-resistant windows to prevent damage from wind-borne debris.

2.7.5 Resilient and Sustainable Communities

Today, civil engineering continues to be the driving force of sustainable and resilient communities worldwide. Through civil engineering, communities are able to withstand and recover from external stresses and shocks, such as natural disasters, economic downturns, and social unrest, while also maintaining a high quality of life and minimizing their impact on the environment. This is achieved through building reliable infrastructure and transportation systems. In that regard, well-planned and maintained infrastructure and transportation systems enable communities to respond to and recover from disasters.

Civil engineering supports communities to prioritize environmental protection and conservation, reducing their carbon footprint and promoting eco-friendly practices. Mitigating environmental degradation is a key aspect of civil engineering because materials used in civil engineering are extracted from nature in huge amounts, and this greatly impacts the ecosystem. If not carefully conducted, lots of flora and fauna can be lost.

Civil engineering projects, such as road construction and dam building, can lead to environmental degradation, including deforestation, habitat destruction, and water pollution. That is why environmental impact assessments are conducted to identify potential environmental risks and develop strategies to mitigate them.

In the same way, environmental pollution can be disastrous; for example, leaving construction sites open for a prolonged period of time can expose societies to negative consequences. Civil engineering projects can generate significant amounts of waste, including construction waste, which can contribute to pollution and environmental degradation. Therefore, civil engineering strives to ensure that negative consequences are greatly minimized. Thus, the impact of civil engineering is felt.

Civil engineering is sensitive to climate change, and measures are promoted to ensure that carbon emissions are reduced from all activities that surround infrastructure development. The production of cement, for example, a critical component of concrete, is a significant contributor to greenhouse gas emissions, which exacerbate climate change. So, civil engineers devise mechanisms to reduce greenhouse gases not only from cement but also from other materials.

The impact of civil engineering on the society and environment is felt in almost everything because it deals with directing great sources of power in nature for the convenience of humans. For example, clean energies are now tapped to power communities all over the world. Without it, many communities would live in darkness, and economic growth would be minimal.

The extraction and processing of raw materials, such as sand, gravel, and cement, required for civil engineering projects can lead to resource depletion and habitat destruction. Therefore, civil engineers are now coming up with designs that incorporate sustainable design principles, such as energy efficiency, renewable energy systems, and green building materials, to reduce the environmental impact of their projects. This could minimize the extraction of raw materials.

In the same way, mimicking nature is the way forward to minimize environmental degradation and build resilient and sustainable communities. That means incorporating green infrastructure, such as green roofs, rain gardens, infiltration systems, and green walls, to reduce stormwater runoff, improve air quality, and mitigate the urban heat island effect is what civil engineers prioritize.

Finally, a collaboration between civil engineers, policymakers, and stakeholders, as well as education and awareness-raising efforts, has greatly promoted sustainable practices and reduced the negative impacts of civil engineering projects on the society and the environment.

2.8 MIND MAPPING QUESTION

The design flaw dilemma—You're a senior engineer at a prestigious construction firm, responsible for designing a critical component of a high-profile project. During a routine inspection, you discover a critical design flaw that compromises the structural integrity of the building. If left unaddressed, the flaw could lead to catastrophic consequences, including loss of life. However, you also know that

- The project is already behind schedule and over budget.
- Correcting the design flaw will require significant resources, potentially leading to financial ruin for the company.
- You are the lead engineer on the project, and your professional reputation will be severely damaged if the flaw is exposed.

You have two options:

- Correct the design flaw, that is, inform your superiors and the client about the issue, potentially saving lives but risking the company's financial stability and your own professional reputation.
- Keep quiet, that is, ignore the design flaw, hoping that it won't cause any harm, but potentially putting lives at risk and violating your professional ethics.

What would you do?

2.9 CONCLUSION

All engineers exercise their obligations aiming for the greater good, serving and saving humanity in all their endeavors. In that regard, they follow ethical codes. While delivering professional expertise, the engineer mediates between a philosopher and a working mechanic, hence requiring the need to have theoretical knowledge and skills to interpret the workings of both! Therefore, it is essential to learn and grasp the key attributes of civil engineering and management, the four principles of management, and 5 Ms of management as these form the underpinning logic to deliver successful projects! History has consistently reminded us of prominent figures such as Leonardo da Vinci and Karl von Terzaghi and others who are of great significance in the history of civil engineering. Their remarkable talents were exercised professionally for the good of humanity. It does no harm when we take inspiration from them!

NOTES

1 A designer focuses on creating and developing designs, while a principal designer takes on a leadership role, overseeing the entire design process, ensuring alignment with project goals and client needs, and managing a team of designers.
2 How can you build a hospital in just a few weeks? | Institution of Civil Engineers (ICE).

REFERENCES

1. Attributes for professionally qualified membership. Retrieved from: https://www.ice.org.uk/join-ice/attributes-for-professionally-qualified-membership
2. Successful Professional Reviews for Civil Engineers 4th Edition, by Patrick Waterhouse (Author), H. MacDonald Steels (Author), Page 20. Accessed 3/August/2015. https://www.emerald.com/books/book/17579/Successful-Professional-Reviews-for-Civil
3. Construction Regulations 2015 (Design and Management). https://www.legislation.gov.uk/uksi/2015/51/contents/made
4. Philosophy Foundation. What is philosophy? Retrieved from: https://www.philosophy-foundation.org/what-is-philosophy Accessed 5/May/2025.
5. Mitcham, C. (1998). The Importance of Philosophy to Engineering. *Teorema: Revista Internacional de Filosofía*, 17(3), 27–47. https://www.jstor.org/stable/43047298
6. Pinto J. K. (2007). *Project Management: Achieving Competitive Advantage* (p. 241). Pearson Education Inc., Upper Saddle River, New Jersey.
7. Marsh J. H. (2015). Quebec Bridge Disaster. Retrieved from: https://www.thecanadianencyclopedia.ca/en/article/quebec-bridge-disaster-feature
8. Bennett M.D.,E.I.T.,A.M.ASCE.(2023).The Johnstown Flood of 1889: A Catastrophe of Civil Engineering (Part 5) | Geo-Institute. Retrieved from: https://www.geoinstitute.org/news/johnstown-flood-1889-catastrophe-civil-engineering-part-5
9. Ren R. (2023). Ahead of His Time: Leonardo da Vinci's Contributions to Engineering. Bartlett School of Environment, Energy and Resources, University College London, London,

WC1E, 6BT, United Kingdom. *Journal of Education, Humanities and Social Sciences.* 21. https://drpress.org/ojs/index.php/EHSS/article/view/13025/12671

10. Leonardo DaVinci (1478). *Codex Atlanticus. https://ambrosiana.it/en/discover/codex-atlanticus-leonardo-da-vinci/codex-atlanticus/*

11. Fisk, P. (2016). What is innovation? Connecting the dots, the ones other people miss. Retrieved from: https://www.peterfisk.com/2016/02/what-is-innovation-connecting-the-dots-the-ones-other-people-miss/

12. CoolMath4Kids. Connect the dots brain teaser. Retrieved from: https://www.coolmath4kids.com/brain-teasers/connect-dots. Accessed 5/May/2025.

13. Heydenreich, L. H., Dibner, B., & Reti, L. (1980). Leonardo The Inventor. McGraw-Hill Book Company, England.

14. Google Arts & Culture. Leonardo: Navigation and Hydraulic Engineering. Exhibition by Museo Nazionale della Scienza e della Tecnologia, Leonardo da Vinci, Via San Vittore 21, Milano. Retrieved from: https://artsandculture.google.com/story/4AVBRMG1X5GLLQ. Accessed 28/June/2025.

15. USC Illumin. (n.d.). Leonardo da Vinci: The Engineer. Retrieved from: https://illumin.usc.edu/leonardo-da-vinci-the-engineer. Accessed 28/June/2025.

16. Scientists Prove Leonardo da Vinci's 500-Year-Old Bridge Design Actually Works. Jason Daley – Correspondent, October 16, 2019. Retrieved from: https://www.smithsonianmag.com/smart-news/da-vincis-unbuilt-super-long-istanbul-bridge-would-have-worked-180973356/

17. Kemp, M., et al. (1987). Inventions of Nature and the Nature of Inventions, in: Ed. Paolo Galluzzi, *Leonardo da Vinci Engineer and Architect.* Montreal Museum of Fine Arts, Montreal.

18. How can you build a hospital in just a few weeks? | Institution of Civil Engineers (ICE). Retrieved from: https://www.ice.org.uk/news-views-insights/inside-infrastructure/how-can-you-build-a-hospital-in-just-a-few-weeks

Civil Engineering Technology

3

3.1 INTRODUCTION

People always hold discussions surrounding technology and engineering and their differences. Some argue that one defines the other, while others say that they are interrelated. Many people keep asking themselves what came first: technology or engineering, a classic chicken-and-egg problem.

The age-old question has sparked intense curiosity everywhere around the globe—which came first, technology or engineering? Did innovative engineering solutions give rise to new technologies or did emerging technologies drive the development of engineering disciplines? These are questions that always come to mind for novices, young engineers, and those in the making or training.

The debate has been raging on for a long time: what's the origin story of technology and engineering? Did engineering lay the groundwork for technological breakthroughs, or did technological advancements pave the way for engineering innovations?

This above inquiry frames the question in a way that highlights the intriguing and complex relationship between technology and engineering. The distinction between technology and engineering has sparked ongoing discussions for long. Some believe that engineering is the foundation of technological advancements, while others see technology as the catalyst for engineering innovation. However, the consensus is that technology and engineering are intertwined, with each playing a vital role in shaping the other.

So, what came first, technology or engineering? The answer is not a simple one. Technology and engineering have evolved together over time, with each influencing the other in a cyclical process, just as engineering and mathematics have developed in parallel throughout the course of history, again,

just as the relationship between science and engineering is iterative, with each field informing and advancing the other, reciprocally. Scientists' discoveries, for example, often spark new engineering innovations, which in turn drive further scientific inquiry. The fact remains that scientists' contributions are instrumental in shaping the development of engineering and advancing engineering technologies.

However, if we look at the history of human innovation and invention, we can see that early humans developed simple tools and techniques (technology) to solve problems and improve their lives. As these tools and techniques became more complex, the need for systematic design, development, and optimization (engineering) arose. Engineering disciplines emerged to address these needs, driving further technological innovations. These innovations, in turn, enabled new engineering solutions, and so on.

So, in a sense, technology came first (early tools and techniques). Engineering emerged as a response to the need for more complex and sophisticated technological solutions—engaging in continuous improvement. The two have since co-evolved, with each driving advancements in the other. So, at large, STEM disciplines have evolved in parallel, advancing the knowledge and understanding of the fundamental concepts such as **power, force, energy, and work** that have driven humanity to greater heights. Those are the foundational concepts of many modern technologies.

It is important to note, however, that the actual relationship between technology and engineering is more complex and interdependent. We learn about the Industrial Revolution and how the steam engine revolutionized the world and eased the production of goods and services. Was the steam engine an engineering invention or a technological device? The steam engine was both an engineering invention and a technological device.

Engineering invention	• The steam engine was a novel solution to a specific problem, converting thermal energy into mechanical work. • It involved innovative design, experimentation, and optimization of components like cylinders, pistons, valves, and crankshafts. Engineers such as James Watt, Thomas Newcomen, and Richard Trevithick developed and improved the steam engine, applying principles from mechanics, thermodynamics, and materials science. Not forgetting Leonardo da Vinci who had designed a much simpler form of Watt's engine that operated by flywheel and crank, although James Watt is credited with inventing the modern steam engine, the answer to a more modern steam engine was contained within Da Vinci's earlier drawings. See Section 2.5.1.5.3.
Technological device	• The steam engine was a technological device in the sense that it was a practical application of scientific knowledge,[a] harnessing the power of steam to perform tasks. • It was a key component in the Industrial Revolution, enabling mass production, transportation, and mechanization. • The steam engine embodied technological advancements in materials, manufacturing, and precision engineering, making it an essential tool for industry and transportation.

[a] Scientists influenced technological innovations and inventions.

In essence, the steam engine was an engineering invention that became a technological device, transforming the industry and society. Its development required both innovative engineering solutions and technological expertise to create a practical, efficient, and powerful machine.

Engineering is a *process*, while technology is a *product*. The primary difference is that engineering is the process of creating solutions, while technology refers to the tools, methods, and products that result from that process. So, when we say that someone is an engineer, it is because they have specialist knowledge to design, construct, operate, and maintain systems. In that process of doing those activities, they engineer products that can be referred to as technology. They can also use existing or new technologies in the engineering process to come up with new and advanced technologies.

Civil engineering technology, therefore, is that technology used in the civil engineering industry, that is, the tools, techniques, methods, and products, used to offer civil engineering services or products.

3.1.1 Staying Ahead of the Curve: Why Civil Engineers must Embrace Emerging Technologies

The primary answer to the question 'why must civil engineers embrace emerging technologies?' is that if you don't embrace emerging technologies, you cannot practice effectively, solving solutions for the modern world—a constantly changing world.

You need to know about the latest technologies shaping the world, from the hardware to software technologies, and find ways to integrate them into your work to safely, reliably, and sustainably offer up-to-date solutions to civil engineering problems.

Civil engineering is constantly evolving. In the 21st century, this means the profession is becoming more interdisciplinary. So, addressing the grand challenges of today and tomorrow starts with civil engineering research. Are you good at research? That is the question that haunts every constantly evolving civil engineer. Can you do the appropriate research without staying ahead of the curve—embracing new and advancing technologies? The answer is 'No.' You must be up to date to offer excellent services.

Whether at planning, design, construction, operation, or maintenance stage of a project, you need to conduct appropriate research and interpret the findings to share recommendations

insightfully! You cannot do this research without staying ahead of the curve—embracing emerging technologies to research and develop practical innovative solutions. Staying informed about new technologies can help you stay competitive in the industry and attract clients.

New technologies can lead to innovative solutions, improved designs, and more efficient construction methods. Therefore, a civil engineer taps into new technologies to innovate practical engineering solutions that improve designs and offer efficient construction techniques that comply with changing regulations and standards.

In Section 2.2.9, the need for continuous professional development was shared. Continuous learning is essential for professional growth and maintaining licensure or certification. Therefore, staying up to date with emerging technologies cannot be overemphasized—it is a must to ensure that the engineer understands and applies these technologies efficiently and effectively in their practice.

Advancing technologies can improve construction speed, reduce costs, and enhance project management. This can reduce the construction time and drastically reduce costs! Staying ahead of the curve is beneficial in this way because the engineer knows that the elements of time and cost are inseparable twins, especially in construction engineering. Therefore, having knowledge of emerging technologies that can help reduce costs or construction time is a great idea. As a civil engineer, be in the know, keep learning, and be up to date with emerging technologies so as to make lasting improvement in your work.

Also, new technologies can provide more sustainable and environmentally friendly solutions, reducing the environmental impact of construction projects.

3.1.2 Engineering versus Technology

To illustrate the difference, let's consider the design and development of a new advanced pump. In that case, engineering is the design and development of the pump. Technology includes the materials, software, and systems used in that pump, as well as the manufacturing processes and maintenance procedures. In summary, engineering is the process of creating solutions, while technology refers to the tools, methods, and products that result from that process. While engineering and technology are closely related and often used interchangeably, there is a subtle distinction between them. See Table 3.1.

TABLE 3.1 Difference between engineering and technology

NO.	ENGINEERING	TECHNOLOGY
1	Refers to the systematic application of scientific knowledge and mathematical principles to design, build, and maintain structures, machines, and systems.	Refers to the application of scientific knowledge and engineering principles to create practical solutions, products, and services.
2	Focuses on the development of solutions to specific problems, often involving the creation of new products, processes, or systems.	Encompasses the tools, techniques, and methods used to achieve specific goals or solve real-world problems.
3	Emphasizes the use of theoretical frameworks, modeling, and analysis to optimize performance, efficiency, and safety.	Includes the development, deployment, and maintenance of physical systems, infrastructure, and software.
4	Typically involves the application of established principles and methods to create new solutions.	Often involves the integration of multiple disciplines, such as engineering, computer science, and social sciences.

FIGURE 3.1 30 m³ water reservoir tank under construction, Kisoro UG.

3.1.2.1 Case study 3.1: Water tank design and construction, Kisoro UG

This case study involves a small 30 m³ water reservoir tank for a small community under construction in Kisoro UG, 2025, sited on dwarf walls in Figure 3.1. The tank measures 2.44 m by 3.66 m by 3.66 m. Each hot dip-galvanized pressed steel panel measures 1.22 m (height) by 1.22 m (width) manufactured according to BS 1564:1975 (specification for pressed steel sectional rectangular tanks). See also case study 9.1. BS EN 10025:2004 Grade S275 is a British Standard that specifies the steel used in **pressed steel tank panels** [1]. The 'S' indicates a steel grade, and the '275' refers to the **minimum yield strength** in MPa, which is 275 N/mm². S275 is often used in the construction of pressed steel tank panels because it offers a good balance of strength, machinability, and weldability. These panels are often hydraulically cold-pressed or hot dip-galvanized steel panels and may have standard thicknesses such as 4, 5, or 6 mm or more depending on the tank size. The thickness varies according to the position of the panel. Thus, the weight per panel differs.

Nominal capacity of the tank:

$$2.44 \times 3.66 \times 3.66 = 32.685 \text{ m}^3$$

This tank was designed to serve a community estimated to consume 30 m³, thus factoring in a safety factor of 1.09 (see Section 7.6.9.9). As a design or construction engineer, determining accurately the panel specifications needed for all faces of the rectangular (cuboidal) tank is crucial as it directly impacts the design of the dwarf walls, including their quantity (to meet tank capacity), dimensions, and thickness. It is

a must to consult appropriate standards, codes, and manufacturer specifications prior to design and specification of the tank capacity and materials for use in construction.

Specifically, the dwarf walls must be configured to support the tank's centered placement, ensuring precise alignment and optimal structural integrity. For example, you can see in Figure 3.2 that the panel joints precisely align with the dwarf walls—center to center of the dwarf walls. See Figures 3.3 and 3.4 demonstrating how standards guide the engineer on how to configure their designs. A key takeaway is that standards serve as a guiding framework for design and decision-making processes in civil engineering projects, ensuring consistency, safety, and quality. So, the engineering process is guided by the existing or advancing technologies.

The standards guide the manufacturer about the required specification for the yield strength of the panels, that is, specifying the thickness of the side, roof, and floor panels. This information would help the engineer during the design process. The bearing capacity of the soil and associated parameters would influence the type of foundation. This can be determined following applicable standards and conducting geotechnical investigations—for example, the dynamic cone penetrometer (DCP) test is commonly done in accordance with a procedure contained in BS 5930:2015+A1: 2020 for trial pit excavations. BS 5930 provides guidelines for conducting ground investigation of construction sites [2]. Thus, the bearing capacity of the soil would indicate if the soil strata at a certain depth would be able to carry the load on the proposed foundation type—for example, raft or strip foundations are common for reservoir tanks sited on dwarf walls.

To emphasize this, technology, codes, and standards of practice significantly influence engineering decisions and outputs. Civil engineers must stay continuously up to date with

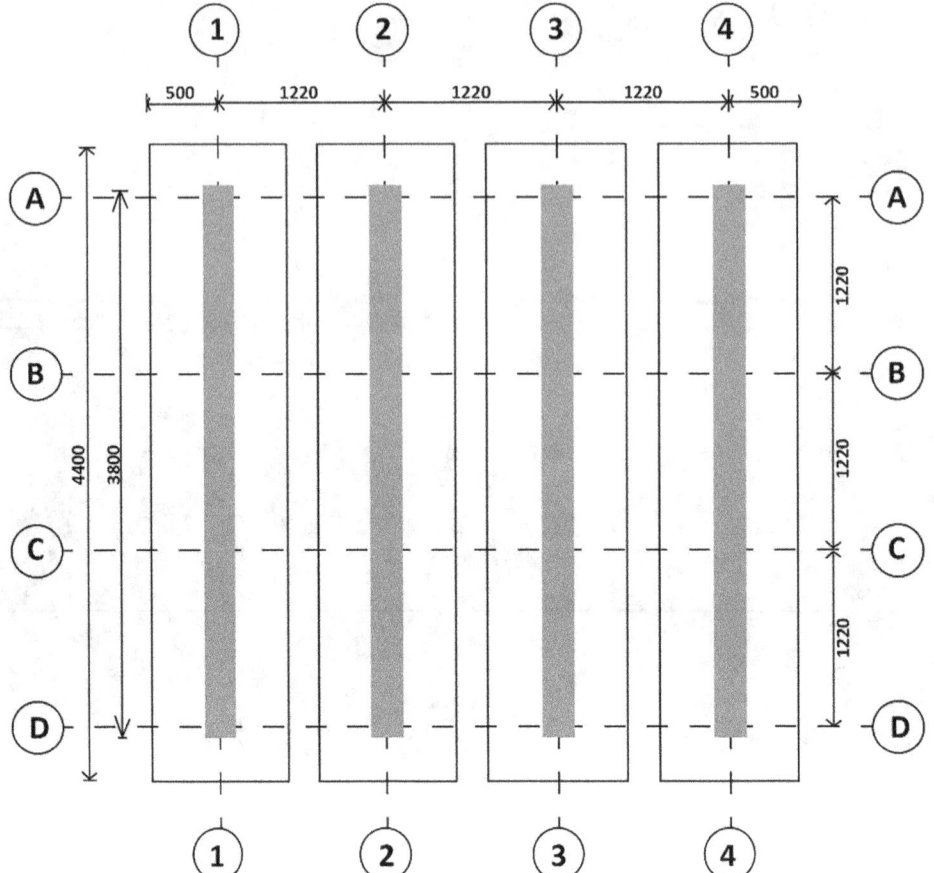

FIGURE 3.2   Foundation layout.

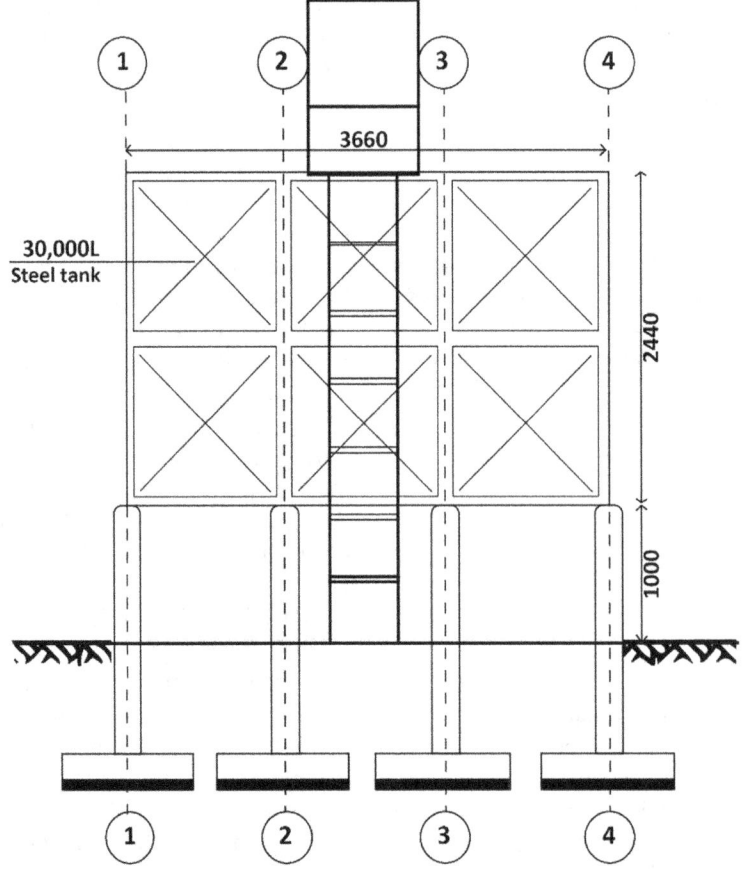

FIGURE 3.3   Side elevation showing support pedestals.

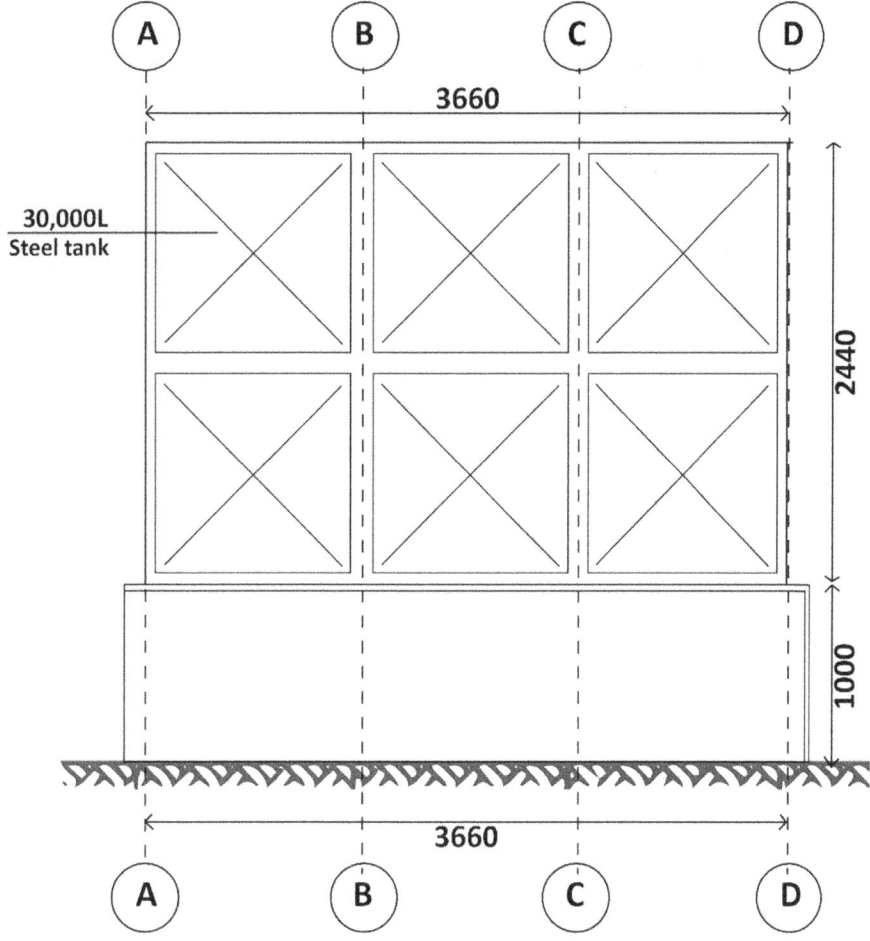

FIGURE 3.4 Side elevation showing support pedestals.

civil engineering technology, products, and systems. It is also important to understand the local codes and standards of the region to practice effectively. By knowing local industry standards, manufacturer standards, and regulations, engineers can initiate conceptual designs from a well-informed perspective. Furthermore, their engineering science must align with what the codes, standards, and technology specify. This is now a true engineering process.

3.2 TECHNOLOGY, CODES, AND STANDARDS OF PRACTICE

A **code** is a set of mandatory rules or requirements that define the minimum acceptable level for something, which may be enforced by law when a government, state, federal, and local authority or country adopts it into their legislation, making it enforceable within their jurisdiction—becoming a legally binding code.

In its entirety, as far as construction and civil engineering are concerned, a code is a set of guidelines or rules that provide a framework for designing, constructing, or operating something. Codes are typically voluntary, and compliance is not always mandatory, except when adopted by the government, making them a **legally binding code**. Codes are often developed by industry associations, government agencies, organizations, or experts in a specific field. They provide recommended best practices, but may not be enforceable by law.

Note: A code is a set of rules that experts in the field recommend people to follow; it is a model. Although it is not a law, it can be adopted into law. A code tells you what needs to be done, but it doesn't explain how it should be done. Some examples of codes include International Building Code, American Society of Mechanical Engineers (ASME) Boiler and Vessel Code, and AWS D1.1.

A **standard of practice** is a set of guidelines or best practices that detail *how* to achieve the requirements set by a code, often not legally enforceable but considered a benchmark for quality within a profession or industry. It is a set of technical definitions, specifications, and guidelines. They function as instructions for designers, manufacturers, operators, or users of equipment. If you are building something, a standard tells you about the materials, process, designs, structure, etc. In brief, standards tell you *how* to do something.

Standards are usually created by individual companies, organizations, or countries. They are not legalized. A standard develops into a code when it is adopted by a set of government bodies and gets legalized. Some examples of standards include American Society for Testing and Materials (ASTM) International standards and ISO standard.

It is essential to note that sometimes a standard does not define explicitly all parts of how something should be done—in other words, it is not a specification. Standards describe minimum requirements and do not contain all of the engineering and administrative information normally contained in specifications. The standards usually contain options that must be evaluated by the user of the standard. Until each optional feature is specified by the user, the product or service is not fully defined.

In summary, a code tells you 'what to do,' and a standard of practice explains 'how to do it.' See Table 3.2. To illustrate this further, consider the Routledge 2023 game-changing book (further reading 1) *Integrated Drainage Systems Planning* *and Design for Municipal Engineers* by Patrick Ssempeera, which sets new standards (in tech and engineering) on how to approach drainage design, development, and maintenance in the modern world. This is because a sustainable world needs more sustainable drainage systems integrated with gray infrastructure. It showcases how engineers contribute to the refinement of codes and standards—which is one of their typical roles. That is typically what professional engineers do—setting or contributing to new standards of practice!

TABLE 3.2   Code versus standard

NO.	CODE	STANDARD
Model	Model that is adaptable by law. A government agency can adopt a **code as a regulation,** making it mandatory and enforceable by law. A **regulation** can reference a code, requiring compliance with specific provisions or standards outlined in the code.	Set of technical definitions, specifications, and guidelines.
Key considerations	Clarifies what needs to be done.	Clarifies how something should be done.
Enforceability	Can be adopted into law. And once adapted into law, codes are typically legally enforceable, meaning non-compliance can result in penalties, say, **'building codes'** enacted through a statutory instrument by a **government agency** in compliance with primary legislation. Non-compliance to industry codes such as **'Code of Ethics'** for civil engineers issued by the ICE for which members must comply is enforceable by the institution once there is breach of the **rules** contained therein by a member. For example, the Institution of Civil Engineers, United Kingdom, and ASCE specify the code of professional conduct for members. The institutions take the conduct of members very seriously. Being a member of the ICE, for example, means that a civil engineer has proved that they can work to the highest possible **standards** and to abide by this ethical code of conduct. The code spells out what a member must abide to. Therefore, it is a set of rules to follow while rendering professional expertise. **Ethical codes of practice** still have a significant influence on industry practices through not being statutory codes.	Is not legalized. Standards of practice are usually not legally mandated.
Level of detail	Codes provide basic requirements. It is a rule or rules to follow. Once adopted into law, a code sets out regulations.[a]	Standards of practice offer more detailed instructions and recommended procedures.
Application	Codes are often used in construction, safety regulations, and other areas where strict compliance is necessary.	Standards of practice are commonly used in professional fields such as medicine, engineering, and accounting to ensure quality of work.
Examples	**Examples** include 2024 International Building Code (IBC)[b] and 2024 International Fire Code [3,4].[c]	**Examples** include the following: American Society for Testing and Materials (ASTM) International standards (such as ASTM D6938[d]) and ISO standard (such as ISO 9001: 2015 Quality Management[e]) [5,6].

[a] **Note:** Rules can be established by various entities, such as organizations, institutions, or individuals. Regulations, however, are typically created by government agencies or legislative bodies. Regulations often carry the force of law, while rules might not always have the same level of legal authority.

[b] The 2024 IBC is a **model code** developed by the ICC. This code applies to all buildings except detached one- and two-family dwellings and townhouses up to three stories. It provides guidelines for designing, constructing, and occupying buildings safely and sustainably. **Note:** The ICC develops and maintains a family of 15 model codes. As a model code, the 2024 IBC is not directly enforceable by law. However, the IBC can become enforceable when local governments, states, or countries adopt it as their building code, making it enforceable within their jurisdiction. Governments can also incorporate the 2024 IBC into their legislation, making it a legally binding code. https://codes.iccsafe.org/content/IBC2024P1.

[c] The 2024 IFC® contains **regulations** to safeguard life and property from fires and explosion hazards. Topics include general precautions, emergency planning and preparedness, fire department access and water supplies, automatic sprinkler systems, fire alarm systems, special hazards, and the storage and use of hazardous materials. https://codes.iccsafe.org/content/IFC2024V1.0.

[d] ASTM D6938 is a standard **test method** for 'in-place density and water content of soil and soil-aggregate by nuclear methods.' This standard provides guidelines for using **nuclear gauges** to measure the density and water content of soils and soil-aggregates in the field. The test method is commonly used in geotechnical engineering, construction, and quality control applications. https://www.instrotek.com/products/mc-3-elite.

[e] ISO 9001:2015 is an international standard that outlines the requirements for a Quality Management System (QMS), designed to help organizations ensure that they meet customer and regulatory requirements, while continuously improving their processes and overall performance.

3.2.1 Evolution of Codes and Standards

The journey of standards and codes in civil engineering, particularly, traces back to early efforts aimed at ensuring public safety and regulating construction practices. Initially, codes focused on structural integrity and basic safety measures, reflecting the challenges and available materials. Over time, as engineering knowledge expanded and construction techniques evolved, these standards grew in complexity and comprehensiveness.

In 1754 BCE, King Hammurabi of Babylon—the sixth king of the First Babylonian Dynasty, ruling for 42 years—established the Code of Hammurabi, which is one of the world's oldest written legal codes. It is the earliest known building code, named as 'Code of Hammurabi' (circa 1754 BCE), which specified standards for construction, materials, and safety. The code shared universal problems such as collapse of buildings, oxen getting loose and trampling fields, and neighbors squabbling, much as they do today [7].[1] Today, Babylon exists as an archeological site in central Iraq.

During the first Industrial Revolution, the need for standardized practices and safety protocols grew. In the late 19th and early 20th centuries, organizations like the Institution of Civil Engineers, ASME, and the ASTM emerged to develop and publish standards for various industries.

Following World War II, there was a significant push for standardization and codification in various fields, including construction, manufacturing, and transportation. This led to the development of modern building codes, such as the **International Building Code (IBC)** and the International Residential Code (IRC).

The advent of computers and digital technologies has revolutionized the development, dissemination, and implementation of codes and standards. Electronic platforms, online databases, and digital tools have made it easier to access, update, and enforce codes and standards.

As global trade and collaboration increase, there is a growing need for harmonized codes and standards. Organizations like the International Organization for Standardization (ISO) and the International Electrotechnical Commission (IEC) play a crucial role in developing global standards that facilitate international trade, innovation, and cooperation.

The ISO story began in 1946 when delegates from 25 countries gathered in London to discuss the future of standardization. A year later, on 23 February 1947, ISO officially came into existence. In this post-war era, the founding members saw international standards as a key to the world's reconstruction efforts [8].[2]

Modern standards are closely tied to technological advancements, environmental considerations, and global collaboration.

The future of civil engineering standards will be shaped by ongoing technological innovation and societal needs. Artificial intelligence (AI) and machine learning (ML) tools are revolutionizing civil engineering standards. Digital technologies like building information modeling (BIM) and advanced analytics are revolutionizing how standards are applied throughout the project lifecycle. Today, codes and standards continue to evolve in response to emerging trends, such as digital transformation and industry 4.0, sustainability and climate change, cybersecurity and data protection, globalization and trade agreements, and technological innovations and disruptions.

Civil engineering codes are usually developed and updated in response to innovations in civil engineering technology, construction techniques, environmental considerations, changes in policy, and so on. New editions are published after a specified period of time to cope with the latest innovations and technologies.

Real-time monitoring and predictive maintenance are becoming integral to maintaining infrastructure resilience and optimizing operational efficiency. These are made easier by integrating AI tools and ML technologies. Additionally, as climate change impacts intensify, standards will continue to evolve to incorporate sustainable practices and resilience against natural hazards.

3.2.2 Role of Standards and Codes

- Standards provide a common language, facilitate the integration of emerging technologies, and help ensure that efforts to mitigate climate change are effective, consistent, and measurable [9].[3]
- Standards and codes play a crucial role in various aspects of society, ensuring safety, quality, and consistency. They help societies to establish minimum acceptable levels for products, processes, and services, providing a common language and framework for different stakeholders. In essence, they serve as guidelines and regulations that help protect the public, facilitate trade, and improve efficiency.
- Standards and codes are the 'rules of the game' used when devising and implementing economic and financial policies. Countries that 'play by the rules' tend to have better economic performance and greater financial stability, so it's in their interest to abide by internationally recognized standards and codes.

3.2.3 Selected Civil Engineering Standards

This section outlines select few civil engineering standards adopted for material development for illustrative purposes. It typically provides how standards have evolved throughout history for these selected products or processes. The list provides some basic examples of select civil engineering codes and standards for illustrative purposes to show how they evolved, why they were adopted into civil engineering, and what keeps them going and getting refined over time.

The section shares interesting scenarios of how the evolution of a standards influenced further research and development in the civil engineering field—furthering engineering knowledge. Therefore, it is not a bad idea for a civil engineer to learn about standards as much as they could. This cannot be overemphasized. It is self-evident. This is because keeping up to date and constantly evolving as a civil engineer requires a check on how emerging technologies influence codes and standards of practice and how you can extend this knowledge to refine civil engineering designs, processes, systems, and products.

You must realize that technology has influenced civil engineering throughout history, and a constantly evolving civil engineer must appreciate the role of new and advancing technology in their practice, especially in the development of codes and standards of practice. Technology and technological devices always emerge from almost every corner and not necessarily from engineers alone—so long as the technology is essential and makes lives better and the society welcomes it. The engineer, being a creative person and a problem-solver, always taps these technologies to refine processes and systems and that is why they keep up to date with emerging technologies. As they develop new codes and standards of practice, engineers borrow technologies that society socially objectivizes and recommend them for adoption, making processes and systems work even much better for the greater good.

In that regard, engineers are not only smart, but they also know how to work on borrowed ideas to refine new or existing processes, systems, products, and technologies! Engineers realize that time is limited and we have to accomplish so much; that is why they leverage teamwork. Mathematics as a tool that engineers harness, in particular, has never been relevant to engineers than today in the AI era! Thus, engineers, today, borrow and integrate ideas, but not working things from first principles—the next time you try first principles, you will be stuck in a rut—basically failing at innovation and entrepreneurship.

3.2.3.1 Sewer pipelines and culverts

What are the reasons behind the standardization of concrete culvert sizes to typically measure 300, 600, 900, 1,200, and 1,500 mm diameters, and what factors contribute to the scarcity of other sizes?

As a skilled engineer, you design a sewer line, properly evaluating the applicable hydraulics, hydrology, and structural requirements. However, despite your precise calculations, you're ultimately forced to round your specifications or output to the nearest available culvert sizes/standards dictated by industry standards, introducing a degree of approximation to your otherwise exacting design. Why is this so?

Culvert sizes evolved from traditional brick and stone arches, which were built in increments of 300 mm (approximately 1 foot). This led to the wide adoption of standard sizes, 300, 600, 900, 1,200, and 1,500 mm diameters, which are multiples of 300 mm. At that time, there was no much engineering science!

People developed technological products trying to ease human life by devising practical simpler ways to bridge structures, make people and goods move from one place to another, and convey water from one side of the obstacle to the other without much conscious scientific or engineering effort!

Since then, the industry has adopted their standard use. As engineering knowledge advanced, it found a culvert (technological product) already present in those sizes because of the reason advanced earlier. So, engineering knowledge continued to refine the specifications and the types and hydraulics of culverts.

Today, we can say that a culvert is an engineered solution that utilizes materials and design principles to convey water across an obstacle. It allows water to flow under a road, railroad, or other obstacle, typically consisting of a pipe or an enclosed channel. It is designed to convey water from one side of the obstacle to the other, while maintaining the integrity of the road or railroad or any other structure above—for example, consider jacked culvert pipes under buildings.

As engineering knowledge advanced, typically the hydraulics, such as **Manning's equation** that expresses flow velocity with channel geometry, the understanding of how the roughness of pipes influences the velocity of flow, and the overall hydraulic efficiency of the sewer system in question came into picture! That was around the 1890s.

Talk about concrete pipes, steel pipes, and uPVC pipes—all have different roughness factors. See Section 7.6 for the role of coefficients. Standards were drawn in response to this breakthrough! Thus, the engineer was able to influence standards of practice for culverts—focusing more on quality, pipeline efficiency, safety, durability, and reliability.

In the same way, the strength of culverts was seen from another perspective influencing the longevity of the sewer pipeline system. Culvert strengths were defined at an ultimate level based on the design intent! By the engineer understanding the structural aspects of sewers, specifically how much load (both internal and external) a pipe culvert can withstand before failing, they specify the correct materials to use for a task:

- See BS 9295:2020 Guide to the structural design of buried pipes [10].
- See BS EN 1295-1-1997 [11] that recommends a factor of safety (FOS) of 1.25 for general use and a value of 1.5 for reinforced pipes with a 67% proof load.

All these were efforts of a constantly evolving civil engineer! Now, advancements in technology, for example, allow testing materials possible, applying scientific and engineering knowledge!

- See BS 5911-1:2021 [12]—TC Concrete pipes and ancillary concrete—Unreinforced and reinforced concrete pipes (including jacking pipes) and fittings with flexible joints (complementary to BS EN 1916:2002) [13].
- BS EN 1916 specifies requirements and describes test methods for precast concrete pipes and fittings, unreinforced, steel fiber and reinforced, with flexible joints and nominal sizes not exceeding DN 1750 or WN/HN 1200/1800, for which the main intended use is the conveyance of sewage, rainwater, and surface water under gravity or occasionally at the low head of pressure in pipelines that are generally buried. The scope includes pipes (collectively referred to as 'jacking pipes') intended to be installed by pipe jacking, micro-tunneling, or other trenchless technology [12].

This part of BS 5911 specifies complementary requirements to those specified in BS EN 1916 for unreinforced and reinforced concrete pipes and fittings, with nominal sizes not exceeding DN 1500 for circular pipes with base and WN/HN 800/1200 for egg-shaped pipes. Requirements are also specified for reinforced concrete circular trench and jacking pipes with nominal sizes greater than DN 1750, but not exceeding DN 3000. All these standards evolve as extensions and improvements of the already existing technological systems—focusing on advancing the right procedures for conducting tasks and quality, reliability, and durability, among other aspects.

In Section 2.2.5, we described applying and extending new and existing technologies as a prerequisite for a competent engineer, i.e., how you have applied codes and standards of practice and how you have led the refinement of these codes and standards. Nevertheless, culverts of differing sizes escaping the standard 300 mm increment are becoming not uncommon because of the advancing technology and materials science.

3.2.3.2 uPVC/HDPE pipes pressure nominal (PN) numbers

What are the reasons behind the standardization of uPVC/HDPE pipes pressure nominal (PN) numbers to typically 6, 10, 16, and 25 and what factors contribute to the **scarcity of other ratings**?

You've probably seen that the standard sizes for water pipes, such as high-density polyethylene (HDPE) pipes manufactured in standard sizes of 20, 25, 32, 40, 50, and 63 mm with pressure ratings ranging from PN 6 (6 bars—approximately 87 psi or 600 kPa, for low-pressure applications, such as water supply and drainage), PN 10 (10 bars—approximately 145 psi or 1,000 kPa for medium-pressure applications, like heating and cooling systems), PN 16 (16 bars—approximately 232 psi or 1,600 kPa for higher-pressure applications, like industrial processes and steam systems), and PN 25 (25 bars—approximately 363 psi or 2,500 kPa for high-pressure applications, like hydraulic systems and high-pressure pipelines). Why not other sizes and ratings?

During engineering design for a pipeline, the engineer may arrive at any pipe diameter and pressure rating. However, they must approximate the design to the nearest standard size and pressure rating, considering the FOS in design.

It is essential to note that manufacturers choose these standard sizes or values for pipes on the basis of **preferred numbers**. In industrial design, preferred numbers (also referred to as preferred values or preferred series) serve as standardized guidelines for selecting specific product dimensions that fall within predetermined design limitations.

ISO 17:1973 provides a guide to the use of preferred numbers and of a series of preferred numbers [14]. This international standard helps organizations and industries adopt a harmonized approach to numerical values, promoting consistency, simplicity, and cost savings. The standard introduces a series of preferred numbers, such as the Renard series (R5, R10, and R20) and the geometric series.

ISO 17:1973 provides a framework for selecting preferred numbers and a series of preferred numbers, applicable to various fields, including engineering, manufacturing, and quality control. The standard aims are to establish a logical and consistent system for numerical values, reduce the number of different values used, and facilitate international cooperation and standardization.

Why are preferred numbers chosen?

- They are selected in a way that when a product is produced in various sizes, they will be approximately evenly distributed on a **logarithmic scale**. See Section 7.8 for the role of logarithms in engineering science.
- Employing them enhances the likelihood of interoperability among components designed independently by various individuals or teams. In essence, it's a strategic approach to standardization, applicable within organizations or industries, and is generally beneficial in industrial settings (except when vendor lock-in or planned obsolescence is intentionally pursued).

Pipe diameters and pressure numbers (ratings) mentioned earlier are selected and manufactured specifically according **Renard's Preferred Numbers**. The PN system was adopted to harmonize pressure ratings across Europe and eventually globally, facilitating international trade and standardization. The chosen values are practical and easy to work with, allowing for simple calculations and conversions.

By choosing these specific values, the PN system provides a clear, consistent, and practical way to designate pressure ratings, making it easier for manufacturers, engineers, and users to select and work with suitable components.

Renard's preferred numbers are a series of numbers used in engineering to minimize error when replacing numbers with the closest Renard number. The numbers are part of a geometric sequence, with each number a multiple of the previous one. The series is named after French military engineer Colonel Charles Renard, who proposed it in 1870. Examples of Renard's series include the following:

- R5 : The series based on the 5th root of 10, is approximately 1.585
- R10 : The series based on the 10th root of 10 is approximately 1.259.
- R20 : The series based on the 20th root of 10 is approximately 1.122.
- R40 : The series based on the 40th root of 10 is approximately 1.059.

Therefore, the pipe pressure ratings PN 6, PN 10, PN 16, and PN 25 follow Renard's series R5. R5 is a set of preferred numbers that are spaced in **a geometric progression**, with a ratio of the following:

$$10^{\left(\frac{1}{5}\right)} \approx 1.585$$

This series is commonly used for pipe pressure ratings, among other applications. The HDPE pipe sizes of 20, 25, 32, 40, 50, and 63 mm follow Renard's series R10. Renard's series R10 is a set of preferred numbers with a geometric progression ratio of the following:

$$10^{\left(\frac{1}{10}\right)} \approx 1.259$$

This series is often used for pipe and tube sizes. Some of the advantages of using Renard's numbers is that it helps reduce the number of sizes or values needed for product manufacturing. It also simplifies the design and manufacturing of a product, enhancing efficiency and cost-effectiveness. Thus, it improves interchangeability and compatibility, overall. UPVC manufacturing pipes in sizes of 110, 125, 140, 160, etc. are also a perfect example of preferred numbers in action.

Pipes manufactured to ISO standards

Other than Renard's series, the PN standards were chosen based on a combination of several factors. The PN system originated in Europe, where the bar unit was already widely used. The PN system is based on the metric system, which is used internationally. The bar unit is a metric unit of pressure, making it a natural choice.

In SI units, Pascal (Pa) is the unit of pressure or stress, defined as follows: 1 Pascal (Pa) = 1 Newton per square meter (N/m²). In other words, Pascal is a measure of the force (in Newtons) applied per unit area (in square meters). Pascals are commonly used to express pressures in various fields, such as physics, engineering, and materials science. To break it down further:

- 1 Newton (N) is the unit of force, equivalent to 1 kilogram-meter per second squared (kg·m/s²).
- 1 square meter (m²) is the unit of area.

So, 1 Pascal (Pa) represents a force of 1 Newton applied to an area of 1 m²:

- 1 bar ≈ 14.5 psi (psi = pound per square inch)
- 1 Pa ≈ 0.000145 psi
- 1 bar = 100,000 Pa.

The PN values are chosen to provide a safety margin, allowing for fluctuations in pressure and ensuring that components can withstand unexpected surges. The PN system was adopted to harmonize pressure ratings across Europe and eventually facilitating international trade and standardization globally. By choosing these specific values, the PN system provides a clear, consistent, and practical way to designate pressure ratings, making it easier for manufacturers, engineers, and users to select and work with suitable components.

3.2.3.3 Steel reinforcement bar (rebars) sizes

Steel reinforcement bars (rebars) are manufactured typically to the sizes of Y8, Y12, Y16, Y20, Y25, and Y32 mm. Why? Steel bars, also known as reinforcement bars or rebars, are solid, long metal bars which are generally used in construction projects in order to reinforce structures such as buildings, dams, and bridges. Their types are thermo mechanically treated (TMT) bars, hot rolled bars, cold-rolled steel bars, and mild steel bars.

TMT bars are additionally earthquake-resistant. They are, therefore, perfect for usage in areas where seismic activity is common. They come in different sizes; steel bars are 8, 10, 12, and 16 mm in size. TMT bars are extensively utilized in the construction industry. Some of the common uses of TMT bars are as follows: marine structures, high-rise buildings, industrial structures, and bridges and heavy-duty infrastructure operations. Steel bars in the diameters of 8, 10, 12, and 16 mm are most widely utilized. The answer is the same: Renard's Preferred Numbers.

3.2.3.4 Reservoir tank steel panels

What are the reasons behind the standardization of hot dip-galvanized pressed steel tank panels or cold-pressed steel panels to typically be manufactured measuring 1.0 m by 1.0 m or 1.22 m by 1.22 m or 0.9 m by 0.9 m or 0.5 m by 0.5 m and what factors contribute to the scarcity of other sizes? Why not 2 m by 2 m, for example?

See Figure 3.1 for a reservoir tank with hot dip-galvanized pressed steel panels. The dimension of 1.22 m × 1.22 m (4 feet × 4 feet) for a tank panel is a common standard in the industry manufactured according to BS 1564:1975 (specification for pressed steel sectional rectangular tanks), and it's based on several factors [15]. BS 1564:1975 specifically specifies 1.2 m by 1.2 m panels. The primary factor considered is the standardization of transportation:

- Standardization for transportation. The panels of these sizes can be easily transported on standard pallets or trucks, reducing transportation costs and logistical challenges. For example, the 1.219 m length is a common standard in many industries, particularly in North America, often referred to as the 'standard GMA pallet' size (48 inches × 40 inches). The UK pallet typically measures 1,200 mm × 1,000 mm (47.2 inches × 39.4 inches). The lesson to take forward is that the transportation standard influenced the tank panel standard! Fitting materials in containers or trucks for transportation is the primary factor.

Other factors include the following:

- Modularity and manufacturing efficiency. Using a standard size such as 1.22 × 1.22 m allows for modular design and construction. Panels can be easily connected and arranged to form larger tanks. The size of 1.22 × 1.22 m is efficient for manufacturing processes as it allows for optimal use of materials and minimizes waste.
- Ease of installation and structural integrity. The size of 1.22 × 1.22 m makes it easier to handle and install panels on site as they can be lifted and placed by a single person or a small team. The size also provides a good balance between structural strength and weight, ensuring the panels can withstand the required loads and pressures.

While these factors might not be universally applicable, they contribute to the widespread adoption of this standard size in the tank industry.

3.2.3.5 Public stand post (PSP) yields (tap stands yields)

The standard tap stand flow is typically 0.225 lps (13.5 L/min) for community rural water supply schemes, as specified in standard manuals for several jurisdictions. Such a tap will adequately serve a population of 200–230 persons. Such a standard is derived from conducting market/appropriate research to come up with consumption rates in rural communities. That is why it is advisable that engineers need to have the skills to conduct appropriate/market/formative research in design and development of civil engineering projects. Although manuals can dictate the standard tap stand flow, the engineer may decide to increase or decrease, as circumstances may dictate.

Where tap stand intends to serve only just a few households, the flow can be reduced a bit, and conversely the flow may be increased for a more densely populated area (a double or triple-faucet tap stand may also be built). The specification of 0.225 lps (13.5 L/min) in the manual for a given jurisdiction stands as a standard benchmark used by designers. Engineers keep constantly revising and updating manuals and standards as technology advances. Modern PSPs may no longer serve the population in some modern villages when restricted to delivering 13.5 L/min. The lesson to take forward is that standard and codes keep changing with advances in technology, and it is the role of the engineer to take this forward.

3.2.3.6 Horsepower (HP) as a standard

Horsepower (HP) as a standard unit of measurement is used in the English system. You have probably seen trucks bearing the label HP, say 320 HP. HP is a unit of power, originally defined by James Watt to compare the power of steam engines to the power of horses.

What is horsepower?
The concept of 'horsepower' originated in the 18th century—coined in 1782 by Scottish inventor James Watt, who developed a unit of measurement to compare the output of his steam engines to the work done by horses. Watt's goal was to demonstrate that his engines were more powerful and efficient than the traditional horse-powered systems used in industries like coal mining.

In 1828, Thomas Tredgold related civil engineering to the concepts of James Watt, who had demonstrated that his engines were more powerful and efficient than the traditional horse-powered systems used in industries like coal mining. Thomas identifies a civil engineer as one who is engaged in the art of directing the great sources of power in nature for the use and convenience of man. Thus, the engineering fraternity socially objectivized the HP as a unit of measurement for power in the English system and the Watt, later, in the SI system.

Watt is best known today for the creation of the steam engine. His contraption was a roaring success from its conception, soon becoming a focal part of the coal mining process. Horsepower (HP) is a unit of power, originally defined by James Watt to compare the power of steam engines to the power of horses. It's calculated as follows:

$$1 \text{ horsepower (HP)}$$

$$= 33,000 \text{ foot} - \text{pounds per minute} \left(\text{ft} - \text{Ib} / \text{min} \right)$$

$$1 \text{HP} = 550 \text{ foot} - \text{pounds per second} \left(\text{ft} - \text{Ib} / \text{s} \right)$$

$$1 \text{HP} = 745.7 \text{ Watts}$$

In essence, HP represents the rate at which work is done or energy is transferred. Since its inception by James Watt, it's commonly used to measure the power output of engines (cars, trucks, aircraft, etc.), motors (electric, hydraulic, etc.), pumps and compressors, and turbines. The term 'horsepower' was coined to help people understand the power of steam engines in relation to the power of horses, which were commonly used for work at the time. Today, it's a widely accepted unit of measurement for power.

James Watt derived the concept of HP from observing horses at work. In the late 18th century, Watt was trying to find a way to explain the power of his steam engines in terms that would be relatable to potential customers, who were familiar with the power of horses. At the time, horses were commonly used for tasks like pumping water, grinding grain, and hauling loads.

Watt observed that a strong draft horse could lift a load of about 33,000 pounds to a height of 1 foot in 1 minute. He used this as a basis for his calculation. See Figure 3.5. It illustrates the four concepts of force, work, energy, and power.

FIGURE 3.5 Horse raising 33,000 pounds to a height of 1 foot per minute.

- **Force:** A push or pull that causes an object to change its motion or shape. The string applies a force (pull) to the load.
- **Work:** The transfer of energy from one object to another through a force applied over a distance. Work is done lifting the load through a distance/height of 1 foot.
- **Energy:** The ability to do work. This is intrinsic. The horse has the ability to do work!
- **Power:** The rate at which work is done or energy is transferred or converted. The horse could lift a load of about 33,000 pounds to a height of 1 foot in 1 minute. How quickly? 1 minute. Thus, you can evaluate the efficiency.

Watt defined one HP as the power required to lift 33,000 pounds by 1 foot in 1 minute, which is equivalent to about 746 watts. This was a marketing strategy because it allowed people to easily understand the power of his steam engines in terms of the familiar power of horses. So, while horses did not directly contribute to the calculation, their capabilities inspired Watt to create a relatable unit of measurement that has since become a standard in the industry!

You can relate this to the same way economists such as Adam Smith, in his 1776 publication, *The Wealth of Nations*, built on slave labor to develop economic theories such as the labor theory of value—in this case, slaves acted as experiments, opening eyes on how political economies could function much better! The philosophy of absolute advantage, for example, must have been directly inspired by slave laborers who worked on infrastructure projects such as roads, railways, and canals and on farms. It is quite surprising that we have no standard unit along those lines, 'the slave unit' per se.

To convert 33,000 foot-pounds per minute (ft-lb/min) to a more metric unit,
1 foot-pound (ft-lb) = 1.356 joules (J)
1 minute = 60 s

So,

33,000 ft-lb/min = 33,000 × 1.356 J/ft-lb/60 s ≈ 745.7 watts (W)

Now, to express this in kilograms, we can use the fact that 1 W is equal to 1 joule per second. 1 kW (Kilowatt) is equal to 1,000 W. Since 1 W = 1 Joule/s, we can also express 1 kW as 1,000 Joules/s. We can also use the conversion factor 1 joule = 0.102 kilograms-force meter (kg-f·m):

745.7 W ≈ 745.7 J/s ≈ 76.28 kgf·m/s.

To simplify, we can say:

33,000 ft-lb/min ≈ 76.28 kilogram-meters per second (kg·m/s)

This conversion is an approximation as the original unit (ft-lb/min) is not a standard metric unit. The unit 'foot-pounds per minute' (ft-lb/min) is a part of the Imperial System of units, also known as the British Imperial System or the English system.

This system was widely used in the United Kingdom and its former colonies, including the United States. The Imperial System is based on traditional units such as the following:

- **Length:** inches, feet, yards, and miles.
- **Weight:** pounds, ounces, and tons.
- **Volume:** fluid ounces, cups, pints, and gallons.
- **Power:** HP (derived from foot-pounds per minute).

Although the Imperial System is still used in some jurisdictions, especially in the United States and the United Kingdom, the metric system (SI units) has become the standard system of measurement in most countries around the world and scientific applications due to its decimal-based consistency and ease of conversion.

3.2.4 Selected Civil Engineering Codes

These can be legally enforceable and non-enforceable codes. Once a government agency adopts a **code as a regulation**, it becomes mandatory and enforceable by law. For example, in many jurisdictions, **building codes**, in some jurisdictions referred to as building regulations (for example, in the United Kingdom), are enacted into law and become the primary legislation, outlining specific requirements for building design, construction, and safety. Other examples include health codes and traffic codes. These are statutory codes enacted into law by a legislative body, such as a parliament or congress. Other types of codes include the regulatory codes, industry codes, and model codes.

Regulatory codes, on the other hand, are codes created by regulatory agencies or government departments to implement or enforce laws. Regulatory codes can be statutory instruments if they are created under the authority of a primary law. Sometimes, they are enacted as statutory instrument, which is a type of law or regulation created by a government agency or official under the authority of a primary law or statute, say a minister. These instruments provide detailed rules, regulations, or procedures to implement or enforce the primary law.

Industry codes are codes of practice or standards developed by industry associations, professional bodies, or other organizations. The 'ethical code of conduct for civil engineers' issued by the Institution of Civil Engineers (ICE) is an excellent example. Industry codes may not be statutory instruments but have significant influence on industry practices.

Model codes are sample codes or guidelines developed by organizations or experts to provide a framework for others to follow. The 2024 IBC is a **model code**. Model codes are not typically statutory instruments.

Civil engineering codes can fall into the above categories, and building codes are a perfect example to discuss because they are the commonest enforceable codes, worldwide. These are regulations that set out the standards to which all buildings and other structures must conform. They cover design, construction, maintenance, repair, and alteration. They specify the minimum requirements necessary to safeguard the health, safety, and welfare of building occupants and the public as well as other impacts on the environment, efficiency, community, and so on.

3.2.4.1 International Building Code

IBC is a model code developed by the International Code Council (ICC). Located in Country Club Hills, Illinois, the ICC has a significant presence in the US building code community. The ICC is a nonprofit organization that develops and publishes model building codes and standards, widely used in the United States and around the world. The organization works with state and local governments, building professionals, and industry stakeholders to develop and implement building codes and standards.

The IBC is a model code developed by the ICC. Federal states and local governments adopt the IBC as part of their building codes. Once adopted, the IBC becomes enforceable by local building authorities, such as building departments or code enforcement agencies. The IBC is not a federal law in the United States, but rather a model code that is adopted and enforced at the state or local level.

In the United States, being a federal republic, building codes are generally enacted, adopted, and enforced at the state or local level. The IBC developed by the ICC is not a federal law in any US jurisdiction. Instead, federal states and local governments choose to adopt and enforce this IBC model code, often with modifications, as part of their own laws and regulations.

3.2.4.1.1 2024 Ohio Building Code, United States

2024 OBC is based on the IBC 2021 model codes, with amendments and additions.[4] Also, the OBC of 2017 used the 2015 IBC as its foundation [16].

The 2024 Ohio Building Code, based on the IBC 2021 [30], establishes minimum standards for the design, construction, and safety of buildings and structures within Ohio. It covers a wide range of topics, including construction types, fire and smoke protection, interior finishes, fire protection systems, means of egress, accessibility, interior environment, and energy efficiency. The code also addresses construction in flood hazard areas and requires site accessibility plans.

3.2.4.1.2 2019 Building codes, Uganda

The national building (building standards) code (S.I. No. 51), the building control (accessibility standards for persons with disabilities) code (S.I. No. 52), the national building (standards for electrical installations in buildings) code (S.I. No. 58), the national building (standards for mechanical installations in buildings) code (S.I. No. 60) [17],[5] and the national building (structural design) code were influenced by international best practices and standards, such as the IBC, given the country's efforts to modernize its infrastructure and regulations.

These statutory instruments (S.I.) were issued by the Minister of Uganda—a unitary republic. In exercise of the powers conferred on the Ugandan Minister responsible for building works by Section 46 of the Building Control Act, 2013, and in consultation with the National Building Review Board of Uganda, these codes were enacted. In this case, the Building Control Act, 2013, is the primary legislation, and the S.I. are secondary legislations.

3.2.5 How Does Technology Shape Codes and Standards and Vice Versa? The Engineer's Role

Many times, people debate the role of an engineer and technician in code and standards development. However, this should be undebatable because if a technician is at i, an engineer is always at $i+1$—note that this is used in a literal sense. Engineers typically take the lead in developing and overseeing the code development process due to their advanced technical knowledge and experience. For example, engineers with expertise in structural/civil engineering, mechanical engineering, and electrical engineering often play a significant role in developing building codes (see Section 3.2.4) due to their technical expertise. Technicians may also provide input and support. The process involves both technicians and engineers. However, code development committees often comprise a range of stakeholders, which may include engineers, architects, builders, contractors, code officials, and other experts.

The engineer plays with technology to define and develop codes and standards effectively. The key governing parameters are quality, safety, reliability, economy, efficiency, effectiveness, durability, strength, operability, and maintainability of the products and systems whose codes and standards are being defined.

The engineer's role in the in the development of codes and standards of practice is unparalleled! Engineers' involvement ensures that codes and standards are grounded in real-world experience, reflecting the complexities and challenges of actual projects. Throughout history, the civil engineer has played a significant role in the development of codes and standards of practice in the construction industry and associated industries.

As outlined earlier, a code sets out rules of what should be done and a standard spell out of how to do it. For example, the plumbing code of the United Kingdom (UK plumbing regulation, the Water Supply (Water Fittings) Regulations (1999), was enacted in 1999) is part of the Building Regulations, which are outlined in the Building Regulations (2010) and its amendments. On the other hand, standards set out how to achieve the outputs and outcomes, generally using available technologies, for example:

- BS 7593:2019 [18] is a standard for plumbing that includes requirements for system maintenance, such as water testing and inhibitor re-dosing. The standard ensures awareness of potential problems and of remedies that maintain efficiency and maximize the life of heating and cooling systems. It covers sealed cooling systems; treatment of water to overcome scale and corrosion; cleaning and flushing of systems; and advice on the use of in-line filters.
- BS EN 806-4:2010 (Specifications for installations inside buildings conveying water for human consumption) specifies requirements and gives recommendations for the installation of potable water installations within buildings [19]. It also covers pipework outside buildings but within the premises in accordance with BS EN 806-1:2000 specifications for installations inside buildings conveying water for human consumption. General. BS EN

806-4 applies to new installations, alterations, and repairs.

- BS EN 806-1 specifies potable water installation requirements and gives recommendations on the following: design, installation, alteration, testing, maintenance, and operation. BS EN 806-1 also covers the system of pipes, fittings, and connected appliances installed for supplying potable water.

Note also that scientific facts influence technological innovations and engineering principles to design, manufacture, and operate engineering systems. For example, nondestructive tests conducted in construction works utilize technologies built on physics principles. Some of typical examples are ultrasonic pulse velocity machine, the rebound hammer, and the nuclear density gauge (Figure 3.6). These tools are used to measure concrete strength/quality (the first and second machines) and the last one used to measure the soil density, asphalt density, and concrete density. These are technological tools! They work based on several physics principles.

Once designed and built, engineers can specify these technological devices in codes and standards, for example, the machines used in nondestructive testing to determine those specific material properties—to ensure that quality is achieved. And we can use them to further engineering and scientific research. Therefore, technology shapes codes and standards at the same time codes, and standards shape advancing technology by furthering research and development in the right way benchmarking on already existing breakthroughs. Standards and codes ensure that things are done the right and safest way, achieving the desired quality, durability, and reliability. This reciprocity is vital and must be grasped by an ever-evolving civil engineer. For the evolving civil engineer, things work as if it were a neural network!

FIGURE 3.6   Nuclear density machine. (Image credit: Shutterstock.)

The two technological tools, ultrasonic pulse velocity (UPV) machine and the rebound hammer, are specified in nondestructive tests, specifically BS 1881: Part 202: 1986 (for the rebound hammer) [20] and BS 1881: Part 203: 1986 (for the UPV machine) [21]. They provide recommendations on methods and devices which can be used to determine the concrete strength and quality. On the other hand, the nuclear density machine (Figure 3.6) is suitable for standards such as ASTM D6938, D2950, D7013, D7759, C1040, and AASHTO T310. Specifically, ASTM D6938 provides standard test methods for in-place density and water content of soil and soil-aggregate by nuclear methods (shallow depth) [22].

3.2.6 AI and ML Advancing Civil Engineering Standards and Codes

AI has emerged as a disruptive force in almost everything shattering traditional norms and redefining industry standards. AI and ML play pivotal roles across multiple domains of civil engineering, including structural health monitoring, predictive maintenance, earthquake engineering, and environmental sustainability.

AI refers to the simulation of human intelligence processes by machines, particularly computer systems. These processes include learning (the acquisition of information and rules for using the information), reasoning (using rules to reach approximate or definite conclusions), and self-correction [23]. AI encompasses a range of subfields, including robotics, natural language processing, vision, and expert systems. One fundamental goal of AI is to develop systems which can perform tasks that typically require human intelligence, such as speech recognition, decision-making, and visual perception. On the other hand, ML, a subset of AI, involves the use of algorithms and statistical models that enable computers to perform tasks without explicit instructions. Instead, ML systems concentrate on identifying patterns in data and learn from them [24].

AI and ML are revolutionizing civil engineering projects by enhancing accuracy, efficiency, and decision-making. These technologies allow engineers to automate repetitive tasks, optimize processes, and provide predictive insights to improve project outcomes. For example, in civil engineering, AI algorithms can predict potential risks, detect structural weaknesses early, and optimize resource allocation, significantly improving project management and reducing costs [25]. Similarly, ML is used in predictive maintenance for real-time monitoring and early failure detection, which helps optimize repair schedules—thus reducing downtime and improving safety. AI in construction comes in many forms which include the following:

a. **Computer vision:**
   This enables machines to recognize and interpret visual data, useful for monitoring construction sites. Computer vision is a field of AI that uses ML and neural networks to teach computers and systems to derive meaningful information from digital images, videos, and other visual inputs—and to make recommendations or take actions when they see defects or issues.

b. **Machine learning:**

This allows systems to learn from data and improve over time. ML, as it appears (learning), is a type of AI, focused on enabling computers and machines to imitate the way that humans learn, to perform tasks autonomously, and to improve their performance and accuracy through experience and exposure to more data. This enables machines to recognize and interpret any form of data as they get exposed to more data useful for survey research and conducting technical audits,

c. **Natural language processing:**

This helps machines understand human language, useful for project management tools. Natural language processing (NLP) is a subfield of computer science and AI that uses ML to enable computers to understand and communicate with human language—which can be very helpful on construction sites, automating tasks, and humans issuing instructions to machines and computers.

d. **Robotics:**

Robots can perform specific construction tasks autonomously, improving speed and safety. Construction robots are a subset of industrial robots used for building and infrastructure construction at site. The robot technology is transforming the industry by automating tasks like bricklaying, demolition, and concrete work; reducing labor costs; and enhancing quality control.

Some of the challenges AI systems present to civil engineering and construction industry and other engineering fields is that they must adhere to the same safety and regulatory standards as human-operated systems in civil engineering. However, the current standards in several jurisdictions do not fully account for the complexities introduced by AI, around the world. For example, safety codes and building regulations in many jurisdictions do not address the complexities and challenges of AI-driven structural assessments or autonomous machinery. Therefore, there is an urgent need for regulatory bodies in several jurisdictions to update these standards to ensure AI systems meet the necessary safety and performance criteria.

3.2.6.1 *AI redefining civil engineering codes and standards*

AI is redefining civil engineering standards and codes in several ways, which include design optimization, predictive maintenance, structural health monitoring, automated code compliance, and risk analysis. With the incorporation of AI in civil engineering, standards are becoming more efficient, sustainable, and resilient, leading to improved infrastructure and better communities:

- **Research and development (R&D):** AI accelerates the discovery of new materials and optimizes existing ones for improved performance through research and development, thus advancing materials science. In this process, AI supports continuous learning and can potentially update standards and guidelines based on practical performance data and emerging trends.

- **Design optimization:** AI algorithms optimize designs for structural integrity, sustainability, and cost-effectiveness. Several codes and standards such as design codes, structural codes, and mechanical and plumbing codes are being reshaped by AI to cater for this reality. With AI, engineers can harness vast amounts of data to optimize designs and predict performance, automate complex processes, streamlining construction and reducing costs, analyze and mitigate risks with uncanny precision, and create intelligent systems that adapt and learn. In the same scheme of things, AI optimizes building energy consumption and reduces environmental impact, ensuring that a highly energy efficient and sustainable building is developed.

- **Predictive maintenance:** AI is playing a crucial role in predictive maintenance. AI-powered sensors monitor infrastructure health, predicting potential failures and reducing maintenance costs. Through a proactive approach that uses data analysis and technology, AI-powered technologies can easily and quickly anticipate equipment failures using sensors before they occur, enabling timely maintenance and reducing the risk of unplanned downtime.

- **Code compliance automation:** AI automates code compliance by checking designs against building codes and regulations, thus streamlining the approval process. AI enables seamless data exchange between different systems and software that could enable the automation of code compliance. In that case, different systems, devices, applications, or organizations are enabled to work together seamlessly, exchanging information and using the information exchanged, thus enabling the automation of code compliance possible.

- **Effective site management:** AI enhances construction site management, optimizing workflows, and reducing errors. AI assists engineers in complex tasks, freeing them to focus on high-value work and enable smooth workforce augmentation. This includes deploying robots, computer vision, and NLPs at construction sites. In the same way, AI identifies potential risks and hazards, enabling proactive mitigation strategies especially at construction sites and during the design process, thus engaging in risk analysis and construction site planning and management.

3.3 SCIENTIFIC LAWS THAT HAVE ADVANCED CIVIL ENGINEERING TECHNOLOGY

Many scientific laws have advanced civil engineering technology. The most famous ones are Sir Isaac Newton's laws of motion, Robert Hooke's law of elasticity, and the law of conservation of energy. The conservation of energy cuts across all engineering disciplines and acts as a link from one discipline to another.

These laws cannot be underestimated in driving the civil engineering industry throughout history alongside other scientific laws and principles. They are heavily applied in research and development, pioneering lots of civil engineering technologies.

3.3.1 Sir Isaac Newton (1643–1727)

Sir Isaac Newton (Figure 3.7) discovered calculus together with German scientist, Gottfried Wilhelm Leibniz (1646–1716). Ever since then, calculus has been widely applied in civil engineering and manufacturing. Civil engineers now use calculus to advance civil engineering technology and civil engineering itself.

Calculus is about infinitesimal changes. Originally called infinitesimal calculus or 'the calculus of infinitesimals,' it has two major branches: differential calculus and integral calculus. Differential calculus cuts something into small pieces to find how it changes. Integral calculus joins (integrates) the small pieces together to find how much there is.

Invented by Sir Isaac Newton (Figure 3.7), the basic principle of integration is to add up all the tiny contributions that make up a larger quantity. Integrals work by summing up an infinite number of small elements to find the whole. In other words, we can say that integral calculus is defined as the study of the area beneath a curve. Integral calculus helps us find the area under curves and accumulate quantities. For a single variable, specifically, the **definite integral** is represented mathematically as follows:

FIGURE 3.7 Sir Isaac Newton. (Image credit: Sir Isaac Newton; Sir Godfrey Kneller's 1689 portrait of Isaac Newton, owned by the Earl of Portsmouth, captures the renowned scientist at the pinnacle of his intellectual prowess, just before he transitioned to his role at the Royal Mint in London [26].)

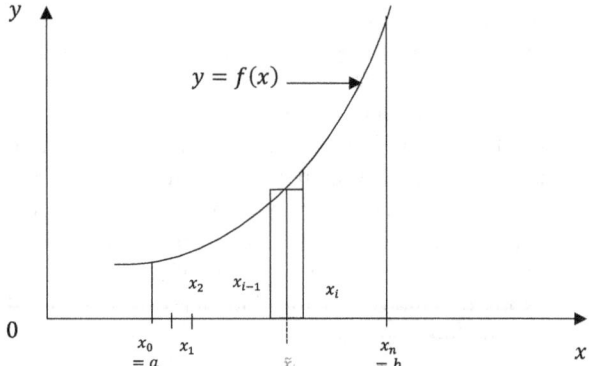

FIGURE 3.8 Definite integral as an area.

$$\int_a^b f(x)\,dx = \lim_{\substack{n \to \infty \\ all\ \Delta x_i \to 0}} \sum_{i=1}^n f(\tilde{x}_i)\Delta x_i \qquad (3.1)$$

This formula calculates the area between the curve $f(x)$ and the x-axis from point a to point b, where $a = x_0 < x_1 < x_2 < \cdots < x_{n-1} < x_n = b$, $\Delta x_i = x_i - x_{i-1}$, and $x_{i-1} \le \tilde{x}_i \le x_i$. Geometrically, we can interpret this integral as the area between the graph $y = f(x)$, the x-axis, and the lines $x = a$ and $x = b$, as illustrated in Figure 3.8. The double and triple integrations are founded on this basic principle.

3.3.1.1 Newton's laws of motion

Newton's groundbreaking work on forces, as outlined in his famous publication *Philosophiæ Naturalis Principia Mathematica* in 1687, established the cornerstone of classical mechanics. So, modern understanding of force as a vector quantity with magnitude and direction was developed by Sir Isaac Newton. He introduced the three laws of motion, which describe how forces affect the motion of objects. These are as follows:

- **The first law:** also called the 'law of inertia,' states that an object at rest remains at rest and an object in motion remains in motion, unless acted upon by an external force.
- **The second law:** force is equal to the rate of change of momentum ($F = ma$).
- **The third law:** for every action, there is an equal and opposite reaction.

These laws are the cornerstone of civil structure, primarily because civil structures deal with loads and how to have these loads attain equilibrium so that the structure is safe and stable! For example, the first law supports the understanding of the stability of a structure in a sense that if there is an unresolved force acting on the structure, the structure cannot be in equilibrium nor stable! See Figure 3.9. It is essential to note that the science of classical mechanics, sometimes called Newtonian mechanics, formed the scientific basis of much of modern engineering.

All forces must be in equilibrium to have a stable structure. Statically determinate structures can be analyzed by equilibrium equations. And that's basically the work of civil engineers. If a road pavement is failing, there must be a certain force resulting from factors perhaps not accounted for, causing

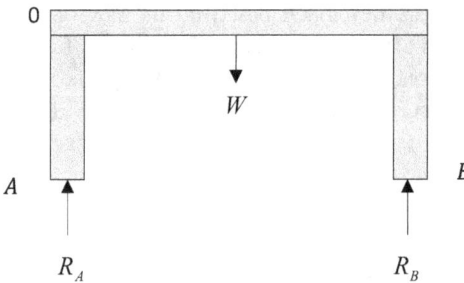

FIGURE 3.9 Statically determinate structures illustration.

the failure. Consider a simply supported beam whose weight, W, is non-negligible resting on two supports, A and B.

For the system to be in equilibrium,

$$\sum F = 0 \tag{3.2}$$

$$W = R_A + R_B \tag{3.3}$$

$$\sum M = 0 \tag{3.4}$$

The beam is designed in a way that it cannot restrain horizontal movement or a moment! It can only restrain vertical movement downward due to the two supports. Here, the beam will remain in equilibrium and stable until when some push or pull comes into contact with the beam either horizontally, clockwise or anticlockwise around a designated axis or pivot point, or from below the beam in-between supports A and B.

Imagine that the beam is bolted onto the supports, A and B, then it becomes fixed. Thus, it is able to resist rotational and horizontal movement.

Note:

A moment is a measure of the turning or rotational effect of a force around a pivot point or axis. It is a vector quantity, characterized by magnitude (amount of turning effect) and direction. Moments are calculated as the product of the force and the distance from the pivot point to the point where the force is applied (level arm). This advance in Newton's discoveries and physics generally led to widespread applications in pulley system design and development, cranes, conveyor belts, and bridge design and development, and much more.

In terms of an object in motion, a flowing river remains in its course unless an obstacle such as a dam (a control) is built in its course interrupting the flow. The civil engineer has really benefited from Newton's discoveries to the extent of learning how to train rivers, commonly labeled as 'river training' in the technical sense. The second law is widely applied in water resource engineering, fluid mechanics, hydraulics, and many other fields throughout civil engineering. Force is equal to the rate of change of momentum, mathematically expressed as follows:

$$F = ma - m\frac{\Delta V}{\Delta t} \tag{3.5}$$

Specifically, in hydropower generation, Newton's second law is applied. This is because the velocity of flow directly impacts the power generated by a turbine in a power dam as higher flow velocity translates to more kinetic energy in the water, which in turn drives the turbine blades with greater force, resulting in increased electricity production; essentially, the faster the water flows, the more power the turbine can generate. Water is diverted from a stream into a pipeline, where it is directed downhill and then through the turbine (flow). The vertical drop (head) in the pipeline creates pressure at the bottom end. The pressurized water discharging from the end of the pipe creates the force that drives the turbine. The turbine in turn drives the generator where electrical power is produced. Therefore, more flow or more head produces more electricity [27].

The **third law**, 'For every action, there is an equal and opposite reaction,' is applied throughout civil engineering to design stable foundations and structural members by understanding that a structural member to perform satisfactorily must be able to resist the deformation due to the imposed load. If for example, a 50 kN force is axially applied to a structural member, the member should be in position to resist 50 kN for it to reliably perform its intended function. This law is applied in the development of a rebound hammer (Schimdt hammer) alongside other laws. See Section 3.3.4.1. This law tells us that if the structural member is deformed, it means it fails to counteract a force, causing the deformation. That means the reactive force is lower than the active force. The design of support systems applies this law to the fullest!

Active forces, also known as applied forces, are exerted by an external agent or object on another object. Reactive forces, also known as reaction forces, are forces that arise in response to an active force. They are the forces exerted by an object on the external agent or object that is applying the active force.

- Example: The force exerted by the ground on the building foundation is a reactive force, and the building's weight is an active force.

It is essential to note that the whole field of civil engineering can be summarized in those **three** laws! Everything works tending toward those three universal laws.

3.3.2 Robert Hooke (1635–1702)

Robert Hooke's law states that the force applied to an object is directly proportional to the object's displacement from its equilibrium position. It's a principle of physics that describes the elastic behavior of materials. Hooke's law, law of elasticity, was discovered by the English scientist Robert Hooke in 1660.

It states that for relatively small deformations of an object, the displacement or size of the deformation is directly proportional to the deforming force or load.

In other words, when a force is applied to an object, that either stretches or compresses it, the resulting deformation will be directly proportional to the magnitude of the force. This linear relationship between force and deformation holds true as long as the material remains within its elastic limit. This means it can return to its original shape once the force is removed. Mathematically stated as follows:

$$F \propto x$$

$$F = kx \tag{3.6}$$

where k is the proportionality constant known as the spring constant [28].

This simple equation (3.6) provides a quantitative understanding of how different materials respond to mechanical forces and serves as a foundation for analyzing the behavior of springs, solid objects, and other elastic systems. It is an underpinning principle in materials science.

Hooke's law is widely applicable and has become a cornerstone of engineering and mechanics, influencing various fields such as material science, structural design, and even biomedical engineering.

The advent of Young's modulus, E, was inspired by Robert Hooke's law. Thomas Young (1773–1829) is credited with identifying Young's modulus as an important material property, but the concept was developed earlier by Leonhard Euler (1707–1783) and Giordano Riccati (1709–1790)—both inspired by Robert Hooke.

Note that Hooke's law applies within the elastic limit of an object. This means that the object returns to its original shape and size when the force is removed. See Chapter 5 for yield strength and ultimate strength. This law has provided a foundation for many civil engineering principles, particularly, applicable in the yield strength determination for civil engineering materials such as metals. Therefore, in its advanced sense, the mathematical formula for Hooke's law is as follows:

$$\sigma = E\varepsilon \qquad (3.7)$$

where σ is the stress, E is the modulus of elasticity or Young's modulus, and ε is the strain.

With its elegant simplicity, Hooke's law offers a basic framework for understanding how objects respond to external forces, paving the way for advancements in countless technological applications.[6]

It has enabled the development of more advanced materials and structures, such as high-strength materials. Through understanding the relationship between stress and strain, engineers are able to develop materials with improved strength and durability. Hooke's law has enabled the design of more complex structural systems, such as suspension bridges and high-rise buildings. It has also enabled cutting-edge research in materials science.

Hooke's law is extensively applied in civil engineering. For example, it is specifically applied in structural engineering to calculate the deflections and stresses in beams, columns, bridge structures, and other structural elements under different loads, ensuring their safety and stability.

Hooke's law is widely applicable in testing materials to determine the mechanical properties of materials, such as Young's modulus, shear modulus, and Poisson's ratio. Hooke's law helps engineers design structures that can withstand seismic forces and deformations.

Overall, Robert Hooke's law has played a crucial role in advancing civil engineering technology, enabling engineers to design and analyze structures with greater precision and accuracy.

3.3.3 The Law of Conservation of Energy

The law of conservation of energy, one of the fundamental laws of physical sciences and engineering, changed the whole game about how engineers today approach 'work and energy.' Several techniques were developed based on the law of conservation of energy and associated concepts. This law is fundamental in relating the concepts of force, work, energy, and power across several engineering sub-disciplines. Modern civilization is possible because people have learned how to change energy from one form to another and then use it to do work.

The law of conservation of energy states that the total energy of an isolated system remains constant. In simpler terms, energy cannot be created or destroyed; it can only be changed from one form to another. Energy can be converted from one form to another but cannot be created or destroyed.

In a closed system, that is, a system that is isolated from its surroundings, the total energy of the system is conserved. This principle applies to closed systems where energy can only enter or leave the system. See examples below:

- When a stick of dynamite explodes, the chemical energy is transformed into kinetic energy, heat, and sound. In case you add up all the energy forms released during the explosion (kinetic energy, potential energy, heat, and sound), it matches the decrease in chemical energy from the dynamite combustion.
- The second example is the food you eat, which contains chemical energy. The body stores this energy until you use it as kinetic energy during work or play. See Section 6.3 for the gym and civil engineer.
- The stored chemical energy in coal or natural gas and the kinetic energy of water flowing in rivers can be converted to electrical energy, which can be converted to light and heat.

Several principles, methods, and techniques used in several engineering fields are derived from the principle of energy conservation in a system. Through equation formulation, manipulation of variables, and mathematical advancements, modern techniques were developed, which serve as alternative techniques to already existing methods, hence advancing several civil engineering technologies because of understanding the functioning of different complex civil engineering systems much better!

Throughout all the fields of civil engineering, energy techniques are deployed, and they offer numerous advantages over other techniques. Energy techniques are versatile and widely applicable across various branches of civil engineering.

It is essential to know that energy is a fundamental physical quantity that is conserved in all physical processes—energy is conserved, meaning it cannot be created or destroyed, only converted from one form to another. Therefore, this universality makes energy techniques applicable to various civil engineering disciplines. This is because many physical laws, such as Newton's laws, widely adopted in civil engineering, and which are the primary laws that advanced civil engineering technology, can be formulated in terms of energy. This allows energy techniques to be applied to a broad range of problems in civil engineering.

In that regard, energy techniques often enable the simplification of complex problems by focusing on energy balances rather than detailed geometric or kinematic analyses, especially in structures. Energy techniques abstract away from specific details, such as material properties or geometric complexities, allowing for more general and widely applicable solutions.

The advantage of energy techniques is that they can be applied to nonlinear and dynamic systems. They are particularly useful for analyzing nonlinear systems, where traditional linear analysis methods may not be applicable.

Energy techniques can be used to analyze dynamic systems, including those with time-dependent loading, vibrations, or other dynamic effects.

Because of their interdisciplinary connections to other disciplines, relying on the principles from thermodynamics, mechanics, and physics, energy techniques are widely applicable to various civil engineering disciplines. This versatility triggers several civil engineering innovations and technological advancements. We acknowledge everyone who contributed to their development in several spheres of engineering and scientific research!

The versatility of energy techniques stems from their ability to simplify complex problems, abstract away from specific details, and provide a universal framework for analyzing a wide range of civil engineering systems. There is no doubt that energy techniques have advanced civil engineering technology throughout history.

3.3.3.1 Work–energy theorem

The work–energy theorem (WET) states that the net work done by the forces on an object equals the change in its kinetic energy (KE). Starting with the work–energy theorem and Newton's second law of motion, we can say that the net work done on the object is mathematically expressed as follows:

$$W = \Delta K \tag{3.8}$$

where

W: net work done on the object
ΔK: change in kinetic energy

Note: Work is done when a force is applied to an object, causing it to move. The force applied can be a push or a pull. The WET has had an enormous impact on the development of various energy techniques throughout civil engineering and other fields. The WET can also express work done as follows:

$$W = \int F(r)\,dr \tag{3.9}$$

Consider Figure 3.10 for the derivation of the kinetic energy formula.

In Figure 3.10, m is the mass of the object, r is the distance moved by the object, and V is the velocity. Relate this with Figure 3.5. Now, using Newton's calculus,

$$W = \int F(r) \times dr \tag{3.10}$$

$$\Delta KE = \int ma(r) \times dr$$

$$\Delta KE = m \int \frac{dV}{dt} \times dr$$

$$\Delta KE = m \int \frac{dr}{dt} \cdot dV$$

$$\Delta KE = m \int_{V_0}^{V} V \cdot dV$$

$$\Delta KE = \frac{1}{2}mV^2 - \frac{1}{2}mV_0^2 \tag{3.11}$$

Since the object is initially in the condition of rest, the initial kinetic energy is zero. Thus, we get the equation (3.12) for kinetic energy as follows:

$$KE = \frac{1}{2}mV^2 \tag{3.12}$$

The WET is very helpful in analyzing situations where a rigid body should move under several forces. A rigid body cannot store the potential energy in its lattice due to its rigid structure, and it can only possess kinetic energy. The WET is a fundamental principle in civil engineering, helping analyze and predict the behavior of objects under various forces and energy transfers—very helpful in structural engineering as several concepts are developed or extend from the WET.

The WET is primarily used to analyze situations where a force acting on an object causes a change in its kinetic energy, allowing you to calculate the work done by that force without needing detailed information about the motion's acceleration or time involved; this is particularly useful in scenarios with variable forces or complex motion paths, where applying Newton's second law directly might be difficult.

The WET relates the energy transferred to an object through work to the resulting change in the object's kinetic energy. It applies to various types of forces, including conservative and non-conservative forces. The **WET** has several implications:

- Energy transfer: Work done on an object transfers energy to it, increasing its kinetic energy.
- Kinetic energy change: The change in kinetic energy is directly proportional to the work done.
- Force and displacement: Work is done when a force is applied over a displacement, resulting in a change in kinetic energy.

3.3.3.2 Power

Power is the rate at which work is done or energy is transferred or converted. See Figure 3.5. The concepts of efficiency relate to power. How quickly a machine or system does work or converts energy? In other words, what is the rate at which work is done or energy is transferred with respect to time? Whenever 'rate' is

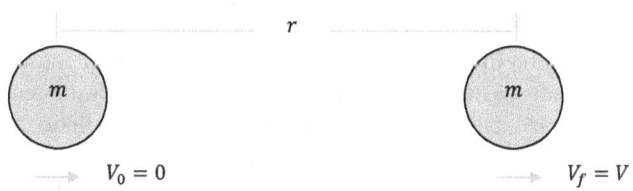

FIGURE 3.10 Derivation of kinetic energy.

mentioned in science or engineering, it is referred to Newton's differential calculus. Power is, thus, the measure of how quickly energy is transferred or converted or how quickly work is done. Mathematically, this is represented as the derivative of work (W) with respect to time (t), which is expressed as follows:

$$P = \frac{dW}{dt} \tag{3.13}$$

Note:

- 1 Watt = 1 Joule/second (metric system).
- 1 HP = 745.7 Watts (English system).

In simpler terms, power is the work done over a certain period of time, often measured in watts (W) (metric units). This shows how Newton and colleagues advanced our understanding of the rate at which work is done, by inventing calculus, and much more.

3.3.3.3 Energy techniques/methods in civil engineering

Since energy techniques are quite powerful, they are relied on analyzing stresses in structures—borrowing on the principle that energy can be converted from one form to another but cannot be created or destroyed, civil engineers formulate equations. The principle of conservation of energy states that the total energy of an isolated system remains constant over time. In some instances, the assumption is made that some energy is converted to other forms, and in case of an isolated system, it is negligible. This is a key concept when analyzing isolated systems. The most important thing in energy techniques is 'no energy is destroyed or created.'

Energy techniques, also known as energy principles or energy methods, are mathematical approaches used to analyze and solve problems in various fields, including physics, engineering, and mechanics. Energy techniques include the WET mentioned earlier, which states that the work done on an object is equal to the change in its kinetic energy. Thus, the kinetic energy equation is given in (3.12), where m is mass and V is velocity. So, the work–energy equation is simply given as follows:

$$W = \Delta KE \tag{3.14}$$

where W is the work done and ΔKE is the change in kinetic energy—this is a crucial technique in structural analysis.

Potential energy (PE) is the energy that an object possesses due to its position, state, or configuration. It is mathematically expressed as follows:

$$\text{PE} = mgh \tag{3.15}$$

where m is the mass, g is the acceleration due to gravity, and h is the height.

Therefore, neglecting other energies in a system and considering the energy conservation principle, the total energy (E) in a system, which is the sum of kinetic and potential energies, is given as follows:

$$E = \text{KE} + \text{PE} \tag{3.16}$$

And in that case, from the energy conservation equation, $\Delta E = 0$ where ΔE is the change in total energy. The virtual work is also a technique used to find the equilibrium position of a system by considering the virtual work done by external forces. Other energy equations include the strain energy and thermal energy.

Thermal energy

$$Q = mc\Delta T \tag{3.17}$$

where m is the mass, c is the specific heat capacity, and ΔT is the temperature change. These energy techniques and equations are fundamental tools for analyzing and solving problems in various civil engineering fields. They are expanded to cater for various configurations and systems. In Section 5.3.1.3, we see that, if an axial member of length L, cross-sectional A, and modulus of elasticity, E, is transmitting a force F undergoes a temperature change ΔT, the strain energy U is given as follows:

$$U = \frac{1}{2}\frac{F^2 L}{AE} + F\alpha L\Delta T \tag{3.18}$$

where α is the coefficient of linear expansion of the material.

3.3.3.3.1 Energy techniques in structural engineering

Structural systems comprise statically determinate and indeterminate systems. A large number of structural engineering projects usually combine a large system of simple elements in a complex and often highly statically indeterminate structure. The deflection theory is usually adopted to determine the complete support reactions.

After the reactions are determined, a comprehensive analysis of the structure is performed. A deformation analysis may also be necessary to perform even when the structure is statically determinate because deflections at various points of the structure may pertain to the overall design requirements.

Note that the deflection analysis which uses geometric approaches and superposition techniques are advantageous for relatively simple systems, but with complex systems, these approaches become difficult and cumbersome.

Although these geometric approaches and superposition techniques have far-reaching success in solving structural problems, they can be aggravating at some point. Energy techniques are used as alternatives. The advantage energy techniques present is that they do not utilize the overall geometry, and in practice, they can actually simplify the analysis.

Energy techniques, also known as energy methods or energy principles, are powerful tools used in structural engineering including stress analysis. These techniques focus on the energy associated with the deformation of a structure, rather than analyzing the entire geometry of the deformed structure.

By applying energy principles, engineers and researchers can: simplify complex problems, avoid detailed geometric analysis, focus on the energy dissipated or stored in the system, and obtain approximate solutions or bounds for the solution.

Have you noticed the advanced seismic base isolation technology in buildings? These steel joints function as **dampers and energy dissipators**: during seismic activity, they flex,

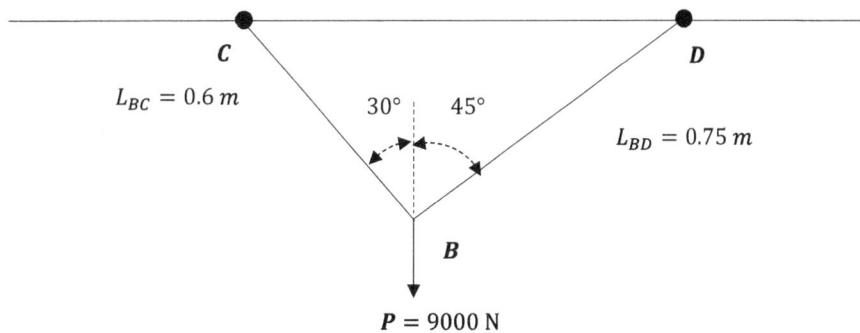

FIGURE 3.11   Cables BC and BD are angled 30° and 45° from vertical.

twist, and absorb the ground motion, effectively decoupling the superstructure from the tremors and preserving the building's structural integrity.

Example 3.1 for illustrative purposes

Cables BC $(0.6$ m$)$ and BD $(0.75$ m$)$, shown in Figure 3.11, form angles of 30° and 45°, respectively, to the vertical. Determine the vertical and horizontal displacement of point B if a vertical force P of 9,000 N is applied at B. The cables can be considered of solid steel $\left(E = 0.2068 \text{ N/mm}^2\right)$ each with a cross-sectional area of 130 mm^2.

Solution

The first step is to draw free-body diagrams. See Section 8.5.1. The second step is to determine the forces in each cable. From Figure 3.12, equilibrium of forces in the x and y directions result in the following:

$$F_{\text{BC}} = 6,586.61 \text{ N}$$

$$F_{\text{BD}} = 4,658.14 \text{ N}$$

Vertical displacement

See Figure 3.13. To determine the vertical drop of point B, we equate the total work done by the force P to the sum of the work P performs on each cable. Assuming a linear force displacement process, the work done by gradually applying the force P to 9,000 N is as follows:

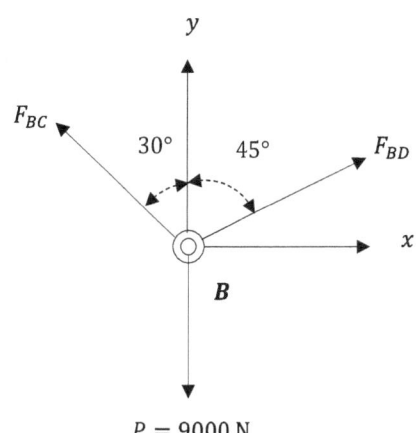

FIGURE 3.12   Free-body diagram for Figure 3.11.

$$W = \frac{1}{2} P \left(\delta_B \right)_V = 4500 \left(\delta_B \right)_V \quad (3.19)$$

The work applied to each cable is as follows:

$$W_{\text{BC}} = \frac{1}{2} F_{\text{BC}} \delta_{\text{BC}} = \frac{1}{2} \frac{F_{\text{BC}}^2 L_{\text{BC}}}{\text{AE}}$$

$$= \frac{1}{2} \frac{\left(6,586.61 \text{ N}\right)^2 \left(600 \text{ mm}\right)}{\left(130 \text{ mm}^2\right)\left(0.2068 \times 10^6 \text{ N/m}^2\right)}$$

$$= 484.12 \text{ N.mm} = 0.48412 \text{ J}$$

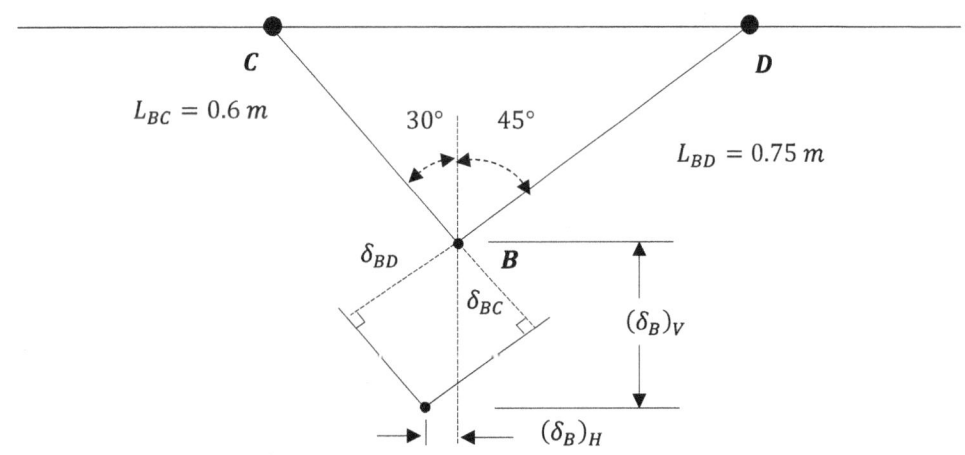

FIGURE 3.13   Determination of the vertical drop of point B.

$$W_{BD} = \frac{1}{2} F_{BD} \delta_{BD} = \frac{1}{2} \frac{F_{BD}^2 L_{BD}}{AE}$$

$$= \frac{1}{2} \frac{(4{,}658.14 \text{N})^2 (750 \text{ mm})}{(130 \text{ mm}^2)(0.2068 \times 10^6 \text{N/m}^2)}$$

$$= 302.67 \text{ N.mm} = 0.30267 \text{ J}$$

The total work is the sum of the work applied to each cable, and so

$$W = W_{BC} + W_{BD} \qquad (3.20)$$

$$4{,}500(\delta_B)_V = 0.48412 + 0.30267$$

$$(\delta_B)_V = 1.75 \times 10^{-4} \text{ m} = 0.175 \text{ mm } (\approx 1.69 \times 10^{-3} \text{ in})$$

Note: One joule is equal to the amount of work done when a force of 1 Newton displaces a body through a distance of 1 meter in the direction of that force. Note carefully that units are always added when performing structural calculations, primarily. This is to avoid discrepancies and to facilitate unit conversions. In this case, you will notice that the product:

$$\frac{1}{2} \cdot \frac{(6{,}586.61 \text{ N})^2 (600 \text{ mm})}{(130 \text{ mm}^2)(206.8 \times 10^6 \text{N/mm}^2)}$$

gives

$$0.48412 \text{ J}$$

See Section 7.6.7 for conversion factors and/or coefficients.

Horizontal displacement

To determine the horizontal deflection of point B, an artifice must be used to avoid the fact that no work is done by P moving horizontally.

Solution

Since there is no horizontal force at point B, there is no work due to the horizontal deflection $(\delta_B)_H$. However, it is possible first to place an exceedingly small horizontal force H at point B where the force is so small that it causes negligible deflections (see Figure 3.14). Then, when the vertical force P is placed on the structure, the constant horizontal force H will do work. The work due to H is

$$W_B = H(\delta_B)_H \qquad (3.21)$$

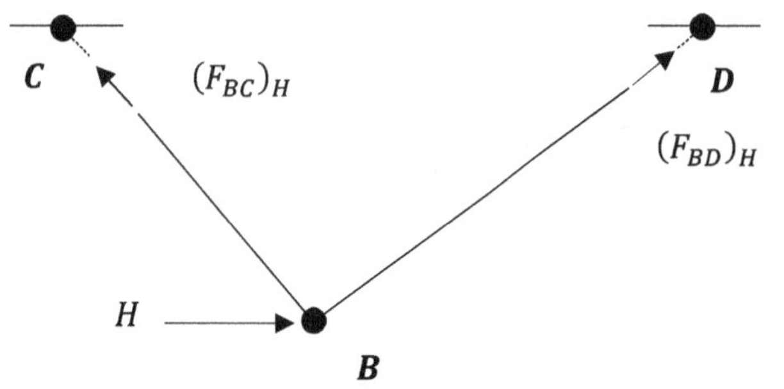

FIGURE 3.14 Determination of the horizontal deflection of point B.

Note: The $\frac{1}{2}$ is omitted since the force H remains constant throughout the deflection $(\delta_B)_H$.

The forces in the cables BC and BD due to H alone are found by static analysis and are as follows:

$$(F_{BC})_H = 0.732H \qquad (3.22)$$

$$(F_{BD})_H = -0.897H \qquad (3.23)$$

Do not be concerned here with a compressive force in a cable. Recall that H is actually an artificial force and is basically infinitesimal in value. The additional work obtained from these forces on the cables arises from the deflections caused by the actual applied force. The additional work terms are as follows:

$$(W_{BC})_H = (F_{BC})_H \delta_{BC} = (0.732H)\delta_{BC} \qquad (3.24)$$

$$\delta_{BC} = \frac{F_{BC} L_{BC}}{AE}$$

$$= \frac{6{,}586.61 \text{ N} \times 600 \text{ mm}}{(130 \text{ mm}^2)(0.2068 \times 10^6 \text{N/m}^2)} = 0.147 \text{ mm} \qquad (3.25)$$

Therefore,

$$(W_{BC})_H = (F_{BC})_H \delta_{BC}$$

$$= (0.732H)(0.147 \text{ mm}) = 0.108H \text{ mm} \qquad (3.26)$$

and

$$(W_{BD})_H = (F_{BD})_H \delta_{BD} = (-0.897H)\delta_{BD} \qquad (3.27)$$

$$\delta_{BD} = \frac{F_{BD} L_{BD}}{AE}$$

$$= \frac{4{,}658.14 \text{ NN} \times 750 \text{ mm}}{(130 \text{ mm}^2)(0.2068 \times 10^6 \text{N/m}^2)} = 0.130 \text{ mm} \qquad (3.28)$$

Therefore,

$$(W_{BD})_H = (F_{BD})_H \delta_{BD}$$

$$= (-0.897H)(0.130 \text{ mm}) = -0.117H \text{ mm} \qquad (3.29)$$

Since the forces $(F_{BC})_H$ and $(F_{BD})_H$ are present before deflections, the work expression will not contain the $\frac{1}{2}$ term.

The work due to H equals the work H performs on the elements of the system:

$$W_H = (W_{BC})_H + (W_{BD})_H \qquad (3.30)$$

$$H(\delta_B)_H = 0.108H \text{ mm} - 0.117H \text{ mm} \qquad (3.31)$$

$$(\delta_B)_H = -0.009 \text{ mm} \qquad (3.32)$$

Note:

- The negative sign indicates that the deflection is in the direction opposite H.
- The procedures followed above avoid analyzing the overall geometry of deformation structure.

Overall geometry of deformation of the structure

Now, let's analyze the overall geometry of the deformation structure. We already determined that

$$F_{BC} = 6{,}586.61 \ \text{N}$$

$$F_{BD} = 4{,}658.14 \ \text{N}$$

Now let's imagine the pin at B is removed and cables BC and BD are allowed to stretch under internal forces F_{BC} and F_{BD}, respectively. Then, by rotating the cables, they can be 'rejoined.' This must be where point B actually displaces to. In case the rotations are very small, the rotation of BC and BD can be approximated by a perpendicular movement, as shown in Figure 3.13. The elongations of each cable are as follows:

$$\delta_{BC} = 0.147 \ \text{mm}$$

and

$$\delta_{BD} = 0.130 \ \text{mm}$$

The deflection of point B can be obtained by developing the relationships between δ_{BC}, δ_{BD}, and $(\delta_B)_V, (\delta_B)_H$ as follows:

$$(\delta_B)_V = \frac{\delta_{BC}}{\cos 30} + (\delta_B)_H \tan 30 \qquad (3.33)$$

and

$$(\delta_B)_V = \frac{\delta_{BD}}{\cos 45} - (\delta_B)_H \tan 45 \qquad (3.34)$$

Substituting δ_{BC} and δ_{BD} yields

$$(\delta_B)_H = 0.0124 \ \text{mm}$$

Substituting this into either of equations (3.32) or (3.33) gives

$$(\delta_B)_V = 0.1769 \ \text{mm}$$

Energy techniques are quite powerful, and they are greatly applied to the field of stress analysis. These techniques can be complex and not so straightforward. It is worth noting that geometric and superposition approaches are generally more practical on simple single-element systems. The major energy techniques include Castigliano's first theorem, the complementary energy theorem, and special applications of the complementary energy theorem that is, Castigliano's second theorem applied to linear systems and the virtual load method. Other energy techniques include Rayleigh's technique and the Rayleigh ritz technique. Find these in advanced structural engineering books. Their applications slightly vary with statically determinate and indeterminate structural systems. See further reading 2.

3.3.3.3.2 Energy techniques in water resources engineering

In water resource engineering, energy techniques play a crucial role, particularly in processes like desalination, water distribution, and renewable energy developments. Dams and other water infrastructure can be designed to generate hydroelectric power, providing a renewable energy source while also serving water resource management purposes.

Water distribution systems can be designed to recover energy from water flow, for example, by using micro-hydro systems or 'pump-as-turbine' technologies. In the same scheme of things, optimizing water distribution networks, including pipe sizes and pump configurations, can improve energy efficiency and reduce energy consumption.

How about drainage and flood control/defense systems? Energy equations are at play—very helpful in hydraulics system design and optimization. The famous Bernoulli's equation, formulated by the Swiss mathematician and physicist Daniel Bernoulli 1738, is versatile and can be applied in drainage design and water distribution systems. The development of several energy techniques in water resource engineering borrows ideas from Bernoulli's theoretical equation and associated principles.

Bernoulli's principle states, 'The pressure of a fluid (liquid or gas) decreases as its velocity increases.' As the speed of a fluid increases, the pressure within the fluid decreases, and vice versa. This fundamental principle is a cornerstone of fluid dynamics and is widely applied in various fields, including aerospace engineering, chemical engineering, and hydraulics. In terms of energy, Bernoulli's principle can be defined as follows: 'The sum of the kinetic energy (related to velocity) and potential energy (related to pressure) of a fluid remains constant.' Mathematically, this can be is expressed as follows:

$$P + \frac{1}{2}\rho v^2 + \rho g h = \text{constant} \qquad (3.35)$$

where

P = pressure
ρ = fluid density
v = fluid velocity
g = acceleration due to gravity
h = height above a reference level

Note: Bernoulli's equation is only applicable for incompressible fluids under inviscid flow. This energy-based definition highlights the conversion between kinetic energy (velocity) and potential energy (pressure) as the fluid flows.

Open-channel flow problem

Imagine water flowing in an open channel (Figure 3.15). Total head H:

$$H = Z + E \qquad (3.36)$$

where

$$E = y + \alpha \frac{V^2}{2g} \qquad (3.37)$$

Equation 3.37 is a simplification of Bernoulli's equation: specific energy equation. **Note:** The coefficient on the velocity (see Section 7.6) head terms arises from the integration of **Euler's**

Open-channel flow problem

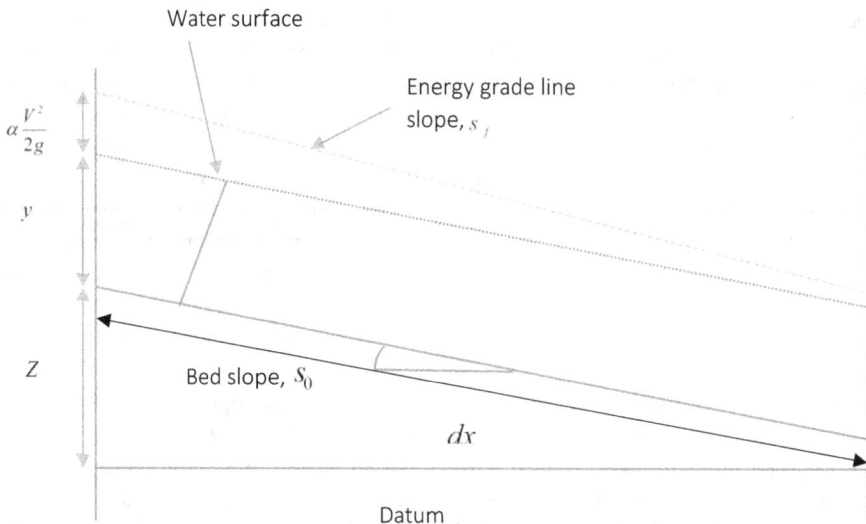

FIGURE 3.15   Derivation of energy in open-channel flow.

differential equation over the cross-sectional area of flow and reflect the non-uniform velocity distribution, which exists in the flow of real fluids.

Differentiating (3.36) with respect to x

$$\frac{dH}{dx} = \frac{dZ}{dx} + \frac{dE}{dx}$$ (3.38)

The slope of the energy grade line and channel bed are downward in the direction of x, as shown in Figure 3.15:

$$\frac{dH}{dx} = \text{Friction slope}, S_f$$

$$\frac{dZ}{dx} = \text{Bed slope}, S_0$$

$$\Rightarrow -S_f = -S_0 + \frac{dE}{dx}$$

$$\Rightarrow \frac{dE}{dx} = S_0 - S_f$$ (3.39)

Equation 3.39 is the equation for non-uniform flow, which is used to locate the longitudinal surface water flow profile. The sign of $\left(S_0 - S_f\right)$ can vary.

Also,

$$\frac{dE}{dy} \cdot \frac{dy}{dx} = S_0 - S_f$$

$$\frac{dy}{dx} = \frac{S_0 - S_f}{\left(dE / dy\right)}$$ (3.40)

These are termed 'energy equations,' which work hand in hand with 'momentum equations' to solve complex hydraulics problems. See further reading [1] for more information.

3.3.3.3.3 Energy techniques in highway engineering

The concept of **strain energy** is applied in pavement design and analysis. Strain energy (Section 5.3.1.3) is a fundamental concept in mechanics of materials and is used to analyze the behavior of pavements under various loads, especially in fatigue analysis. Of course, with some little modifications, the same equations applied in structural engineering apply to the analysis of the rigid and flexible pavements to determine deformations to axle loading.

Flexible pavements

Energy equations are used to calculate the stress and strain responses of flexible pavements (Figure 3.16) under traffic loading, helping engineers and designers optimize pavement thickness and material selection. In a flexible pavement, the surface layer is primarily designed accounting for safety, vehicle speeds, and avoiding skidding and sliding, among other factors. However, the road base and sub-base are structural layers designed to carry loads, and sub-grade requires the proper analysis of strain and stresses induced in those layers. See Figure 3.16, a section of road pavement structure showing the layers. The development of several road pavement technologies, ranging from geotextiles/geocells used in soil stabilization (especially geocells in base layer), soil reinforcement materials, and many others, has things to do with energy dissipation as vehicles pass over the constructed road systems.

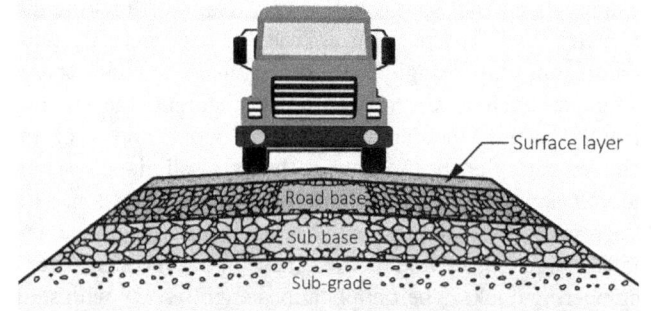

FIGURE 3.16   Flexible pavement layers.

FIGURE 3.17   Neoloy® geocell.

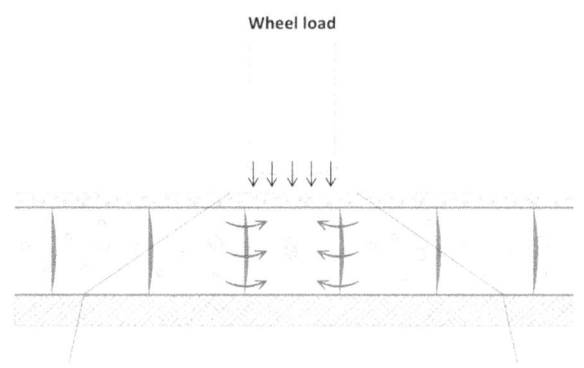

FIGURE 3.18   Geocell-reinforced soil.

Figure 3.17 showcases a geocell, neoloy® type, manufactured by PRS Geo-technologies.[7] The geocell material such as the neoloy® resists the lateral pressure exerted by the infill materials resulting in tensile forces within the geocell [29]. When the vehicle passes over the geocell-reinforced base layer (Figure 3.18), the geocell distributes the load laterally, reducing the stress concentrations (see Section 6.3.3.7).

As a design engineer, specifying the depth of the geocell-reinforced base layer would take into consideration the height of the geocell. Geocells are available in a range of heights, typically including 50, 75, 100, 120, 150, and 200 mm. The specific height chosen depends on the application and the required level of soil stabilization and confinement. These calculations, of course, with tensile forces building up in the base layer, energy equations cannot be escaped.

Rigid pavement

Energy principles are applied to analyze the behavior of rigid pavements, such as concrete slabs, under traffic loading and environmental conditions.

3.3.3.3.4 Energy techniques in geotechnical Engineering
Energy equations are used to analyze the interaction between soil and highway structures (soil–structure interaction), such as foundations and retaining walls. Energy principles are applied to analyze the stability of slopes and embankments

along highways, ensuring they can withstand natural and external forces.

What about earthquake engineering? Earthquake-resistant structures rely on energy dissipators to resist the impact of earthquakes on structures. Energy dissipators are used to absorb and dissipate the energy generated by earthquakes, reducing the stress on the structure. Energy dissipators include base isolators, damping systems, and shock absorbers. These help absorb seismic energy, reduce structural vibrations, and minimize damage to the building.

3.3.4 Typical Civil Engineering Technologies at the Peak of Scientific Laws

This section outlines how the civil engineering industry and various technologies have successfully enabled to integrate Newton's laws, energy conservation principles, and Hooke's law to advance civil engineering technologies and other technologies, generally. Sections 3.3.4.1–3.3.4.3, particularly, describe technologies developed through a proper understanding and application of scientific principles to improve the quality of concrete—the chief civil engineering material.

3.3.4.1 The rebound hammer (Schmidt hammer)

Rebound hammer test conducted using the rebound hammer is a nondestructive testing method of concrete, which provides a convenient and rapid indication of the compressive strength of the concrete. BS 1881: Part 202: 1986 gives recommendations on the use of rebound hammers for testing the hardness of concrete. It describes the areas of application of rebound hammers, their accuracy, the calibration procedure, the procedure for obtaining a correlation between hardness and strength, the conditions of the concrete and its testing which influence results, the method of testing concrete *in situ*, and the reporting of results.

The rebound hammer, also called a Schmidt hammer, consists of a spring-controlled mass that slides on a plunger within a tubular housing. The operation of rebound hammer is shown in Figure 3.19. When the plunger of rebound hammer is pressed against the surface of concrete, a spring-controlled mass with constant energy is made to hit the concrete surface to rebound back. The extent of rebound, which is a measure of surface hardness, is measured on a graduated scale. This measured value is designated as rebound number (rebound index). A concrete with low strength and low stiffness will absorb more energy to yield in a lower rebound value.

Key points to note

- Spring-controlled mass (Hooke's law in action).
- Constant energy and energy absorption (energy conservation in action).
- Active and reactive forces (Newton's third law in action).

Twelve readings are usually sufficient to obtain a reliable estimate of the surface hardness at one location on a structure,

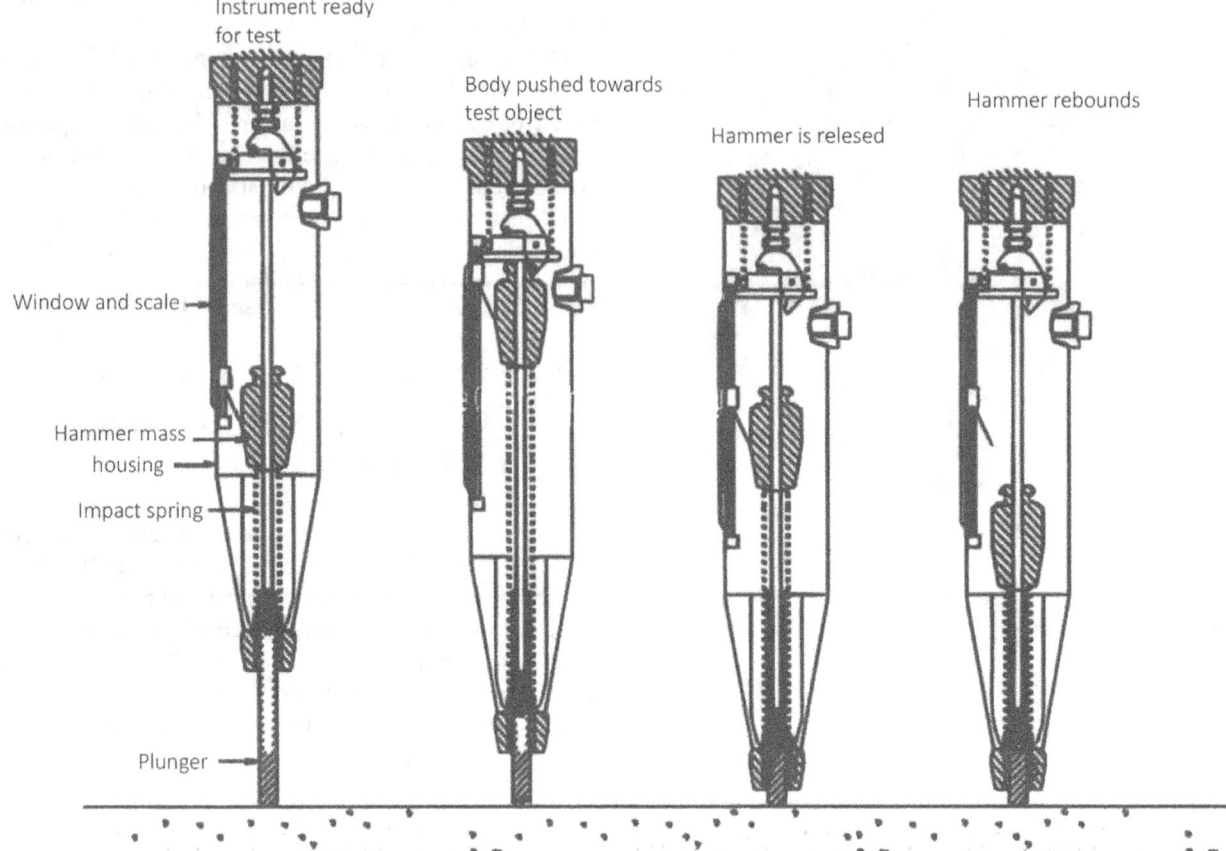

FIGURE 3.19   Rebound hammer.

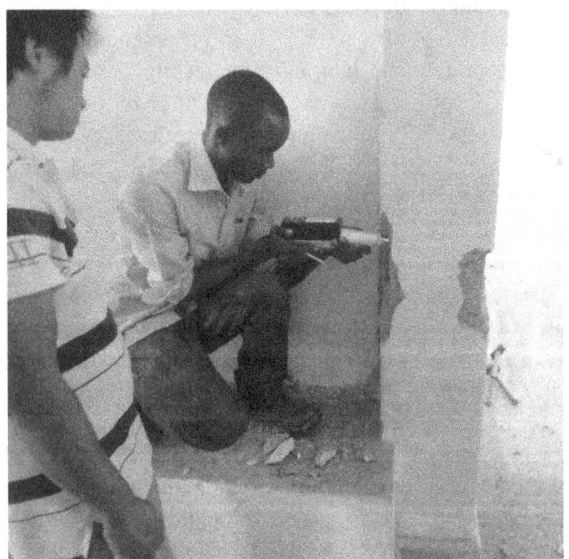

FIGURE 3.20   (a) Rebound hammer testing column concrete strength. (b) Rebound hammer readings.

say a column. Sometimes cubes are used as the specimens. And where cubes are used as the specimens, take nine readings using the rebound hammer on each of the two side faces accessible in the compression machine. Figure 3.20a shows a rebound hammer used to test column concrete strength. Figure 3.20b gives the reading of the rebound hammer. Then, construct a correlation curve from the mean rebound number and strength for each test specimen. The equation for this curve can be determined by any standard curve fitting procedure. See Section 7.4.5.2 (curve fitting).

The confidence in the test results can be improved by combining hardness testing with measurements of UPV, as described in BS 1881: Part 203 (see also BS 1881: Part 201)—Section 3.3.4.2. Usually, the Schmidt hammer test is conducted to supplement observations/results by the PUNDIT (Section 3.3.4.2).

3.3.4.2 UPV testing machine

Also named portable ultrasonic nondestructive digital indicating tester ('PUNDIT'), it is the equipment used to conduct UPV tests. A UPV machine is a nondestructive testing device that uses high-frequency sound waves to measure the velocity of ultrasonic pulses through a material, such as concrete. These waves have frequencies above 20 kHz,[8] which is beyond the human hearing range. Because high-frequency sound comes with a short wavelength, it is the first to fade as sound travels over a great distance or through a dense surface. The UPV machine uses the principle of ultrasonic wave propagation to measure the velocity of sound waves through a material.

UPV machines offer several advantages, including nondestructive testing, fast and easy measurements, high accuracy and precision, and ability to test a wide range of materials. UPV machines are commonly used to evaluate concrete quality and detect defects; assess the condition of bridges, buildings, and other structures; and in materials science to research and develop new materials. A UPV machine or PUNDIT conducts an ultrasonic pulse velocity test, which is an *in situ* nondestructive test to check the quality of concrete and natural rocks. In this test, the strength and quality of concrete or rock are assessed by measuring the velocity of an ultrasonic pulse passing through a concrete structure or natural rock formation.

This test is conducted by passing a pulse of ultrasonic wave through concrete to be tested and measuring the time taken by pulse to get through the structure. Higher velocities indicate good quality and continuity of the material, while slower velocities may indicate concrete with many cracks or voids.

BS 1881: Part 203: 1986. British Standard, Testing concrete: Part 203. Recommendations for measurement of the velocity of ultrasonic pulses in concrete. See Table 3.3.

The technique generally used is the measurement of ultrasonic wave group velocity from the transit time of a pulse between separate transmitting and receiving transducers through a known path distance in the concrete. This provides a measurement of the mean ratio of **elastic stiffness** to **density** along the path and has been found to be a useful index of concrete quality.

The velocity is a function of the composition, degree of compaction, maturity, and free water content, which are inherent in concrete products and structures. Under certain conditions and for properly defined ranges of material, useful correlations may be established between the velocity and properties such as the modulus of elasticity and strength.

What does the velocity measurement indicate? The velocity of the ultrasonic pulse is related to the material's density, modulus of elasticity, and porosity. By measuring the velocity, the UPV machine can detect defects—lower velocities may indicate defects, such as cracks or voids. Higher velocities

typically indicate higher-quality materials. UPV can be used to monitor the development of concrete strength over time—determining concrete maturity.

All ultrasonic pulses passing through concrete are attenuated by an amount depending on their frequency and the properties of the concrete. This attenuation is not easy to measure, and techniques based on this aspect of pulse propagation have not been widely used, but helpful additional information may be obtained under certain circumstances. Specialist literature should be referred to.

Measurement of the velocity of ultrasonic pulses of longitudinal vibrations passing through concrete may be used for the following applications:

- determination of the uniformity of concrete in or between members.
- detection of the presence and approximate extent of cracks, voids, and other defects.
- measurement of changes occurring with time in the properties of the concrete.
- correlation of pulse velocity and strength as a measure of concrete quality.
- determination of the modulus of elasticity and dynamic Poisson's ratio of the concrete.

How UPV machine works

Two transducers (a transmitter and a receiver) are placed on the surface of the material, typically on opposite sides of the area being tested—direct transmission approach, Figures 3.21 and 3.22, or semi-direct transmission approach, Figures 3.23 and 3.24.

A pulse of longitudinal vibrations is produced by an electro-acoustical transducer, which is held in contact with one surface of the concrete under test. After traversing a known path length L in the concrete, the pulse of vibrations is converted into an electrical signal by a second transducer.

Ultrasonic pulse transmission: The transmitter sends a high-frequency ultrasonic pulse through the material. The receiver detects the pulse after it has traveled through the material.

Electronic timing circuits enable the transit time T of the pulse to be measured. The machine measures the time it takes for the pulse to travel from the transmitter to the receiver

TABLE 3.3 Criteria for UPV Tests [BS 1881, 1986].

UPV [M/S]	CONCRETE QUALITY
>4,500	Excellent
3,500–4,500	Good
3,000–3,500	Medium
Below 3,000	Doubtful

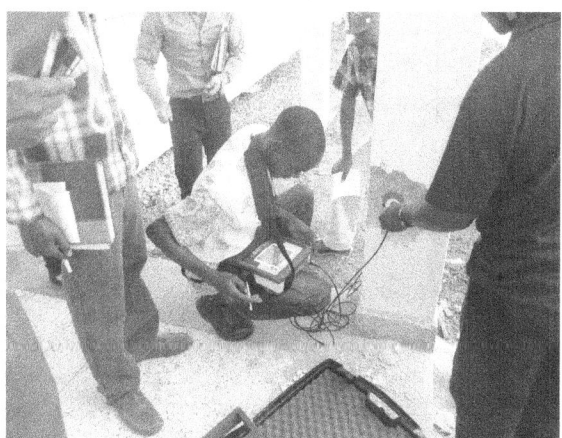

FIGURE 3.21 Testing a column with UPV machine—direct transmission approach.

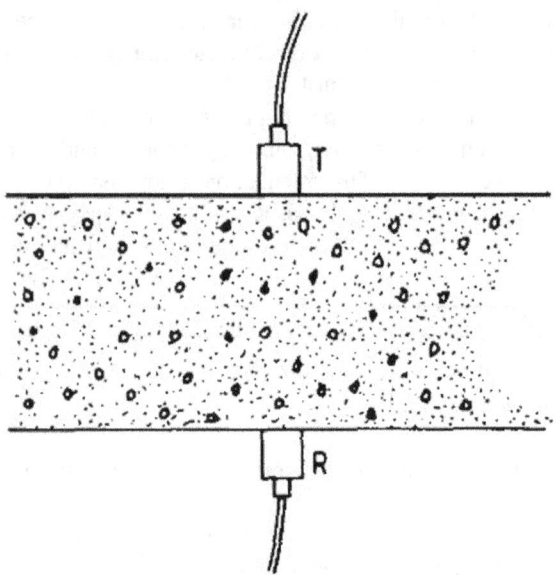

FIGURE 3.22 Direct transmission approach illustration.

FIGURE 3.23 Testing a column with UPV machine—semi-direct transmission approach.

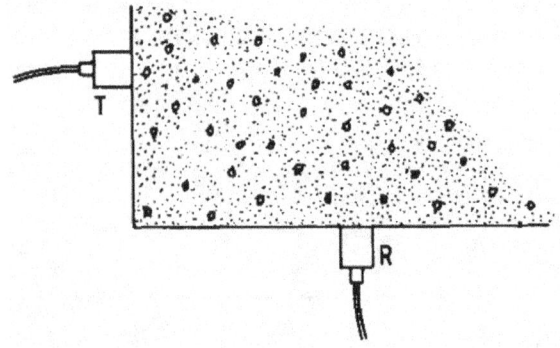

FIGURE 3.24 Semi-direct transmission approach illustration.

(called the 'transit time'). The machine then calculates the velocity of the ultrasonic pulse using the transit time and the known distance between the transducers. The pulse velocity v (in km/s or m/s) is given by the following:

$$v = \frac{L}{T} \tag{3.41}$$

where

- L is the path length;
- T is the time taken by the pulse to traverse that length.

A pulse of vibrations of ultrasonic rather than sonic frequency is used for two reasons:

- to give the pulse a sharp leading edge.
- to generate maximum energy in the direction of propagation of the pulse.

When the pulse is coupled into the concrete from a transducer, it undergoes multiple reflections at the boundaries of the different material phases within the concrete. A complex system of **stress waves** is developed, which includes both longitudinal and shear waves propagating throughout the concrete.

3.3.4.3 Advanced nuclear density equipment machine

An advanced nuclear density gauge, also known as the nuclear densometer, is a nondestructive testing device that uses nuclear technology to measure the density of materials, such as soil, concrete, or asphalt. It is commonly used in road construction to measure densities of compacted sub-grade, sub-base, base, and pavement layers. The nuclear density machine (Figure 3.6) is suitable for standards such as ASTM D6938, D2950, D7013, D7759, C1040, and AASHTO T310.

An advanced nuclear density gauge utilizes radioactive materials to measure the density and moisture content of materials such as soil and pavement. The gauge emits radiation (gamma or neutron) which interacts with the material being tested. By analyzing the amount of radiation that passes through or is reflected back, the gauge determines the material's density and moisture content. Gamma radiation is used to measure density, and neutron radiation is used to measure moisture content.

Paul Ulrich Villard (1860–1934), a French chemist and physicist, discovered gamma rays in 1900 while studying the radiation emanating from radium. He identified gamma rays as a type of electromagnetic radiation that is highly penetrating and has a shorter wavelength than X-rays. His discovery contributed to the development of a civil engineering technological instrument—the nuclear densometer—by finding applications in the density determination of materials such as asphalt pavements!

The gauge uses a small amount of radioactive material, typically Cesium-137 (Cs-137) or Americium-241 (Am-241), to emit gamma rays. These gamma rays interact with the surrounding material, causing some to be scattered or absorbed. Cesium-137 (Cs-137) isotope emits gamma radiation, which is used to determine the density of materials. Americium-241 (Am-241) isotope emits neutrons, which are used to measure moisture content.

The density of the material is directly proportional to the extent of gamma ray scattering. As the gamma rays interact with the material's atoms, the denser the material, the more gamma rays are scattered. By measuring the intensity of the scattered gamma rays, the gauge can calculate the material's density.

Density testing verifies that the material has been properly compacted, which is essential for preventing settlement, deformation, or collapse. This is because density influences the material's ability to support loads. Accurate density measurements help engineers design structures that can withstand expected loads.

The machine measures moisture content. Moisture content significantly affects the behavior of materials like soil and pavement. Excess moisture can lead to reduced strength, increased settlement, or other issues. Moisture content impacts the material's compatibility and stability. Optimal moisture content ensures proper compaction and prevents future problems. Engineers exercise professional judgment and approach issues from more than one angle while using the densometer so that they arrive at the best solution for the civil engineering project.

Advanced nuclear density gauges offer several advantages, including **nondestructive testing**, high accuracy and precision, fast measurement times, and ability to measure density in a variety of materials used in construction. The densometer gauges are commonly applied in geotechnical and structural engineering to measure soil density, moisture content, asphalt density, and concrete density. The gauge is also used to control the quality by accurately measuring the density of materials in manufacturing processes.

Measurement process

The gauge has three main components:

- **Source rod:** which is inserted into the soil being tested. The tip of this rod contains a radioactive element.
- **Sensors:** which is located at the back of the gauge. The sensors detect the amount of radiation coming from the source rod.
- **Display/controls:** this is where the gauge user inputs the proctor value and sets the testing depth.

Procedure

- The tester hammers the pin into the soil. The tester is the person conducting the test.
- The tester places the gauge over the hole created with the pin.
- The tester pushes the source rod into the hole.
- The tester begins the test and steps back. The radioactive source emits gamma rays into the surrounding material. The gamma rays interact with the material's atoms, causing some to be scattered or absorbed.
- The sensors in the rear of the gauge begin to record how much radiation they are exposed to. The sensors are usually a scintillator or a Geiger counter, which measure the intensity of the scattered gamma rays. The scintillator material converts radiation into light, while Geiger counter converts radiation into an electric pulse.
- The gauge calculates the material's density based on how much or how little radiation the sensors detect—the intensity of the scattered gamma rays.

- The tester then returns to the gauge and divides the test density by the proctor density to determine the compaction percentage. Finally, the tester reports the compaction percent to the engineer or contractor.

Nuclear densometer modes of testing

There are two different modes of testing that a nuclear densometer can perform, that is, direct transmission mode and backscatter mode. In the direct transmission mode (Figure 3.25), the sensor which is located in the back of the nuclear density gauge measures the amount of gamma radiation emitted by the source rod. The less radiation detected by the sensor, the higher the material's density. This is why it is important to select the correct depth of penetration of the source rod as a shallower depth will allow for more radiation to reach the sensor than if it were deeper in the same material.

In the backscatter mode (Figure 3.26), the source is in the same plane as the sensor. The shielding within the gauge means that radiation emitted by the source must first be deflected by the material before reaching the sensor. This means that the more the radiation detected by the sensor, the higher the material's density. An important note to make is that in the backscatter mode (Figure 3.26), the results are heavily weighted to the top few inches of material, where compactive efforts are strongest.

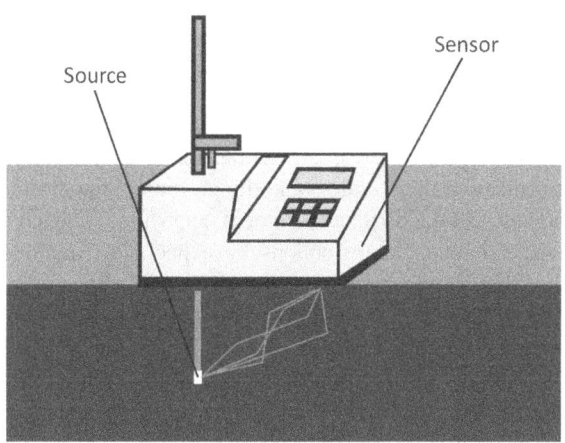

FIGURE 3.25 Direct transmission mode.

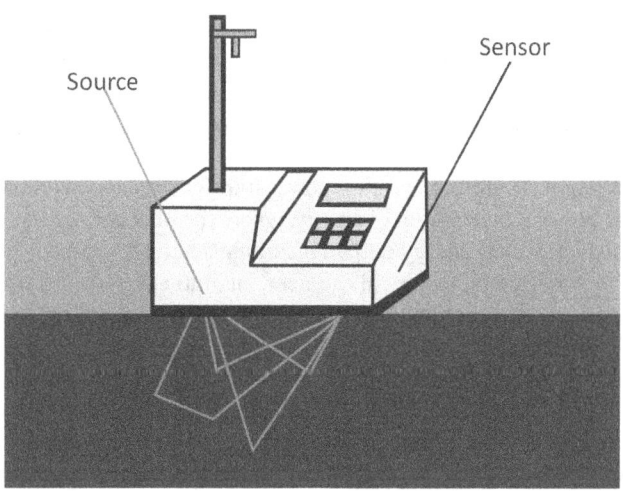

FIGURE 3.26 Backscatter mode.

A nuclear densometer also measures the moisture content. The source rod uses two different types of radiation to provide complete measurements, that is, gamma radiation and neutron radiation. The gamma radiation is used to measure density, and the neutron radiation is used to measure moisture content. Thus, neutrons are released from the source and emanate out, colliding with hydrogen molecules within water or organic material within the soil. When a neutron collides with hydrogen, its speed is reduced, and further collisions with hydrogen no longer reduce the particle's speed. The higher the number of slowed particles detected by the sensor, the more hydrogen is in the material, and hydrogen constitutes approximately 11.11%–11.19% hydrogen by mass, with the remaining percentage consisting of oxygen. This is based on the molar masses of hydrogen (H) and oxygen (O) in a water molecule (H_2O).

Water is not the only thing found in soils that can contain hydrogen. Hydrogen is also found in organic material present in soil. In case the material being tested feels dry but the gauge is indicating a high moisture content reading and low air voids, it is organic. Organic materials are not suitable for construction, neither, in most cases. An alternative method for detecting organics in the soil being tested is smelling it and verifying if the soil is black. If the soil has a strong odor, high gauge moisture content but feels dry, and is black, it is inevitable that it contains organic material.

3.3.4.4 Electromagnetic theory and wireless power transfer, 19th century

This section, although appears smuggled in the civil engineering technology Chapter 3, is essential for civil engineers and other professionals to understand the profound impact of electromagnetism on global advancement, and shows, particularly, how the scientists' contributions have been instrumental in shaping the development of engineering and advanced engineering technologies including civil engineering technologies we use today.

In wireless power transfer, energy is transmitted from a transmitter to a receiver through electromagnetic fields. The energy is converted from electrical energy to electromagnetic energy at the transmitter and then back to electrical energy at the receiver.

Wireless power transfer is a practical application of electromagnetic theory, enabling the convenient and safe transfer of energy without the need for wires or cables. While there are losses in the system due to factors like resistance, radiation, and inefficiencies, the total energy remains conserved. The energy that is not transferred to the receiver is typically lost as heat or radiation. Therefore, wireless power transfer systems do follow the law of conservation of energy, where the total energy input equals the total energy output, including any losses.

An inventor, electrical engineer, and futurist, Nikola Tesla (1856–1943) contributed practically to electromagnetism and wireless power transfer. His designs for AC motors and generators paved the way for efficient long-distance power transmission. Invented in 1891, the Tesla coil is a type of resonant transformer that produces high-voltage, low-current electricity—still used today in applications such as radio transmitters and medical equipment. Tesla developed the coil in **1891,** before conventional iron-core transformers were used to power things such as lighting systems and telephone circuits.[9]

Tesla's experiments with wireless power transmission demonstrated the potential for electricity to be transmitted wirelessly over long distances. His work on resonance and electromagnetic induction laid the groundwork for modern wireless charging technologies. The first system that could wirelessly transmit electricity, the Tesla coil was a revolutionary invention by Nikola Tesla. Early radio antennas and telegraphy used the invention, but variations of the coil can also do other things such as shoot lightning bolts, send electric currents through the body, and create electron winds.

Something of great interest here is that transmission of electrical energy without wires was observed by early inventors and experimenters, but lack of a coherent theory attributed these phenomena ambiguously to electromagnetic induction. Nikola Tesla came a bit later. Did you know that Nikola Tesla developed his ideas benchmarking breakthroughs in earlier scientific theories and experimental validations—the 19th century saw many developments of theories and counter-theories on how electrical energy might be transmitted.

The early discoveries included André-Marie Ampère who discovered, in 1826, the connection between electric current and magnetism, laying the groundwork for understanding electromagnetic phenomena. In 1826, André-Marie Ampère discovered a connection between current and magnets. Michael Faraday, in 1831, described the law of induction, which explains how a changing magnetic field induces an electric current in a conductor. Michael Faraday described in 1831 with his law of induction the electromotive force driving a current in a conductor loop by a time-varying magnetic flux.

Theoretical foundations were laid by James Clerk Maxwell who formulated the Maxwell's equations in 1860s, unifying electricity and magnetism into electromagnetism. He predicted the existence of electromagnetic waves, which can propagate without wires. Maxwell's theory of electromagnetism, articulated in his 'a dynamical theory of the electromagnetic field' in 1865, unified electricity and magnetism into a single framework, demonstrating that electric and magnetic fields travel through space as waves at the speed of light. This theory proposed that light is an electromagnetic wave, an undulation in the same field that causes electric and magnetic phenomena. Maxwell showed that the speed of propagation of electromagnetic waves, which was the same as the speed of light, and came to the conclusion that Electromagnetic waves and visible light are similar, hence formulating a set of partial differential equations that form the foundation of classical electrodynamics, electric circuits, and classical optics along with Lorentz force law. These fields highlight modern communication and electrical technologies. A concise explanation of these phenomena would come from the 1860s Maxwell's equations by James Clerk Maxwell, establishing a theory that unified electricity and magnetism to electromagnetism, predicting the existence of electromagnetic waves as the 'wireless' carrier of electromagnetic energy.

John Henry Poynting defined the Poynting vector and Poynting's theorem in 1884, describing the flow of electromagnetic energy and enabling analysis of wireless power transfer systems. Around 1884, John Henry Poynting defined the Poynting vector and gave Poynting's theorem, which describe the flow of power across an area within electromagnetic radiation and allow for a correct analysis of wireless power transfer systems.

This was followed on by Heinrich Rudolf Hertz's 1888 experimental validation of Maxwell's theory, which included the evidence for radio waves. Heinrich Rudolf Hertz experimentally validated Maxwell's theory, demonstrating the existence of radio waves and confirming the predictions of electromagnetic wave propagation.

These pioneers laid the foundation for our understanding of electromagnetic energy and its applications, including wireless power transfer and radio communication. Their work has had a lasting impact on the development of modern technology. The discoveries and theories developed by scientists, that is, Ampère, Faraday, Maxwell, Poynting, and Hertz, laid the groundwork for engineers such as Nikola Tesla to design and develop invent/innovative technologies.

Scientists' research has helped engineers understand the underlying laws of physics, such as electromagnetism, which informs the design of electrical systems. And a number of electrical systems and technologies are used in civil engineering, terming them civil engineering technologies. These discoveries enabled engineers to create new technologies, such as wireless power transfer, radio communication, and electrical power distribution. Engineers were able to optimize and refine existing technologies, making them more efficient, reliable, and cost-effective, borrowing scientists' findings.

Nikola Tesla and other inventors borrowed several ideas from scientists to invent cool technologies including the Tesla coil and AC motors and generators, which garnered them a number of patents. Today, ideas are in the open due to global advancements. You can literally obtain ideas from all disciplines to innovate or invent whatever you are capable of. As a civil engineer, therefore, if you don't show how you manage to borrow ideas from a number of spectrums, you may not be able to convince others that you have something worth to advance because ideas are in the open—books, articles, online resources, public discussions, technologies we interact with, and etc. and can be understood. The future, even now, promises engineers integrating AI technologies and systems to develop a sustainable world. Following codes and standards of practice means that your innovation or invention, as an engineer, is safe, reliable, durable, up to date, technologically sound, conforms to current industry norms, and complies with the law.

From this chronology of events, you can easily see that an engineer, as defined in Section 2.3.3, must have the ability to play such games as Nikola Tesla objectively while advancing the society. Engineers integrate ideas collected from scientists and practically bring them to life with their skills. Nikola Tesla was a genius who has inspired engineers from different walks of life—he masterfully played games to bring ideas to life—crafting and extending knowledge that scientists failed to devise practical applications. Today, engineers are even smarter, by the time you know your ideas are already borrowed and products are taking shape around the world.

3.4 SYSTEMS OF UNITS OF MEASUREMENT

There are several systems of units of measurement which include the International System of Units (SI), Imperial System, and the United States Customary System (USCS). The International System of Units is based on the metric system. It uses units such as meters (m), grams (g), liters (L), and seconds (s). The modern International System of Units is built upon seven base units: meter, kilogram, second, ampere, kelvin, mole, and candela. It is widely used in science, technology, and international trade.

The Imperial System originated in the United Kingdom and uses units such as inches (in), feet (ft), yards (yd), pounds (lb), and gallons (gal). It is still used in some countries, including the United States and the United Kingdom.

The USCS is similar to the Imperial System, but with some differences. It is used in the United States and some other countries.

Converting between different systems of units is often necessary and can be done using conversion factors.

Standardized units of measurement are essential for technology development, enabling precision, collaboration, innovation, calibrating equipment, and interoperability. Standardized units ensure that components and systems from different manufacturers work together seamlessly. Specifically, interoperability is critical in fields like electronics, telecommunications, and transportation. In civil engineering, metric units like pascals (Pa) for pressure and kilograms per cubic meter (kg/m³) for density support the development of new materials and their applications.

3.4.1 Evolution of Systems of Units of Measurement

The evolution of measurement systems has progressed from early, localized systems based on readily available references such as body parts to the standardized, internationally recognized International System of Units. This development was driven by the need for consistent and accurate measurements in trade, science, and industry, especially as societies grew more complex and interconnected.

Early civilizations relied on localized unites such as body parts (e.g., cubits, feet, and palms) and other easily accessible items as standards for measurement. Form this description, it is easy to tell how the standard measurement of length in the English system, that is, 'foot' originated!

In Section 3.2.3.6, we saw how James Watt derived the concept of HP from observing horses at work and found a way to explain the power of his steam engines in terms that would be relatable to potential customers, who were familiar with the power of horses. Thus, the units of measurement varied from place to place, leading to inconsistencies in trade and other transactions. The units were often specific to fields, such as dry goods or cloth, with no connection between them.

Later, the development of standardized systems was influenced by the increase in trade. As trade expanded, the need for standardized and universally accepted units became crucial. The metric system, developed in France during the age of enlightenment, introduced a decimal-based system, with the meter and kilogram as base units.

The 1875 Metre Convention established international prototypes and a body to oversee the system. The Metre Convention established the International Bureau of Weights and Measures (BIPM), promoting global consistency in measurements. Ever since then, many countries around the world have undergone the metrication process! The modern International System of Units is built upon seven base units: meter, kilogram, second, ampere, kelvin, mole, and candela. The SI system (metric system) is the most widely used system of measurement globally, particularly in science, technology, and international trade.

The metric system's use of **base 10** simplifies conversions and calculations. **Note:** The concepts of base units (e.g., meter and kilogram) and derived units (area or volume) are based on the base units. Recent advancements in science led to the redefinition of some SI base units in terms of fundamental physical constants. Thus, metrology continues to evolve as scientists refine measurement techniques and redefine units.

It is essential to note that the development of a nation relies heavily on its ability to understand and apply standardized measurement systems, such as the metric system, to facilitate smooth trade and commerce. However, not all countries have fully adopted the metric system, and some still use a mix of measurement units. For a nation's development to be sound, it is important that it consistently applies and promotes the use of standardized measurement systems. Level of adoption of the system, consistency and education, and awareness of citizens are indicators of a sound country's metrication process. While many emerging African countries have officially adopted the metric system, in practice, the adoption is often incomplete or inconsistent, with various degrees of continued use of traditional or other measurement systems—and they actually must define what metrication means for their growth and development.

3.4.2 Imperial Units

The Imperial System defines force (pound) as the weight of a specific mass, which remains constant regardless of the gravity. In contrast, the metric system assumes mass remains constant. So, the key differences are as follows:

- Imperial System: force (pound) is defined by weight.
- Metric system: mass (kilogram) is constant.

The acceleration due to gravity (32.2 ft/s²) is rarely needed in calculations. The standard units are as follows:

LENGTH	
1 mile	1,760 yards
1 furlong	220 yards
1 yard (yd)	3 feet
1 foot (ft)	12 inches
1 inch (in)	1/12 foot
Volume	
1 cubic yard	27 cubic feet
1 cubit foot	1/27 cubic yards
1 cubic inch	1/1,728 cubic feet
Area	
1 acre	4,840 sq. yd
1 square mile	640 acres
1 sq. ft	144 sq. in
1 sq. in	1/144 sq. ft
1 sq. yd	9 sq. ft
	(Continued)

LENGTH	
Weight	
1 ton	2,240 pounds
1 hundredweight (cwt)	112 pounds
1 stone	14 pounds
1 pound (lb)	16 ounces
1 ounce	1/16 pound
Capacity	
1 bushel	8 gallons
1 gallon	4 quarts
1 quart	2 pints
1 pint	½ quart
1 fl. oz	1/20 pint
Nautical measure	
1 nautical mile	6,080 feet
1 cable	600 feet
1 fathom	6 feet

3.4.3 Metric System

The International System of Units is the most widely used measurement system globally. It's an absolute system that relies on fundamental units of mass, length, and time. This system ensures consistency across different locations. The SI system assumes that regardless of the location, mass is an unchanging property and that the force varies with location, the unit of force, the Newton (N), depends on the gravitational acceleration at a specific location. On Earth, gravity accelerates objects at approximately 9.81 m per second squared $\left(m/s^2\right)$. The SI system uses the following basic units:

LENGTH	M	METER
Time	s	second
Temperature	K	
Unit of plane angle	rad	
Luminous intensity	cd	candela
Quantity/substance	mol	mole (6.02×10^{23} particles of substance (Avogadro's number))
Mass	kg	

The most commonly prefixes in engineering are as follows:

giga	G	1 000 000 000	1×10^9
mega	M	1 000 000	1×10^6
kilo	K	1 000	1×10^3
centi	c	0.01	1×10^{-2}
milli	m	0.001	1×10^{-3}
micro	μ	0.000001	1×10^{-6}
nano	N	0.000000001	1×10^{-9}

The base units and the prefixes listed above imply a system of supplementary units which forms the convention for noting SI measurements, such as the Pascal for measuring pressure where $1\ Pa = 1\ N/mm^2$. The unit of measurement for pressure, the Pascal (Pa), is named after Blaise Pascal.

The Pascal is defined as one Newton per square meter (N/m²), and it is used to measure pressure, stress, and modulus of elasticity. The International System of Units adopted the Pascal as the standard unit of pressure in 1971, in recognition of Pascal's contributions to the study of fluid mechanics and pressure. So, every time you see 'Pa' used to measure pressure, you know that it's a tribute to the pioneering work of Blaise Pascal!

Typical metric units

MASS OF MATERIAL	KG
Density of the material	kg/m³
Load per unit length	kN/m
Bending moment	kNm
Wind loading	kN/m²
Bulk density	kN/m³
Weight/force/point load	kN
Distributed load	kN/m²
Earth pressure	kN/m²
Stress	N/mm²
Torque	kNm
Modulus of elasticity	kN/mm²
Floor area	m²
Volume of material	m³
Reinforcement spacing	mm
Reinforcement area	mm² or mm²/m
Deflection	mm
Moment of inertia (of an area)	cm⁴ or mm⁴
Span or height	m
Section area	cm² or mm²
Section dimensions	mm
Section modulus	cm³ or mm³
Radius of gyration	cm or mm

3.4.4 Conversion Factors

See Section 7.6.7. As mentioned, engineers working on international projects need an understanding of SI and English units. Given the dual use of SI and English units in some jurisdiction, quick and easy conversion between the two systems is essential. The following is a selection of useful conversion factors.

Volume	1 mm³	0.000061 in³
	1 m³	35.32 ft³
	1 m³	1.308 yd³
Area	1mm²	0.00153 in²
	1 m²	10.764 ft²
	1 m²	1.196 yd²
Mass	1 kg	2.205 lb.
	1 tonne	0.9842 tons
Length	1 mm	0.03937
	1 m	3.281 ft
	1 m	1.094 yd
Density	1 kg/m³	0.06242 lb/ft³
Force	1 N	0.2248 1bf
Line loading	1 kN/m	68.53 lbf/ft

(Continued)

	1 kN/m	0.03059 tonf/ft
Moment	1 N/m	0.73761bf ft
Modulus of elasticity	1 N/mm²	145lbf/in²
	1 kN/mm²	145 032 lbf/in²
Section modulus	1 mm³	0.000061 in³
	1 cm³	0.061 in³
Second moment of area	1 mm⁴	2.403 x 10⁻⁶ in⁴
	1 cm⁴	2.403 x 10⁻² in⁴
Temperature	X °C	[(1.8X+32)]° F
Stress and pressure	1 N/mm²	145 lbf/in²
	1 N/mm²	0.0647 tonf/ in²
	1 N/m²	0.0208 lbf/ft²
	1 kN/m²	0.0093 tonf/ft²

3.4.5 Measurement of Angles

There are two common systems for the measurement of angles.

3.4.5.1 International system

The radian is a constant angular measurement equal to the angle subtended at the center of any circle, by an arc equal in length to the radius of the circle (see Figure 3.27). It is commonly used for the measurement of plane angles in mechanics and mathematics:

$$\pi \text{ radians} = 180° \text{ (degrees)}$$

$$1 \text{ radian} = \frac{180°}{\pi} = \frac{180°}{3.1416} = 57°17'44''$$

This is because $\pi \approx 3.14$ radii wrap halfway around a circle (Figure 3.28).

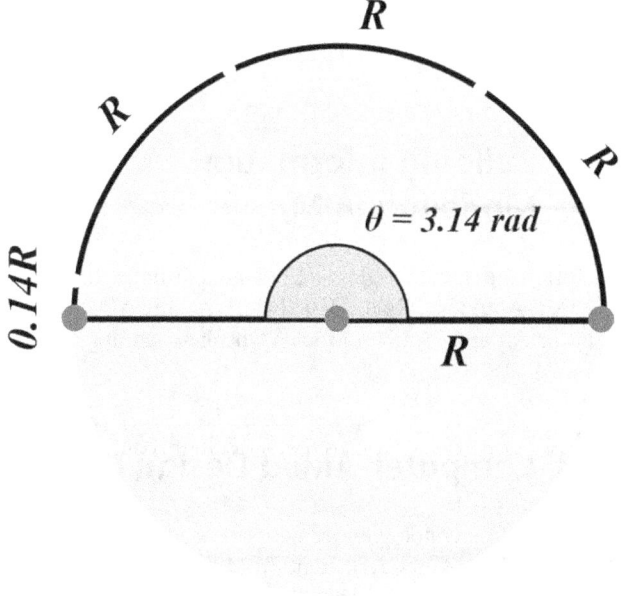

FIGURE 3.27 $\pi \approx 3.14$ radii.

3.4.5.2 English system

The English or sexagesimal system is universal:

- 1 right angle = 90° (degrees)
- 1° (degree) = 60′ (minutes)
- 1′ (minute) = 60″ (seconds)

Equivalent angles in degrees and radians and trigonometric ratios

Angle θ in radians	0	$\frac{\pi}{6}$	$\frac{\pi}{4}$	$\frac{\pi}{3}$	$\frac{\pi}{2}$
Angle [1] in degrees	0°	30°	45°	60°	90°
$\sin\theta$	0	$\frac{1}{2}$	$\frac{1}{\sqrt{2}}$	$\frac{\sqrt{3}}{2}$	1
$\cos\theta$	1	$\frac{\sqrt{3}}{2}$	$\frac{1}{\sqrt{2}}$	$\frac{1}{2}$	0
$\tan\theta$	0	$\frac{1}{\sqrt{3}}$	1	$\sqrt{3}$	∞

3.5 CIVIL ENGINEERING SOFTWARE TECHNOLOGIES

The industry employs several software technologies to speed up the planning, design, construction, operation, and management of civil engineering structures. These software technologies are developed by teams of engineers and computer scientists. They are developed by coding and applying scientific knowledge, computer science, and mathematics. They have their own architecture.

3.5.1 Structural Analysis and Design

The following technologies are adopted around the world for structural analysis and design: Autodesk Revit, STAAD.Pro, ETABS, SAP2000, RISA-3D, PROKON, and Tekla Structures, among several others.

3.5.2 Building Information Modeling (BIM)

The following technologies are adopted around the world for BIM: Autodesk Revit, Graphisoft ArchiCAD, Trimble Navisworks, and Solibri Model Checker, among several others.

3.5.3 Computer-Aided Design (CAD)

The following technologies are adopted around the world for CAD: Autodesk AutoCAD, Bentley MicroStation, ZWCAD, and BricsCAD, among several others.

3.5.4 Geotechnical and Foundation Engineering

The following technologies are adopted around the world for geotechnical and foundation engineering: PLAXIS, DEEPEX, FLAC, GeoStudio, and SETTLE3D, among others.

3.5.5 Transportation Engineering

The following technologies are adopted around the world for transportation engineering: Autodesk Civil 3D, Bentley In Roads, Transoft AutoTURN, and Road Estimator, among several others.

3.5.6 Hydrology and Hydraulic Engineering

The following technologies are adopted around the world for hydrology and hydraulic engineering: HEC-HMS, HEC-RAS, Autodesk Storm and Sanitary Analysis, EPANET, and Bentley WaterGEMS, among several others.

3.5.7 Construction Management and Estimating

The following technologies are adopted around the world for construction management and estimation: Autodesk Construction Cloud, Procore, PlanGrid, Bluebeam Revu, and Sage Estimating, among several others.

3.5.8 Surveying and Mapping

The following technologies are adopted around the world for surveying and mapping: Autodesk AutoCAD Civil 3D, Trimble TerraModel, Bentley MicroStation, and Carlson Survey, among several others.

3.5.9 AI Tools for Civil Engineering

- **Autodesk forma:** This generates multiple design options instantly and optimizes structural designs for sustainability and cost-effectiveness.
- **Autodesk civil 3D:** This streamlines infrastructure design and documentation with AI-enhanced features like corridor modeling and grading optimization.

3.6 MIND MAPPING QUESTIONS

1. What are the key factors that influence the development of effective codes and standards in civil engineering?

2. How do sustainability, engineering, and policy intersect, and what opportunities arise from these connections?

3. What are the potential barriers and enablers to adopting codes and standards through government legislation, and how can engineers navigate these complexities?

4. What are the emerging trends, challenges, and innovations that could impact codes and standards in your region or civil engineering industry, generally?

5. What legislative, regulatory, or contractual mechanisms could be used to make professional code/s of conduct legally enforceable?

3.7 CONCLUSION

With all the insight delivered in this chapter, it is clearly obvious that a civil engineer is trained to think out of the box and around the box—never trained to cram!

Unlocking curiosity is the work of a civil engineer, and what you learn in class is introductory, aiming to shape you into a graduate who knows how to think beyond borders to innovate and develop practical solutions for the industry by harnessing existing, new, and advancing technologies, but not forgetting the extraction of appropriate scientific principles from different fields to shape new and advancing civil engineering technologies. This is made possible by drawing lessons from different fields, learning how to conduct appropriate research, and embracing interdisciplinary learning! You definitely standout!

Learning as though, you were, a neural network is the new normal skill in civil engineering. This is because the essentials of a civil project are interdependent.

Sir Isaac Newton's laws, Robert Hooke's law, and the conservation of energy principle worked or work together to advance civil engineering technology. Without those scholars, many technologies would not be invented. Those are primary scientific laws that advanced civil engineering technology. Everything else is secondary! They dedicated their time to study the forces of nature, and civil engineers learnt how these forces can be harnessed to direct the great sources of power in nature for the use and convenience of humans.

The question is 'What have we done to challenge these geniuses in the modern day?' It looks like whatever we do today derives from the concepts of energy or refines them or it forms the underpinning principle. The profound question of the modern world is energy in spiritualism!

Studied by primarily by scientists, the following concepts have been at the forefront of advancing civil engineering technology:

• **Work**: The transfer of energy from one object to another through a force applied over a distance.
• **Energy**: The ability to do work.
• **Power**: The rate at which energy is transferred or converted.
• **Force**: A push or pull that causes an object to change its motion or shape.

NOTES

1 Hammurabi's code. Was it Just? The DBQ Project, 2011.
2 https://www.iso.org/news/2017/02/Ref2163.html. ISO celebrates 70 years.
3 https://standards.ieee.org/wp-content/uploads/2023/10/Emerging-Technology-Standards-and-Sustainability-Policy-Brief.pdf. Emerging technology, standards, and sustainability.
4 Ohio Building Code 2024. Retrieved from: https://up.codes/viewer/ohio/ibc-2021.
5 Building Codes – NBRB. Retrieved from: https://nbrb.go.ug/resources/building-codes/.
6 Hooke's Law: Definition, Formula, Graph, Applications. Retrieved from: https://byjus.com/gate/hookes-law/.
7 https://www.prs-med.com/.
8 1 Hz (Hertz) is equal to: 1 cycle per second. In other words, if something vibrates or oscillates at a frequency of 1 Hz, it completes one cycle or oscillation every second. 20 Hz is equal to 20 cycles per second. This frequency is within the range of human hearing and is typically perceived as a low-pitched sound or rumble. The frequency of a wave refers to the number of cycles or vibrations of the wave that occur in a given unit of time, often in Hertz (Hz).
9 Wireless Electricity? How the Tesla Coil Works | Live Science. https://www.livescience.com/46745-how-tesla-coil-works.html.

REFERENCES

1. BS EN 10025:2004 Grade S275.
2. BS5930:2015 + A1: 2020.
3. https://codes.iccsafe.org/content/IBC2024P1
4. https://codes.iccsafe.org/content/IFC2024V1.0
5. MC-3 Elite™- Nuclear Density Gauge | InstroTek, Inc. https://www.instrotek.com/products/mc-3-elite
6. ISO 9001: 2015 Quality Management.
7. Hammurabi's code. Was it Just? The DBQ Project, 2011.
8. https://www.iso.org/news/2017/02/Ref2163.html. ISO celebrates 70 years.
9. https://standards.ieee.org/wp-content/uploads/2023/10/Emerging-Technology-Standards-and-Sustainability-Policy-Brief.pdf. Emerging technology, standards, and sustainability
10. BS 9295:2020 Guide to the structural design of buried pipes.
11. BS EN 1295-1-1997.
12. BS 5911-1:2021 – TC Concrete pipes and ancillary concrete.
13. BS EN 1916:2002.
14. ISO 17:1973 Guide to the use of preferred numbers and of series of preferred numbers.
15. BS 1564:1975 Specification for pressed steel sectional rectangular tanks.
16. Ohio Building Code 2024. Retrieved from: https://up.codes/viewer/ohio/ibc-2021
17. Building Codes – NBRB. Retrieved from: https://nbrb.go.ug/resources/building-codes/
18. BS 7593:2019.
19. BS EN 806-4:2010 Specifications for installations inside buildings conveying water for human consumption.
20. BS 1881: Pat-t 202: 1986. British Standard. Testing concrete. Part 202. Recommendations for surface hardness testing by rebound hammer.

21. BS 1881: Part 203: 1986. British Standard. Testing concrete. Part 203. Recommendations for measurement of velocity of ultrasonic pulses in concrete

22. ASTM International. (2017). Standard test methods for in-place density and water content of soil and soil-aggregate by nuclear methods (shallow depth) (ASTM D6938-17ae01). Retrieved from: https://store.astm.org/d6938-17ae01.html. Accessed 06/07/2025.

23. Russell, S. J., & Norvig, P. (2020). *Artificial Intelligence: A Modern Approach* (4th ed.). Scientific Research Publishing, USA. https://www.scirp.org/reference/referencespapers?referenceid=3614787.

24. Goodfellow, I., et al. (2016). *Deep Learning*. MIT Press, Cambridge, MA.

25. Pan, Y., & Zhang, L. (2021). Roles of Artificial Intelligence in Construction Engineering and Management: A Critical Review and Future Trends. *Automation in Construction*, 122, 103517. https://doi.org/10.1016/j.autcon.2020.103517

26. Newton by Kneller | Lines of thought. Retrieved from: https://exhibitions.lib.cam.ac.uk/linesofthought/artifacts/newton-by-kneller/

27. Gatte, M. T., & Kadhim, R. A. (2012). Hydro Power. https://doi.org/10.5772/52269. Retrieved from: https://www.intechopen.com/chapters/40550

28. Hooke's Law: Definition, Formula, Graph, Applications. Retrieved from: https://byjus.com/gate/hookes-law/

29. PRS Geo Technologies. (n.d.). *Cellular confinement systems. PRS Geo Technologies*. Retrieved January 8, 2026, from https://www.prs-med.com/prs-neoloy-geocells/choosing-a-tough-cell/cellular-confinement-systems/

30. 2021 International Building Code (IBC). https://codes.iccsafe.org/content/IBC2021P2

FURTHER READING

1. Ssempeera, P. (2023). *Integrated Drainage Systems Planning and Design for Municipal Engineers* (1st ed.). CRC Press. https://doi.org/10.1201/9781003255550

2. Budynas, R. G. (1999). *Advanced Strength and Applied Stress Analysis.* 2nd ed. McGraw-Hill, USA.

Forces in Civil Engineering

4

4.1 INTRODUCTION

A force is a push or pull that an object experiences as a result of interacting with another object. When two objects interact, a force is exerted on each of the objects. The force is no longer felt by the two objects when the interaction finishes. Only when there is interaction or induction between two objects, forces by the objects actually exist.

A force can be categorized as either a contact force or a field force [1]. A contact force develops in between two bodies touching each other. On the other hand, a field force develops between bodies that acts through space. A complete specification of the action of a force must include its magnitude, direction, and a point of application, in which case it is treated as a fixed vector [2].

When an object pushes or pulls on another, the former object applies a force to the other object. As an example, a dog pushing a door to open is applying force to the door. Similarly, a cat pulling a doll by its handle exerts force on the handle.

This chapter primarily introduces the reader to the concept of force, load, and stress. The civil engineer's primary role is to design, construct, operate, and maintain structures that are safe, durable, operable, and reliable to safely carry loads throughout the lifecycle phases. Therefore, civil engineers account for all applicable forces in the development of infrastructure.

Force is scientifically described as an action that causes or tends to cause a change in the state of rest or motion of a body. It is any agency capable of producing an acceleration of an unsupported body. Force is equal to the rate of change of momentum (we can also say that how fast momentum changes dictates the magnitude of the force). The unit of force is Newton (N) in SI units and pounds (lb.) in the English system. Forces are all around us, and understanding them is essential for building, designing, and predicting the behavior of various civil engineering systems and structures.

4.2 QUICK INSIGHTS

- A **water supply pipeline** is designed to convey a specific volume of water, and its carrying capacity must be sufficient to resist the internal and external forces acting on the pipeline, ensuring its structural integrity and preventing failure.

- A **retaining wall** is designed to resist forces due to the soil mass it retains, hydrostatic forces (induced by pore water), and surcharge loads. See Figure 4.1.

- A **building** is designed to be able to withstand both dead and live loads. Dead loads may be permanent loads such as weight of walls, floors, roofs, beams, columns, windows, and doors. See Figure 4.2. Live loads may be temporary (variable) such as people (occupants), furniture, snow, wind loads (lateral forces), seismic loads (earthquake loads), and temporary construction loads (e.g., scaffolds), among others.

- A **road pavement** structure is designed to withstand and resist the stresses and strains induced by various loads, including traffic, environmental, and other external forces, to ensure the structural integrity and durability of the pavement layer structure.

- A **reservoir water tank** is designed to withstand hydrostatic forces, tensile, and bending forces. That means the tank panels in the case of pressed steel panels (for example, for a cuboidal tank) must meet the standard flexural strength, and their yield strength must be robust to carry water loads without cracking or failing entirely.

- A **tower** (e.g., steel tower) is designed to be stable enough with stiff materials to resist the forces induced by wind loads, seismic forces, dead loads, and live loads (e.g., a water reservoir tank) that the tower carries. That means it must be designed considering aerodynamic forces (wind loads), seismic forces, and all the dead and live weight the tower must carry.

4.3 THE NEED TO UNDERSTAND AND ADDRESS FORCES

Civil engineers need to understand and address different types of forces to design, develop, and operate infrastructure systems. Understanding the different types of forces and their effects on materials and structures is fundamental for engineers and designers to ensure safety and efficiency in the construction and operation of various civil engineering projects.

DOI: 10.1201/9781003561620-4

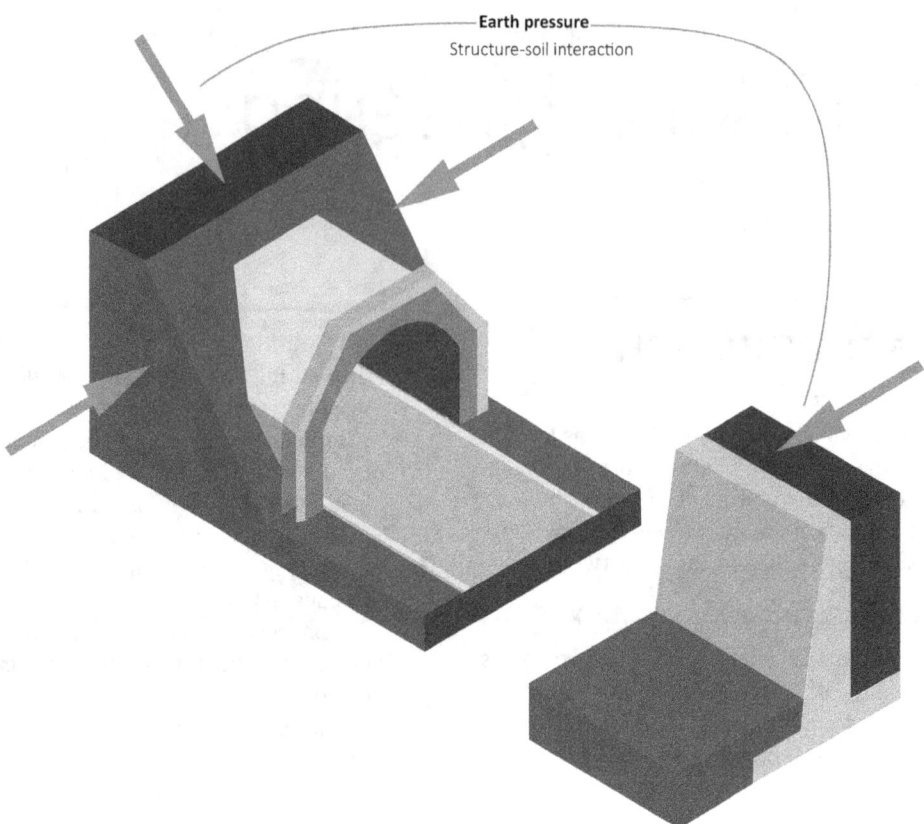

FIGURE 4.1   Earth pressure acting on retaining wall.

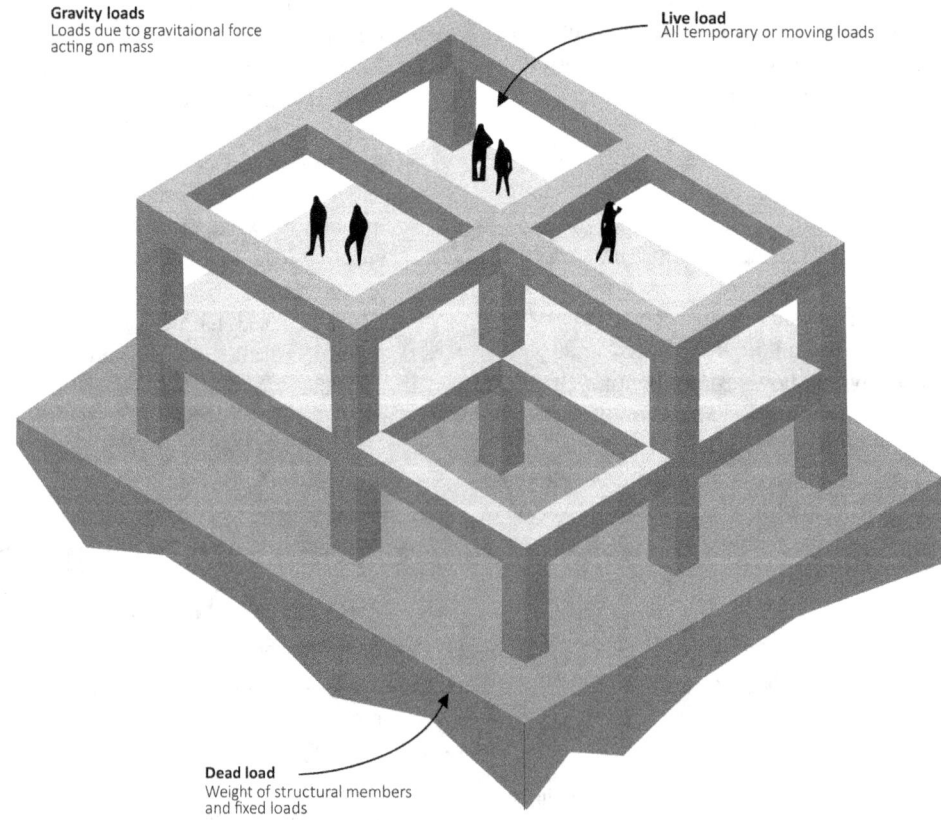

FIGURE 4.2   Dead and live loads on a building.

Engineers optimize design to minimize material usage, meet sustainability goals, and maximize efficiency by considering forces like friction and torque. To prevent material failure, engineers account for forces such as tension, compression, shear, and bending forces, among others.

Ensuring structural integrity and safety when designing infrastructure systems is the one of the core roles of civil engineers; thus, they develop structures that resist gravity forces, wind, and seismic forces to prevent collapse or damage and ensure public safety.

To protect against natural hazards, civil engineers design structures that withstand extreme scenarios such as earthquakes, hurricanes, and floods. Engineers also consider hydrodynamic forces to prevent damage to waterways and surrounding environments, thus preventing environmental damages. Also, civil engineers are required to meet performance requirements, build reliability in civil engineering systems, and therefore, need to account for all forces, and must ensure structures perform as intended, withstanding various forces while maintaining functionality.

Note: In the design of civil engineering structures, for example, water systems, some forces may be considered negligible, such as friction forces and boundary forces. The point to learn here is that even though force evaluation is key, in some cases, we tend to disregard some forces and assume that they are negligible. This comes with design intuition in many instances.

objects, or live load. Force covers a broader range of actions, including friction, air resistance, and tension.

A force can be a push or pull, and it has both magnitude and direction, which makes it a vector quantity. On the other hand, load is often the result of a force, such as gravity, and is typically measured as a weight.

Force is an interaction that changes the motion of an object, while load is the weight or pressure borne by a structure.

Forces can be either temporary or constant. For instance, the force applied to a ball when it's kicked is temporary, lasting only as long as the foot is in contact with the ball. Load often refers to a continuous pressure or force exerted on an object, like the weight of furniture, people, sacks of coffee on a floor, and other movable objects. Both concepts, force and load, are essential in civil engineering, but they are used to describe different phenomena.

According to Newton's second law, forces are the driving factor behind the acceleration of masses. Loads, in construction and engineering, need to be carefully calculated to ensure safety and structural integrity. **Note:** while all loads are forces, not all forces are considered loads.

Forces are experienced through various actions like magnetism or electricity, while load is specifically the force due to weight or an external pressure on a structure or an automobile. This difference is essential for understanding the practical applications of each term in different contexts, such as physical movements as opposed to structural engineering.

4.4 LOADS VERSUS FORCES

A civil engineer need to clearly understand the two terms 'loads' and 'forces.' These words are used interchangeably, but there is a difference in how they must be used depending on the context. Loads are external actions (wind, gravity, or seismic) applied to a structure. Forces are internal actions (axial, shear, and moment) within a structure.

Therefore,

- A force is internal, and a load is external.
- Forces induce loads on a structure (**in conceptual form**). Therefore, active (external) forces are treated as loads in civil engineering.

Illustration,

Wind load (external) → induces → **Force (internal)**

→ **induces** → **Load (on structural component)**.

In the design of structures, loads are calculated as the forces a structure must support. These loads include the weight of the structure itself, known as dead load, and the weight of movable

4.5 OUTLINE OF FORCES THAT CIVIL ENGINEERS ENCOUNTER

Civil engineers deal with various forces because their primary responsibility is to design, build, and maintain an infrastructure that is safe, durable, and functional. Thus, civil engineers create resilient, sustainable, and reliable infrastructures that serve communities and support economic development worldwide.

Civil engineers consider forces outlined in Sections 4.5.1–4.5.14, which translate into loads to ensure the safety, durability, and functionality of infrastructure projects. These forces are inherent in the natural environment and affect the behavior of structures, materials, and systems.

4.5.1 Gravity

This is the downward force that affects structures, foundations, and materials. Civil engineers consider gravity as a primary load that must be accounted for in the design of buildings, bridges, roads, and other infrastructure. Figure 4.3 illustrates the gravity force (downward force). Did you know that snow loads are gravity loads? The forces they cause act downward due to the weight of the snow.

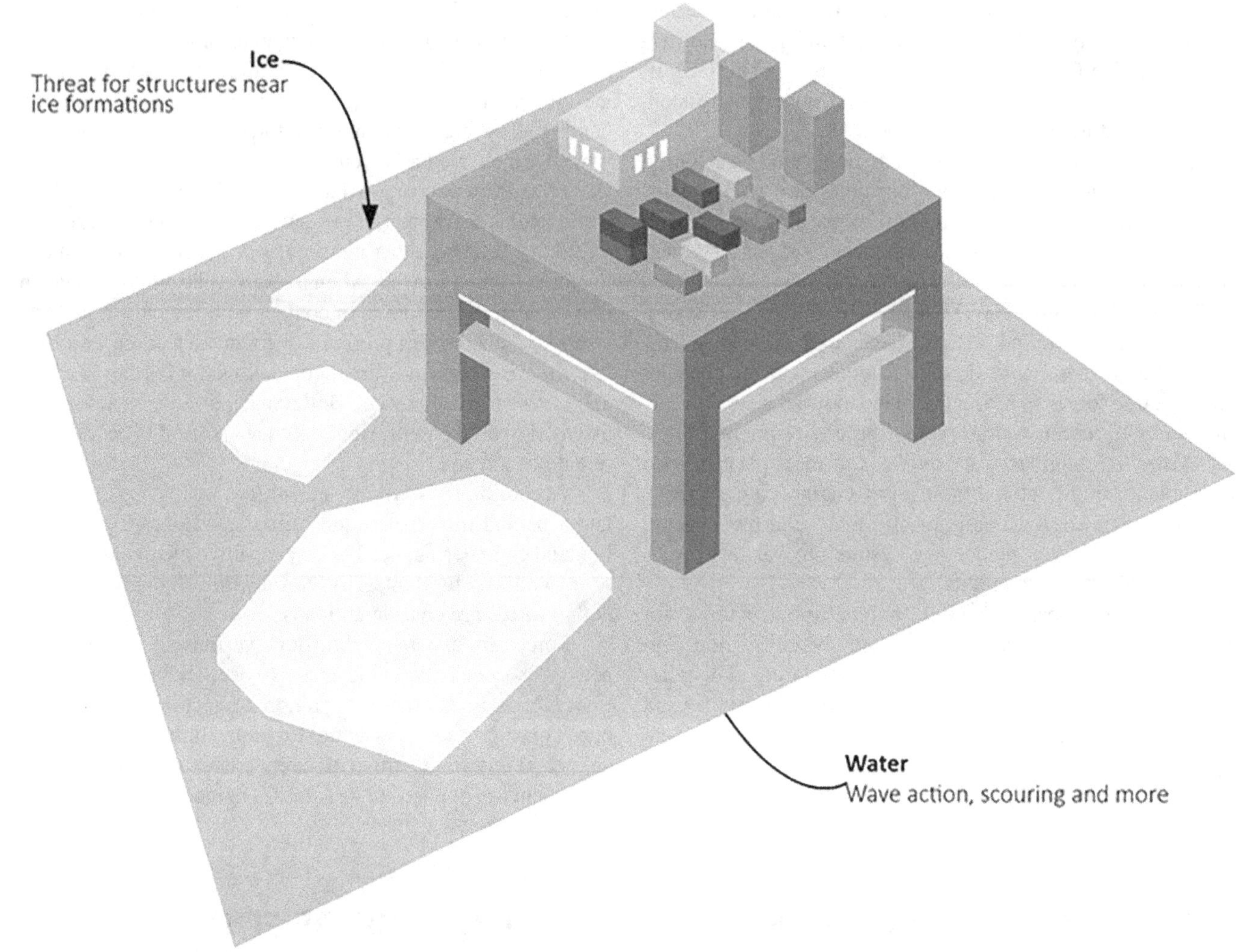

Ice
Threat for structures near
ice formations

Water
Wave action, scouring and more

FIGURE 4.3   Gravity force.

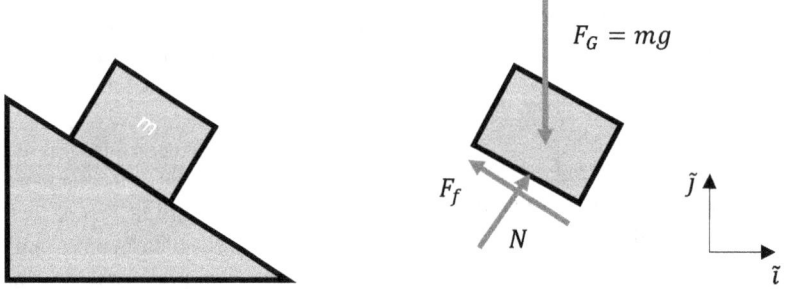

FIGURE 4.4   Friction force.

4.5.2 Friction

Friction is the force that opposes motion between surfaces, which is crucial for structural stability. As far as civil engineering structures are concerned, friction force is encountered in several ways, for example, foundations—that is, friction between the foundation and soil helps transfer loads to the ground, preventing settlement and sliding.

The friction between tires and pavement surfaces ensures safe vehicle braking and cornering. The friction between bridge decks and piers helps resist horizontal loads like wind and seismic forces.

Friction force is the silent protector of civil engineering structures, ensuring safety, stability, and strength. Figure 4.4 illustrates friction force, F_f. Friction force is the unseen guardian of motion, opposing the slide, slip, or slide between surfaces. In civil engineering, this force is the backbone of stability, ensuring structures stand tall and strong.

4.5.3 Tension

Tension, also termed the tensile force, tends to stretch materials like cables or beams. This is a force that tends to elongate or stretch an object. It occurs when two forces act away from each other along the same straight line. Tension is responsible for pulling materials apart and is typically observed in structures like cables, ropes, and springs.

When a tensile force (Figure 4.5) is applied to an unloaded material (Figure 4.6), it produces a stress corresponding to the applied force, causing the cross-section to become thinner and the length of the material to elongate.

The stretching forces acting on the material are referred to as tensile force, which has two components: tensile stress and tensile strain. This indicates that the material being acted upon is under tension, and that the forces are attempting to stretch it.

4.5.4 Compression

This is a force that tends to shorten or squash an object. In other words, it squeezes structural elements such as columns or foundations. It occurs when two forces act toward each other along the same straight line. Compression is responsible for pushing materials together and is commonly observed in columns, beams, and other load-bearing elements. See Figure 4.7.

4.5.5 Shear

Shear causes one material layer to slide past another layer along a parallel plane. See Figure 4.8. In other words, shear

FIGURE 4.5 Tension force.

FIGURE 4.7 Compressive force.

FIGURE 4.6 Unloaded member.

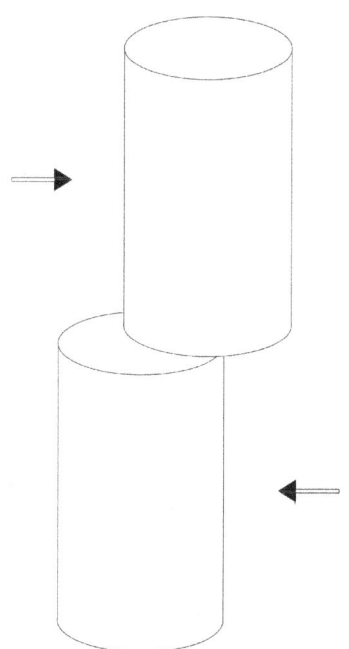

FIGURE 4.8 Shear force.

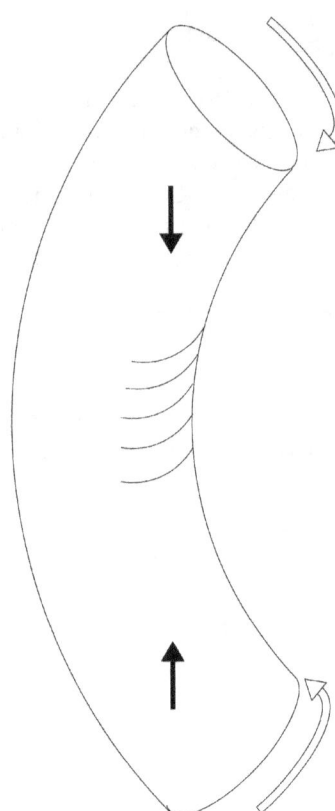

FIGURE 4.9 Bending force.

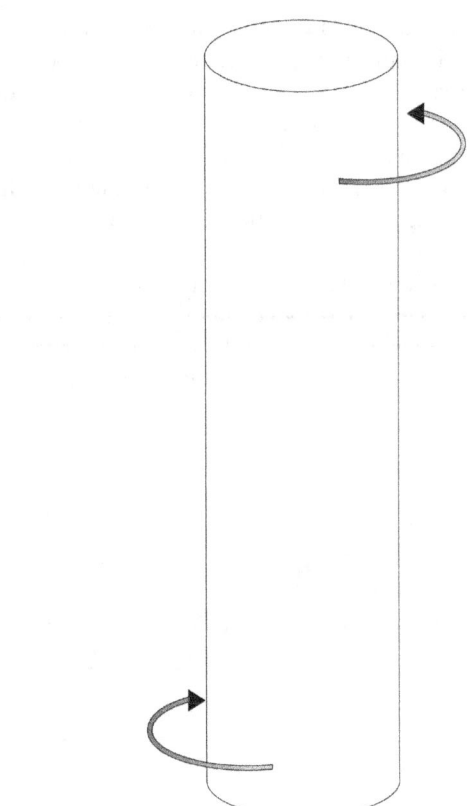

FIGURE 4.10 Torsion force.

force causes deformation of materials or structural elements by sliding or rotating parts. It occurs when two forces act in opposite directions, parallel to each other, but not along the same line. Shear forces are common in bolts, rivets, and other fasteners that hold materials together.

4.5.6 Bending

This is the force that causes curvature in beams or structures. Bending is a combination of tension and compression that causes an object, usually a beam or any other structural member, to experience internal stresses that create concave and convex regions. It happens when an external force is applied perpendicular to the longitudinal axis of the object. Bending is very much considered in the design of bridges, buildings, and other structures. Figure 4.9 illustrates a bending force.

4.5.7 Torque (Torsion)

Torque, also called torsion force (Figure 4.10), is the rotational force that affects shafts, foundations, and structures. In other words, it is a force that causes an object to twist along its longitudinal axis. It occurs when forces act in opposite directions, causing a twisting motion. Torsional forces are significant in shafts, propellers, and other rotating components.

4.5.8 Seismic Forces

These are forces generated by earthquakes, affecting structural design. It's caused by the ground motion and shaking that

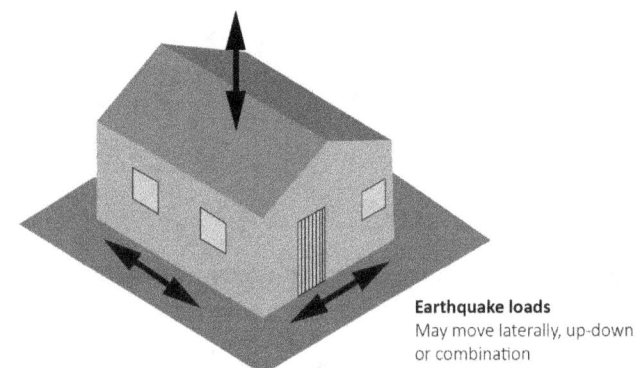

Earthquake loads
May move laterally, up-down or combination

FIGURE 4.11 Seismic force.

occurs during seismic activity. Seismic waves are mechanical waves of acoustic energy that travel through the Earth or another planetary body. They can result from an earthquake, volcanic eruption, magma movement, a large landslide, and a large manmade explosion that produces low-frequency acoustic energy.

Seismic forces (Figure 4.11), generated by earthquakes, are dynamic forces that act on civil engineering structures, primarily causing horizontal vibrations and inertia forces. These forces are a critical consideration in structural design, aiming to ensure buildings and infrastructure can withstand earthquakes and minimize potential damage or collapse. Clearly, more mass implies higher inertia force, meaning that lighter buildings sustain the earthquake shaking better. The inertia force experienced by the roof is transferred to the ground via the columns, inducing forces in columns. Seismic forces' impact on structures and earthquake-resistant structural design is a very active area of research for civil engineers today!

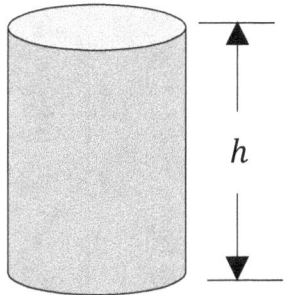

FIGURE 4.12   Hydrostatic force.

4.5.9 Hydrodynamic Forces

These are forces exerted by a fluid in motion on an object or structure such as dams or bridges. These forces are caused by the velocity and momentum of the fluid. These forces present a very active area of research for civil engineers!

4.5.10 Hydrostatic Forces

Hydrostatic force is the force exerted by a fluid (liquid or gas) on an object due to the pressure the fluid generates when it is at **rest or stationary**. Consider a stationary fluid in a cylindrical container in Figure 4.12. Hydrostatic force can be calculated using the following formula:

$$F = P.A = h\rho g.A \qquad (4.1)$$

where

> F: hydrostatic force
> P: fluid pressure
> A: cross-sectional area
> h: height of the fluid
> ρ: density of the fluid
> g: acceleration due to gravity

The center of pressure is a point on the immersed surface where the resultant hydrostatic pressure force acts.

4.5.11 Geotechnical Forces

These are various forces acting on soil and rock masses, influencing their behavior and stability. These forces can be internal (such as gravity and pore water pressure) or external (like loads from structures, seismic activity, or groundwater flow). The forces are related to soil mechanics and affect foundations and earthworks.

4.5.12 Thermal Forces

The forces caused by temperature changes, affecting material expansion and contraction. These forces are catered for in structures by allowing for expansion joints. See Figure 4.13.

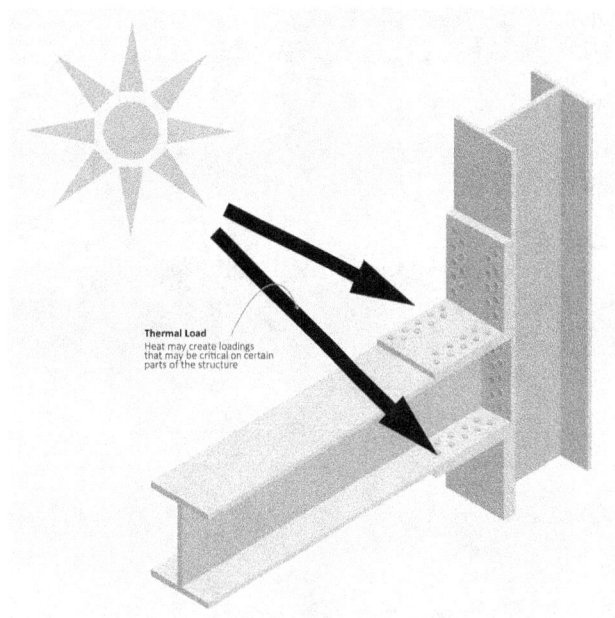

FIGURE 4.13   Thermal force.

4.5.13 Aerodynamic Forces

An aerodynamic force is a force exerted on a body by the air in which the body is immersed and is due to the relative motion between the body and the gas. For example, wind loads are a type of aerodynamic force that acts on structures, buildings, bridges, towers, and other objects due to wind flow. These forces can cause pressure on surfaces of the structures, which can lead to forces and moments. Wind can also create suction or partial vacuum or negative pressure on some surfaces.

Shear forces can also occur when wind loads create a force parallel to a surface or a structural element, causing it to deform or fail by sliding or rotating. Pressure acts normal to the surface, and shear force acts parallel to the surface. Both forces act locally. The net aerodynamic force on the body is equal to the pressure and shear forces integrated over the body's total exposed area.

Wind loads arising from wind flow speeds can be detrimental to civil engineering structures, especially towers, skyscrapers, and bridges. Basic wind speed is an essential parameter used for conversion into wind loads on building structures. Wind speeds are categorized and considered based on factors like location, building type, and the desired level of risk tolerance.

Imagine building towers or skyscrapers at **Barrow Island** which currently holds the Guinness World Record for the highest recorded wind speed not associated with a tornado. During 1996's Tropical Cyclone Olivia, 253 mph (407.164 km/hour) winds were clocked by an unmanned weather station on this portion of Western Australia's northwest coast. The most common categories include basic wind speeds (used for initial design calculations), ultimate design wind speeds (considering potential for extreme events), and gust wind speeds (accounting for rapid fluctuations in wind). These wind speeds are used to calculate wind loads, which are then factored into structural designs to ensure safety and durability.

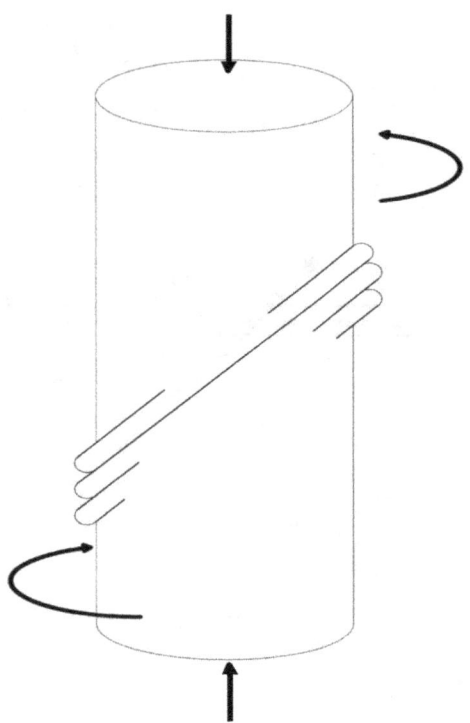

FIGURE 4.14 Combined loading.

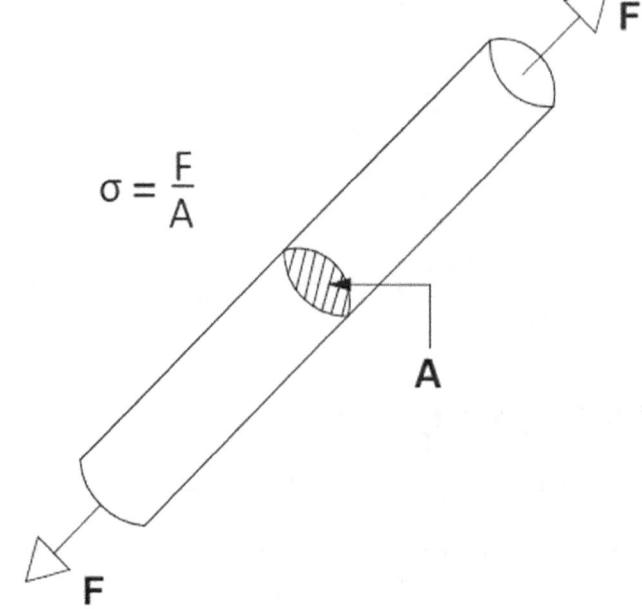

FIGURE 4.15 illustrating the concept of stress.

where

σ: stress
P: force (N)
A: cross-sectional area (m^2)

Note: Stress is measured in Pascals, $1\,\text{Pa} = 1\text{N/m}^2$. We may, therefore, say that there are five primary types of forces and stresses in civil engineering: tensile forces, compressive forces, flexural stress, shear forces, and torsion. Different types of stresses that materials can experience due to external forces are discussed in Sections 4.6.1–4.6.4.

4.5.14 Combined Loading

Combined loading (Figure 4.14) involves the simultaneous application of multiple forces on a structure. It is not an explicit force, but instead it is a combination of several forces that act as a whole on a structure. These forces can combine tension, compression, bending, shear, and torsion. In real-world situations, structures are often subject to various combination of loads, and engineers need to consider all these forces when designing to ensure structural integrity.

4.6 WHAT IS STRESS IN CIVIL ENGINEERING?

Stress is defined as the **internal-force distribution** along a specific surface orientation at a uniquely distinct point within a structural element experiencing applied loads. You will notice that instead of using forces such as compressive force or tensile force, in many cases, compressive **stress or tensile stress** is used. The reason is provided in Section 5.5.1.

In simple terms, stress is **'force per unit area'** and is the product of the applied force F (or P) over the cross-sectional area. In a much more general sense, the concept of stress might be used to understand the situation at any point inside a solid. Stress could also be utilized to estimate when a material will fail. Figure 4.15 illustrates the concept of stress. Stress formula is written as follows:

$$\sigma = \frac{\text{Force}}{\text{Cross}-\text{sectional area}} = \frac{F}{A} \qquad (4.2)$$

4.6.1 Axial Stress

Axial stress occurs when a force is applied parallel to the length of an object, causing it to stretch or compress. It's a uni-axial stress, meaning that the force acts along a single axis. It is, therefore, the result of a force acting perpendicular to an area of a body, causing the extension or compression of the material. **Example:** A rope being pulled apart or a column being compressed.

4.6.2 Tensile Stress

Tensile stress is a type of axial stress that occurs when an object is stretched or pulled apart. It's a stretching force that tries to elongate the material. **Example:** A rubber band being stretched or a wire being pulled.

4.6.3 Compressive Stress

Compressive stress is also a type of axial stress, but it occurs when an object is squeezed or compressed. It's a squeezing force that tries to shorten the material. **Example:** A brick being compressed or a column being crushed.

4.6.4 Shearing Stress

Shearing stress occurs when a force is applied parallel to the surface of an object, causing it to deform by sliding or rotating. It's a tangential force that tries to change the shape of the material. **Example:** A piece of paper being cut or a bolt being sheared.

4.7 UNDERSTANDING LOADS IN CIVIL ENGINEERING

Understanding and accurately calculating applicable loads is essential for designing safe and durable civil engineering structures. From a 3-D perspective (Figure 4.16), loads encountered by civil engineers affect structures from different planes, that is, the vertical, horizontal, and transverse planes. Engineers need to understand this and visualize the applicable forces when designing systems:

- The vertical plane divides/indicates the civil engineering structure into the front and back or left and right sections.
- The horizontal plane divides/indicates the structure into top and bottom sections.
- The transverse plane (or lateral plane) divides/indicates the structure into left and right sections, perpendicular to the longitudinal axis.

From the above explanation, vertical loads can be dead or live loads imposed/acting on a structure; dead loads include the weight of the structure itself, while live loads include weight of people, furniture, and movable objects. The vertical loads can be gravity loads such as snow loads (Figure 4.17), that is, weight of snow accumulation. Surcharge loads (vertical) are additional loads imposed on a structure or soil, applicable in retaining wall development and foundation design, buried pipelines, etc.

Lateral loads such as wind loads (forces exerted by wind) and seismic loads (forces caused by earthquakes) are always encountered by civil engineers. Soil pressure/geotechnical forces are lateral pressures exerted by soil. The

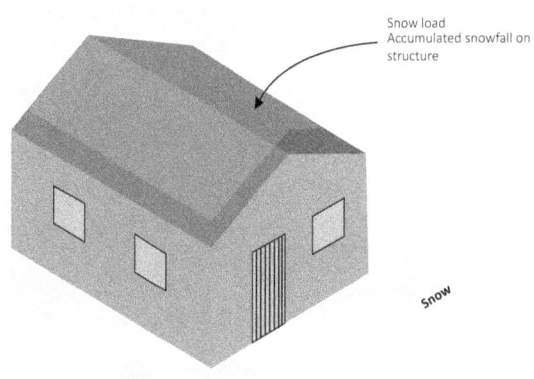

FIGURE 4.17   Snow loads.

hydrostatic loads (lateral) are forces exerted by water at rest. Hydrodynamic loads act from all angles, that is, across three planes; they are forces exerted by moving water. Impact loads are sudden forces caused by impact or collision. Thermal loads are forces caused by temperature changes. Dynamic loads are forces caused by vibration, oscillation, or other dynamic effects. Figure 4.18 illustrates point loads, distributed triangular loads, and uniformly distributed loads. Civil engineers have to differentiate all those terminologies listed above.

4.8 MIND MAPPING QUESTIONS

1. Explore forces on a dam, skyscraper, road pavement, buried pipeline, and tunnel.
2. What are the different types of external forces that can act on a bridge, and how do they impact its stability and design?

4.9 CONCLUSION

Civil engineers deal with forces in the design, construction, operation, and maintenance of structures. Their primary role is to design and develop safe structures for public use.

In that regard, they strive to ensure that all forces are accounted for in their work because once significant forces are neglected or failed to be accounted for, catastrophic events are bound to happen.

A civil engineer must ensure that all forces that are likely to impact the structure are envisioned straightaway because civil engineering structures are large systems whose reliability must be ensured right from the start. Failure to do so can have great impacts on systems because remodeling halfway or after project development can be costlier.

In the design and construction, dead and live loads must be considered without neglect. This is in line with safety considerations. It is important to note that safety considerations

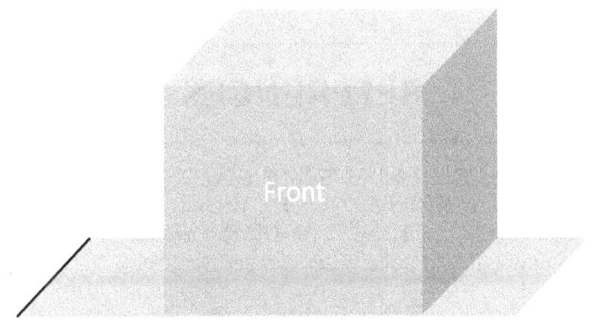

FIGURE 4.16   3D representation of structure.

Point load
P

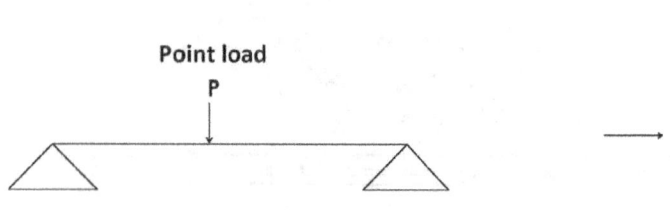

Unifromly distributed load
Q

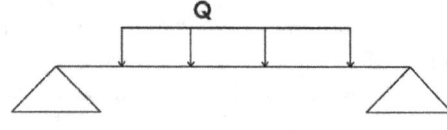

Triangular distributed load

W

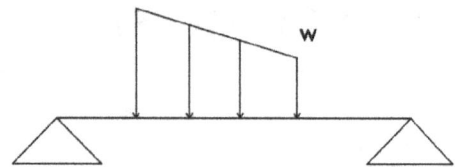

FIGURE 4.18   Point and distributed loads.

are the key cornerstones of a successful civil engineering proj-ect, and no compromise should ever be made on safety.

Many structures may be impacted by combined forces, and all of them must be accounted for to design, develop, or maintain a safe structure to perform the intended function. In summary,

- **Axial stress:** Force applied along the length of an object (stretching or compressing).
- **Tensile stress:** Force stretching an object.
- **Compressive stress:** Force compressing an object.
- **Shearing stress:** Force causing an object to deform by sliding or rotating.

These stresses can occur individually or in combination, depending on the loading conditions. Understanding these concepts is essential in engineering, materials science, and architecture to design and analyze structures and materials:

- **Force** is an external push or pull acting on an object.
- **Load** is a specific term for the forces acting on a structure or system, often used in engineering to specify the forces that cause stress.
- **Stress** is the internal resistance of a material to an external force, measured as force per unit area.
- While all loads are forces, not all forces are consid-ered loads.

REFERENCES

1. Costanzo, F., Plesha, M. E., & Gray, G. L. (2010). *Engineering Mechanics: Statics & Dynamics*. McGraw-Hill, New York.
2. Meriam, J. L., & Kraige, L. G. (1998). *Engineering Mechanics: Statics* (4th ed.). John Wiley & Sons, New York.

The Science of Materials

5

5.1 INTRODUCTION

It is essential to know what kind of materials are suitable for a civil engineering project. Usually, civil engineering structures are designed to carry loads. These loads can vary in magnitude. The study of the science of materials is important because it informs the engineer about the most suitable materials to adopt taking into consideration sustainability issues, cost, safety, and durability.

> In the civil engineering context, a material in its primary sense, is a substance or combination of substances having specific physical, chemical, and mechanical properties, used in the construction, operation, maintenance, or repair of infrastructure, buildings, and other engineered systems.

In the civil engineering industry, two primary attributes govern the study of the science of materials, that is, temperature and pressure. These influence the physical, chemical, and mechanical properties of materials. Temperature variations influence the overall integrity of the material or structure, and they exert forces on structural members. Temperature changes are responsible for thermal expansion; materials contract or expand with temperature changes, and this affects their dimensions and structural integrity, overall. Thus, temperature fluctuations in construction or building materials can cause stresses to build up in the materials. In terms of the strength and durability, temperature influences the material strength in such a way that some materials weaken or become brittle at extreme temperatures. For example, asphalt concrete loses its binding properties and becomes soft and prone to deformation at temperatures above 50°C. Also, uPVC pipes become brittle and prone to cracking at temperatures above 60°C. Note that temperature-induced changes, for example, in a concrete setting, impact its properties, and materials absorb moisture differently at different temperatures, thus affecting material properties.

The combined temperature and pressure effects have interesting facts. For example, sustained loads and high temperatures cause material deformation over time. Engineers, therefore, plan, design, and construct structures to understand how will the structure stand permanently without creeping (gradual deformation) bearing dead and live loads with temperature fluctuations over the design period?

Material degradation is accelerated by two primary factors: pressure and temperature, for example, corrosion. Due to repeated loading and unloading cycles, intensified by temperature fluctuations, materials are fatigued, influencing chemical reactions such as concrete hydration and degradation. This reduces the material lifespan. Therefore, while other attributes are important, the primary attributes used by engineers are temperature and pressure while studying and specifying materials. These modify and influence chemical, mechanical, and physical properties of materials right from the manufacturing process, through the installation, operation, and maintenance of infrastructure systems.

It is, thus, a technical axiom that the control of pressure and temperature in all civil engineering applications is crucial, whether in planning, design, operation, or maintenance of a civil engineering structure, and this applies throughout the lifecycle phases of the project. Therefore, the understanding of how these key attributes will impact the structure is key in all lifecycle phases.

Pressure is the force exerted per unit area on an object or surface. In its basic sense, it is a measure of the normal force (perpendicular force) applied to a surface, divided by the area of that surface. Sometimes, the force that deforms or interacts with a material or structure may not be normal. It may be at an angle to the surface. The material response to that force is termed as 'stress.' So, stress can be generated by pressure or temperature fluctuations. In summary, pressure is an external force, while stress is the internal response of a material to that force. Therefore, these external forces acting on a civil engineering structure or material induce stresses within the structure or material.

Temperature is a measure of the average kinetic energy of particles in a substance or system. It is a fundamental physical property that describes the thermal energy of a system. In simpler terms, temperature is a measure of how hot or cold a material is. Temperature significantly impacts the strength of materials, and it's a crucial factor in the study of materials and structures.

Temperature and pressure play crucial roles in civil engineering as they affect the behavior and properties of substances, materials, and systems. These are the two primary factors that cause the deformation of materials.

Materials are selected based on how well they resist these factors to remain durable and safe for the purpose—not to deform. Several terminologies and concepts are built on those two primary attributes. Factors based around these two primary attributes in the study of civil engineering materials have much importance. As a civil engineer, it is important to grasp the fact that temperature and pressure are the primary factors whose impact on structures and materials presents further analysis and understanding of our built environment at both the micro and macroscopic scales.

Sample cases

- **Imagine a fluid flowing through a buried pipeline!** The temperature at which it flows influences the type of pipe material, and the quantity of flow influences the internal force exerted on the pipe

DOI: 10.1201/9781003561620-5

walls, that is, pressure. Moreover, the temperature of the fluid influences its viscosity, and the material (i.e., pipe material) that withstands the buildup of stresses (both internal and external) due to the pressure in the pipeline and backfill loads (or external forces) and temperature at which it moves will be selected. That means the material with enough strength to counteract the estimated forces must be selected. Thus, water, gas, sewage, oils, and other hydrocarbons have varying pipeline materials, sizes, and strengths. All other factors that influence the choice of the type of pipe to use, say HDPE, UPVC, GI, and ductile iron, are secondary.

- **Imagine building a storied structure in South Sudan, Juba, where temperatures can go up to 40°C, or in furnace Creek in Death Valley, California, USA, where temperatures can be above +50°C!** The temperature impacts the characteristics of structural members, and the loads the structure will carry will influence the sizes, type, etc. Still, temperature and pressure are the primary factors. Materials expand and contract with temperature fluctuations. For example, metals like steel are known to perform relatively well in a temperature range from −20°C to 50°C. That's why we study thermal expansion of materials. And when designing bridges and railways, we account for thermal expansion, providing expansion joints to cater for cooling and expansion from time to time.

- **Imagine proper concrete curing while minimizing shrinkage!** Concrete curing process involves controlling the temperature and moisture levels to promote proper hydration and hardening of the concrete. During curing, concrete is more prone to shrinkage due to the ongoing hydration reactions and moisture loss. Proper curing techniques, such as maintaining adequate moisture levels, can help minimize shrinkage and reduce the risk of cracking. The primary factors that contribute to concrete shrinkage are evaporation and hydration. However, secondary factors include temperature, humidity, cement content, aggregate type, and mix design. Shrinkage is a natural process in concrete, and it can lead to cracking if not properly managed. Proper curing helps reduce shrinkage. Proper curing involves maintaining a suitable environment for the concrete to set and harden, typically by, keeping the concrete moist, **controlling temperature**, and preventing excessive evaporation. It is essential to note that temperature influences all the factors that contribute to shrinkage.

- **Imagine building an overhead storage reservoir tank to store clean and safe water!** The two primary items are the volume of water in the tank and the temperature at which water will be stored. The former gives the hydrostatic forces to act on the walls of the tank and later provides the required temperatures (thermal conditions). So, when deciding the materials to use to build the reservoir and tower, we must find those that can withstand the tensile and compressive forces—with enough tensile and flexural strength. The loading on the tank tower exerts forces into the structural members. Their internal responses (stresses) to resist deformation due to the loading must be robust.

- **Imagine building an asphalt concrete road!** The two primary factors that affect the pavement strength determination are the vehicle loads exerting pressure on the pavement and the temperature at which asphalt is laid and compacted and operated—whether it allows to achieve the desired compaction—strength! And whether the area's climate would support the longevity of the asphalt road! Viscosity–temperature relations are key concepts in the design, analysis, evaluation, and construction process for the road pavement, that is, 'viscosity–temperature relationship of binder—bitumen.' The air-void content in asphalt concrete is found to decrease with an increase in compaction temperature. If the air-void content is too high, it allows for air intrusion and water. The higher the binders' viscosity, the greater the resistance to compaction. Therefore, temperature is always monitored. Similarly, the road base and sub-base materials are compacted (pressure exerted) to achieve the desired strength to carry the vehicular traffic. Other parameters and analyses are secondary.

- **Imagine manufacturing steel rebars!** In industrial applications, every civil engineering material, for example, composites, metals, and alloys, is manufactured with two key attributes: modification in pressure exerted on the constituent materials and temperature fluctuations or variations in the industrial processes. Iron ore is smelted to manufacture rebars at 1000°C. An alloy of carbon and iron makes steel; we have thermo-mechanically treated (TMT) bars, hot-rolled bars, cold-rolled steel bars, and mild steel bars; all of these are manufactured by controlling temperatures.

- **Pressed steel tank panels' manufacturing**. The manufacturing process for hydraulically cold-pressed panels and hot dip-galvanized (HDG) panels differs in the sense that hydraulically cold-pressed panels are formed without heat, whereas HDG panels involve a hot-dipping process for coating, showcasing the role of temperature and pressure in the manufacturing of civil engineering materials!

The above examples provide an insight into how several factors are built on the two primary factors that modify or deform the physical, chemical, and mechanical characteristics of materials, that is, pressure and temperature. Engineering concepts such as stress, strain, Young's modulus, Poisson's ratio, viscosity, chemical composition, entropy, and density start where the study of pressure and temperature boils down, as far as civil engineering materials are concerned.

Note: Physical properties refer to observable characteristics of a material. Examples of physical properties include color, melting point, and density. Chemical properties are characteristics of a substance or material that may be observed when it participates in a chemical reaction. Examples of chemical properties include flammability, toxicity, chemical stability, reactivity, corrosion resistance, oxidation resistance, and heat of combustion.

Mechanical properties describe how a material reacts to external forces like pushing, pulling, or twisting. Mechanical properties are studied under strength of materials, which is a crucial topic for civil engineers, necessitating an outline of the reasons why every civil engineer must study the strength of materials provided in this chapter. Altogether, the study of physical, chemical, and mechanical properties of materials is essential for proper and reliable design of civil engineering structures.

Materials that can be deformed significantly before breaking are called ductile materials. Some examples of ductile materials are many metals and plastics. The opposite of ductile is brittle. Brittle materials don't deform much before breaking. Instead, they maintain their shape right until they break. **Note:** When designing or developing a civil engineering structure, two primary material considerations arise:

- How will temperature fluctuations affect the structure and materials and how will the induced stresses due to temperatures changes be dealt with?
- How will the pressures or external forces applied to the structure induce stresses in the structure and materials and how will the induced stresses due to external forces be dealt with?

From buildings, to earth-retaining structures, to water tanks, to bridges, to road pavements, to towers, to hydraulic structures, etc., these are the primary questions as far as materials to use in the project are concerned. Other factors such as environmental factors follow the two primary attributes.

5.2 WHY STUDY THE SCIENCE OF MATERIALS?

Civil engineers study the science of materials to be able to select appropriate materials for a project. Without knowing how materials behave under different conditions, civil engineers cannot select the best materials for the job.

The science of materials primarily focuses on the study of the structure, composition, and properties of various materials. This includes their physical, chemical, and mechanical characteristics. It explores how materials behave under different conditions, such as temperatures, pressure, and environmental exposure.

Understanding the properties and behavior of materials helps civil engineers design and build structures that are durable, safe, and resistant to various loads and environmental conditions.

Through the study of material science, civil engineers are able to understand the physical and mechanical properties of materials. Thus, they can design structures optimized for specific loads and environmental conditions. Hence, they minimize unnecessary material usage and cost.

Because civil engineers primarily deal with forces that induce loads on structural members, they must study the science of materials to understand how they behave when loaded or exposed to different environmental conditions. Without knowing how they behave, engineers can design and build structures that put too much load on the materials. Poorly built structures can buckle, twist, and the entire structure can collapse, killing people inside it. In that regard, civil engineers work to make structures more effective, efficient, and sustainable. Therefore, understanding the properties and behavior of materials helps civil engineers design and build structures that are safe, durable, and resistant to various loads and environmental conditions, achieving structural integrity.

Knowledge of material behavior is crucial to prevent structural failures by selecting materials with adequate strength and resistance to factors like corrosion, fatigue, and impact. Materials that are not good enough to bear loads exist, that is, those that cannot withstand pressures and those that can. Also, some can corrode easily, and some cannot.

Knowledge of material science enables civil engineers to select the most suitable materials for specific applications. By understanding how different materials behave under stress and load, civil engineers can choose the right materials for various construction projects, such as concrete for buildings, steel for bridges, or asphalt for roads. This ensures the safety, durability, and functionality of structures, considering factors like strength, stiffness, and resistance to environmental conditions.

Studying materials science enables civil engineers to perform quality checks (quality control) on construction materials to ensure they meet design specifications.

Civil engineers need to understand what the materials are made of. From the science of compounds and atoms, civil engineers need to understand the chemical composition of materials to be able to specify the right materials for the job. A civil engineer need to differentiate between a mixture, compound, and a solution as several materials used in construction fall under those categories. Understanding mixtures such as a concrete (composite mixture), alloys, blend, suspension, emulsion, and colloid is essential for a civil engineer.

Understanding compounds such as cements, aggregates, polymers, elastomers, bituminous compounds, lime compounds, and thermosets and solutions such as adhesives, coatings, sealants, bonding agents, grouts, and mortars is essential for a civil engineer. All these are used in civil engineering and construction. It is important to note that these materials behave differently under different conditions, such as **temperatures**, **pressure**, and **environmental exposure**. During construction, these materials are mixed as appropriate, forming solutions or mixtures, to develop or beautify structures, and it essential to recognize that pressure and temperature are crucial drivers in modifying or influencing/shaping the desired outcome.

By studying the science of materials, civil engineers are able to differentiate between a composite and an alloy, additives and admixtures, and when to use/specify them because they are some of the common materials used in construction.

Civil engineers need to select appropriate construction methods and be familiar with material properties and behavior to make decisions on construction techniques, ensuring that materials are used efficiently and effectively.

Understanding material properties can help engineers choose environmentally friendly materials with reduced environmental impact. Studying material science allows civil engineers to explore new, sustainable, and innovative materials, contributing to the development of more resilient and environmentally friendly infrastructure.

By understanding material degradation and failure mechanisms, civil engineers develop strategies for maintenance, repair, and rehabilitation of structures.

5.3 TEMPERATURE EFFECTS ON MATERIALS

Temperature greatly affects civil engineering materials, that is, metals, concrete, ceramics (including bricks and masonry), fiber composites, polymers, bituminous materials, timber, and glass. These are the primary classes of materials. Temperature affects all these materials in different ways, influencing their properties, behavior, and performance. Understanding these effects is crucial for designing and constructing safe, durable, and efficient buildings and infrastructure.

Temperature effects on materials include thermal expansion and contraction, loss of strength and ductility, corrosion, hydration, shrinkage, moisture absorption, durability losses, degradation, freeze–thaw damage, and moisture and fungal damage.

Metals such as steel and aluminum are affected by thermal expansion and contraction. Metals expand and contract with temperature changes, which can cause structural movements and stresses. Temperature affects metal strength and ductility. Some metals, like steel, lose strength at high temperatures, while others, like aluminum, become weaker at low temperatures. Temperature and humidity can influence corrosion rates in metals.

Glass expands and contracts with temperature changes, which can cause structural stresses. Glass is also susceptible to thermal shock, which can cause it to shatter or crack. Temperature and humidity can influence the chemical durability of glass.

Ceramics including bricks and masonry are affected by thermal expansion. Like metals, ceramics expand and contract with temperature changes. Ceramics can absorb moisture, leading to expansion and potential cracking. Ceramics can be susceptible to freeze–thaw damage, especially if saturated with water.

On the other hand, temperature changes affect the hydration process, which can impact concrete strength and durability. Concrete shrinks as it cools, which can cause cracking and structural issues. Concrete's resistance to freeze–thaw cycles is critical in cold climates. In the same way, the curing process can be affected by temperature changes.

Fiber composites are affected by temperature. For example, high temperatures can cause the matrix material (e.g., polymer) to degrade, reducing the composite's strength and stiffness. Some fibers (e.g., glass) can degrade at high temperatures, affecting the composite's properties. Fiber composites can absorb moisture, leading to degradation and reduced performance.

High temperatures can cause polymers to degrade, losing their strength and stiffness. Polymers can become brittle and lose their impact resistance below their glass transition temperature. Polymers can absorb moisture, leading to degradation and reduced performance.

For bituminous materials, temperature affects the viscosity of bituminous materials, influencing their flow and application properties. Also, bituminous materials can soften and lose their strength at high temperatures. Bituminous materials can oxidize at high temperatures, leading to degradation and reduced performance.

Temperature affects the moisture content of timber, influencing its strength, stiffness, and durability. Timber shrinks as it dries, which can cause structural movements and stresses. Temperature and moisture can influence the activity of insects and fungi that can damage timber.

5.3.1 Thermal Expansion

Metals and ceramics, for example, expand and contract with temperature changes, which can cause structural movements and stresses. Glass, too, expands and contracts with temperature changes, which can cause structural stresses.

It is important to recognize the fact that when an object such as a metal is heated or cooled, its length changes by an amount proportional to the original length and the change in temperature. The linear thermal expansion is the change in the length of an object when heated and can be expressed as

$$\Delta l = l_0 \alpha (t_1 - t_0) \tag{5.1}$$

where

Δl: change in object length (m, inches)
l_0: initial length of object (m, inches)
α: linear expansion coefficient (m/m°C, in/in°F)[1]
t_0: initial temperature (°C, °F)
t_1: final temperature (°C, °F)

Therefore, the final length of the object can be calculated as

$$l_1 = l_o + \Delta l = l_0 + l_0 \alpha (t_1 - t_0) \tag{5.2}$$

where

l_1 = final length of object (m, inches).

The linear thermal expansion coefficient (α) is a material property that describes the change in the length of a material in response to a change in the temperature. It is the ratio of the change in length (Δl) to the original length (l_0) of a material, per unit change in temperature (Δt). See Table 5.1 for linear expansion coefficients for common products [1]. Linear expansion coefficients for most materials vary with temperature.

5.3.1.1 Worked example [5.1] for illustrative purposes

An **I-beam** made of structural steel is designed for temperatures ranging from −30°C to 50°C. If a beam's length is 6 m when assembled at 20°C, what is the shortest final length of

TABLE 5.1   Linear expansion coefficients for common products

PRODUCT	LINEAR TEMPERATURE EXPANSION COEFFICIENT (α) $(10^{-6}m/(m°C))$
ABS (acrylonitrile butadiene styrene) thermoplastic	72–108
ABS—glass fiber-reinforced	31
Acetal—glass fiber-reinforced	39
Acetals	85–110
Acrylic	68–75
ALLVAR Alloy 30 (negative thermal expansion)	−30
Alumina (aluminum oxide, Al_2O_3)	8.1
Aluminum	21–24
Aluminum nitride	5.3
Amber	50–60
Antimonial lead (hard lead)	26.5
Antimony	9–11
Arsenic	4.7
Bakelite, bleached	22
Barium	20.6
Barium ferrite	10
Benzocyclobutene	42
Beryllium	12
Bismuth	13–13.5
Brass	18–19
Brick masonry	5
Bronze	17.5–18
Cadmium	30
Calcium	22.3
Caoutchouc	66–69
Cast Iron Gray	10.8
Celluloid	100
Cellulose acetate (CA)	130
Cellulose acetate butynate (CAB)	96–171
Cellulose nitrate (CN)	80–120
Cement, Portland	11
Cerium	5.2
Chlorinated polyether	80
Chlorinated polyvinylchloride (CPVC)	63–66
Chromium	6–7
Clay tile structure	5.9
Cobalt	12
Concrete	13–14
Concrete structure	9.8
Constantan	15.2–18.8
Copper	16–16.7
Copper, Beryllium 25	17.8
Corundum, sintered	6.5
Cupronickel 30% (constantan)	16.2
Diamond (carbon)	1.1–1.3
Duralumin	23
Dysprosium	9.9
Ebonite	70
Epoxy—glass fiber-reinforced	36
Epoxy, cast resins, and compounds, unfilled	45–65
Erbium	12.2
Ethylene ethyl acrylate (EEA)	205

(Continued)

TABLE 5.1 (*Continued*) Linear expansion coefficients for common products

PRODUCT	*LINEAR TEMPERATURE EXPANSION COEFFICIENT* (α) ($10^{-6}m/(m°C)$)
Ethylene vinyl acetate (EVA)	180
Europium	35
Fluoroethylene propylene (FEP)	135
Fluorspar, CaF_2	19.5
Gadolinium	9
German silver	18.4
Germanium	6.1
Glass, hard	5.9
Glass, plate	9.0
Glass, pyrex	4.0
Gold	14.2
Gold—copper	15.5
Gold—platinum	15.2
Granite	7.9–8.4
Graphite, pure (carbon)	4–8
Gunmetal	18
Gutta percha	198
Hafnium	5.9
Hard alloy K20	6
Hastelloy C	11.3
Holmium	11.2
Ice, 0°C water	51
Inconel	11.5–12.6
Indium	33
Invar	1.5
Iridium	6.4
Iron, cast	10.4–11
Iron, forged	11.3
Iron, pure	12.0
Kapton	20
Lanthanum	12.1
Lead	29
Limestone	8
Lithium	46
Lutetium	9.9
Macor	9.3
Magnalium	23.8
Magnesium	25–26.9
Magnesium alloy AZ31B	26
Manganese	22
Manganin	18.1
Marble	5.5–14.1
Masonry, brick	4.7–9.0
Mercury	61
Mica	3
Molybdenum	5
Monel metal	13.5
Mortar	7.3–13.5
Neodymium	9.6
Nickel	13.0
Niobium (columbium)	7
Nylon, general purpose	50–90
Nylon, glass fiber-reinforced	23

(*Continued*)

TABLE 5.1 (Continued)   Linear expansion coefficients for common products

PRODUCT	LINEAR TEMPERATURE EXPANSION COEFFICIENT (α) $(10^{-6}m/(m°C))$
Nylon, Type 11, molding and extruding compound	100
Nylon, Type 12, molding and extruding compound	80.5
Nylon, Type 6, cast	85
Nylon, Type 6/6, molding compound	80
Oak, perpendicular to the grain	54
Osmium	5–6
Palladium	11.8
Paraffin	106–480
Phenolic resin without fillers	60–80
Phosphor bronze	16.7
Plaster	17
Plastics	40–120
Platinum	9
Plutonium	47–54
Polyacrylonitrile	70
Polyallomer	92
Polyamide (PA)	110
Polybutylene (PB)	130–139
Polycarbonate (PC)	65–70
Polycarbonate—glass fiber-reinforced	21.5
Polyester	124
Polyester—glass fiber-reinforced	25
Polyethylene (PE)	108–200
Polyethylene (PE) —high molecular weight	108
Polyethylene terephthalate (PET)	59.4
Polyphenylene	54
Polyphenylene—glass fiber-reinforced	36
Polypropylene (PP), unfilled	72–90
Polypropylene—glass fiber-reinforced	32
Polystyrene (PS)	70
Polysulfone (PSO)	55–60
Polytetrafluoroethylene (PTFE)	112–135
Polyurethane (PUR), rigid	57.6
Polyvinyl chloride (PVC)	54–110
Polyvinylidene fluoride (PVDF)	128–140
Porcelain, industrial	4
Potassium	83
Praseodymium	6.7
Promethium	11
Quartz, fused	0.55
Quartz, mineral	8–14
Rhenium	6.7
Rhodium	8
Rock salt	40.4
Rubber, hard	80
Ruthenium	9.1
Samarium	12.7
Sandstone	11.6
Sapphire	5.3
Scandium	10.2
Selenium	37
Silicon	3–5

(Continued)

TABLE 5.1 (*Continued*)   Linear expansion coefficients for common products

PRODUCT	LINEAR TEMPERATURE EXPANSION COEFFICIENT (α) $(10^{-6}m/(m°C))$
Silicon carbide	2.77
Silver	19–19.7
Sitall	0.15
Slate	10
Sodium	70
Solder lead—tin, 50% –50%	25
Speculum metal	19.3
Steatite	8.5
Steel	10.8–12.5
Steel stainless austenitic (304)	17.3
Steel stainless austenitic (310)	14.4
Steel stainless austenitic (316)	16.0
Steel stainless ferritic (410)	9.9
Strontium	22.5
Tantalum	6.5
Tellurium	36.9
Terbium	10.3
Terne	11.6
Thallium	29.9
Thorium	12
Thulium	13.3
Tin	20–23
Titanium	8.5–9
Topas	5–8
Tungsten	4.5
Uranium	13.4
Vanadium	8
Vinyl ester	16–22
Vulcanite	63.6
Wax	2–15
Wedgwood ware	8.9
Wood, across (perpendicular) to grain	30
Wood, fir	3.7
Wood, parallel to grain	3
Wood, pine	5
Ytterbium	26.3
Yttrium	10.6
Zinc	30–35
Zirconium	5.7

Notes:

- $10^{-6}m/m°C = 1\mu m/m°C$.
- m/m = meter per meter, in/in = inches per inches.
- Most values for are for temperature 25°C(77°F) The span in the values may be caused by the variation in the materials themselves—or by the variation in the sources used.
- Superficial expansion: The amount by which a unit area of a material increases when the temperature is increased by one degree is called the coefficient of superficial (area) expansion.
- Cubic expansion: The amount by which a unit volume of a material increases when the temperature is raised by one degree is called the coefficient of cubic expansion.
- $t_K = t_c + 273.16$.
- $t_R = t_F + 459.67$.
- 1 in (inch) = 25.4 mm.
- 1 ft (foot) = 0.3048 m.

the beam at minimum temperature −30°C and the longest final length of the beam at maximum temperature 50°C?

Solution

The shortest final length of the beam at minimum temperature −30°C can be calculated as

$$l_1 = 6 + 6(0.000012 \text{ m/m}°C)(-30°C - 20°C) = 5.9964 \text{ m}$$

$$(5.3)$$

Therefore, the contraction is 3.6 mm.

The longest final length of the beam at maximum temperature 50°C can be calculated as

$$l_1 = 6 + 6(0.000012 \text{ m/m}°C)(50°C - 20°C) = 6.00216 \text{ m}$$

$$(5.4)$$

Therefore, the expansion is 2.16 mm.

5.3.1.2 Worked example [5.2] for illustrative purposes (thermal stress)[2]

A steel bar (Figure 5.1) of square cross-section is constrained to just fit between two fixed supports A and B when the temperature $T_1 = 60°C$. If the temperature is increased to $T_2 = 120°C$, determine the average normal stress developed in the bar. ($\alpha = 1.2 \times 10^{-6}/°C$); $E_{\text{steel}} = 200$ GPa.

Schematic

Solution

In no external load conditions, the stress is caused by the force developed at the support due to thermal expansion. Suppose support A is removed and the temperature increased by 60°C as shown in Figure 5.2.

Extension caused

$$\delta_T = \alpha(\Delta T)l \qquad (5.5)$$

$$\delta_T = 12 \times 10^{-6} \times 60°C \times 2 \text{ m} = 1.44 \times 10^{-3} \text{m} \ (1.44 \text{ mm}) \qquad (5.6)$$

Now, the force required to push the bar back to its original position at support A is computed as follows:

$$P = \frac{\delta_T EA}{l} \qquad (5.7)$$

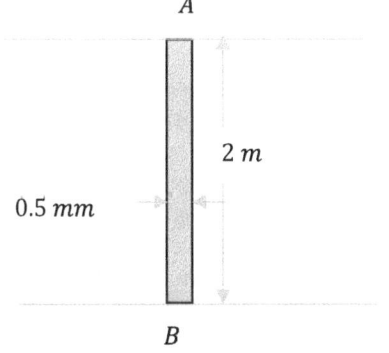

FIGURE 5.1   Steel bar of square cross-section.

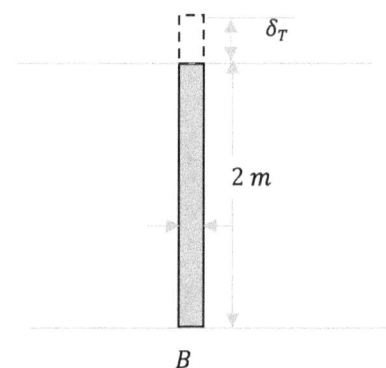

FIGURE 5.2   Steel bar after removing support A and heating to 60 °C.

$$P = \frac{1.44 \times 10^{-3} \times 200 \times 10^6 \times (0.5 \times 10^{-3})^2}{2} = 0.036 \text{ kN}$$

$$(5.8)$$

Therefore, the average normal stress (compressive stress (force)) developed in the bar is

$$\sigma_c = \frac{P}{A} = \frac{0.036}{(0.5 \times 10^{-3})^2} = 144 \text{ MPa} \qquad (5.9)$$

5.3.1.3 Case notes on thermal expansion

Generally, from Example 5.1, a 3.6 mm (contraction) and 2.16 mm (expansion) change might seem insignificant, but it can cause significant stresses in the structure, especially those with high thermal expansion coefficients. Tiny changes in dimensions can create stress concentrations, which can lead to material failure, cracks, or other defects. A stress concentration point is a localized area in a structure at which the stress is significantly higher than in the surrounding material. See Section 6.3.3.7. Considering cumulative effects, small changes can accumulate over time, leading to significant deviations from the intended design. This can compromise the structural integrity of buildings, towers, bridges, or other infrastructure. As seen from Example 5.2, the thermal expansion in bridge structures can cause thermal cracks, leading to failures.

Note: In real practice, structural steel, such as that used in I-beams, is typically designed to perform well within a temperature range of −20°C to 50°C (−4°F to 122°F). However, some steel grades and specialized coatings can extend this range: Low-temperature applications, between −50°C and −20°C (−58°F and −4°F), and high-temperature applications, between 50°C and 100°C (122°F and 212°F) or even higher.

As a practical application, expansion joints are placed in railroads (Figure 5.6) to cater for thermal expansion, typically referred to as 'expansion gaps' or 'thermal expansion gaps.' Expansion joints provide space for this movement, preventing track buckling or warping. In short, rails contract at low temperatures and experience tensile stress in high temperatures. This could lead to heat kinks, which force the track out of gauge and could cause derailments if preventive measures are not taken. So, to deal with thermal expansion, prior to installation, the rail is heated to its rail neutral temperature and then cooled as it is laid. Additionally, various technologies utilizing wayside devices are implemented to detect rail faults [2].

FIGURE 5.3   Qian'an–Caofeidian railway, PRC. (Image credits: Richard Li Chengzhe.)

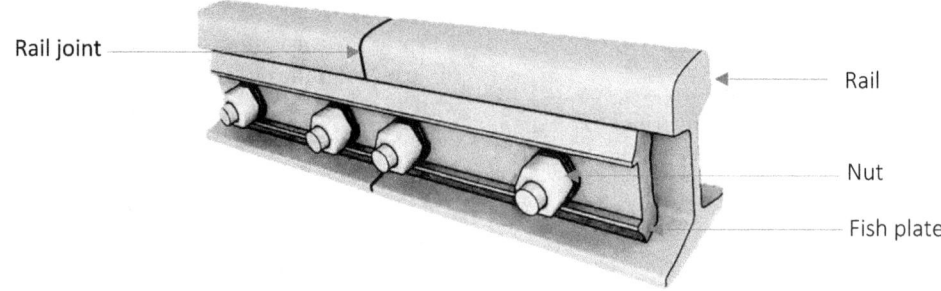

FIGURE 5.4   Rail track component.

Expansion joints consist of a series of interconnected steel plates that allow the tracks to expand and contract without causing any significant damage. By incorporating expansion joints, the stress on the tracks is substantially reduced as the expansion is absorbed by these flexible sections. Rail tracks (Figure 5.3—Qian'an–Caofeidian railway, PRC) are typically made from steel as it contains a unique combination of properties, that is, high strength, durability, cost-effectiveness, and perhaps low maintenance.

Expansion joints are designed to accommodate thermal expansion and contraction of concrete structures due to temperature changes. Their purpose is to allow for movement between adjacent concrete sections, prevent damage from thermal expansion and contraction, and also maintain the structural integrity of the structure. Expansion joints are typically wider (about 1–2 inches) and spaced farther apart (e.g., 50–100 feet) compared to **contraction joints**.

Other joints that cater for temperature fluctuations in civil engineering systems include contraction joints that control cracking in concrete caused by shrinkage and dilation joints that pass through each floor, except the foundations, and divide the entire building into smaller sections. They are designed to

safely absorb the expansion and contraction of various construction materials, absorb vibrations, and allow the building to properly accommodate earth movements caused by earthquakes. In structures, deflections due to temperature changes are not uncommon. If an axial member that is transmitting a force F undergoes a temperature change ΔT, the strain energy U is as follows:

$$U = \frac{1}{2}\frac{F^2 L}{AE} + F\alpha L\Delta T \qquad (5.10)$$

where α is the coefficient of linear expansion of the material, A is the cross-section of the member, E is Young's modulus, and L is the length of the member.

A rail joint is a track component that connects two pieces of rail, and it provides room for rail expansion and contraction due to the changes in temperature. See Figure 5.4. The mechanical rail joint is designed to allow a gap variation and, as such, has two well-defined limits of this gap.

The minimum gap of a mechanical rail joint is null, and that is usually achieved without exerting shear force on any of the joint bolts. The maximum **expansion gap** of a mechanical

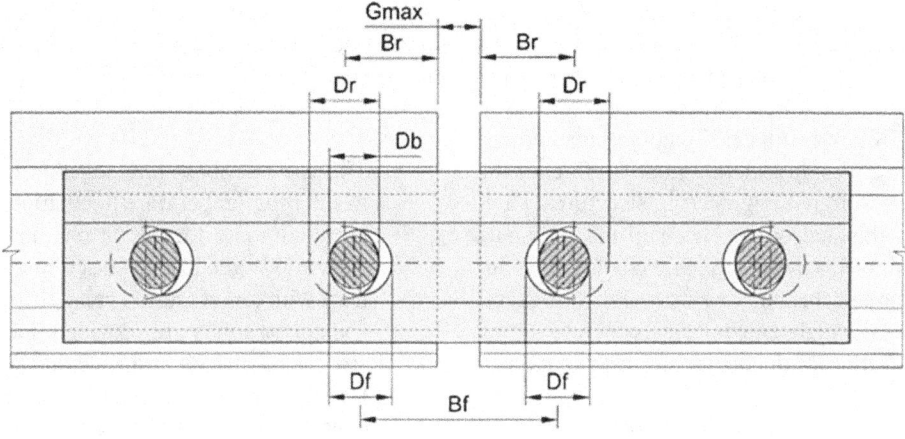

FIGURE 5.5   Rail track component expansion gap.

rail joint is dependent on the position of the joint holes and on the diameter of the holes and bolts of the joint [3]. This maximum gap shown in Figure 5.5 is reached when the rails contract due to temperature decreases and are starting to exert shear forces on the joint bolts. The maximum expansion gap G_{max} can be computed as follows:

$$G_{max} = Bf + Df + Dr - 2Db - 2Br \qquad (5.11)$$

where

- Bf: the distance between the centers of the middle holes of the fishplate.
- Df: the fishplate hole diameter.
- Br: the distance from the end of rail to the first rail hole center.
- Dr: the rail joint hole diameter.
- Db: the joint bolt diameter.

For a mechanical joint, there will always be a relation between the joint holes and bolt diameters, that is, Db <Df<Dr. The maximum gap G_{max} is designed for the longest rail used on the jointed track to avoid the shearing of the joint bolts when the rail contracts—it is an important parameter in the analysis of the jointed track behavior due to rail temperature variations.

5.3.2 Viscosity–Temperature Relationship of Binder Materials Such as Bitumen

In road construction, the air-void content in asphalt concrete (bituminous material) is found to decrease with an increase in the compaction temperature. If the air-void content is too high, it allows for intrusion of air and water. The higher the viscosity of binders, the greater is the resistance to compaction. Therefore, temperature must be regulated. Typically,

- The laying temperature for hot mix asphalt (HMA) ranges between 280°F and 320°F (138°C and 160°C), and compaction temperature ranges between 220°F and 280°F (104°C and 138°C). It is

essential to note that factors influencing the life of pavements including load, climate-related factors including precipitation and temperature, and the effect of temperature on water in the pavement are the principal causes of pavement degradation over time. The stripping mechanism can be exacerbated by a combined presence of high temperature, water, and traffic loading [4].

- The laying temperature for warm mix asphalt (WMA) ranges between 200°F and 250°F (93°C and 121°C), and compaction temperature ranges between 150°F and 200°F (66°C and 93°C).

When the desired temperatures are not achieved, for example, at low temperatures, asphalt concrete may not be compacted properly, leading to a weak and porous pavement. And at very high temperatures, the asphalt concrete may become too fluid, leading to over-compaction and formation of a dense, impermeable pavement. Overall, pavements that are not compacted at the correct temperature may age more quickly (garnering premature failures), leading to a shorter lifespan. High temperatures can cause the asphalt binder to oxidize, leading to a harder, more brittle pavement that is prone to cracking. Low temperatures can cause the asphalt binder to become too viscous, making it difficult to achieve proper compaction, resulting in increased air voids.

The goal of compacting asphalt concrete is to densify the asphalt mixture and reduce the air-void ratio so as to create a more solid and stable structure. A well-compacted asphalt layer transfers the load to the surrounding material, thus reducing the pressure on the pavement surface. This helps minimize **the** risk of deformation and damage.

In summary, the stress buildup in well-compacted pavement helps counteract the pressure from wheels by distributing the load, transferring the stress, and creating a state of stress equilibrium. This results in improved load-carrying capacity, reduced deformation, and increased pavement life.

Generally, for bituminous materials, the viscosity–temperature relationships must be observed when being applied for civil engineering applications. Temperature affects the viscosity of bituminous materials, influencing their flow and application properties. Bituminous materials can soften and lose their strength at high temperatures. Softening point is thus a key parameter. Bituminous materials can also oxidize at high temperatures, leading to degradation and reduced performance.

It is also important to note that gravel roads in tropical African countries produce heavy dust as compared to roads in other countries with different climates. Temperatures and climatic conditions of the place, in general, greatly influence the design and development of civil engineering structures. Civil engineers understand this and devise mechanisms to handle adverse climatic conditions. Adverse climatic conditions influence the hydrologic models that engineers develop. For example, in regions of extremely high temperatures, concrete setting time is monitored for better results, and materials expand and contract.

Climate of the area is very important in design of structures from roads, to pipelines, and to buildings. Engineers, thus, collect climatological data such as temperature, wind speed, humidity, and rainfall (mm) of the area to support their models. Thus, they can draw relationships between rainfall received in an area and temperature, which is complex and influenced by various factors, including evaporation, atmospheric circulation, topography, and human activities.

5.3.3 Pipeline Design

Temperature plays a vital role in the design, construction, and operation of oil and gas or water pipelines. Understanding its effects on fluid properties and incorporating temperature considerations into pipeline design and operation are essential for efficient and safe transportation of the fluids. It is crucial to note that pipeline systems are designed to accommodate thermal expansion to prevent damage. During the operating stage of the pipeline flow, there is pressure and temperature inside the pipeline. The resultant forces due to the pressure and the difference between temperature inside the pipeline and the surrounding fluid are created, which are to be contained within the tolerances, and the pipelines tend to expand rapidly and longitudinally. The pipeline design is greatly affected by the 'pressure,' 'volume,' and 'temperature' properties of the fluid being handled by the pipeline.

5.3.4 Conclusion

Because temperatures cause expansion and contraction of materials, it must be controlled or monitored to achieve the desired quality of the material or type for civil engineering use. When using materials in construction, we must be sensitive about the effects of temperature changes. In the development of alloys, temperatures are monitored. In the extraction of iron from the iron ore and the development of steel alloys of different material components, the temperature is monitored. Bituminous materials and oil refineries rely on temperature control and monitoring to achieve the desired output. Therefore, whatever industrial application or civil engineering application we need, we consider temperature effects to come up with the desired output, the durability, reliability, etc. In real-world applications, exposure to extreme temperatures can affect the material's strength, ductility, and corrosion resistance, for example, steel. Engineering concepts such as stress,

strain, Young's modulus, Poisson's ratio, viscosity, chemical composition, entropy, and density start where the study of pressure and temperature boils down, as far as civil engineering materials are concerned.

- The use of rollers in bridges (see Section 8.4.1) is to cater for temperature fluctuations. Roller supports are commonly located at one end of long bridges. This allows the bridge structure to expand and contract with temperature changes.
- Expansion joints are provided in concrete and steel structures such as bridges, highway pavements, and buildings to accommodate thermal expansion and contraction, as well as other movements, to prevent damage and ensure the structural integrity of the structure. Materials expand when heated and contract when cooled. Expansion joints allow for this movement, preventing stresses that can lead to cracks, damage, or even structural failure.
- Binder materials such as bitumen are very sensitive to temperature in performing their role.
- In the manufacturing/production process for civil engineering materials, temperature is a key factor. Did you know that temperature affects the setting rate of concrete?
- Pipeline systems are designed to accommodate thermal expansion to prevent damage.

5.4 PRESSURE EFFECT ON MATERIALS

In its simplest form, pressure is defined as force per unit area on a surface of an object. All civil engineering structures are designed and built to withstand pressures induced on them. As mentioned earlier, it is important to note that all other concepts in the study of the mechanics of materials are developed from the study of pressure. It is, therefore, important to note that pressure and stress (the dominant concept) are related but distinct concepts. Pressure is an external force, whereas stress is an internal response to that force.

5.4.1 Key Differences between Pressure and Stress

Pressure and stress, though related concepts, have interesting differences to note:

- Pressure is an external force, while stress is an internal response to that force.
- Pressure is a scalar, while stress is a tensor—meaning it has both magnitude and direction.
- Pressure is typically measured on a surface, while stress is measured within the material. Stress is a measure of the material's internal resistance to external forces, like tension, compression, or shear. For example, a retaining wall experiences external

earth pressure, but the material of the retaining wall also experiences internal stress due to that pressure. A bridge may be subjected to external wind pressure, but the structural members also experience internal stress due to the load. See the concept of load and force in Section 4.4.

Note: Because of the above explanations, materials are selected based on how they behave when stresses build up within them as they counteract different kinds of loads. Hence, we study the science of materials and strength of materials dwelling more on the different types of stresses and their causes. Let's now dive into why we study the science of materials.

5.5 STRENGTH OF MATERIALS

The field of strength of materials deals with forces and deformations that result from their acting on a material. Strength of materials (also called mechanics of materials) calculates stresses and strains in structural members. It focuses on analyzing how solid objects respond to external forces. We can also say it deals with mechanics of deformable bodies, structural analysis, etc. **Note:** Stresses develop in materials and structures primarily due to pressures exerted on materials or temperature fluctuations/variations in the materials.

Structures are designed and analyzed using stress and strain analysis principles to ensure safety and durability. The field (strength of materials) has numerous applications in civil engineering. It is applied in buildings, bridges, dams, and many more. Strength and stress are often mistakenly used interchangeably, with some books even using the same symbols for both. To clarify the difference, we must distinguish between the terms stress and strength:

> **Important:** Stress is defined as the internal-force distribution along a specific surface orientation at a uniquely distinct point within a structural element experiencing applied loads. Strength denotes some kind of limit and is more complex to define—it is based on the design intent. Typically, strength is considered as the highest stress level that a material can tolerate before failure occurs.

Understanding the differences between the three mechanical properties of materials, that is, strength, stiffness, and hardness, is foundational in civil engineering:

- **Hardness** measures a material's resistance to *surface deformation*. For some metals, like steel, hardness and tensile strength are roughly proportional.
- **Stiffness** is an indicator of the tendency for an element to return to its original form after being subjected to a force. In other words, it is a material's ability to resist deformation under an applied load. It

is a measure of how much a material will stretch or compress when subjected to a given force.

- **Strength** measures how much stress can be applied to an element before it deforms permanently or fractures. It denotes some kind of limit.

5.5.1 Example [5.3] for Illustrative Purposes: A 1.0-in-Diameter Aluminum Rod Loaded in Pure Tension

Let's consider a 1.0-in-diameter aluminum rod loaded in pure tension (Figure 5.6). If the tensile force attains a peak magnitude of 13 kips (57.83 kN) prior to catastrophic failure, we can assert that the maximum tensile capacity of the aluminum rod, subjected to the specified loading conditions, is 13 kips. Here, we recognize that this definition of strength is conditional upon the specific loading configuration, as well as the dimensions, geometry (shape and size), and material properties of the particular test specimen employed.

Note that dividing the force (13 kips) by the original cross-sectional area provides a size-independent measure of the maximum tensile stress reached before fracture. **Note:** The definition of stress in this case is based on the initial area differentiating it from the true stress, which is based on the final reduced area at fracture. This is the standard engineering practice.

Initial cross-sectional area of the aluminum rod:

$$A = \pi \frac{(1.0)^2}{4} = 0.785 \text{ in}^2 \tag{5.12}$$

Maximum tensile stress before fracture:

$$\frac{13}{0.785} = 16.561 \text{ kpsi} \tag{5.13}$$

Important: The ultimate tensile strength is now characterized by the maximum tensile stress value, rather than the maximum force, attained at the point of fracture. This value is typically listed in engineering and materials reference texts or manuals and, by convention, represents the ultimate tensile strength S_U (force per unit area). It is typically assumed that when strength is quantified in terms of a maximum achievable stress value, size effects become negligible [5].

However, it is important to note that although this is normal practice, in certain cases, it is far from the reality. This is even more exciting when the size is small. Overall, for a given material, the strength in force per unit area decreases as the size increases. For example, if the aluminum rod in discussion was 0.5-in in diameter, the maximum tensile stress would be 66.3 kpsi (294.92 kN)! However, it is important to note again that this effect diminishes rapidly as the size increases. Thus,

FIGURE 5.6   1.0-in-diameter aluminum rod loaded in pure tension.

in most situations, size effects are neglected, especially when the loading is static and not dynamic.

Strength theories are relied on which relate a general state of stress to the strength values obtained from simple tests of the material. This minimizes the amount of experimentations necessary to test each new loading case—some standard static tests performed on materials include pure tension, compression, bending, and torsion. Thus, this approach (relying on strength theories) answers the following question: what strength is associated with a structure with a more complex geometric and/or loading condition?

Thus, the design civil engineer must understand the many modes of failure and their limits primarily for ductile and brittle materials, which are the chief materials used for structural purposes. As outlined in Chapter 7, civil engineers develop and modify governing equations for a particular assignment to simplify designs. These are modified by the **design factor** resulting in the sound design equations used in civil engineering applications. See Section 7.6.9.9 for design factor (factor of safety). These strength theories are relied on to design structural members such as beams, columns, and slabs primarily forming statically indeterminate ductile structures for which the analysis of elastic behavior of ductile materials applies.

5.5.2 Common Structural Members

The structural members that civil engineers primarily deal with are beams, columns, and slabs, see Figure 5.7 (structural elements of a building). These three elements work together to form a structural system, providing support, stability, and functionality to buildings and other structures.

5.5.2.1 Beams

This is a horizontal or sloping structural element that supports loads from above mostly perpendicular to their axis and transfers them to supporting elements like columns or walls. Examples are floor beams, roof beams, and header beams (above doors or windows). Beams resist bending and shear forces and can be made of various materials like wood, steel, or concrete.

Beams mostly act in bending because they are designed to support transverse loads acting perpendicular to their longitudinal axis. These applied loads will bend, deflect, or displace the beam depending on the geometry of the structure, shape, material properties, supports, load patterns, etc.

These loads cause deformation of the beam that is usually expressed in terms of deflection. The deflection describes the change of the beam from its original neutral surface (before loading) to its neutral surface after the deformation has occurred. There are numerous methods to compute beam deflections, which results in well-known formulas used in the industry.

5.5.2.2 Columns

A column is a vertical structural element that transmits compressive loads from above to a foundation or supporting element. Columns provide support and stability to buildings, resisting axial compressive forces and sometimes bending moments. Examples are building columns, pillars, and piers (supporting bridges or structures).

5.5.2.3 Slabs

This is a flat, horizontal, or sloping structural element made of concrete, wood, or other materials. Slabs provide a flat surface for floors, roofs, or ceilings and can be supported by beams, columns, or walls. Examples are floor slabs, roof slabs, ceiling slabs, and foundation slabs (like a house's foundation).

5.5.3 Why Civil Engineers Study Strength of Materials?

Strength of materials, also known as mechanics of materials, deals with the analysis of the internal forces and stresses that deformable materials undergo when subjected to external loads, such as tension, compression, bending, and torsion. It aims to predict the behavior of materials under these loads and determine their ability to withstand various types of stress.

The study of strength of materials is crucial for civil engineers because it enables structural analysis. Therefore, civil engineers are able to conduct reliable and safe design and construction of structures. Through structural analysis, knowledge of the strength of materials enables civil engineers to analyze the behavior of structures under various loads and stresses, ensuring that they are safe and durable.

By combining knowledge of material properties (science of materials) and structural behavior (strength of materials), civil engineers can design and construct structures that meet performance requirements and safety standards—reliably performing as expected throughout their design lifespan.

Civil engineers need to learn and grasp the chemical and physical properties of materials such as the density of materials.

Civil engineers need to understand and apply the primary/chief mechanical properties of deformable materials, that is, Young's modulus (E), shear modulus (G), Poisson's ratio (v), area moment of inertia (I), structural stiffness, coefficient of linear thermal expansion (CLTE), and flexural rigidity (EI). Throughout the analysis of the strength and stability of materials and structures, the aforementioned are used and manipulated predominantly as they form the basis of further analysis and understanding of how structures behave under various loading conditions.

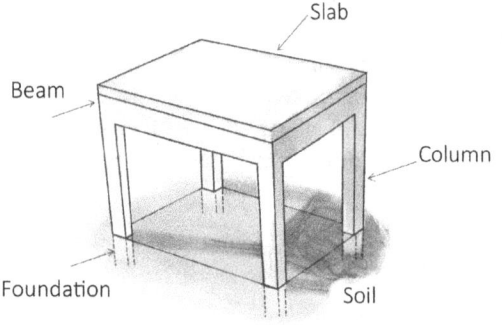

FIGURE 5.7 Structural elements of a building.

5.5.3.1 Young's Modulus, E

Young's modulus (E), also known as the modulus of elasticity, is a measure of the stiffness of a solid material (**deformable**)—it describes the relationship between stress and strain. It is given as the ratio of stress to strain within the proportional limit of the material. It is an important parameter in defining the **stress–strain curve** (Figure 5.8) of a material and is used to calculate the **stiffness** of materials. It is the parameter used to determine which material is stiffer than the other. The unit of elastic modulus (Young's modulus) is **megapascal (MPa) or gigapascal (GPa)**.

   Young's modulus plays a crucial role in structural design by providing engineers with a measure of a *material's stiffness,* allowing them to predict how much a material will undergo deformation under a given load, which is essential for selecting appropriate materials and designing safe and efficient structures such as bridges, buildings, and components that can withstand specific stresses without undergoing excessive deformation.

   Young's modulus is a critical parameter in the design of structures—it helps engineers predict how materials will behave under various loads. Thus, it helps in selecting materials. Engineers rely on Young's modulus to choose materials that can withstand the expected loads and stresses in a particular application.

   A higher Young's modulus indicates a stiffer material, making it better suited for applications requiring *rigidity*, such as steel in construction, while a material with a lower Young's modulus, such as rubber, is desirable for flexible applications. Therefore, by considering Young's modulus in the design process, engineers can create structures that are safe, efficient, and cost-effective. The following are some ways Young's modulus is used in design.

   The traditional stress–strain curve is primarily used to describe the behavior of materials under tensile forces. Modulus of elasticity is one of the most important concepts of material science and engineering. It is the property of material that measures the stiffness to elastic deformation under stress.

   In 1676, Robert Hooke discovered that the force (F) required to stretch or compress a spring is directly proportional to its displacement (x) from its equilibrium position. Mathematically, this is expressed as follows:

$$F = kx \tag{5.14}$$

   where k is the spring constant.

After 131 years, in 1807, Thomas Young expanded on Hooke's law to describe the *behavior of deformable solids*. He defined Young's modulus (E) as the ratio of stress (σ) to strain (ε) within the proportional limit of a material. Mathematically, this is expressed as follows:

$$E = \frac{\sigma}{\varepsilon} = \frac{\text{Rise}}{\text{Run}} = \text{slope of stress-strain curve} \tag{5.15}$$

Thus, Young's modulus (E) is a measure of a material's stiffness or resistance to deformation. In Figure 5.8, Young's modulus is the slope of the linear part of the stress–strain curve for a material under tension. The connection between Hooke's law and Young's modulus is pretty straightforward. Hooke's law describes the behavior of a spring, while Young's modulus describes the behavior of a solid material. *However, both concepts rely on the idea of proportionality between force (or stress) and displacement (or strain).*

   Hooke's law laid the foundation for Young's modulus by introducing the concept of proportionality between force and displacement and providing a mathematical framework $F = kx$ that could be adapted for *deformable solids such as ductile materials.*

   Note: Ductile materials can be deformed significantly before breaking, whereas brittle materials break with little to no deformation. Examples of ductile materials include metals and plastics. Brittle materials don't undergo much deformation before breaking. Instead, they keep their shape right until they break. Glass is a typical example of a brittle material.

   Thomas Young built upon Hooke's work by extending the concept to deformable solids, rather than just springs. He defined Young's modulus (E) as a material property, rather than a specific spring constant. Therefore, Hooke's law provided the foundation for Young's modulus by introducing the concept of proportionality and a mathematical framework that could be adapted for deformable solids.

   In Figure 5.8, the **elastic region** is where the material can return to its original shape after an external force or stress is removed. In other words, elastic materials can stretch, compress, or deform, but they will return to their original state once the force is removed. In the **plastic region**, the material undergoes permanent deformation without breaking. Plastic materials can be molded, shaped, or deformed, and they will retain their new shape even after the external force is removed.

- **Point a**, proportional limit, is the maximum stress at which the stress–strain relationship remains linear, meaning that the material will return to its original shape when the stress is removed. Note: Allowable stress a material can withstand is always specified within this limit.
- **Point b**, elastic limit (yield point), is the maximum stress that a material can withstand without undergoing permanent deformation. It's a critical point on a stress–strain curve, marking the boundary between elastic and plastic behavior.
- **Point c**, lower yield point, is the lowest stress level reached after the initial sharp decrease in stress,

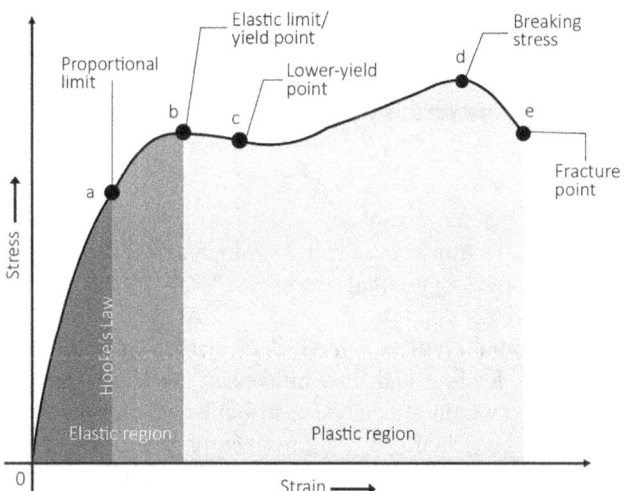

FIGURE 5.8 Stress–strain curve for a deformable solid.

marking the point where a material begins to undergo plastic deformation with a minimal increase in applied stress, essentially signifying the onset of stable plastic deformation within the material; it is typically observed in materials like mild steel that exhibit a distinct 'yield point phenomenon' with an upper yield point followed by a sudden decrease to a lower yield point.

- **Point d**, breaking stress (ultimate tensile strength (UTS)), is the maximum stress that a material can withstand before failing or breaking.
- **Point e**, fracture point, is the point at which a material fails or breaks under stress, resulting in a complete loss of its load-carrying capacity.

Region b–d is referred to as strain hardening, while d–e is the necking region. Strain hardening occurs in the plastic region of the stress–strain curve, after the yield point. In this region, the material becomes stronger and more resistant to deformation.

The necking region occurs after the UTS point on the stress–strain curve. In this region, the material's stress-carrying capacity decreases, and it eventually fails.

Note: Strain hardening is a strengthening mechanism, while necking is a weakening mechanism. Strain hardening occurs in the plastic region, while necking occurs after the UTS point. Strain hardening increases the material's strength, while necking leads to material failure.

5.3.3.2.1 Analyzing stress and strain

Young's modulus is used to analyze the stress and strain in structures. By knowing Young's modulus of a material, civil engineers can calculate the stress and strain in a structure and ensure that it can withstand the expected loads without failing:

$$E = \frac{\sigma}{\varepsilon} \tag{5.16}$$

5.3.3.2.2 Calculating deformation and deflection

Young's modulus is used to calculate the deflection and deformation of structures under load, typically beams and columns. This is crucial in ensuring that structures can withstand external forces without excessive deformation or failure. To calculate the deflection $y(x)$ of the cantilever beam of length L, point-loaded with load, P, at the extreme end, we use the following equation:

$$y(x) = \frac{PL^3}{3EI} \tag{5.17}$$

where E is Young's modulus and I is the area moment of inertia.

5.3.3.2.3 Determining structural stiffness

Young's modulus is a measure of a material's stiffness. Stiffness is the ability of a structure to withstand bending, twisting, compression, or stretching without losing its original shape. By knowing Young's modulus of a material, civil engineers can determine the structural stiffness of a component or system, which is essential in designing structures that can resist deformation and vibration.

5.3.3.2.4 Optimizing structural shapes

Young's modulus is used to optimize the shape of structures for maximum stiffness and minimum weight. By knowing Young's modulus of a material, engineers can design structures that are efficient and cost-effective.

5.3.3.2.5 Designing for buckling and instability

Young's modulus is used to predict the buckling and instability of structures, particularly slender columns and beams. Establishing Young's modulus of a material, engineers can design structures that are resistant to buckling and instability.

5.3.3.2.6 Accounting for temperature effects

Note: Young's modulus can vary with temperature. Engineers use this information to design structures that can withstand temperature fluctuations and ensure that they remain stable and functional. See Section 5.3.1.2 worked example [5.2] for illustrative purposes (thermal stress).

5.5.3.2 Shear modulus (G)

The term 'shear modulus of elasticity,' also called the 'modulus of rigidity,' is a measure of a material's resistance to transverse deformations, essentially how much it can resist shearing forces without permanent deformation, and is calculated by dividing shear stress by shear strain, represented by the symbol 'G' and measured in Pascals (Pa). Young's modulus is related to the shear modulus of elasticity (G) by the following equation:

$$E = 2G(1+v) \tag{5.18}$$

where E is Young's modulus, G is the shear modulus, and v is Poisson's ratio.

5.5.3.3 Poisson's ratio, v

Poisson's ratio (v) is the ratio of transverse contraction (or expansion) strain to longitudinal extension strain in the direction of stretching force. Poisson's ratio (v) is a dimensionless quantity that describes the lateral strain response of a material to a longitudinal tensile loading. In other words, it measures how much a material will deform sideways when stretched or compressed. See Figure 5.9. Tensile deformation is considered positive, and compressive deformation is considered negative:

$$v = \frac{\varepsilon_t}{\varepsilon_l} = \frac{\text{Lateral strain}}{\text{Longitudinal strain}} = \frac{-(\Delta w/w)}{(\Delta l/l)} \tag{5.19}$$

where

- v is Poisson's ratio
- ε_t is the transversal/lateral strain, and
- ε_l is the longitudinal strain.

Poisson's ratio predicts lateral deformation of a material under axial loading and also influences the stress distribution. Poisson's ratio affects the distribution of stresses within a material, particularly in the presence of holes, notches, or other stress concentrators. Thus, it affects material stiffness, with higher values indicating greater stiffness. For most materials, the value of Poisson's ratio lies in the range of 0–0.5.

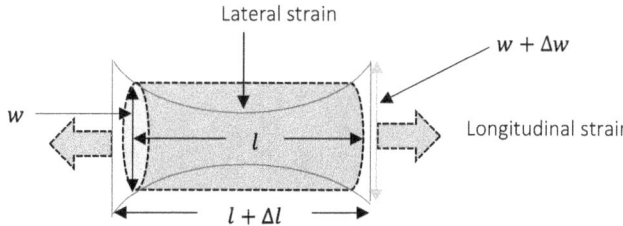

FIGURE 5.9 Poisson's ratio illustration.

TABLE 5.2 Poisson's ratios for different materials

MATERIAL	VALUES
Concrete	0.1–0.2
Cast iron	0.21–0.26
Steel	0.27–0.30
Rubber	0.4999
Gold	0.42–0.44
Glass	0.18–0.3
Cork	0.0
Copper	0.33
Clay	0.30–0.45
Stainless steel	0.30–0.31
Foam	0.10–0.50

A few examples of Poisson's ratios for different materials are given in Table 5.2.

5.5.3.4 Linear coefficient of thermal expansion

The CLTE, also known as the coefficient of thermal expansion (CTE), is a material property that measures how much a material expands when heated. The CLTE describes how much a material's length changes for each degree change in temperature. It's a key factor in designing structures that can withstand temperature changes.

5.5.3.5 Structural stiffness

This is the ability of a structure to resist deformation or deflection when subjected to an external load. The most common types of structural stiffness are bending stiffness (flexural stiffness) and torsional stiffness. *Stiffness* is the ability of a *structure* to withstand bending, twisting, compression, or stretching without losing its original shape.

5.5.3.6 Flexural rigidity (EI)

The product of Young's modulus (E) and moment of inertia (I) is known as **flexural rigidity**. This value is used to describe the resistance of a beam or other structural members to bending. **Note:** The area moment of inertia (I) is a geometric property of the material. That means flexural rigidity varies with I.

Flexural rigidity is a specific type of structural stiffness that deals with a structure's resistance to bending. You have probably seen structural members with sections of different shapes: L-section, I-section, U-section, hollow sections, circular section, and triangular section. All these sections influence the flexural rigidity of the structural member. The material properties can be the same, but because of the differences in geometrical properties, the flexural rigidity differs. In addition, do you ever wonder how topology can transform the strength of materials?

Different cross-sectional shapes have varying moments of inertia, which affect the flexural rigidity. Even if the material properties are the same, the differences in geometrical properties (i.e., cross-sectional shape) can result in varying flexural rigidities. In Section 7.5.1, we notice how the product of variables communicates important information.

In Figure 5.10, the log deflects more when the mass is placed on it with the cross-section's length oriented horizontally, compared to when the breadth is horizontal. The log's stiffness and resistance to bending depend on its cross-sectional shape and orientation. When the length is horizontal, the log is more prone to bending, resulting in greater deflection. In contrast, when the breadth is horizontal, the log is stiffer and more resistant to bending, resulting in less deflection. This is the primary reason why more depth is provided to the beam than its width. It is about the area moment of inertia about an axis. The area moment of inertia about an axis, also called the '**second moment of area**,' is a measure of how the area of a cross-section is distributed around a specific axis, essentially indicating how resistant a shape is to bending when a moment (force applied at a distance)

FIGURE 5.10   100 kg mass loaded on the same log but when the log is placed differently.

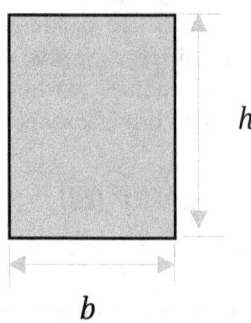

FIGURE 5.11   Log cross-section.

is applied about that axis; it is calculated by summing up the product of each small area element and the square of its perpendicular distance to the axis of interest, expressed mathematically as

$$\int y^2 dA \tag{5.20}$$

where y is the perpendicular distance from the axis to the area element dA and the integral is taken over the entire area of the cross-section. See Figure 5.11.

$$I_{xx} = \frac{bh^3}{12} \tag{5.21}$$

$$I_{yy} = \frac{b^3h}{12} \tag{5.22}$$

5.5.3.7 Worked examples for illustrative purposes

These worked examples share insights on how the chief mechanical properties of materials outlined in Section 5.5.3 are used to **design safe structural systems**. They are relied on to compute the level of stress that the structural system can withstand without failure. Because structural systems can be loaded differently in shear, tension, compression, torsion, and/or bending, they must be evaluated for each different scenario/loading configuration.

The primary factors that cause stresses in deformable bodies are temperature, pressure, and environmental conditions. These worked examples illustrate the basic application of the chief mechanical properties of materials emphasizing that variations in loading patterns/configurations, temperature, and environmental conditions are responsible for causing deflections, translations, and rotations in structural systems. Did you

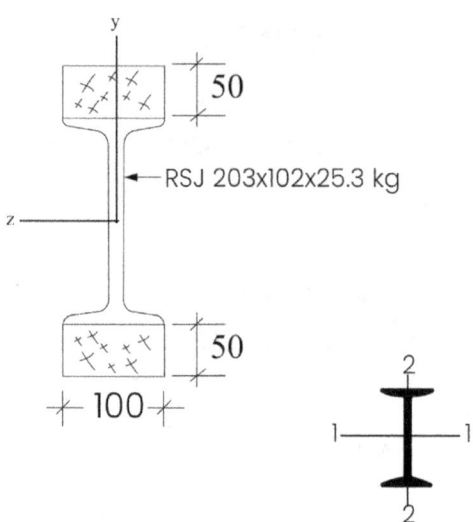

FIGURE 5.12   I-beam.

know that Young's modulus can vary with temperature? Yet, it measures the stiffness of materials!

These worked examples are convergent in nature. However, despite being convergent, they share insights, bearing in mind that the convergent and divergent approaches are reciprocal systems or approaches, with each influencing the other and that they can be modified accordingly, to generate ideas. Structural systems are predominantly designed through divergent approaches.

5.3.3.2.1 Example [5.4]: Uniformly loaded steel I-beam reinforced with wood beams

A simple beam (Figure 5.12) that is 5 m long supports a uniform load of intensity q. The beam is constructed of a rolled steel joist, RSJ 203 × 102 × 25.3 kg section reinforced with wood beams that are securely fastened to the flanges. The wood beams are 50 mm deep and 100 mm wide. The modulus of elasticity (E) of steel is 20 times that of wood. If the allowable stresses in the steel and wood are 110 MPa and 8.2 MPa, respectively, what is the allowable load, q_{all}? (disregard the weight of the beam). The properties of the joist sections (RSJs) are given in Table 5.3.

Note: Axes 1-1 and 2-2 are centroidal principal axes.
Solution

Transforming the section into one made of entirely of steel, since steel is stiffer than wood, the width of the wood must be reduced to an equivalent width of steel:

$$n = \frac{E_{wood}}{E_{steel}} = \frac{1}{20} \tag{5.23}$$

TABLE 5.3   Properties of the joist sections (RSJs)

DESIGNATION	MASS PER METER	AREA OF THE SECTION	DEPTH OF THE SECTION	WIDTH OF THE SECTION	THICKNESS		AXIS 1-1			AXIS 2-2		
					WEB	FLANGE	I	S	R	I	S	R
MM	KG	CM²	MM	MM	MM	MM	CM⁴	CM³	CM	CM⁴	CM³	CM
203×102	25.3	32.3	203.2	101.6	5.8	10.4	2294	226	8.43	163	32.0	2.25
178×102	21.5	27.4	117.8	101.6	5.3	9.0	1519	171	7.44	139	27.4	2.25
152×89	17.1	21.8	152.4	88.9	4.9	8.3	881.1	116	6.36	86.0	19.3	1.99
127×76	13.4	17.0	127.0	76.2	4.5	7.6	475.9	74.9	5.29	50.2	13.2	1.72
102×64	9.7	12.3	101.6	63.5	4.1	6.6	217.6	42.8	4.21	25.3	8.0	1.43
76×51	6.7	8.5	76.2	50.8	3.8	5.6	82.58	21.7	3.12	11.1	4.4	1.14

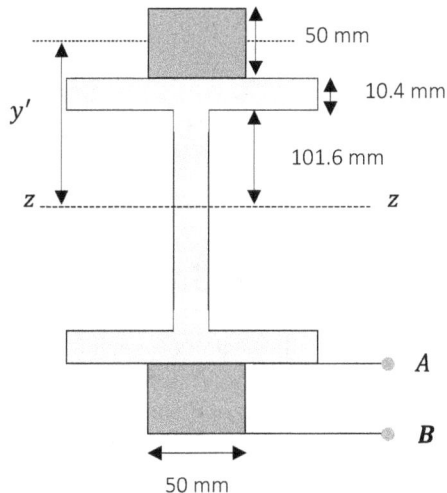

FIGURE 5.13 Transformed section of I-beam in Figure 5.12.

So,

$$b_{steel} = nb_{wood} = \frac{100}{20} = 50 \text{ mm} \qquad (5.24)$$

Transformed section, Figure 5.13.

$$I_{z\text{-}z} = 2,294 + 5 \times \frac{5^3}{12} + 5^2 \times (10.16 + 10.4 + 2.5)^2 = 7,038.33 \text{ cm}^4 \qquad (5.25)$$

Bending stress at A:

$$\sigma = \frac{Mc}{I} \qquad (5.26)$$

$$\sigma_A = \frac{M \times (101.6 + 10.4) \times 10^{-3}}{7,038.33 \times 10^{-8}} \qquad (5.27)$$

where σ_A is the maximum stress in steel.
So,

$$M = \frac{110 \times 10^3 \times 7,038.33 \times 10^{-8}}{(101.6 + 10.4) \times 10^{-3}} = 69.13 \text{ kNm} \qquad (5.28)$$

Bending stress at B:

$$\sigma_{B_{wood}} = n\sigma_{B_{steel}} \qquad (5.29)$$

$$\sigma_{B_{wood}} = \frac{1}{20} \times \frac{M(101.6 + 10.4 + 50) \times 10^{-3}}{7,038.33 \times 10^{-8}} = 8.2 \times 10^3 \text{ kPa} \qquad (5.30)$$

$$M = 71.25 \text{ kNm}$$

Maximum allowable bending moment = 69.13 kNm, see Figure 5.14.

So,

$$\frac{ql^2}{8} = 69.13 \qquad (5.31)$$

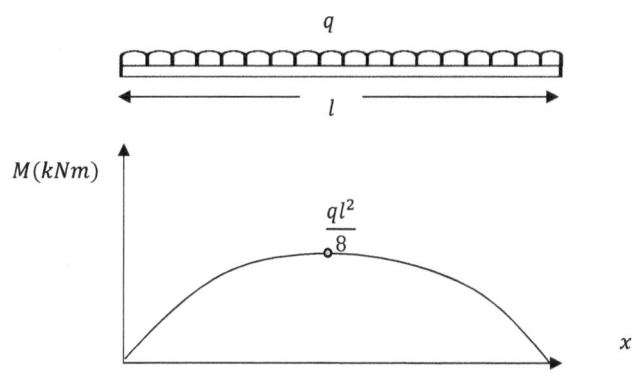

FIGURE 5.14 Maximum allowable bending moment.

$$l = 5$$

Therefore,

$$q = 22.12 \frac{\text{kN}}{\text{m}} \qquad (5.32)$$

5.3.3.2.2 Example [5.5]: Hollow steel shaft
A hollow steel shaft ACB (5.15) of external diameter 50 mm and internal diameter 40 mm is held against rotation at ends A and B. Horizontal forces P are applied at the ends of the vertical arm. Determine the allowable value of the forces P if the maximum permissible shear stress in the shaft is 55 MPa.

Solution

See Figure 5.16 **(inner and outer diameters of the shaft)**
Torque at C

$$T_C = 0.4P \text{ kNm}$$

$$J = \frac{\pi}{32}\left(D_1^4 - D_2^4\right) \qquad (5.33)$$

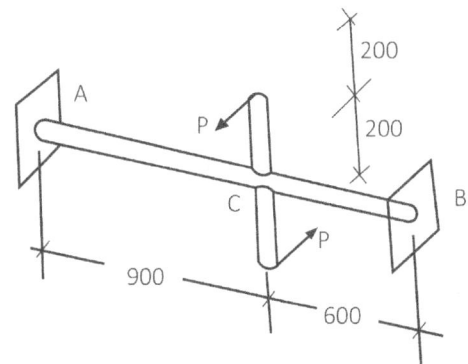

FIGURE 5.15 Hollow steel shaft.

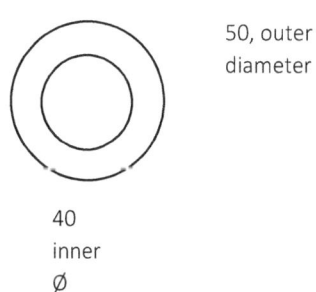

FIGURE 5.16 Inner and outer diameters.

$$J = \frac{\pi}{32}\left(0.05^4 - 0.04^4\right) \tag{5.34}$$

$$J = 3.623 \times 10^{-7}\,\text{m}^4$$

Hollow shaft free-body diagram, Figure 5.17.
Equilibrium requires the following:

$$T_A + T_B = T_C \tag{5.35}$$

$$T_A + T_B = 0.4P \text{ kNm} \tag{5.36}$$

Compatibility

$$\varnothing_{A/C} = \varnothing_{B/C} \tag{5.37}$$

and

$$\varnothing = \frac{TL}{JG} \tag{5.38}$$

So,

$$\frac{T_A \times 0.9}{JG} = \frac{T_B \times 0.6}{JG} \tag{5.39}$$

So,

$$T_A = \frac{0.6}{0.9}T_B \tag{5.40}$$

Substituting in equation (5.36):

$$\frac{0.6}{0.9}T_B + T_B = 0.4P \text{ kNm} \tag{5.41}$$

$$T_B = 0.24P \text{ kNm}$$

$$T_A = 0.16P \text{ kNm}$$

$T_B > T_A$, so maximum shear occurs in BC.

So,

$$\tau_{\max} = \frac{T_c}{J} \tag{5.42}$$

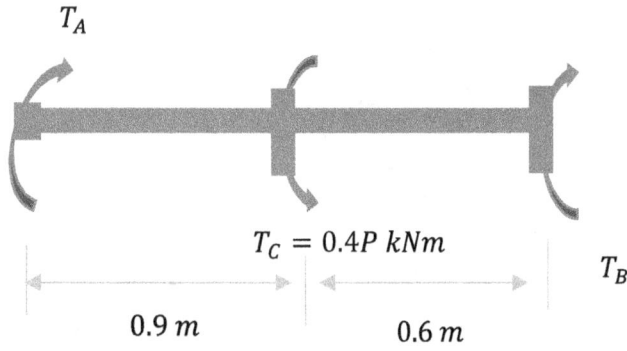

FIGURE 5.17 Hollow steel shaft FBD

$$55 \text{ MPa} = \frac{0.24P \text{ kNm} \times 0.025 \text{ m}}{3.623 \times 10^{-7}\,\text{m}^4} \tag{5.43}$$

So,

$$P = 3.32 \text{ kN}$$

5.5.4 Materials and Structures

A structure is the overall system or assembly, while a material is the substance used to build it. See steel tower analysis model for water reservoir tank in Figure 5.18—it is an analysis model for the 100 m³ water reservoir tank—12-m-high tank tower support. The tower contains the structures and structural members, that is, beams, columns, ties, and struts, are the steel materials. Structures are designed to perform specific functions, while materials have inherent properties that influence their behavior. Structures can be made from multiple materials, and materials can be used in various structures. Thus, a structure is the overall arrangement of components, while a material is the substance used to construct those components. To illustrate the difference further, consider the two examples: a bridge and a car.

- A bridge (structure) can be made from steel (material), concrete (material), or a combination of both.
- A car (structure) can have a body made from steel (material), aluminum (material), or carbon fiber (material).

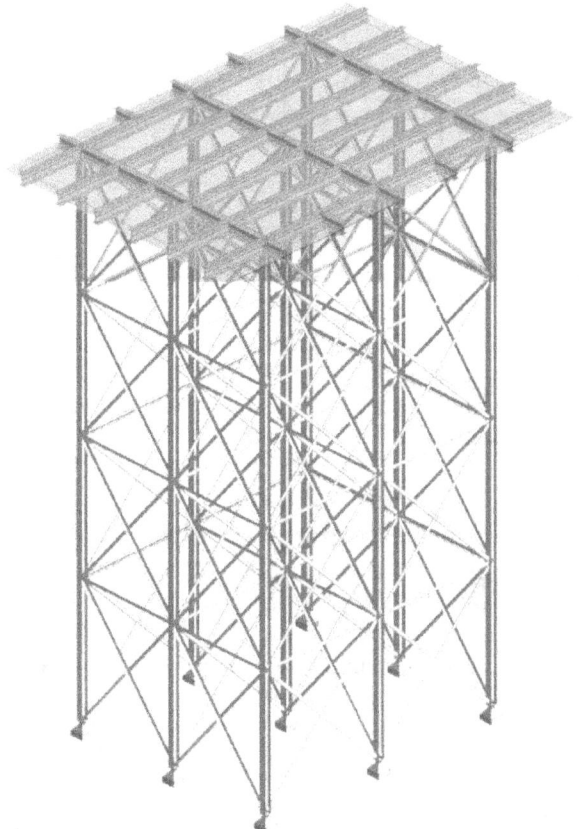

FIGURE 5.18 Steel tower model.

5.5.5 Stability of Structures

The earlier sections explained the stiffness and strength of materials. It is now important to recognize that stability applies to structures.

A structure is an assembly of materials in equilibrium used to carry loads.

Generally, stability of a structure refers to the ability of a structure to recover its equilibrium or resist sudden changes, dislodgment, or overthrow. A stable structure should maintain stability under any conceivable loading conditions, irrespective of the type or location of the load. To achieve this, the load-carrying capacity of the individual members must be evaluated while assembling the structure. As outlined in Section 8.4.5, redundant members play a crucial role to ensure the safety and stability of structures. These redundant members are illustrated in the steel tower model, Figure 5.18.

5.5.6 Roadmap for Specifying Materials for a Civil/Structural Project

Specifying materials for use on a project is a back-and-forth approach juggling between cost and durability. The material must be strong enough to achieve the desired reliability. Going through a rigorous approach where calculations are performed based on available standards, codes of practice, and regulations, the project specifications are drawn in accordance with safety, sustainability goals, reliability, durability, operability, and maintainability. Specifying the quality of material to use on a project must meet the minimum strength parameters for the job. That means the material must be safe to carry loads and unhazardous.

5.5.6.1 Typical example [5.6]—for illustrative purposes

This is a basic example to illustrate how a specification may be arrived at. The two-bar truss ABC shown in Figure 5.19 has a pin support at points A and C which are 2.0 m apart. Members AB and BC are steel bars, with the pin connected at joint B. The length of bar BC is 3.0 m. A sign weighing 5.4 kN is suspended from bar BC at points D and E, which are located 0.8 m and 0.4 m, respectively, from the ends of the bar. Determine the required cross-sectional area of the bar AB and the required diameter of the pin at support C if the allowable stresses in tension and shear are 125 MPa and 45 MPa, respectively. (**Note:** The pins at the supports are in double shear. Also, disregard the weights of members AB and BC.)

Solution

Free-body diagram (Figure 5.20) for the two-bar truss.

Assumptions:

- Neglecting self-weight.
- This a 2-D problem.

Taking moments about C:

$$F_{x,\,A} \times 2 = 5.4 \times 1.7 \tag{5.44}$$

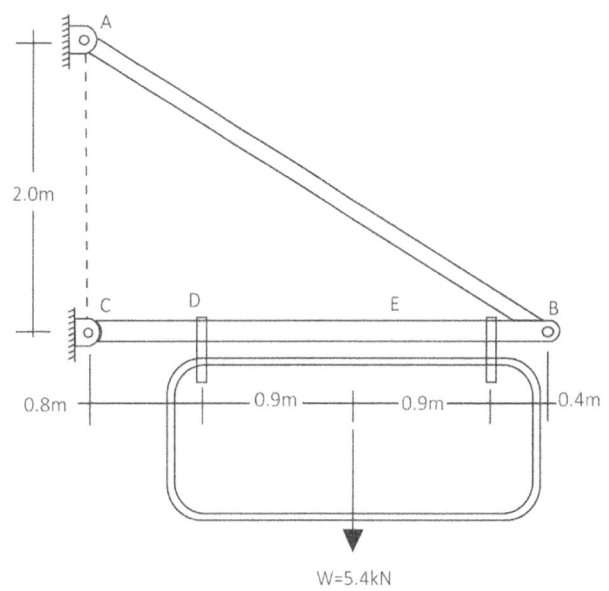

FIGURE 5.19 Two-bar truss.

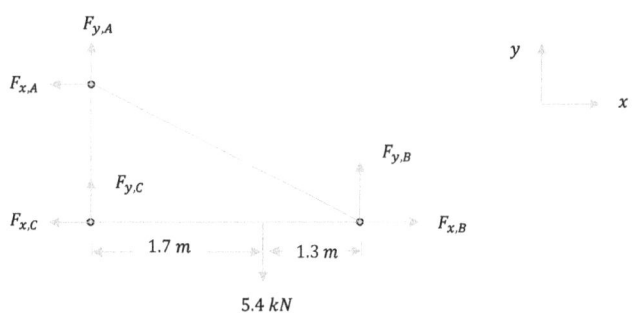

FIGURE 5.20 Two-bar truss FBD.

$$F_{x,\,A} = 4.590 \text{ kN} \tag{5.45}$$

Dismembering at B; taking equilibrium of rod BC (Figure 5.21):

Taking moments at C:

$$F_{y,\,B} \times 3 = 5.4 \times 1.7 \tag{5.46}$$

$$F_{y,B} = 3.06 \text{ kN} \tag{5.47}$$

$$\sum F_y = 0; \tag{5.48}$$

$$F_{y,\,C} + F_{y,\,B} = 5.4 \tag{5.49}$$

$$F_{y,\,C} = 5.4 - 3.06 = 2.34 \text{ kN} \tag{5.50}$$

For the entire body,

$$\sum F_x = 0; \tag{5.51}$$

$$F_{x,\,A} + F_{x,\,C} = 0 \tag{5.52}$$

$$F_{x,\,C} = -4.590 \text{ kN} \tag{5.53}$$

(i.e., in opposite direction)

$$\sum F_y = 0; \tag{5.54}$$

FIGURE 5.21 Rod BC, FBD.

$$F_{y, A} + F_{y, C} = 5.4 \tag{5.55}$$

$$F_{y, A} = 5.4 - 2.34 = 3.06 \text{ kN} \tag{5.56}$$

Internal forces

Internal force in AB:

$$F_{AB} = \sqrt{\left(F_{y, A}\right)^2 + \left(F_{x, A}\right)^2} \tag{5.57}$$

$$F_{AB} = \sqrt{3.06^2 + 4.59^2} = 5.5165 \text{ kN} \tag{5.58}$$

(tension)

Internal axial force in BC:

$$F_{x, C} = 4.59 \text{ kN} \tag{5.59}$$

(compression)

Internal shear force in BC:

$$F_{y, C} = 2.34 \text{ kN} \tag{5.60}$$

(shear)

Section sizing

AB

$$\sigma = \frac{F}{A} \tag{5.61}$$

$$A_{AB} = \frac{F_{AB}}{\sigma_{all}} = \frac{5.5165 \text{ kN}}{125 \text{ MPa}} = 44.13 \text{ mm}^2 \tag{5.62}$$

Pin at C,

Resultant force at support C,

$$F_c = \sqrt{\left(F_{x, C}\right)^2 + \left(F_{y, C}\right)^2} = \sqrt{(4.59)^2 + (2.34)^2} = 5.152 \text{ kN} \tag{5.63}$$

Since the support is in double shear, shear force in the support road,

$$\frac{5.152}{2} = 2.576 \text{ kN} \tag{5.64}$$

Again, the required area of the pin,

$$A_{pin} = \frac{F_C}{\sigma_{all}} = \frac{2.576 \text{ kN}}{45 \text{ MPa}} = 57.24 \text{ mm}^2 \tag{5.65}$$

(cross-sectional area due to shear)

$$A = \frac{\pi D^2}{4} \tag{5.66}$$

So,

$$D = \sqrt{\frac{4A}{\pi}} = \sqrt{\frac{4 \times 57.24}{\pi}} = 8.53 \text{ mm} \tag{5.67}$$

So, we take a 10 mm diameter pin/bolt, the nearest standard pin, and consider a margin of safety. Thus, this becomes a specification for the size of the pin required for the project. This can be rounded to the nearest existing standard as per given jurisdiction, say M10. 'M' stands for metric.

Several points to note are as follows:

- Pins (bolts) are manufactured in standard sizes. Bolts are threaded fasteners with external male threads. They mate with nuts, which have female— meaning, internal—threads. Both the bolt and the nut grip the materials being fastened, creating a bolt joint, with the nut also preventing axial movement.
- Test materials should be checked if they meet the required compressive and tensile strength. In this example, the allowable stresses in tension and shear at support C are 125 and 45 MPa, respectively. Therefore, the required materials, that is, bolts (pin) and members AB and CB, must meet the required strength. That's why the materials (pin and members) are tested to ascertain whether they are able to carry the required loads.
- It is important to note that in construction, it is not only about achieving the required sizes, that is, required cross-sectional areas AB, BC, and required diameter of pins, but also the material being able to withstand the loads is what counts, and this is ascertained through material testing. The most common/ typical materials for such purposes are steel—an alloy of iron and carbon. Very low carbon content is responsible for poorly performing steels, being prone to corrosion, and cracking. Yield strength is improved in steels with a higher carbon content.
- When a pin/connection is in double shear, it means that the member is subjected to two separate shear forces, one on either side of the member, and these forces are acting in opposite directions.

  In other words, the member is being 'sandwiched' between two opposing shear forces, resulting in a doubling of the shear stress on the connection/pin.

- Allowable stress, also known as permissible stress or working stress, is typically within the proportional limit of a material. The proportional limit is the maximum stress at which the stress–strain relationship remains linear, meaning that the material will return to its original shape when the stress is removed.

- Allowable stress is usually set at a value below the proportional limit to ensure that the material remains within the elastic range, where it will not undergo permanent deformation. Note that allowable stress is often calculated as a fraction of the yield strength or UTS of the material, taking into account factors such as safety, reliability, and potential overload conditions. Allowable stress is typically within the proportional limit of a material, ensuring that the material remains safe and reliable under expected loading conditions.

5.5.6.1.1 Analysis
Example 5.6 presents a typical convergent exercise (see Section 9.2.6), leading to specification of a 10-mm-diameter pin for use to suspend a sign weighing 5.4 kN from a two-bar support. The pin was loaded and subjected to the specified loading conditions—loaded in shear. In case it was loaded in a different loading configuration, the scenario would be different. This sheds light on how specifications for materials to use in construction can be arrived at.

5.6 THE NEED TO SELECT APPROPRIATE MATERIALS FOR CIVIL ENGINEERING PROJECTS

Earlier listed are several materials used in civil engineering projects. These include bricks, cement, concrete, aluminum, steel, sand, glass, timber, wood, aggregates, adhesives, bitumen, asphalt, paints, tiles, steel plates, pipes (HDPE, uPVC, GI, ductile iron, soon iron, etc.), marble, water, coarse and fine aggregates, slates, stones, rockfill, geotextiles, geomembranes, and waterproof materials.

Materials selected for civil engineering projects are deemed to meet technical, economic, environmental, safety, aesthetic, regulatory, and social requirements. Materials are specified and selected based on how close they strike a balance between the three pillars of sustainability. The guiding principle of sustainable projects is that materials are selected only when they can strike a balance between the three pillars of sustainable development, that is, social, economic, and environmental. Figure 5.22 shows the three pillars of sustainable development. Thus, designers and contractors primarily select materials that are more sustainable than others as far as the task is concerned.

5.6.1 Strength and Durability

Materials are selected based on strength and durability requirements, that is, how best they resist applied forces and how long they last with applied forces. As an example, cable-stayed bridges exert significant tensile forces on the piers, towers, and foundation, requiring **high-strength reinforcement**.

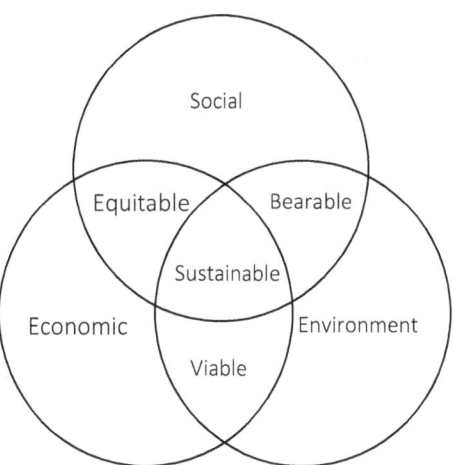

FIGURE 5.22 Sustainable development pillars.

And because of the long spans which increases the structural loads and necessitating stronger reinforcement, high-tensile strength rebars are used. Stay cables of a cable-stayed bridge are post-tensioned to counteract the effect of the bridge dead load. Wind, traffic, and seismic activities create dynamic loads on the bridge that high-strength rebars help resist. The **yield and ultimate strength** for typical HT rebars in cable-stayed bridges ranges between 80–100 ksi (550–690 MPa) and 100–120 ksi (690–830 MPa), respectively.

5.6.2 Reliability

Materials can be selected based on the reliability, operability, functionality, and maintainability. Building materials such as solvents, putties, releasing agents, plaster and special mortars, waterproof materials, and glue cements are purely selected based on reliability—that is, their ability to perform the task effectively and efficiently in the required time period. The desired quality output influences the choice of materials.

The efficiency and effectiveness of a material inform the reliability indicator and ultimately influencing the functionality, overall. In water pipeline construction, HDPE, UPVC, and GI may be selected in different situations based on the desired hydraulic efficiency, the terrain, and pressure ratings necessary to counteract the flow pressures.

A material is said to be operationally and functionally reliable when it meets the users' needs efficiently and effectively in a specified time period. Reliability, durability, safety, and cost-effectiveness are pertinent factors. Why choose steel over concrete? Why opt for slab decks instead of conventional slab system? Such questions must be answered to find the most reliable choice.

5.6.3 Corrosion Resistance

In many cases, materials are selected based on how best they resist corrosion and degradation. Some material alloys are preferred because the combination of various materials in the metal makes them a better choice, thus improving the corrosion resistant properties of the alloy. Coated materials are also preferred because they resist corrosion better than the uncoated ones. For example, choosing hot dip-galvanized steel tank panels is common in reservoir tank construction because thy resist corrosion and rust.

Corrosion mechanism and protection

Iron corrodes when exposed to oxygen and water, forming iron oxide (rust) as per the following equation:

$$4Fe + 3O_2 + 2H_2O = 2Fe_2O_3 \cdot H_2O \qquad (5.68)$$

5.6.4 Cost-Effectiveness

Based on lifecycle cost analysis, some materials are better than others. The initial capital investment of a material may be high, but when you perform a lifecycle cost analysis, because of reduced maintenance and operation costs, the material is better than the chosen alternatives. That means due to long-term maintenance reduction, the material is selected. For example, comparing sustainable drainage systems (SuDS) with gray infrastructure lifecycle costs, SuDS may be favored especially when environmental sustainability is also considered a factor in the analysis. Cost-effective materials may be preferred to those which are not. A cost-effectiveness approach is used to compare the costs of different material options to determine the most economical materials to achieve a desired outcome.

5.6.5 Low-Carbon Footprint

The world is fast-tracking sustainable development practices, and that means low carbon footprint-producing materials are a better choice. Therefore, materials are selected over others when they are of low-carbon footprint, recyclable, reusable, non-toxic, produce minimal environmental impact, and comply with industry regulations.

5.6.6 Safety and Health Concerns

It is believed that safety and human health are the cornerstone of a sound civil engineering project because in any case, these cannot be compromised. In that regard, materials are selected for safety reasons. If an alternative material is hazardous, it cannot be used to safeguard human safety and other living beings. That means materials that mitigate hazards are largely opted. Similarly, structural reliability is a key consideration for material selection. It is important that materials that promote health, safety, and hygiene better than others are opted as this can help prevent accidents or provide fire safety. For example, fire-retardant paints (intumescent paint) may be opted for their ability to prevent the spread of fire. They are typically applied to surfaces such as steel, wood, dry wall, and concrete. Solid blocks are more fire-resistant than perforated cellular bricks of the same thickness due to their thermal mass, less voids, and overall density.

5.6.7 Aesthetics

Materials are selected based on the pursuit of the most aesthetical appearance for a structure, especially the finishing materials such as tiling, painting, stonework, woodwork, compound designing, sustainable drainage systems, and landscaping works, among others. Therefore, aesthetic and social reasons guide the designers while selecting materials basically for the visual appeal, community acceptance, historical preservation, cultural sensitivity, user comfort, accessibility, and social responsibility.

5.6.8 Standards and Regulations Compliance

Materials that meet the regulatory and standardization requirements are prioritized. This means compliance with building codes, adherence to industry standards, meeting certification requirements, permitting regulations, environmental regulations, health and safety regulations, and international standards, among others.

5.6.9 Up-to-Date Materials

Selecting materials based on innovation and advancement in technology is no longer a topic to debate about. Emerging technologies, research and development, material science advancements, innovative applications, improved performance, enhanced functionality, and future-proofing have all provided a platform for better materials.

5.7 MAIN CATEGORIES OF CONSTRUCTION MATERIALS

From elementary chemistry, we know that materials (matter) can take three forms: solid, gas, or liquid. These can either be compounds, solutions, or mixtures. Construction materials are primarily liquids and solids. The main categories of construction materials include composites, alloys, additives, admixtures, binders, concrete and masonry, polymers, and plastics.

5.7.1 Composites

A composite material is a combination of two materials with different physical and chemical properties. When they are combined, they create a material which is specialized to do a certain job, for instance, to become stronger, lighter, or resistant to electricity.

- Reinforced concrete is a composite material. Plain concrete is good in compression but weak in tension; in contrast, steel is good in tension and weak in compression. Composites are also a mixture of two or more elements, but it does not contain metals.
- Geotextiles are composite materials or permeable fabrics used in geotechnical engineering to stabilize soil, prevent erosion, and improve drainage. They are typically made from a combination of materials, such as synthetic fibers (polypropylene,

polyester, or nylon), natural fibers (jute, coconut, or cotton), and other additives (stabilizers, UV protectants, etc.)

- Wood is a composite material. Wood's composition and structure are naturally occurring, making it a complex composite natural material. It consists of components such as cellulose fibers, hemicellulose, and lignin, which work together to yield its structure and properties. Some other natural materials such as stone, soil, and bamboo are natural composite materials.
- Other natural materials include aggregates, sand, gravel, and crushed stone.

5.7.2 Alloys

An alloy is a mixture of two or more elements, at least one of which is a metal, where the resulting material has a combination of properties that are different from those of the individual elements.

A mixture consists of two or more elements, at least one of which is a metal. The elements are combined at the atomic or molecular level. The resulting material has a uniform composition and properties. Examples: steel, brass, and bronze.

Alloys are mixtures of elements at the atomic or molecular level; composite materials are made from distinct substances combined at the macroscopic level. An alloy is a mixture of one or more metals with other elements. An alloy can either be a homogeneous or a heterogeneous mixture. A composite is always a heterogeneous mixture. Examples of alloys are steel (iron–carbon alloy), brass (copper–zinc alloy), bronze (copper–tin alloy), titanium alloy (titanium–aluminum–vanadium alloy), and stainless steel (iron–chromium–nickel alloy).

Most metals used in construction are alloys, while some are pure elements. For example, steel is a pure alloy, aluminum can be pure aluminum (element) or an aluminum alloy (aluminum with other elements like copper, zinc, or magnesium), copper can be pure copper (element) or a copper alloy (copper with other elements like zinc, tin, or nickel), and iron can be pure iron (element) or part of an alloy like steel (iron and carbon).

5.7.3 Admixtures

Admixtures are substances added to concrete, mortar, or other construction materials to enhance their properties, performance, or workability such as air-entraining agents, water-reducing agents, and corrosion inhibitors. TEAIS ES, a Spanish company, manufactures several admixtures (visit: https://www.teais.es).

5.7.4 Additives

Additives are substances added to materials, products, or processes to enhance their properties, performance, or functionality such as plasticizers in plastics, retarders in concrete, and UV stabilizers in polymers. TEAIS ES, a Spanish company, manufactures several additives (visit: https://www.teais.es).

5.7.5 Binders

A binding agent, or binder, is a material used to form materials into a cohesive whole, as a means of providing structural stability. These include cement, lime, gypsum, filler, tar, and bitumen.

5.7.6 Concrete and Masonry

Concrete is a mixture of aggregates (fine and coarse), cement, and water. Concrete and masonry materials are physical combinations of two or more substances—for example, concrete (a blend of cement, water, and aggregates) and mortar (a blend of cement, water, and sand). Soil is also a mixture. When we define concrete and masonry, it chiefly refers to concrete, bricks, blocks, mortar, etc.

5.7.7 Polymers and Plastics

Polymers are large molecules made up of repeating subunits, called monomers. Plastics are a specific type of polymer, typically synthetic, that can be molded and shaped. Plastics are hydrocarbon-based polymeric materials derived from crude oil and natural gas and are a subset of polymers. Therefore, while all plastics are considered polymers, not all polymers are considered plastics.

5.7.8 Other Materials

These include glass, ceramics, and geosynthetics.

5.8 COMMON CIVIL ENGINEERING MATERIALS

The common civil engineering materials include reinforced concrete, composite steel and concrete, structural glass, aluminum, timber, plywood, masonry, stones, ceramics, composites, polymers, plastics, admixtures, and structural steel.

5.8.1 Reinforced Concrete

This is made out of cement, reinforcement steel, coarse and fine aggregates, and water. These components are outlined below:

- **Cement:** Usually, gray is the standard color of cement. However, white cement can also be used to change the mix appearance. It is composed of limestone and clay fired to temperatures of about 1400°C and then ground to a powder. The cement content of a mix affects the strength and finished surface appearance.

- **Aggregate:** These are coarse and fine aggregates—coarse sizes range from 10 to 20 mm, while fine/sand range from 0.075 to 4.4.75 mm—and sand makes up about 75% of the mix volume. Coarse aggregates can be natural dense stone or lightweight furnace by-products.
- **Water:** Water is added to create the cement paste, which coats the aggregate. The water/cement ratio must be carefully controlled as the addition of water to a mix will increase workability and shrinkage, but will reduce the strength if cement is not added.
- **Reinforcement:** Reinforcement normally consists of deformed steel bars. Traditionally, the main bars were typically high yield ($f_y = 460 \text{ N / mm}^2$), and the links are mild steel ($f_y = 250 \text{ N / mm}^2$). However, the new standards on bar bending now allow small-diameter high-yield bars to be bent to the same small radii as mild-yield steel bars [6].
- **Admixtures:** Concrete admixtures are materials added to concrete during mixing to modify its properties, either in its plastic or hardened state. Durability, workability, and setting time can be affected by the use of admixtures.
- **Formwork:** These are temporary works made up of either steel, timber, or plastic mold used to keep the liquid concrete in place until it has hardened. Formwork can account for up to half the cost of a concrete structure and should be kept simple and standardized where possible.

5.8.1.1 Summary of reinforced concrete properties

- Dense, plain concrete is typically 24 kN/m³, while reinforced concrete is 25 kN/m³.
- Poisson's ratio, 0.2.
- Modulus of elasticity, E, 14–41 kN/mm².
- Modulus of rigidity, G, 21 GPa.
- Linear coefficient of thermal expansion, $13–14 \times 10^{-6}/°C$.

5.8.1.2 Concrete behavior under different conditions

Temperature: under high temperatures, concrete can degrade, lose strength, and potentially spall or crack due to thermal expansion and water loss. At low temperatures, for example, during the curing process, concrete can slow down hydration, leading to slower gain in strength. Freezing temperatures can cause damage if water inside the concrete freezes and expands. Temperature plays a role in concrete shrinkage—concrete can shrink as it dries, leading to cracking or stress buildup.

Pressure: concrete is strong under compressive stress, but excessive pressure can lead to crushing or failure. Concrete is weak under tensile stress and can crack or fail if subjected to excessive tensile forces.

Stating: when loaded, concrete can creep or deform under sustained loads, leading to long-term deformation or cracking.

Environmental conditions: exposure to excessive moisture can lead to degradation, erosion, or chemical reactions that weaken the concrete. Concrete can be damaged by exposure to chemicals such as acids, salts, or sulfates, which can react with the cement paste or rebars. That's why appropriate concrete cover is always recommended to mitigate such risks. Prolonged exposure to UV radiation can cause surface degradation and discoloration of concrete. Repeated freezing and thawing can cause damage to concrete, especially if the concrete is saturated with water. Therefore, understanding how concrete behaves under different conditions is important for designing and constructing durable structures. Engineers and builders consider these factors to take steps to mitigate potential issues and ensure the longevity of concrete structures.

5.8.2 Steel Reinforcement Bars

The role of steel reinforcement bars (rebars) is to improve the mechanical properties of the structure as they provide tensile strength, ductility, and crack-growth resistance. Steel is considered a mixture because it is composed of multiple elements such as iron and carbon, which are not chemically bound together, but rather physically combined to form the alloy known as steel; therefore, it is not a compound.

Summary of hot-rolled steel properties

- Density, 78.5 kN/m³
- Tensile strength, 275–460 N/mm² yield stress and 430–550 N/mm² ultimate strength
- Poisson's ratio, 0.3
- Modulus of elasticity, E, 205 kN/mm²
- Modulus of rigidity, G, 80 kN/mm²
- Linear coefficient of thermal expansion, $12 \times 10^{-6}/°C$

5.8.3 Structural Steel

Structural steel is a category of steel specifically designed and manufactured for use in construction, known for its high strength and ability to support heavy loads. It is commonly used in building frames, bridges, and other large structures. Proper use of structural steel in Figure 5.23 enabled engineers to build a non-column balcony at Nanjing railway station. This cantilevered balcony or platform extends outward from the main structure without visible support columns underneath. These balconies are often designed with a specific structural approach to manage the forces and stresses induced in the steel members.

5.8.4 Composite Steel and Concrete

The term 'composite structure' refers to the structure in which different materials such as timber, steel, concrete, and masonry are used together for construction. The most common type of composite construction is the use of steel and concrete to form steel–concrete composite structures. Steel–concrete composite

FIGURE 5.23 Steel structure—non-column balcony at Nanjing railway station. (Image credits: Richard Li Chengzhe.)

structures cover structural elements such as beams, slabs, and columns in which the best structural properties of each material are combined. Designs of composite steel and concrete structures are covered by BS EN 1990–1999 Eurocode 4.

Composite construction is extensively used in bridges, multistory buildings, warehouses, marine structures, and many more. Many applications in the mentioned structures are categorized as beams and girders, floor systems, and column systems. Steel members are susceptible to buckling, while their tensile strength is remarkable. On the other hand, plain concrete members can withstand a large magnitude of compressive force; however, their tensile strength is very low. Therefore, the simultaneous use of steel and concrete allows the structural designers to take advantage of steel and concrete and neutralize each material's weakness by the advantage of the other material.

5.8.4.1 Case study [5.1]: Slab deck systems

The composite-suspended slab system is made from galvanized steel profiles, laid in position, into which concrete is poured. The concrete provides strength, with the steel providing stability. This composite slab system uses substantially less concrete than conventional decking systems, resulting in substantial cost-effectiveness for the client. The additional benefit is that the composite slab system uses less concrete and steel, thus making it substantially less harmful to the environment. This case study is selected to demonstrate a building system that matches modern-day sustainability goals! The slab deck system presents several advantages as compared to the conventional concrete slab systems:

- Lightweight system; much less concrete and steel.
- Exceptional strength, due to T-beam design.

- Exceptional bond between steel and concrete (composite action).
- Ample space in voids for electrical and plumbing services.
- High-quality soft finish.
- 30% cost saving versus conventional system.
- Easy to install.
- Minimal propping required.
- Ideal for complicated shaped slabs.
- Much less harmful to the environment.
- Good sound and temperature insulation.

How it works?

The composite-suspended slab system is manufactured, utilizing galvanized steel profiles, laid in position as panels. These panels once laid into position act as the formwork for wet concrete to be poured over the profiled galvanized panels [7]. See Figures 5.24 and 5.25. The concrete provides strength and, together with the steel, higher stability. The composite-suspended slab system is adaptable and can be used on any design—curved or straight; all that is needed is the exact measurements to tailor the system to your needs. They are used for factories, warehouses, shopping malls, office blocks, and blocks of flats, among others.

5.8.5 Structural Glass

The term 'structural glass' is a term used to describe a frameless glass assembly where the glass is taking an element of the structural load. The structural load that the glass is designed to take will depend on the location and purpose of the glass element.

A structural glass floor, for instance, will be designed to withstand a walk on load across its surface, whereas a frameless

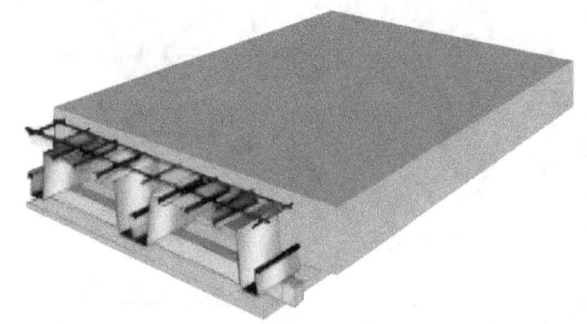

FIGURE 5.24   Slab decking system (T-beam model).

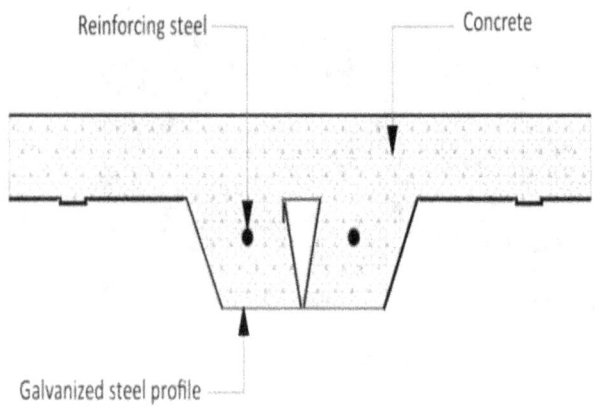

FIGURE 5.25   Cross-section through a typical T-beam.

glass balustrade installation will be designed to withstand the required line loading in a perpendicular direction. Whatever the use or purpose of a frameless glazing element correct structural engineering, glass specification and fixing details are essential in ensuring the installation is safe and secure.

The design of structural glass elements requires a good understanding of how the material behaves when placed under load. Glass is a very strong, but also extremely brittle, material. This key attribute causes it to fail suddenly as it cannot yield, unlike more traditional materials such as steel and timber. This fact presents unique challenges to structural engineers when designing structural elements to be made from glass.

5.8.6 Aluminum

Aluminum is commonly employed in construction because of its inherent advantages of lightness and corrosion resistance. Aluminum is frequently utilized as a substitute for steel in situations where its unique qualities are critical and worth paying for. Since 1886, aluminum has been utilized extensively as an industrial metal. Its unique characteristics are lightweight and sparkling appearance. Aluminum usage in the globe is estimated to be in the range of 20 million tonnes per year.

The present smelting method was invented at this time. When compared to metals like lead, copper, and bronze, which have been used for thousands of years, this suggests that the metal has been available for a shorter time. Aluminum is utilized in a variety of applications, including external facades,

roofs and walls, windows and doors, stairwells, railings, and shelves, among others.

5.8.7 Timber

Typically refers to wood that has been processed for construction or building purposes, such as beams, lumber, or boards. Timber often implies a more specific use or application. Construction-grade timber is specifically selected for structural support in building projects. It undergoes rigorous testing to ensure it meets the necessary standards for strength, stability, and durability.

5.8.8 Wood

Wood is the raw material from trees, including logs, planks, or boards. Wood can be in its natural state or processed for various uses. It offers various advantages for both structural and aesthetic applications. The different types of wood are suitable for different purposes, with hardwoods like oak and maple often used for durability and appearance, while softwoods like pine and fir are favored for framing due to their affordability and ease of use.

5.8.9 Other Common Materials

These include bamboo, plywood (composite material), building elements/fixings and fastenings, masonry (includes brickwork and blockwork—hollow blocks and solid blocks), stones, ceramics, and composites, polymers and plastics, geosynthetics (especially geotextiles and geomembranes), admixtures, and additives.

5.9 MIND MAPPING QUESTIONS

Temperature

1. How do temperature fluctuations affect the durability of concrete?
2. What civil engineering materials expand or contract significantly with temperature changes?
3. How can thermal insulation be optimized in modern building design?

Pressure

1. How does pressure impact the structural integrity of materials such as steel or concrete?
2. What civil engineering materials are best suited for high-pressure applications?
3. How can earth pressure be managed in retaining wall design?

5.10 CONCLUSION

Temperature, pressure, and environmental conditions affect civil engineering materials. The physical, chemical, and mechanical properties of civil engineering materials are shaped by these three factors. The primary questions born in an engineer's mind early in the planning, design, or construction phase of projects revolve around the ability of a material to withstand variations in temperature, pressure, and environmental conditions. The selection of suitable materials for civil engineering purposes primarily answers questions related to durability, reliability, safety, cost, and sustainability. These factors are influenced by the science of materials which benchmarks on the behavior of materials when subjected to the three primary attributes, that is, temperature, pressure, and environmental conditions.

NOTES

1 See the section for the role of coefficients in equations.
2 This example illustrates the effect of temperature changes on statically indeterminate axially loaded structures (thermal stress on axially loaded statically indeterminate members).

REFERENCES

1. Engineering ToolBox. (n.d.). Linear thermal expansion coefficients. Retrieved from: https://www.engineeringtoolbox.com/linear-expansion-coefficients-d_95.html. Accessed 04/07/2025.
2. Worldwide Rails. (n.d.). How do railroads deal with thermal expansion? Retrieved from: https://worldwiderails.com/how-do-railroads-deal-with-thermal-expansion/. Accessed 04/07/2025.
3. Radu, C. (1999). Cai Ferate – Suprastructura Caii (Railway – Track Superstructure) – Course notes, Faculty of Railways, Roads and Bridges – Technical University of Civil Engineering Bucharest.
4. Babu, G. L. S., Kandhal, P. S., Kottayi, N. M., Mallick, R. B., & Veeraragavan, A. (2019). *Pavement Drainage: Theory and Practice* (1st ed.). CRC Press. https://doi.org/10.1201/9781351135948
5. Budynas, R. G. (1999). *Advanced Strength and Applied Stress Analysis* (2nd ed.). McGraw-Hill, New York.
6. Cobb, F. (2009). *Structural Engineer's Pocket Book* (2nd ed.). Routledge, London.
7. Steel, C. (n.d.). Permanent formwork voidcon. Retrieved from: https://www.clotansteel.co.za/portfolio-item/permanent-formwork-voidcon/

FURTHER READING

1. Budynas, R. G. (1999). *Advanced Strength and Applied Stress Analysis* (2nd ed.). McGraw-Hill, New York.
2. Cobb, F. (2009). *Structural Engineer's Pocket Book* (2nd ed.). Routledge, London.

Mimicking the Natural World

6

6.1 INTRODUCTION

The civil engineering industry greatly mimics the natural world. Civil engineering often draws inspiration from anatomy, physiology, and the natural world (environment) to design and develop innovative solutions.

When engineers of all disciplines encounter design dilemmas, whether it's creating better materials, efficient processes, or smarter systems, they usually turn to nature for inspiration. This practice is *generally* referred to as **biomimicry**.

The purpose of this chapter, therefore, is to introduce scholars, students, and professionals in all fields of civil engineering design and development to biomimicry. When practiced in a deep and thoughtful way, biomimicry has the potential to yield sustainable civil engineering designs and/or products. The truth is that civil and other engineers, who presently dream of innovating cool stuff, cannot look farther than nature-inspired products!

Civil engineers greatly mimic the natural world by drawing inspiration from it, comparing their designs, products, and systems with nature. And now that sustainability is a hot topic of the 21st century, civil engineers gear all efforts and spend days and nights trying to figure out how best to reduce the adverse impact on nature by mimicking it as much as possible or applicable—that is the work of a constantly evolving civil engineer, now and in the future.

Analogizing the nature can lead to innovative solutions and sustainable designs. How best can we not antagonize nature and at the same time connect the unconnected? That's the question evolving civil engineers constantly contribute to!

Simply put, biomimicry is the practice of designing and engineering solutions by emulating nature and its biological systems. Biomimicry (literally, imitation of the living) aims to take inspiration from natural selection solutions adopted by nature and translate the principles to human engineering. The biomimicry approach aims to favor 'choices' tested by nature, which had millions of years to understand what works best and what doesn't.

Biomimicry involves studying and mimicking the forms, processes, and functions of living organisms, such as plants, animals, and microorganisms, to develop innovative and sustainable solutions for human challenges. It is a practice that significantly learns from nature and mimics the strategies found in nature to solve human design challenges and find hope. For most challenges we face, nature probably has a solution.

Leonardo da Vinci (1452–1519), one of the most well-known historical figures of high ingenuity, demonstrated to the world how connections can be made, basically connecting the unconnected. His insights were perhaps the first to share a glimpse into biomimicry. He was great for making new connections basically drawing inspiration from nature. His greatest breakthroughs came from connecting the unconnected, and in particular, the insights gained by bringing different perspectives together. His expertise in anatomy helped him create better portraits and sculptures and also helped him make sense of mechanics and engineering. He was a great artist, engineer, and scientist! He defined innovation as connecting the unconnected, drawing inspiration from nature!

Da Vinci explored the human form with the same probing mind with which he investigated all his interests; he also documented his findings in his books. One of his most famous sketches, The Vitruvian Man (Figure 2.2), exemplifies his exploration of the human form and is often interpreted solely as an artistic work. However, it exemplifies his scientific perspective and interest in the human form. As he learned about the human body, his sense of wonder at its perfection grew. In addition to considering the human form as 'the greatest instrument in nature,' Da Vinci also viewed the human body as a machine that can be analyzed and understood [1]. Such an approach is similar to those who study biomedical engineering, a field that focuses on understanding, repairing, or replacing physiological systems in order to improve medicine.

Da Vinci studied human movement and mechanics, envisioning a robot-like device. He drew parallels between human joints and muscles and machine components, recognizing similarities between the two. Within his analytical mind, Da Vinci realized that the joints and muscles of the human body reflect the simple gears and pulleys that make up machines. By analyzing joints and muscles, he saw analogies with mechanical systems, laying the groundwork for future innovations.

He showed how the anatomy can be mimicked to develop systems for the convenience of man. His work in art, engineering, and philosophy is of paramount importance in the study of civil engineering. He is remembered for being a highly creative person. See Chapter 2. It is obvious that civil engineering is a mirror of the history of human beings on Earth.

It is obvious and exciting to observe how the infrastructure that civil engineers develop mimics the natural world. The physical and built environment which includes roads, power plants, railroads, bridges, canals, aqueducts, tunnels, buildings, airports, dams, water supply and distribution systems, sewage systems, airfields, schools, offices, hospitals, and sustainable drainage systems all mimic the natural world directly or indirectly—in whole or in part.

The science behind why things stand and not fail or fall in the natural world is greatly copied and harnessed to develop artificial systems throughout the civil engineering industry. Civil engineers do two things based on nature:

- Drawing inspiration from nature.
- Using nature as an analogy (analogizing nature).

DOI: 10.1201/9781003561620-6

The former encourages, motivates, or incites action due to its positive influence, and the latter leads to clarification and understanding of a civil engineer's work by comparing primarily two things, processes, or ideas—that is, comparing how nature works with civil engineers' innovations. *Doing two things, that is, inspiration and analogizing, is biomimicry!*

In this sense, 'inspiring' is often used to describe natural things that evoke a sense of hope, enthusiasm, or determination. It is about the impact something has on our emotions and actions. Analogy is a comparison between two things, processes or ideas. It helps explain one concept by likening it to another—in this particular case, basically, engineers make an analogy between the workings of nature and those of human societies in a practical sense—comparing between one thing (natural) and another (artificial), typically for the purpose of explanation or clarification or appreciation.

As engineering is a practical field, engineers usually analogize the nature by developing artificial systems that resemble it! Hence, they gain a better understanding of how systems that mimic the natural environment could work much better serving humanity, which is the highest good in the practice of an engineer.

In their literal sense, analogies are commonly used to clarify complex ideas. They involve drawing parallels between two different things. For instance, analogies focus on explaining by comparing, using words like 'like' or 'as.' However, in civil engineering, they compare nature with artificial systems that engineers develop. So, analogizing nature is a practical subject.

Nature's designs are timeless and full of inspiration. An artist draws inspiration from natural forms like dates, kiwis, tomatoes, and fish patterns to create unique architectural wonders that bring serenity to urban spaces—he may not necessarily analogize nature in his works. Have you seen the urban spaces that artists beautify? Inspiration can come from the simplest elements.

Nature's designs bring peace to our environment. Integrating nature into design creates spaces that resonate deeply. In the civil engineering world, particularly, embracing nature-inspired solutions can lead to innovative and sustainable solutions. It is essential to create environments where functionality and esthetics harmonize, inspired by nature's seamless integration of form and function.

Have you heard about **nature-based solutions**? These solutions draw inspiration from nature to solve the problems facing the ecosystem today. Thus, they contribute to a sustainable world.

There is no doubt that civil engineers greatly mimic the natural world by drawing inspiration from it, comparing their designs, products, and systems with nature. And now with the focus on sustainability in the 21st Century, civil engineers gear all efforts and spend days and nights trying to figure out how best to reduce the adverse impact on nature by mimicking it as much as possible or applicable—that is, the work of a constantly evolving civil engineer, now and in the future. In the same scheme of reasoning, a well-schooled engineer raised while living in green environment may have far-reaching ideas mimicking nature than another raised living predominantly in gray infrastructure.

Analogizing nature can lead to innovative solutions and sustainable designs. How best can we not antagonize nature and at the same time connect the unconnected? That's the question evolving civil engineers constantly contribute to! To put things in perspective and make it clearer, we provide two examples of *'blood vessels and pipelines'* and *'tree and column'* henceforth.

6.1.1 Blood Vessels and Pipelines

The flow of fluids through pipelines is similar to the circulatory system, with its network of blood vessels and capillaries.

6.1.1.1 Inspiration

Blood vessels and capillaries in the human body efficiently transport vital fluids (such as blood) throughout the network, supplying oxygen and nutrients to the cells. This natural system inspires the design of pipelines, aiming to transport fluids (in this case liquids or gases) over long distances typically kilometers, often with similar efficiency and reliability or even more. Potable water pipelines are typical examples of such lengthy pipelines that draw inspiration from blood vessels.

6.1.1.2 Analogizing

Analogizing blood vessels to pipeline design and construction involves identifying the underlying principles and mechanisms in blood vessels that can be applied to pipeline design. Some key analogies include the following.

6.1.1.2.1 Network structure
Both blood vessels and pipelines have a branching, hierarchical network structure, allowing for efficient distribution and flow. Engineers study the blood vessel network structure, hierarchy, and branches to mimic it to produce efficient and effective pipeline routes and layouts.

In terms of location, arteries are typically located deeper in the body, closer to organs. The veins are often located closer to the surface, with more visible and accessible pathways. In water pipeline design, for example, pumps are connected to the transmission pipeline, while the reservoir tank typically allows water to flow by gravity through the distribution networks and then to a network of intensification lines and service lines. These imitate arteries and veins. A very good example is two-stage pumping, and pressures need to be boosted to allow water to flow effectively through the distribution networks.

Through arteries, blood flows away from the heart, and through veins, blood flows toward the heart. Underground water is abstracted by pumps (relate this to an organ—the 'heart'), and aquifers keep recharging. The pumps imitating the heart pump water through the transmission pipeline to the reservoir which supply water through the distribution, intensification, and service lines. When water is used, it goes back to the earth and percolates, thereby recharging the aquifers. See Section 6.2.1.3.1 for the detailed analysis of how the pump analogizes the heart.

6.1.1.2.2 Flow regulation
Blood vessels have valves and constriction/dilation mechanisms to control flow; similarly, pipelines use valves, pumps, and compressors to regulate fluid flow. Such valves on

pipelines include air-release valves, control valves, and gate valves. Control valves, gate valves, and air-release valves are installed on pipelines. Veins have one-way valves to prevent blood from flowing backward—the engineer mimics this to produce a non-return valve. Arteries do not have valves, except for the pulmonary artery. Veins need valves to overcome low pressure and gravity, while arteries do not require valves due to their high pressure and continuous flow.

Arteries have thicker walls, are more muscular, and are elastic to withstand higher pressure. On the other hand, veins typically have thinner walls, are less muscular, and are more flexible to accommodate varying blood volumes—this is because veins have a lower level of pressure. So, their walls don't need to be as thick as those of arteries. Have you seen pressure ratings in pipes (say PN 6, 10, 16, and 25)? These pipes are produced with varying thicknesses to withstand pressures. It is typically based on the pressures they transmit that thicknesses are determined. That means selecting pipes of different materials, sizes, and thickness, for example, HDPE, uPVC, GI, or ductile iron depends on pressure they are designed to withstand. This directly compares with blood vessels composed of flexible, durable materials; pipelines often use materials with similar properties, such as flexibility, corrosion resistance, and strength.

6.1.1.2.3 Pressure management

Blood vessels maintain optimal pressure to ensure consistent flow; pipelines are designed to manage pressure drops and surges to prevent damage or loss of flow. Civil engineers install a pressure gauge and switch on the pipeline.

By analogizing the functional principles of blood vessels, pipeline designers can create more efficient, reliable, and adaptable systems for transporting fluids. This **biomimetic approach** has led to innovations in pipeline materials, design, operation, and maintenance, improving the safety and efficiency of fluid transportation. Did you know that we can boost pressure in a water pipeline branching from another by reducing its size?

6.1.2 Tree and Columns

The structure of trees, with their trunks, branches, and roots, inspires the design of columns and support systems in civil engineering. We learn fixed connections from the way branches are connected to the trunk. Figure 6.1 shows the basic tree anatomy.

6.1.2.1 Inspiring

The structure of trees, with their trunks, branches, and roots, inspires the design of columns and support systems in civil engineering. Trees have inspired civil engineers to design and develop columns for centuries. The trees' forms and structures, specifically the trees' trunks, with their sturdy, upright forms, have inspired the shape and function of columns, especially the circular columns.

Columns provide support and stability, just like tree trunks. The top and base of columns are reminiscent of a tree's branching and root systems, respectively. The capital (top

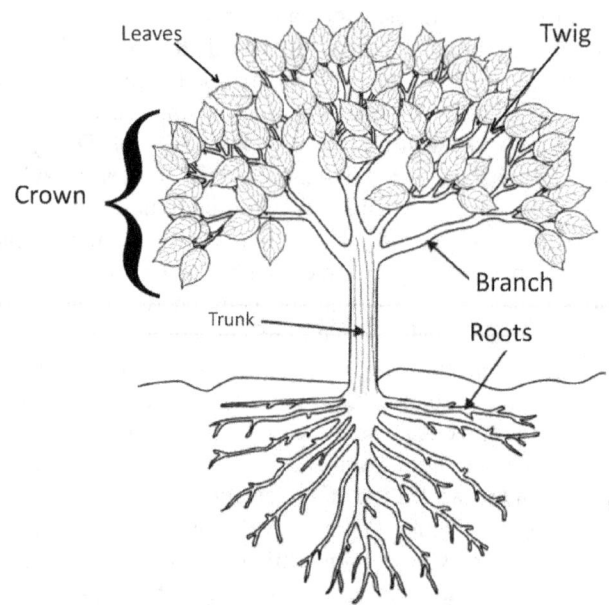

FIGURE 6.1 Basic tree anatomy.

of the tree) distributes the weight, while the base provides a sturdy foundation.

Trees' trunks often taper from a wider base to a narrower top. This tapering shape has been incorporated into column design to create a sense of elegance and stability. Also, the tree bark's texture and pattern have inspired various column finishes, such as fluted or rusticated surfaces.

The proportions of trees, with their balanced ratios of trunk, branches, and leaves, have influenced the proportions of columns and their relationship to the surrounding architecture.

Trees' organic, natural forms have inspired the use of columns in design to create a sense of harmony with nature.

Engineers develop column designs that are both functional and esthetically pleasing by observing and emulating the forms, structures, and proportions of trees. This biomimicry enriches the architecture and design, connecting us to nature's beauty and wisdom.

6.1.2.2 Analogy

Just as a tree's trunk provides support and stability, allowing the tree to grow and flourish, a column in engineering and architecture design provides structural support, enabling the building to stand tall and thrive.

Just as a tree's roots dig deep and its branches reach wide, a column's foundation is strong and its capital (upper part of a tree trunk, just below the branches) reaches upward, connecting and sustaining the surrounding structure. Note that the column base distributes loads to the ground as roots of a tree distribute the loads to the ground.

Did you know that the taller the tree, the deeper the roots? The same applies to buildings—high-rise buildings, typically skyscrapers, usually utilize deeper foundations—typically pile foundations! Did you know that a tree's height is often matched by the depth of its roots? Similarly, in building design, the taller the skyscraper, the deeper and more extensive its foundation—typically utilizing pile foundations that transfer weight to stable soil layers, ensuring stability and structural integrity.

6.1.3 Appreciating Nature in General

Going forward, the analogies in Sections 6.1.1.2 and 6.1.2.2 highlight the inspiration, mimicking, and connection between blood vessels and pipelines and between trees and columns, respectively, demonstrating how nature's forms and structures can inform and enrich civil engineering design.

It is essential to note that civil engineers mimic nature and either *analogize it or simply get inspiration* and work around it to make something completely different. Nature can inspire something without being directly analogized. Therefore, inspiration can come from observing nature:

- **Principles**: Nature's fundamental principles, such as sustainability or efficiency, can guide human endeavors. This is very common in sustainable drainage systems (SuDs). See further reading 1.
- **Forms**: Shapes, patterns, and structures in nature can inspire designs, such as architecture or product design. For example, *domes and skulls*—the shape and structure of domes in architecture—are similar to those of the human skull, providing protection and support.
- **Processes**: Natural processes, such as evolution or adaptation, can inspire problem-solving approaches or innovative methods. How natural systems function greatly inspires civil engineering processes; for example, infiltration basins mimic how water infiltrates into the ground and thus contribute to sustainable management of water resources and ecological sustainability. Transport management systems mimic the central nervous system. Civil engineers design systems that mimic these natural processes.

However, analogizing nature requires explicitly comparing or mapping natural phenomena to human-made concepts or systems—in this case, civil engineering structures. If inspiration from nature doesn't involve making explicit practical connections or comparisons, it's not analogizing. In this context, analogizing implies a deeper connection between the two systems compared, that is, the natural and the artificial, where the artificial is not just a simple copy of the natural but an innovative application of the nature's principles to solve a similar problem or closely related problem.

Analogizing nature can also involve drawing metaphorical connections. This means drawing abstract parallels between natural phenomena and human experiences, like comparing the flow of traffic to the flow of water. Through analogizing nature, systemic analogies may be inspired. This means comparing complex natural systems, like ecosystems or weather patterns, to understand and design complex human systems, like supply chains or social networks.

However, it is important to note that while biomimicry might occasionally involve metaphorical connections or systemic analogies, its core focus is on direct inspiration and application of natural designs or processes to solve human problems, in this case civil engineering problems. In contrast, metaphorical connections and systemic analogies are more abstract and might be used in other fields such as systems thinking, design thinking, and innovation strategies. These approaches might draw inspiration from nature, but they do not necessarily involve direct imitation or application of natural designs like biomimicry does.

6.1.3.1 Biomimicry

Nature's solutions are used to develop innovative technologies, for example, developing water-repellent materials inspired by lotus leaves and designing more efficient wind turbines based on whale fin shapes. These two processes involve emulating the forms, processes, and functions of living organisms to create innovative solutions. This is analogizing nature—typical biomimicry.

It involves both inspiration and analogizing, as well as a deep understanding of the biological systems and mechanisms being emulated. Several steps are undertaken in biomimicry, which include the detailed study of the natural system to understand its structure, function, and performance and converting the natural principles into a technological context, often involving significant innovation and creativity. It also involves testing and validating the **biomimetic solution** to ensure it meets performance, reliability, and functionality requirements.

6.1.3.2 Nature-inspired art

This is a process of creating art pieces that evoke natural forms or emotions without direct comparison. This is not analogizing; it is inspiration. It is common in urban spaces in cities, homes, hotels, and recreation centers, among others—drawing inspiration from natural systems to make unique pieces of art that immerse the human into the feeling that they are in nature.

Nature can inspire without being directly analogized, but analogizing it helps explicitly connect natural principles to human innovations.

Now, that climate change has serious mental health implications than ever, designers are constantly integrating nature-inspired art in their works. Landscaping works and SuDs predominantly utilize this kind of art. Climate change has significant mental health implications that may include anxiety and stress related to environmental uncertainty.

Trauma and grief from experiencing extreme weather events can lead to increased risk of mental health disorders (e.g., depression and post-traumatic stress disorder (PTSD)). So, by integrating nature-inspired art in design and development of engineering solutions, the communities' well-being improves, and the esthetical appearances are therapeutical, drawing people and environment in harmony.

Examples of nature-inspired art in engineering include biomimetic architecture, green infrastructure, eco-friendly materials, public art installations, and sustainable urban planning. By embracing nature-inspired art in civil engineering, we can create solutions that not only serve their purpose but also promote well-being, sustainability, and community harmony.

6.2 MIMICKING ANATOMY, PHYSIOLOGY, AND THE NATURAL WORLD

Civil engineering often draws inspiration from anatomy, physiology, and the natural world to design and develop innovative solutions. In addition to the two examples given in Section 6.1, illustrating 'inspiration' from nature and 'analogizing' nature,

Section 6.2.1 outlines some more examples of how civil engineering mimics the anatomy of living beings. Section 6.2.3 outlines examples of how civil engineering mimics the natural environment.

It is essential to note that anatomical systems and physiological process work jointly or alongside each other to support the body functions. However, as discussed in Section 6.1, the engineer may be inspired by the anatomy but fails to analogize the system as it is. The engineer may exploit in part or whole—implying they can utilize the physiological process in an architecture that does not mimic the real anatomy or define the anatomical-inspired architecture but functioning not as exactly as the corresponding physiological process.

Anatomy studies the 'hardware' (structure), and physiology studies the 'software' (function). Anatomical processes can be considered static and physiological process dynamic. Note that anatomical changes often occur over longer periods, while physiological processes happen rapidly. For example, bone fractures and subsequent healing are structural changes which are anatomical. In comparison, blood clotting and platelet activation are functional responses—a physiological process.

Anatomical processes are related to the structure and organization of body parts. It also involves changes in shape, size, or position of tissues, organs, or systems as it focuses on the morphology (form and structure). The primary anatomical processes include bone growth and development, wound healing and tissue repair, embryological development, and musculoskeletal movements (joint movements).

On the other hand, physiological processes are related to the functions and activities of living organisms involving chemical, electrical, or mechanical changes that maintain homeostasis. It primarily focuses on function and dynamics. The primary examples of physiological processes include nerve impulse transmission, blood circulation and oxygenation, digestion and nutrient absorption, and hormone regulation and metabolism.

A civil engineer has been inspired by nature and analogizes anatomy, physiology, and the natural environment to design and develop systems. From blood flow, skeletal changes, digestion, and homeostasis, the engineer has exploited the nature-inspired wisdom to develop beautiful things to save humanity.

The engineer also mimics the natural environment, that is, the trees and other plants, water resources, lakes and rivers, caves, natural landscapes, and forests, among other natural creations. SuDs are a perfect example of how engineers mimic the natural world to solve drainage and flooding problems, while preserving the ecosystem. Architects also significantly mimic fruits, natural landscapes, and plants to come up with beautiful ideas to design high-rise buildings and/or skyscrapers presently. For example, by cutting an onion and observing the section properties, the architect can plan a building marvelous beyond imagination!

As humans have sought to understand and replicate the natural world much better, illustrations of our discoveries reveal the intricacies of natural creation. The more we appreciate nature, the more we draw closer to the Supreme Being, appreciating the higher power.

Remember, the 1828, Thomas Tredgold's definition of civil engineering: as the general advancement of mechanical science, and more particularly for promoting the acquisition of that species of knowledge which constitutes the profession of a civil engineer being the art of directing the great sources of power in nature for the use and convenience of man. The most crucial part of this definition is '*the art of directing the great sources of power in nature....*' It is now high time we greatly mimic nature and not only stop at harnessing the great sources of power! That will lead to sustainable development of the world.

Today, with AI revolutionizing civil engineering, the time is ripe for innovative, sustainability-driven civil engineers to champion biomimicry—emulating nature's genius to develop eco-friendly solutions—we must tap AI resources and automation to locate beautiful ideas that can be developed, leading to environmental sustainability.

Through merging human ingenuity with AI-powered insights and nature-inspired design, human progress can be harmonized with environmental stewardship, ensuring a thriving planet for future generations.

6.2.1 Anatomy, Physiology, and Civil Engineering

Let us start by defining anatomy.

Anatomy is the study of the structure and organization of living things, such as the arrangement of bones, muscles, and tissues in the human body. Anatomy is thus the study of the structure of the body.

Often, functions of the body which include digestion, respiration, circulation, and reproduction may be of much interest.

Physiology is the study of the functions and processes that occur within living organisms, like the circulatory or nervous systems. Civil engineers develop systems that mimic these processes, such as water supply networks or communication systems.

The body is a chemical and physical machine. As such, it is subject to certain laws. These are sometimes called natural laws. Each part of the body is engineered to do a particular job. These jobs are functions. For each job or body function, there is a particular structure engineered to do it.

Civil engineers apply similar principles to design buildings, bridges, water systems, and other civil engineering structures.

The human body is composed of **12** major systems that work together to maintain overall health and function. Once one of the systems malfunctions, the human body becomes weak, and a drug must be administered to heal the person. The body has to be operated and maintained properly to keep healthy.

The **12** major systems of the body are nervous system, circulatory system, respiratory system, digestive system, endocrine system, integumentary system, muscular system, skeletal system, urinary system, reproductive system, lymphatic system, and immune system. Engineers mimic these systems' design while developing products. Sometimes, the mimicking overlaps—whereby the engineer's design mimics functions and structures or shapes of one, two, or more anatomical systems.

By understanding these anatomical systems' structures and functions, engineers make breakthroughs in sustainable, efficient, and effective designs. They design systems that not only perform reliably but also economically conserve energies.

In the 21st century, many engineering marvels are turning out to be nature-inspired, one of the indicators that the planet, Earth, in which alone life is found, is calling upon us to understand the ingredients of life.

6.2.1.1 Respiratory system

This system transports oxygen into the body and eliminates carbon dioxide through the lungs, trachea, bronchi, and diaphragm.

Civil engineers mimic the respiratory system is several ways, but the most common way is through the suction (vacuum) pump. The vacuum pump is inspired by the human's suction mechanism, and the entire system analogizes the human suction mechanism. The suction mechanism is involved in sucking liquids, say, juice through a straw fall under the respiratory system, specifically the mouth and lungs.

The mouth creates a negative pressure by forming a vacuum with the lips and tongue. The diaphragm contracts and descends, increasing the thoracic cavity volume. The intercostal muscles expand the rib cage, further increasing the lung volume. The lung pressure decreases, allowing air (and liquid) to flow into the lungs (or straw). The combination of these factors creates the suction force that draws liquid up the straw.

Additionally, the following systems contribute to the overall process: the nervous system coordinates muscle contractions and relaxations, and the muscular system controls lip, tongue, and diaphragm movements. This intricate interplay enables us to enjoy our favorite beverages through a straw!

6.2.1.1.1 The suction (vacuum) pump and human suction mechanism

Suction describes the air pressure differential between areas. Removing air from a space results in a pressure differential.

Suction pressure is therefore limited by external air pressure. Even a perfect vacuum cannot suck with more pressure than is available in the surrounding environment.

While sucking liquids through a straw (Figure 6.2), a partial vacuum is created in the oral cavity by retracting the tongue to the back of the mouth. The person creates a vacuum by closing their lips around the straw and sucking. The rear portion of the tongue seals against the roof of the mouth, allowing liquids to be drawn into the front region.

When the oral cavity is full, the tongue relaxes, and fluids flow to the throat to be swallowed. **Note:** sucking creates a vacuum in the drinking straw so that the greater ambient pressure pushes the beverage through the drinking straw. The liquid is able to rise up the straw due to the pressure difference created by the vacuum in the mouth. This mechanism is mimicked to develop vacuum pumps of different types!

Note that, in a vacuum, the air molecules are removed or significantly reduced, creating a region with extremely low air pressure because air pressure is caused by the weight and collision of air molecules. With fewer air molecules, the pressure decreases.

A vacuum pump draws gas particles from a sealed volume in order to leave behind a partial vacuum [2]. The first vacuum pump was invented in 1650 by Otto von Guericke (1602–1686) and was preceded by the suction pump, which dates to antiquity.

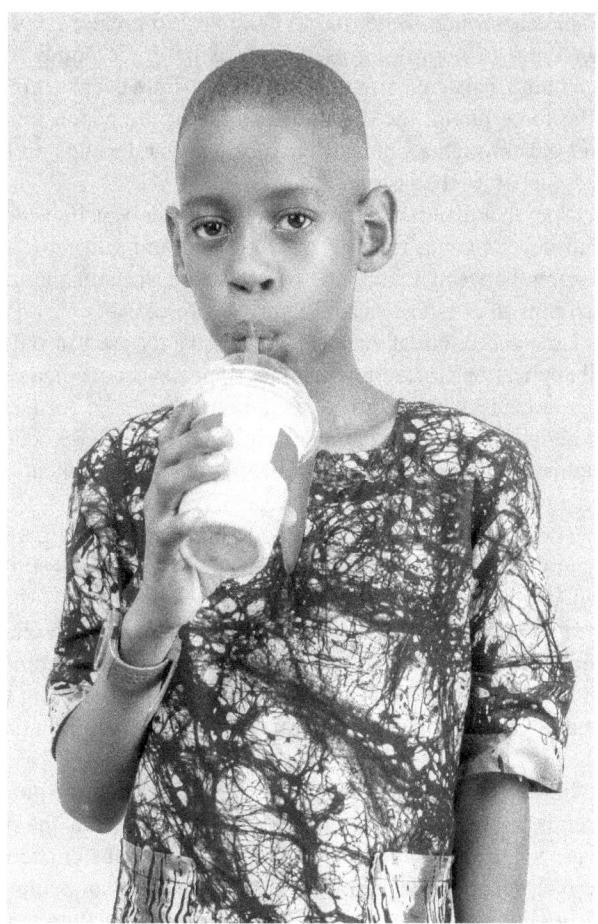

FIGURE 6.2   Mark enjoying an Oreo milkshake through a straw.

A suction pump, specifically, works by atmospheric pressure; when the piston is raised, creating a partial vacuum, atmospheric pressure outside forces water into the cylinder, where it is permitted to escape by an outlet valve.

Note that atmospheric pressure alone can force water to a maximum height, h, of about 34 feet (10 m). See the calculation below:

$$h = \frac{p}{\rho g} \tag{6.1}$$

$$h = \frac{101,325 \text{ Pa}}{1,000 \text{ kg/m}^3 \times 9.81 \text{ m/s}^2} = 10.33 \text{ m} \tag{6.2}$$

Note:

- Atmospheric pressure, p, at the sea level is approximately 1 atm = 101,325 Pascals (Pa).
- The density of water, ρ, is approximately 1,000 kg/m^3.
- Acceleration due to gravity, g, is approximately 98.1 m/s^2.

Therefore, if you try to drink water from a straw at a height more than 10 m, the water won't rise up the straw, and you won't be able to drink it, although, in practice, it's unlikely to drink from a straw at such extreme heights! The suction force created by your mouth won't be enough to overcome the weight of the liquid and the atmospheric pressure pushing down on it.

In cases where water has to be moved to greater heights, other types of pumps are used, such as the force pump. The force pump, particularly, was developed to drain deeper mines. In the force pump, the downward stroke of the piston forces water out through a side valve to a height that depends on the force applied to the piston.

Note that different degrees of vacuum can be achieved, in pump development, ranging from low vacuum with absolute pressures between 1 and 0.03 bars, to high vacuum that can reach pressures as low as a billionth of a Pascal.

Low and medium vacuums are frequently used in industrial applications including vacuum grippers, vacuum cleaners, incandescent bulbs, painting, sandblasting, vacuum furnaces, and negative pressure ventilation. In contrast, higher vacuum systems are essential for specialized laboratory applications such as particle accelerators and reactors.

There are two primary methods for generating a partial vacuum, that is, gas transfer and entrapment. Gas transfer methods include positive displacement vacuum pumps—these use chambers that alternately expand and contract with check or non-return valves to draw in and eject gases—and momentum transfer vacuum pumps—which work by accelerating the speed of gas movement to create a low-pressure region in their wake.

Figure 6.3 shows a reciprocating piston vacuum pump, which is a type of positive displacement pump, and as the piston is rotated, a constant (fixed) amount of gas transfer occurs. The reciprocating vacuum pump mimics the human action of sucking liquids such as juice through a straw—a vacuum pump or specifically, a positive displacement pump with a suction mechanism.

Entrapment vacuum pumps, on the other hand, capture gas molecules through various mechanisms such as:

- Sublimation, where gases are directly transformed into a solid;
- Condensation, where gases are cooled and converted into a liquid or solid state;
- Adsorption, where gas molecules adhere to a surface; and
- Ionization, where gases are ionized and then captured.

Each of the above methods offers specific advantages depending on the requirements of the vacuum system.

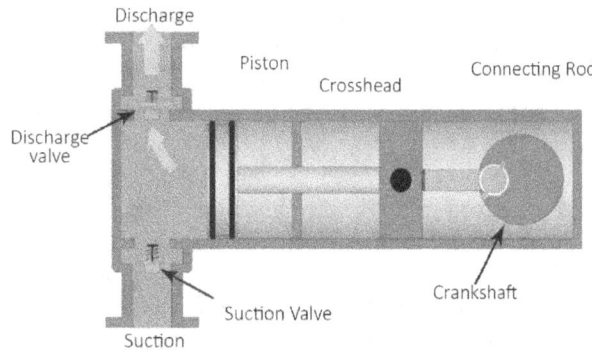

FIGURE 6.3 Reciprocating vacuum pump.

One of the applications of the vacuum pump includes the sewer cleaning pumps/trucks. The sewer cleaning trucks use vacuuming through which all of the dirt is extracted with a vacuuming system. All the solids and water removed this way are then disposed into a sludge tank.

It is important to note that Daniel Bernoulli's principle plays a crucial role in the suction process and vacuum pump operation. According to Bernoulli, the pressure of a fluid (liquid or gas) decreases as its velocity increases. It is possible that the suction mechanism inspired Bernoulli!

When a fluid (liquid or gas) flows through a constricted or narrow passage, its velocity increases. According to Bernoulli's principle, as the velocity increases, the pressure decreases. This decrease in pressure creates a suction effect, which pulls the fluid into the narrow passage. All these principles revolve around conservation of energy, as illustrated in Chapter 3, clearly showing how the law of conservation of energy advanced the civil engineering technology!

Vacuum cleaners, particularly, utilize the principles of suction and Daniel Bernoulli's principle to remove trash and debris. When a fan or impeller located inside the vacuum cleaner is turned on, it rotates at a high speed, creating a region of low air pressure behind it. The fan/impeller accelerates the air, increasing its velocity. According to Bernoulli's principle, the increased air velocity results in a decrease in pressure. This pressure drop creates a suction effect, which pulls air, dirt, dust, and debris into the vacuum cleaner. The sucked-in debris is collected in a dustbin or bag, while the clean air is expelled out of the vacuum cleaner.

6.2.1.1.2 Ventilation and air quality systems

Ventilation and air quality systems in buildings mimic the respiratory system. Heating, ventilation, and air conditioning (HVAC) systems regulate air quality, temperature, and humidity, similar to the respiratory system's function in exchanging oxygen and carbon dioxide. Air purification systems such as high-efficiency particulate air (HEPA) purifiers, activated carbon filters, UV purifiers, ionic purifiers, and ozone generators remove pollutants and contaminants from the air, mirroring the respiratory system's filtering function. Each comes with distinct technology and benefits for improving indoor air quality.

6.2.1.2 Central nervous system

The CNS controls and coordinates body functions, including the brain, spinal cord, and nerves. AI, smart systems/structures, IoT, and sensors mimic the CNS.

You may perhaps have heard the notion that the body systems are part of the brain. There is a growing body of evidence and theoretical frameworks that support this idea. This notion is rooted in the concept of embodied cognition and the extended mind theory.

Although many of the systems and technologies that mimic the CNS do not fall directly in the civil engineering docket, they have found many applications in the civil engineering industry, especially in the modern construction and design techniques/methods.

The world is now tending toward comparing everything with the natural world, particularly living beings, and because the mind is a universe of its own, many smart developments in engineering systems have been recorded—the growing biomimicry and systems thinking justifies this. Scientists and engineers are increasingly drawing inspiration from nature and all living beings to develop sustainable innovative solutions, products, and technologies. The growing interdisciplinary thinking and a profound philosophical perspective on the nature of the human mind are of concern lately.

6.2.1.2.1 Computers and artificial intelligence

Computers are designed to mimic the CNS. For example, artificial neural networks (ANNs) are inspired by the structure and function of the brain's neural networks. ANNs are composed of interconnected nodes (neurons) that process and transmit information. Neurons are the fundamental units of the nervous system. Neurons in the CNS act as information messengers, transmitting signals throughout the body via electrical and chemical signals, enabling communication between the brain, spinal cord, and the rest of the body.

ANNs are a subset of machine learning algorithms inspired by the biological neural networks of the human brain. Composed of interconnected nodes, or 'neurons,' ANNs process information through layers of computation, enabling them to recognize patterns, make predictions, and learn from data.

The brain uses feedback loops to refine its processing and adapt to changing conditions. AI systems, such as recurrent neural networks (RNNs), also employ feedback mechanisms to update their internal state and generate outputs. The brain processes information in parallel through its vast network of neurons. Similarly, computers and AI systems can process multiple tasks concurrently using parallel processing architectures.

The brain's neural networks are organized in a hierarchical manner, with early sensory areas processing basic features and higher areas integrating more complex information. Brian-inspired AI systems, such as convolutional neural networks (CNNs), also employ hierarchical representations to process data. A fundamental goal of AI is to develop machines that can learn and think like humans—basically mimicking everything human! Modern AI models, that is, brain-inspired artificial intelligence (BIAI) models, derive inspiration from the brain neural networks [3].

Figure 6.4 shows the neuron anatomy, which is mimicked by ANNs. In Figure 6.5, ANNs have nodes (often referred to as artificial neurons).

Summarized view of how ANNs work

1. Compare each individual node as its own linear regression model, composed of input data, weights, a bias (or threshold), and an output. The formula is as follows:

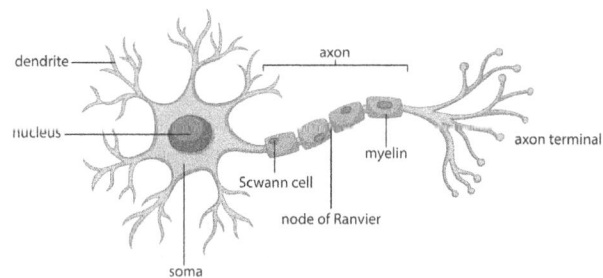

FIGURE 6.4 Neuron anatomy.

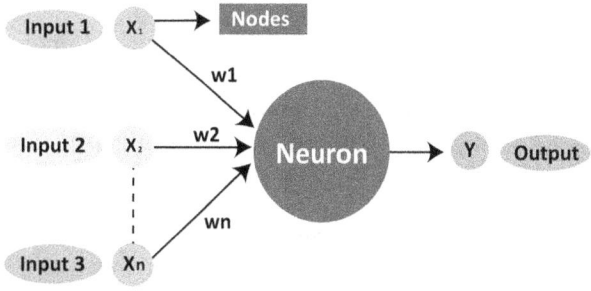

FIGURE 6.5 Artificial neural networks.

- $\sum wixi + \text{bias} = w1x1 + w2x2 + w3x3 + \text{bias}$
- $\text{output} = f(x)$

$$= 1 \text{ if } \sum w1x1 + b >= 0; \ 0 \text{ if } \sum w1x1 + b < 0$$

2. Inputs are weighted: each input is assigned a weight that determines its importance. Once an input layer is determined, weights are assigned. These weights help determine the importance of any given variable, with larger ones contributing more significantly to the output compared to other inputs.

3. Weighted inputs are summed. Inputs are multiplied by weights and combined. All inputs are then multiplied by their respective weights and then summed.

4. Activation function is applied next. Afterward, the output is passed through an activation function, which determines the output. The result is passed through a function that determines the node's output. If that output exceeds a given threshold, it 'fires' (or activates) the node, passing data to the next layer in the network. Node activation: If the output exceeds a threshold, the node 'fires' and sends data to the next layer. This results in the output of one node becoming the input of the next node.

5. Data flows through layers, with each node's output becoming the next node's input. This process of passing data from one layer to the next layer defines this neural network as a feedforward network.

How ANNs mimic biological neurons

- ANN inputs mimic biological dendrites. Just like biological neurons receive signals from multiple dendrites, ANNs receive inputs from multiple sources.
- ANNs weighted summation (weights) mimics biological synapses. Biological neurons weigh the strength of incoming signals, and ANNs use weights to determine the importance of inputs. Synapses are small gaps between two neurons where chemical signals are transmitted.
- ANN activation function mimics the biological firing process. Biological neurons fire when the cumulative signal exceeds a threshold, and ANNs use activation functions to determine the output.
- ANN signal propagation mimics the biological axon. Biological neurons transmit signals to other neurons, and ANNs pass outputs to subsequent layers. The axon is a long, thin extension of a neuron that plays a crucial role in transmitting signals.

6.2.1.2.2 The five senses and sensors

Although the CNS is essential for processing and interpreting sensory information, the five senses are distributed across multiple systems in the body. However, the CNS plays a crucial role in transmitting and processing sensory information from the senses to the brain. Engineers mimic the five senses, that is, sight (vision), hearing (audition), touch, taste (gustation), and smell (olfaction), to design and develop systems that save humanity and make the world a better place to live.

Sight (vision)

Sight (vision) is associated with the visual system, which includes the eyes, optic nerves, and visual pathways in the brain and whole nervous system. Civil engineers mimic the sense of sight in several ways, especially in buildings, for example, installation of cameras in buildings, where cameras capture visual data, just like the eyes. They can detect light, color, and movement.

Automatic sliding doors installed in buildings mimic the sense of vision (sight) and motion detection. They sense the presence or movement of something using sensors like infrared or motion detectors to automatically open and close, eliminating the need for manual intervention. These doors use sensors to detect movement and presence, typically using infrared (IR) sensors. They significantly detect heat and motion, allowing the doors to open and close in response to approaching or departing individuals.

Motion-sensing lights or occupancy-sensing lights or automatic lights are equipped with sensors, usually IR or passive infrared (PIR) sensors, which detect movement or occupancy within a specific range. When the sensor detects motion, it sends a signal to the lighting system, which then turns on the lights. These are perfectly aligned with sustainable development objectives—no energy wastage!

Hearing (audition)

Civil engineers, as the lead project developers of infrastructure systems, usually work with mechanical, electrical, and building service engineers, among others. They, therefore, install several building services and amenities. These services and amenities greatly mimic anatomical systems.

Hearing (audition) is associated with the auditory system, which includes the ears, auditory nerves, and auditory pathways in the brain and whole nervous system. For example, microphones installed in buildings and other infrastructures systems, in this case, mimic the hearing sense—they detect **sound waves**, just like ears. They can capture audio signals, allowing machines to recognize voice commands, detect anomalies, or monitor environmental noise.

Acoustic sensors and speech recognition services are all needed in buildings. And these are services that mimic the sense of hearing. Acoustic sensors measure sound pressure, frequency, and other acoustic properties, enabling machines to analyze and respond to sound patterns, while speech recognition devices enable machines to interpret and understand spoken language, allowing for voice-controlled interfaces and applications. All these applications are applied in the building system, mimicking the human sense of hearing.

Touch

The sense of touch is mimicked in a way that tactile sensors, force sensors, and haptic feedback loop mechanisms are installed in buildings. Touch is associated with the skin and the somatosensory pathway in the nervous system, which includes sensory receptors in the skin, muscles, and joints, as well as somatosensory pathways in the brain. Tactile sensors detect physical contact, pressure, and vibrations, mimicking the way the skin responds to touch. They are used in applications like robotics, prosthetics, and interactive interfaces—systems that are integrated in civil engineering infrastructure design and development.

Force sensors measure the magnitude and direction of forces applied to a surface, allowing machines to respond to physical interactions. Force sensors, or sometimes called load cells, are very helpful in weighbridges! Load cells, a type of transducer, convert the weight of trucks into an electrical signal.

Haptic feedback provides tactile feedback to users, simulating the sense of touch and texture in virtual environments.

Taste (gustation)

Taste (gustation) is associated with the tongue and the gustatory pathway in the nervous system. Taste (gustation) is associated with the gastrointestinal system, specifically the tongue and taste buds. Chemical sensors used to detect chemical compositions and concentrations mimic the way our taste buds respond to different flavors and substances. They are used in applications like food quality control, environmental monitoring, and medical diagnostics. Similarly, gas sensors measure the presence and concentration of specific gases, allowing modern machines to detect and analyze chemical signatures. These systems can be installed in specialized buildings.

Smell (olfaction)

The sense of smell is greatly mimicked in safety systems! Smell (olfaction) is associated with the nose and the olfactory pathway in the nervous system. Smell (olfaction) is associated with the olfactory system, which includes the nose, olfactory receptors, and olfactory pathways in the brain.

The gas sensors installed in buildings used to detect and analyze chemical signatures in the air mimic the way the olfactory system responds to different odors. The electronic nose uses sensor arrays to detect and identify specific odor patterns, allowing machines to recognize and classify different scents.

These sensors and technologies enable machines to perceive and interact with their environment in ways that mimic human senses, allowing for more sophisticated and responsive applications. These systems are applicable in modern and smart buildings.

6.2.1.2.3 Smart systems

In civil engineering, 'smart systems' encompass the integration of advanced technologies like sensors, data analytics, and AI to enhance infrastructure performance, safety, and sustainability. This includes areas like smart structures, smart buildings, and intelligent transportation systems. Section 6.2.1.2.2 outlined how several human senses are mimicked in the development of several smart technologies. These technologies are incorporated in the development smart infrastructure systems. You cannot design or develop smart infrastructural systems and forget about the **five** senses!

Smart systems in civil engineering integrate advanced technologies to create intelligent infrastructure systems that can monitor and respond to various conditions in real time. These systems utilize sensors, data analytics, and machine learning to enhance the safety, durability, and efficiency of civil infrastructure.

Some of the key components of smart systems include sensors, data acquisition, data analytics, machine learning, and automation. Sensors and data acquisition collect data. Automation and control systems respond to data insights by adjusting structural parameters, such as temperature, lighting, or traffic flow. Data analytics and machine learning interpret data to identify patterns, detect anomalies, and predict potential issues.

Applications of smart systems in civil engineering

With the advent of smart systems, many benefits have been realized. We've seen an improvement in safety and a reduction in the number of accidents, for example, with use of traffic management systems. Leveraging smart systems is fast becoming a technique engineers use to envision sustainable projects. In that regard, civil engineers can create more resilient, sustainable, and responsive infrastructure that adapts to changing conditions and user needs. Smart systems lead to enhanced durability and extended lifespan of infrastructure, increased efficiency and reduced maintenance costs, and better decision-making through data-driven insights. Some examples of smart systems in civil engineering include the following:

- Structural health monitoring (SHM) systems detect damage, deterioration, or other changes in infrastructure condition.
- Smart buildings and bridges integrate sensors and automation to optimize energy efficiency, comfort, and safety.
- Intelligent transportation systems (ITS) enhance traffic management, routing, and safety using real-time data analytics.

6.2.1.2.4 Internet of Things

Internet of Things (IoT) refers to the network of physical devices, vehicles, appliances, and other items embedded with sensors, software, and connectivity, allowing them to collect and exchange data. A network of physical devices, vehicles, appliances, and other physical objects that are embedded with sensors, software, and network connectivity allows them to collect and share data.

The IoT is another technology that uses sensors and data analytics mentioned earlier to monitor construction progress, optimize resource usage, and improve safety. The IoT is a crucial component of smart system infrastructure in civil engineering. In the context of smart systems, IoT enables the integration of various devices and sensors to create a seamless and interconnected infrastructure.

The benefits of IoT in smart systems include enhanced data collection and analytics capabilities, improved real-time monitoring and decision-making, increased automation and control, and better integration with other smart systems and technologies. With IoT incorporated into smart system infrastructure, civil engineers can create more connected, efficient, and responsive infrastructure that improves the quality of life for people and enhances the sustainability of urban environments. IoT applications in smart systems are as follows:

- **Remote monitoring and management (RMM) or remote surveillance:** Using IoT-enabled cameras and sensors, business owners can monitor their premises remotely through a mobile app. This allows them to view live footage: check in on their business in real time, thus ensuring that everything is running smoothly. Receive alerts: get notifications when something unusual occurs, such as motion detection or temperature fluctuations. Control devices: remotely control lighting, temperature, and security systems to optimize operations and reduce energy consumption. Monitor analytics: track **key performance indicators** (KPIs) such as customer traffic, sales, and inventory levels. Some popular examples of RMM solutions include the following: **Hikvision**—a surveillance solution that allows remote monitoring and management of cameras and security systems. **Nest**—a smart home and business solution that enables remote monitoring and control of security cameras, thermostats, and other devices. **August**—a smart lock solution that allows business owners to remotely monitor and control access to their premises. These RMM solutions provide business owners with greater flexibility, improved operational efficiency, and enhanced security, allowing them to manage their business remotely and make quick data-driven decisions.
- **IoT in security systems:** IoT security system components include the following: sensors—motion detectors, door and window sensors, glass break sensors, and environmental sensors; cameras—IP cameras, CCTV cameras, and doorbell cameras with motion detection and night vision; alarm systems—IoT-enabled alarm panels that can send alerts to authorities or security personnel; and access control—IoT-enabled door locks, gate controllers, and access control systems.
- **Environmental monitoring:** IoT sensors monitor temperature, humidity, air quality, and other environmental factors to optimize building performance and occupant comfort.
- **Structural health monitoring:** IoT sensors detect changes in structural integrity, such as cracks, vibrations, or stress, to predict potential failures and schedule maintenance.
- **Smart lighting and energy management:** IoT-enabled lighting systems adjust brightness and color based on the occupancy, time of day, and energy usage to optimize energy efficiency.
- **Traffic management and transportation systems:** IoT sensors and cameras monitor traffic flow, optimize traffic signal timing, and provide real-time traffic updates to reduce congestion and improve safety.

6.2.1.2.5 Communication and control systems

These following systems mimic the nervous system:

- Traffic management systems (TMS): a set of technologies and processes that are used to monitor, control, and optimize traffic flow.
- Building automation systems (BAS): a centralized and integrated system of electronic devices designed to monitor and control a building's mechanical and electrical operations, including HVAC, lighting, and security.

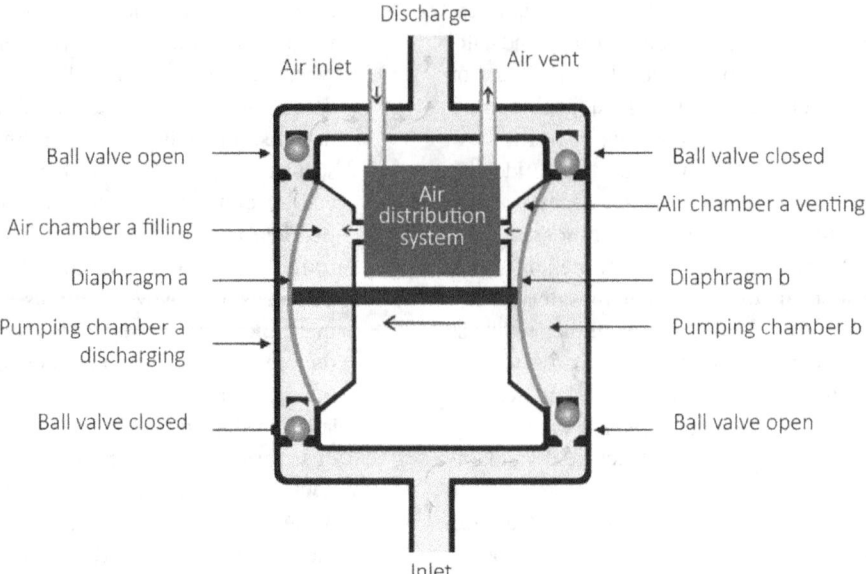

FIGURE 6.6 Diaphragm pump.

TMS use technology to improve traffic flow, safety, and efficiency by addressing congestion, accidents, and other issues through techniques like adaptive signal control, incident detection, and real-time information dissemination.

Similar to the nervous system, these TMS monitor and control traffic flow, adjusting signals and traffic patterns to optimize traffic movement. TMS are systems that use technology to manage traffic flow and reduce congestion. Key components of TMS include traffic detectors, communication systems, software, ITS, and traffic management centers.

Traffic detectors use sensors that collect data on traffic flow, speed, and volume. Communication systems allow for real-time data exchange between traffic management centers and the road network. Then, software processes data and makes decisions about the traffic management strategies.

ITS integrate various technologies to manage traffic flow and improve safety. Traffic management centers (TMCs) are the centralized locations that monitor and control traffic flow.

Examples of TMS technologies include adaptive traffic signal control (ATSC) which adjusts traffic signals in real time based on current traffic conditions and the incident detection TMS which uses cameras and sensors to quickly identify traffic incidents.

BAS integrate various building components, such as lighting control, HVAC, security systems, fire alarm systems, and access control systems, to optimize efficiency and comfort and to create a centralized control system similar to the CNS. BAS automate and control various building functions.

6.2.1.3 Circulatory system (cardiovascular)

This system transports blood, oxygen, and nutrients through the heart, arteries, veins, and blood vessels.

6.2.1.3.1 The heart and the pump

Section 6.2.1.1.1 shows how the vacuum pump analogizes the human suction mechanism (respiratory system). The purpose of all pumps is to convert energy into pressure. Section 3.3.3 explains how/why energy is conserved. We saw that the vacuum pump works by creating a pressure difference, igniting a

suction force, and the fluid is sucked. And it is shown that this suction force cannot raise the fluid to a height more than 10 m, specifically for water (non-compressible fluid).

Engineers have since developed several pumps of different models and working processes to pump fluids to several meter heads. In water supply systems, with different types of pumps available, water can be pumped/raised to heads of more than (>500 m). These include the following: centrifugal pumps, positive displacement pumps, axial flow pumps, industrial pumps, and specialized pumps such as diaphragm pumps and vacuum pumps.

Diaphragm pump (Figure 6.6) mimics the heart's function (Figure 6.7) by using a flexible diaphragm [4] and valves to create a reciprocating motion, drawing in and ejecting the fluid, similar to the heart's rhythmic pumping action. Diaphragm pumps are greatly used to pump liquids and gases and even in medical devices that mimic the heart's function, for example, in cardiac surgery training simulations. For example, Quattroflow® develops and manufactures four-piston laboratory pumps specifically designed for critical applications in the biopharmaceutical industry. Quattroflow pumps are known for their low-friction, low-shear, and low-pulsation operation, making them ideal for applications involving delicate fluids, such as in biopharmaceutical manufacturing [5].

- **Reciprocating action:** The pump moves the fluid in cycles mirroring the heart which contracts (systole) and relaxes (diastole) in a rhythmic manner.
- **Chamber mechanism:** A double-diaphragm pump moves the fluid using alternating chambers similar to how the heart has two main pumping chambers (the left and right ventricles).
- **Diaphragm movement:** In a diaphragm pump, a flexible diaphragm moves back and forth, creating a vacuum and pressure to draw in and eject fluid. This mimics the heart.
- **Valves:** Valves in the pump, like those in the heart, ensure that fluid flows in one direction, preventing backflow. The built-in check valves, integrated directly into equipment or piping systems, control

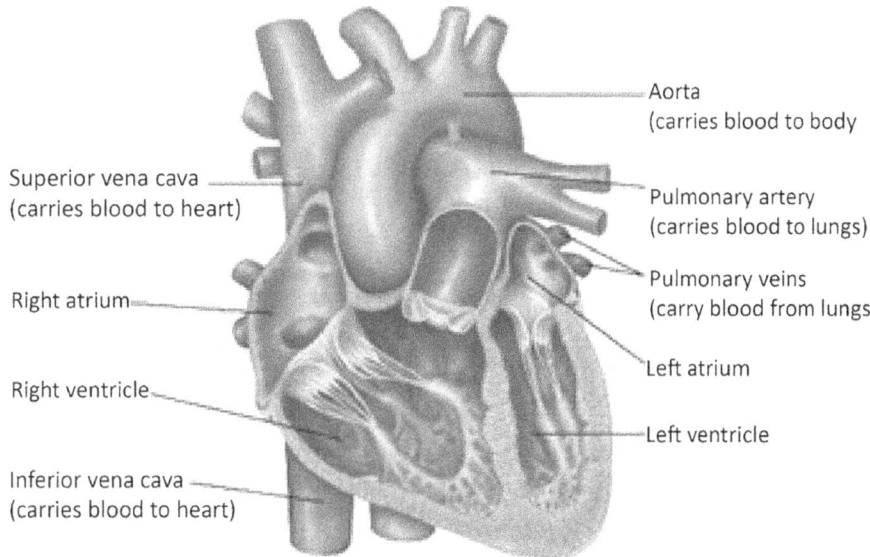

FIGURE 6.7 Human heart.

flow direction, similar to the heart's valves. These built-in check valves are designed to allow the fluid to move in one direction while preventing backflow, which is similar to how the heart's valves ensure blood flows in the correct direction through its chambers.

6.2.1.3.2 Water supply and drainage systems

These systems mimic the system that transports blood, oxygen, and nutrients through the heart, arteries, veins, and blood vessels. Water supply and drainage systems have valves that mirror those of the heart. The flow of fluids through pipelines is similar to the circulatory system, with its network of blood vessels and capillaries. Water distribution networks supply clean water to communities, much like the circulatory system supplying oxygenated blood to the body. Sewer systems collect and transport wastewater away from communities, mirroring the circulatory system's removal of waste products.

6.2.1.4 Digestive system

This breaks down and absorbs nutrients from food through the mouth, esophagus, stomach, small intestine, and large intestines. The digestive system, also known as the gastrointestinal (GI) tract, is a complex process by which the body breaks down food into nutrients, absorbs nutrients into the bloodstream, and eliminates waste products. Also, water, drinks, and sauces act as lubricants, easing food movement through the throat as oil does in machines.

6.2.1.4.1 3D manufacturing and digestive system

3D printing or 3D manufacturing, also known as additive manufacturing (AM), is the construction of a three-dimensional object from a CAD model or a digital 3D model. It can be done in a variety of processes in which the material is deposited, joined, or solidified under computer control, with the material being added together (such as plastics, liquids, or powder grains being fused), typically layer by layer.

Physical products are created from digital designs by layering materials such as metals, plastics, ceramics, and glass. This technology enables the production of complex shapes, structures, and geometries that cannot be produced through traditional manufacturing methods.

The 3D printing process can also be compared to the human digestive system's process that leads to passage of stool. This is because 3D printing conducts layering just as the same way stool layers itself. In 3D printing, layers of the material are deposited. Similarly, the digestive system layers nutrients and waste. 3D printing builds up layers to form an object. The digestive system accumulates waste to form stool. 3D printing extrudes the material through a nozzle. The digestive system extrudes waste through the anus. 3D printing creates a desired shape. The digestive system shapes stool through muscular contractions. The digestive system breaks down food into nutrients, absorbs nutrients into the bloodstream, and eliminates waste products. In the same way, 3D manufacturing uses necessary input materials to produce output.

3D printed houses is the likely future of construction because of the advantages it brings that include faster construction, lower costs, less waste, and higher precision.

6.2.1.4.2 Waste management and recycling systems

The waste management and recycling systems mirror the digestive system's functions. The waste-to-energy plants/ facilities process waste into energy, similar to the digestive system's breakdown of food into nutrients. Similarly, the recycling facilities sort and process recyclable materials, mirroring the digestive system's absorption of nutrients.

6.2.1.5 Endocrine system

The endocrine system produces and regulates hormones through glands like the pancreas, thyroid, adrenal, pituitary, and hypothalamus. The endocrine system is a network of glands and organs that produce and regulate hormone levels in the body. Hormones are chemical messengers that help control various bodily functions. Imbalances or disorders in the endocrine system can lead to various health issues, such as diabetes, thyroid disorders, or growth hormone deficiencies.

6.2.1.5.1 Environmental monitoring and control systems

These include water and air quality monitoring systems which mimic the endocrine system. The air quality monitoring/control systems track pollutant levels, while the water quality monitoring/control systems monitor water quality parameters. These monitoring/control systems are designed to trigger alerts or controls to maintain air and/or water quality, similar to the endocrine system's regulation of hormones.

6.2.1.6 Integumentary system

The integumentary system protects the body from external damage through skin, hair, nails, sweat glands, and oil glands.

6.2.1.6.1 Skin and waterproofing materials

The development of waterproof materials and membranes in civil engineering is inspired by the human skin's ability to protect the body from water and other external elements. Waterproofing materials such as waterproof cements are used to prevent water from penetrating surfaces and structures, thus protecting them from damage and deterioration.

6.2.1.6.2 Building envelope and weatherproofing systems

The building facades and roofing systems mimic the integumentary system. The former systems protect buildings from external environmental factors such as weather, temperature, and humidity, similar to the integumentary system's protection of the body, while the latter shield buildings from weathering, insulation, and waterproofing, mirroring the integumentary system's protective functions.

6.2.1.7 Muscular system

This system allows movement and maintains posture through skeletal muscles, smooth muscles, and cardiac muscle.

6.2.1.7.1 Actuators and muscles

The development of actuators and motors in civil engineering is inspired by the human body's muscles and their ability to generate force and movement. An actuator is a component of a machine that produces force, torque, or displacement, usually in a controlled manner, when an electrical, pneumatic, or hydraulic input is supplied to it in a system.

6.2.1.7.2 Mechanical and robotic systems

The development of cranes and hoists mimics the muscular system. These systems provide mechanical advantage and lift heavy loads, similar to the muscular system's ability to move and support the body. Similarly, in the development of robotic systems, the muscular system is mimicked to perform tasks that require strength, precision, and endurance, mirroring the muscular system's functions.

6.2.1.8 Skeletal system

This system provides structural support and protection to a human through bones, joints, and ligaments. The field of structural engineering particularly mimics the skeletal and muscular systems; see Sections 6.2.1.7 and 6.2.1.8—that is, the musculoskeletal system. See the skeletal system in Figure 6.8.

6.2.1.8.1 Bridges and spines

The structural integrity of bridges can be compared to that of the human spine (Figure 6.9), with its interconnected vertebrae and disks distributing loads and providing flexibility. The bridges mimic the musculoskeletal system of the body.

6.2.1.8.2 Domes and skulls

The shape and structure of domes in the architecture are similar to those of the human skull, providing protection and support. The dome is inspired by the skull.

6.2.1.8.3 Arches and ribcages

The arches in buildings and bridges resemble the ribcage with its curved structure distributing loads and providing support.

6.2.1.8.4 Joints and connections

The design of joints and connections in civil engineering, such as hinges and bearings, is inspired by the human body's joints and articulations. Refer to support and connection types in Chapter 8.

Ball-and-socket joints (Figure 6.10)—for example, hip joint—are a type of synovial joints that allow for a wide range of movements in multiple directions. They are characterized by a rounded bone, called the ball, that moves within a depression in another bone, called the socket. Ball-and-socket joints are synovial joints where the spheroid articular surface of one bone sits within a cup-like depression of another bone.

The ball-and-socket configuration allows for movement with 3 degrees of freedom, which is more than any other type of synovial joint. The depth of the cup and any additional fibrocartilaginous labrum is the major limitation to the extent of motion allowed in any direction.

Engineers mimic the ball-and-socket joint to design ball-and-socket joints in machines—see Figure 6.11. Ball-and-socket joints are mechanical components that allow for pivoting and rotating motion in machines. They are often used in construction equipment and automobiles [6].

Ball-and-socket joints, sometimes known as axial joints, are mechanical bearing components that fit together much like a knee joint and allows pivoting and rotating motion in mechanical applications. They are used in several machines including construction equipment and automotives. Axial bearings are made from either steel or plastic or sometimes both.

> **Hinge joint:** A hinge joint is a type of joint in the body that allows for movement in one plane, primarily bending and straightening. The elbow, knee, ankle,

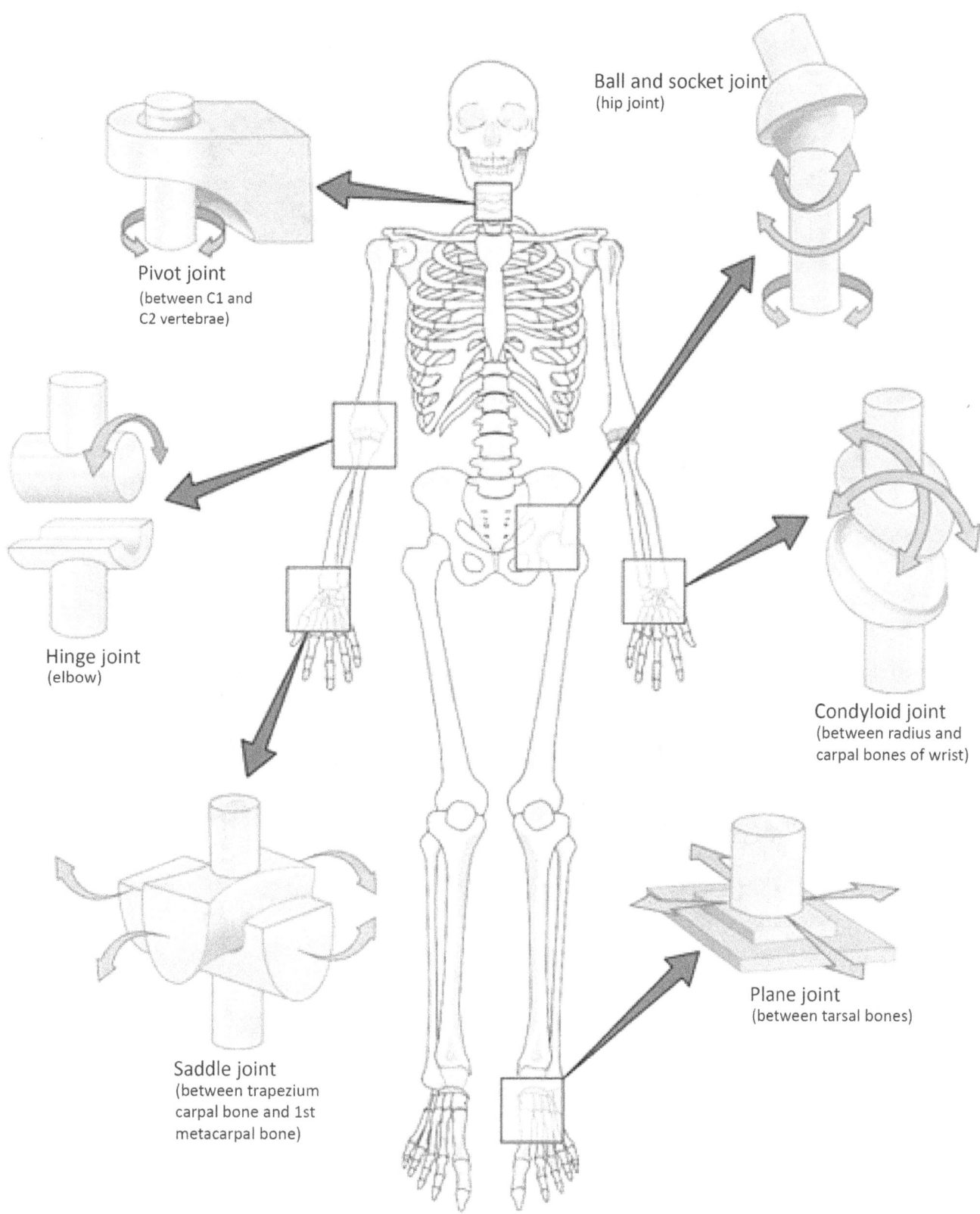

FIGURE 6.8 The skeletal system.

and interphalangeal (IP) joints of the hand and foot are all hinge joints. Engineers mimic hinge joints (Figure 6.12) to design and develop door hinges (Figure 6.13).

Pivot joint: A pivot joint (Figure 6.14) allows bones to rotate around a single axis. In a pivot joint, a cylinder-shaped bone rotates within a ring formed by another bone or ligament. Engineers mimic pivot joints to create mechanisms that allow for rotational movement around a single axis. This is mainly useful in robotics and other engineering applications where controlled rotation is needed, such as in robotic arms or steering mechanisms that can be used in construction.

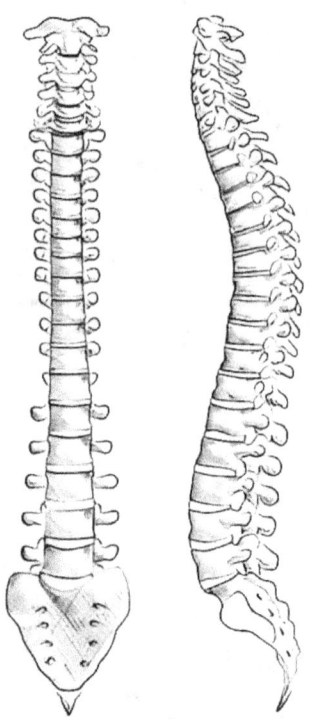

FIGURE 6.9   Human spine.

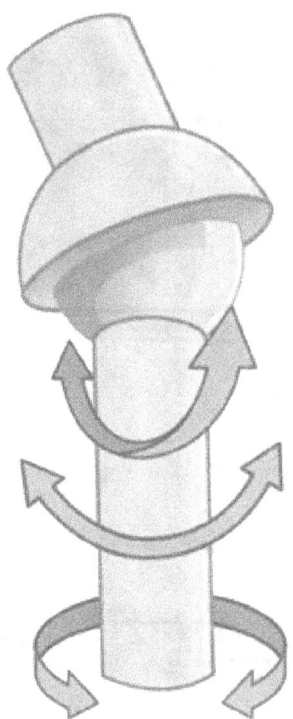

FIGURE 6.10   Ball and socket (e.g., hip joint).

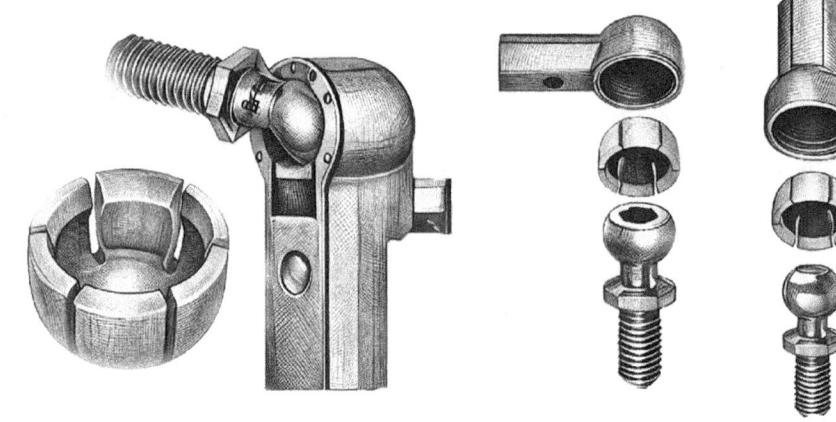

FIGURE 6.11   Ball-and-socket joint in machines.

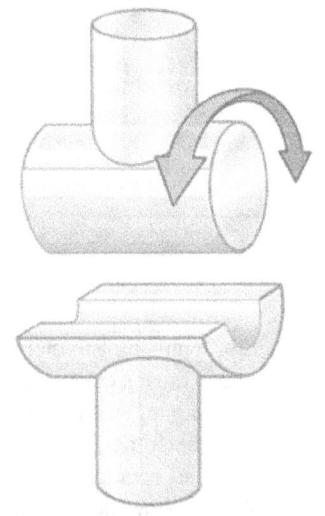

FIGURE 6.12   Hinge joint (e.g., elbow).

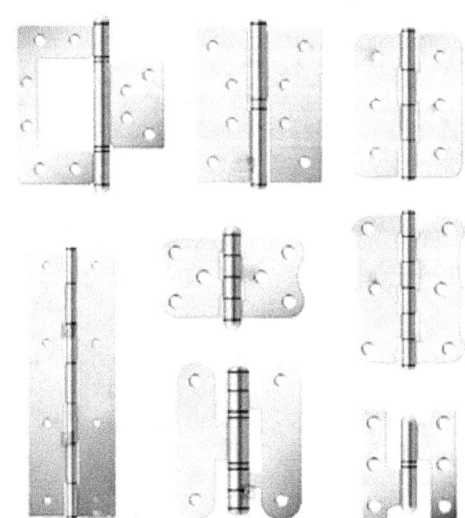

FIGURE 6.13   Door hinges.

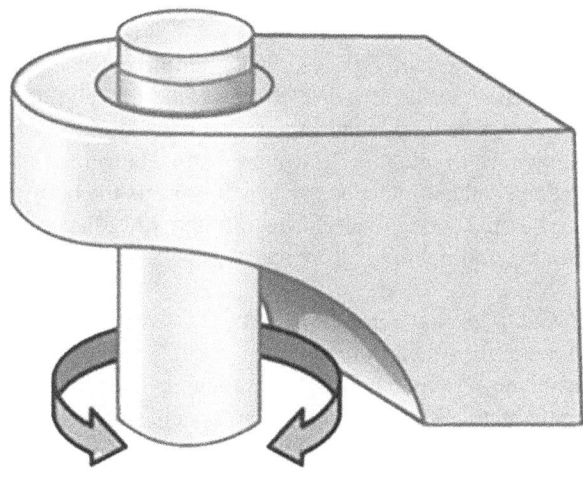

FIGURE 6.14 Pivot joint (e.g., median atlantoaxial joint).

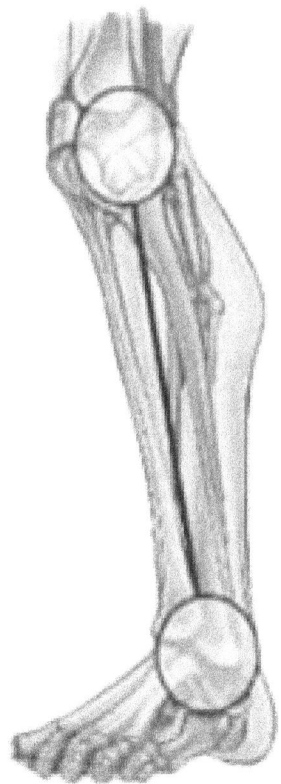

FIGURE 6.15 Knee joint.

Knee joint: The pin connection in structures mimics the knee joint (Figure 6.15). It rotates in a single direction!

6.2.1.8.5 Bones and reinforcement
The use of reinforcement materials like rebar in concrete is similar to the human body's use of bones to provide structural support. See also the cactus tree in Section 6.2.2.3.

6.2.1.8.6 Structural systems
Civil engineering mirroring the human anatomy implies structural engineering because frames really mimic the skeletal system. For example,

- The building frames provide structural support and stability to buildings, similar to the skeletal system's provision of a framework for the body.
- Bridge structures distribute loads and provide structural integrity, mirroring the skeletal system's functions.

6.2.1.9 Urinary system

The urinary system, also known as the renal system, consists of organs that work together to remove waste and excess fluids from the body. The main components of the urinary system are as follows:

- **Kidneys:** These are two bean-shaped organs that filter waste and excess fluids from the blood.
- **Ureters:** These are the two tubes that carry urine from the kidneys to the bladder.
- **Urinary bladder:** A hollow organ that stores urine.
- **Urethra:** A tube that carries urine from the bladder out of the body.

The urinary system removes waste and excess fluids through kidneys, ureters, bladder, and urethra.

6.2.1.9.1 Plumbing system and waste and water treatment units
You may have seen prostate cancer awareness campaigns reading 'in your plumbing' metaphor to highlight the location of the prostate gland and its connection to urinary and reproductive functions. The phrase 'in your plumbing' is a common, easily understood metaphor used in prostate cancer awareness campaigns to help men visualize where the prostate gland is located and how it relates to their urinary and reproductive systems.

This phrase helps men understand that prostate cancer is a condition that can affect their urinary and reproductive functions, and it encourages them to be aware of potential symptoms and seek medical attention, if needed. Therefore, from social psychology, we know that there is no joke without serious meaning—the prostate cancer awareness campaigners already saw a connection between the urinary system and plumbing system! For this discussion, we capitalize on the plumbing system. The plumbing system and waste water treatment units mimic the urinary system. **Note:** the water treatment plants (filtering impurities) mimic the kidney functions!

The kidneys play a crucial role in regulating the volume of bodily fluids, maintaining fluid balance (fluid osmolality), acid–base equilibrium, and controlling electrolyte levels, while also eliminating waste products and removing toxins. The glomerulus is where filtration takes place, with approximately 20% of the blood flowing through the kidneys being filtered. Substances that are typically reabsorbed include water, sodium, bicarbonate, glucose, and amino acids, whereas substances like hydrogen, ammonium, potassium, and uric acid are secreted. See more examples:

- Drain pipes (transports waste) mimic ureters (transports waste).

- Storage tank (holds waste) mimics the bladder (holds urine). Septic tanks treat and dispose of wastewater from buildings, mirroring the urinary system's functions.
- Drain valves (controlling waste flow) mimic the urethra (controls urine flow).

Water purification plants are essential facilities that ensure clean and safe drinking water for communities. These plants use various methods and technologies to remove impurities and contaminants from water sources, such as rivers, lakes, or groundwater. Through the water purification process, undesirable chemicals, biological contaminants, suspended solids, and gases are removed from water. The goal is to produce water that is fit for specific purposes.

Most water is purified and disinfected for human consumption (drinking water), but water purification may also be carried out for a variety of other purposes, including medical, pharmacological, chemical, and industrial applications. The history of water purification includes a wide variety of methods which include physical processes such as filtration, sedimentation, and distillation; biological processes such as slow sand filters or biologically active carbon; chemical processes such as flocculation and chlorination; and the use of electromagnetic radiation such as ultraviolet light. These mirror the urinary system's filtration and removal of waste products.

6.2.1.10 Reproductive system

This is a collection of organs involved in sexual reproduction, encompassing both male and female systems, which work together to produce gametes (sperm and eggs), facilitate fertilization, and nurture a developing fetus. This produces offspring through male (testes, epididymis, vas deferens, and prostate) and female (ovaries, fallopian tubes, uterus, and vagina) organs.

6.2.1.10.1 Manufacturing and fabrication systems
3D printing and additive manufacturing are similar to the reproductive system's creation of new babies/life. These systems create complex shapes and structures. Assembly lines and manufacturing systems produce and assemble products, mirroring the reproductive system's assembly of genetic material.

6.2.1.11 Lymphatic system

This system includes a network of vessels, nodes, and ducts that collect and circulate **excess fluid (lymph)** in the body, and also contains immune cells like lymphocytes. It defends the body against infection and disease through lymph nodes, lymph vessels, spleen, and lymphoid organs.

6.2.1.12 Drainage and water management systems

Stormwater drainage systems mimic the lymphatic system because they collect and transport rainwater and surface runoff, similar to the lymphatic system's removal of waste and toxins. Flood control/defense systems regulate water levels and prevent flooding, mirroring the lymphatic system's functions of collecting and circulating excess fluid (lymph) in the body.

6.2.1.13 Immune system

Your immune system is your body's first-line defense against invaders like germs. It helps protect you from getting sick and promotes healing when you're unwell or injured. You can strengthen your immune system by eating nutritious foods, exercising, and getting enough sleep. It protects the body from pathogens and foreign substances through white blood cells, antibodies, and immune responses.

6.2.1.13.1 Security and surveillance systems
Access control systems are designed to mimic the functions of the immune system. These systems regulate entry and access to secure areas, similar to the immune system's defense against pathogens. The intrusion detection systems also monitor and detect potential security threats, mirroring the immune system's recognition and response to foreign substances. The intrusion detection systems are used to monitor network traffic and detect suspicious activity. Intrusion detection systems (IDS), such as Snort, Suricata, and Zeek, are used to detect and alert on malicious network or system activity, while host intrusion detection systems (HIDS), such as OSSEC, monitor individual hosts for unauthorized access. These systems are integrated in smart infrastructure systems to detect potential security threats, mirroring the immune system's recognition and response to foreign substances. These examples illustrate how various civil engineering areas and technologies mimic the functions and systems of the human body.

6.2.2 The Natural Environment and Civil Engineering

Let us start by defining the natural world/environment. Natural world or natural environment refers to all living and nonliving things that occur naturally on Earth, without human intervention, including plants, water bodies, animals, landforms, air, and climate, essentially encompassing the planet's ecosystem as it exists without significant human alteration; it signifies the 'wild' aspects of the planet, as opposed to manmade structures and environments.

Civil engineers observe and adapt natural processes to design more efficient, resilient, and sustainable infrastructure, such as:

- Imitating the branching patterns of trees to design more efficient transportation networks.
- Studying the flow of rivers to improve water supply and sewerage systems.
- Mimicking the structure of abalone shells to develop stronger construction materials.
- SuDs mimic the natural world—bioretention gardens, infiltration systems, swales, etc.

Civil engineers develop innovative solutions, improve existing designs, and create more sustainable and efficient infrastructure by drawing inspiration from these natural systems and analogizing them. In addition to the example given in Section 6.1.2 (tree and column), additional illustrations of how civil engineers mimic the natural environment are provided henceforth.

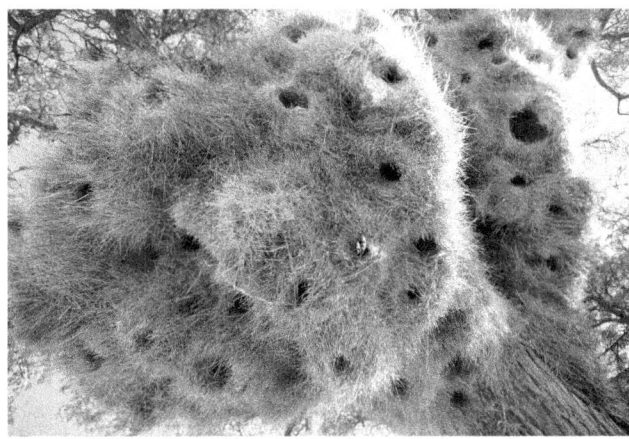

FIGURE 6.16 Sociable weaver nests. Image credit: Shutterstock.

6.2.2.1 Ecosystems and urban planning

The design of sustainable urban ecosystems is inspired by the natural world's ecosystems, with their interconnectedness and balance. Through mimicking the natural world, civil engineering can develop innovative solutions that are efficient, sustainable, and resilient. In urban planning settings, particularly, civil engineers look at natural habitats to design and develop housing units. To build resilient communities, these natural habitats are a source of inspiration.

Today, civil engineers are making use of nature to apply their ingenuity for the convenience of humans and other living beings, while minimizing the harm they would otherwise cause to the environment. That means they harness the properties of engineering science to find solutions that make the world better to live, applying and extending techniques that boost the ecosystem health instead of endangering it. A smart engineer may design to develop a habitat mimicking the sociable weaver's nest (Figure 6.16). Why not?

6.2.2.2 Infiltration systems and the natural percolation process

SuDs, such as the infiltration systems, greatly mimic the natural processes. They are designed to capture and filter stormwater runoff, allowing it to infiltrate into the soil [7]. These systems mimic the natural percolation process, reducing stormwater runoff and alleviating pressure on drainage infrastructure. In the natural percolation process, water moves downward through soil layers, recharging groundwater aquifers. It occurs when precipitation falls on the land, water infiltrates the soil surface and moves downward through soil pores and layers, and water recharges groundwater aquifers.

6.2.2.3 Steel reinforcement in concrete and cactus tree

Reinforced concrete and cactus trees share some interesting similarities:

a. **Strength in resilience:** Cacti are known for their ability to thrive in harsh environments, and reinforced concrete is designed to withstand heavy loads and stresses and also withstands harsh environments such as being buildable in water. Cacti can adapt to different environments, and reinforced concrete can be molded and shaped to suit various architectural needs.

b. **Internal support or reinforcement:** Cacti have a strong internal skeleton that provides structural support, just as reinforced concrete uses steel rebar or fibers to add tensile strength.

c. **Protection from external factors:** Cacti have thick skin to protect themselves from extreme temperatures and predators, while reinforced concrete is often used to create protective barriers against environmental elements.

d. **Durability:** Both cacti and reinforced concrete are known for their long lifespan and resistance to decay. This analogy can inspire innovative approaches to building design, materials science, and sustainability. When engineers study the unique properties of cacti, they may be able to develop more resilient and adaptable reinforced concrete structures that thrive in challenging environments.

6.2.2.4 Artificial drainage systems mirroring rivers

It is obvious that the flow of rivers inspired the engineers to develop efficient drainage systems. Rivers naturally occur differently—some flow on bed rocks, others occur with grassed banks, while others flow on bases with coarse or fine aggregates. The flow velocity varies with slopes and materials with which the river is made of! Therefore, civil engineers observed these factors and engineered artificial drainage systems that could effectively convey stormwater runoff!

6.3 THE GYM AND THE CIVIL ENGINEER

The gym shares similarities with the civil engineering industry. Civil engineers deal with loads, that is, dead and live loads. They also deal with equipment and machinery and the structures they develop. Gym-goers also deal with loads, equipment, and structures. **Dead loads** are permanent and static loads of a structure and its components, while **live loads** are the variable, transient loads imposed by occupancy, such as people, furniture, or vehicles.

Because civil engineers deal with loads just like gym-goers, it is quite exciting to see how civil engineering structures mimic gym-goers and their weightlifting activities, among other activities, often carried out in the gym. This would support knowledge growth and understanding of civil engineering principles by comparing the civil engineering principles and gym activities.

Section 6.2.1 illustrates how engineers mimic the human anatomy, physiology, and natural environment—that is, the plants, water resources, rivers, lakes, and fruits, to design and develop beautiful structures as habitats or esthetical features. In this section, a deeper understanding of principles is provided by comparing dead and live loads that civil engineering deals with, with the gym!

We delve deeper to illustrate anatomically how forces induce loads on structures, just as the same way forces induce loads on gym-goers. A civil engineering structure is designed to withstand a number of loads, and a safe limit is usually considered above which the structure may buckle, collapse in shear, bulge in case of a water tank, surge, sway, or become unstable. Therefore, structures are designed with safe limits above which the structure becomes unsafe—that is, the ultimate limit state (ULS) and the serviceability limit state (SLS).

In the same way, gym-goers carry loads they can handle and perform their activities within safe limits, whether dealing with loads, equipment, or structures. Usually, the gym integrates safety in all operations the same way civil engineers integrate safety rules in practice. The gym-goer, the instructor, and management all play safety roles just like the employer, designer, engineer, staff, workers, and management in any civil engineering function, for example, construction or design process.

Forces induce loads on a civil engineering structure just the same way forces induce loads on a gym-goer (**in conceptual form**). So, forces come first, and loads follow as a consequence, the things we study as civil engineers!

6.3.1 Why Do People Go to the Gym?

People workout for several reasons—and these reasons shed insights into why civil engineers need to maintain and operate civil engineering structures safely and effectively—and at the same time learn why they need to be healthy and fit for work. The motivations behind each individual's exercise habits offer valuable insights into the importance of effective maintenance and operation of civil engineering structures by professionals in the field. We learn a great deal of civil engineering from the gym and about safety, engineering principles, operation, and maintenance; how the body and mind work together; ergonomics; and civil engineering machines and structures that mimic the gym equipment. If you need to grasp civil engineering and develop passion, visit the gym today!

6.3.1.1 Fitness and life goals

The gym helps you see progress through consistency and patience. Fitness goals provide direction, whether it's transforming your body, esthetic improvements, or personal growth. Achieving these goals builds discipline, persistence, and motivation. You can achieve several fitness goals, ranging from complete transformation of your body, esthetic and cosmetic reasons, to personal growth and motivation.

The gym environment fosters collective energy, encouraging you to become your best self. Tracking progress, celebrating milestones, and pushing limits help you evolve and adapt. By setting targets and working toward achieving them, you'll gain a sense of accomplishment and pride. Having a target gives you something to strive for, whether your goal is to gain weight, lose weight, run a marathon, or bench press 120 pounds (54.43 kg). Whether gaining muscle or losing weight, the gym provides a space to build on your potential and strive for continuous improvement. Progress is measurable.

Everyone begins their fitness journey somewhere in the gym. As you work out, you can monitor your advancement. Celebrate each milestone, no matter how minor, and you'll develop a sense of pride, whether you're building strength or increasing mass through bulking. Stanford Medicine found that digitally tracking your health can help you lose weight [8]. Whether it's adding more weight, increasing reps, or trying more challenging exercises, you're constantly evolving and adapting.

6.3.1.2 Maintain mental health

People go to the gym to boost their mental health. The gym plays a significant role in improving and maintaining mental health. Regular gym sessions can create a sense of euphoria, helping combat daily stress and tension by releasing endorphins, also known as **'feel-good'** chemicals. Consistent workouts can induce feelings of elation, reducing anxiety and pressure. Zhang et al. [9] found evidence that exercise is a viable intervention for improving depression and anxiety in young individuals. The study's findings indicate that resistance exercise is particularly effective in treating and preventing depression among young individuals. To achieve optimal benefits, the following recommendations are suggested:

- **Duration:** commit to a training program lasting longer than 6 weeks to experience significant benefits.
- **Frequency:** engage in strength training exercises 3–4 times a week. This type of training involves using resistance to challenge muscles and enhancing strength, endurance, and muscle mass. Various forms of resistance, such as weights, resistance bands, or bodyweight exercises, can be utilized.
- **Session length:** aim for workout sessions that last between 30 and 60 minutes.

When young individuals incorporate resistance training into their routine, they can potentially alleviate symptoms of depression and improve their overall mental health. Post-gym clarity is everything! Coming out of the gym, you just feel good! Your mind feels free of the burden of stress and anxiety, and it helps you take on the day with clear thinking. Generally, the mental health benefits of gym workouts are numerous, including the following:

- Regular gym workouts can boost self-esteem and confidence.
- Gym workouts help release endorphins that act as natural mood lifters and pain relievers, leading to feelings of well-being and reduced stress. Mood enhancement occurs as a result of triggering the release of endorphins.
- Exercise can improve sleep quality by reducing stress, improving mood, and regulating body temperature.
- Exercise provides a mental break from daily worries, promoting relaxation and reducing cortisol levels (stress relief and relaxation). Engaging in gym workouts can be an effective way to relieve stress and promote relaxation by releasing endorphins,

improving mood, and providing a mental break from daily worries.

- The gym provides opportunities for social interaction and community support.

6.3.1.3 Boosting self-esteem

The gym workouts can boost one's self-esteem, and it contributes to a positive self-image and confidence in the following ways:

- **Improved mental health:** When mental health improves, self-esteem tends to follow, creating a more positive self-perception.
- **Accomplishment and progress:** Fitness milestones showcase personal progress and can create a belief in one's capabilities and boost self-esteem.
- **Physical transformation:** Seeing tangible progress in the mirror or with better fitting clothes can give you a boost.
- **Increased energy and health:** Feeling healthier and more energized allows individuals to tackle daily tasks with more enthusiasm and confidence.
- **Body positivity and acceptance:** The gym can help individuals appreciate their bodies for what they can do rather than focusing more on appearance.

6.3.1.4 Maintaining physical health

The gym helps gain and maintain physical health. You gain peak physical strength, conditioning, flexibility, and endurance. Thus, people go to the gym for maintaining the body—gym visits can help you shed some unwanted pounds or gain some healthy weight. Physical health benefits—such as enhanced endurance and stamina—propel people to visit the gym. Your body burns calories faster when you work out in the gym. See the treadmill illustrative example in Section 6.3.3.3.3. Going to the gym offers several physical health benefits that positively impact the aspects of your body and overall well-being. Some key advantages are as follows:

- **Training with weights builds lean muscle:** Muscle tissue is metabolically active, meaning it is burning calories even when you're not exercising. But the gym is more than about building muscle and getting a pump. It is about increased strength and muscle mass.
- **Reduced risk of chronic diseases:** Regular exercise is associated with a decreased risk of various chronic conditions like type **2** diabetes, certain cancers, and stroke.
- **Flexibility and mobility:** Stretching routines relieve muscle tension and increase blood flow. It also reduces the risk of injuries with a full range of motion; in addition, stretching feels so good! It provides improved flexibility and mobility.
- **Bone density:** Weight-bearing exercises, like lifting weights or doing bodyweight exercises, can strengthen bones, reducing the risk of osteoporosis and fractures.
- **Weight management (loss or gain):** The gym helps reduce or gain weight—working out burns fats, which contribute to the weight. On the other hand, gaining weight requires a combination of proper nutrition and a well-structured workout routine—consuming more calories than you burn and focusing on compound exercises, that is, exercises that work multiple muscle groups at once, such as squats, deadlifts, and bench presses.
- **Cardio improves heart health:** The gym offers ways to do cardio like ellipticals, stationary bikes, and treadmills. You can also do cardio-intensive workouts such as **high-intensity interval training (HIIT)**. Cardiovascular fitness implies heart health.

6.3.1.5 Enjoy gym facilities and amenities

The gym has a range of cool stuff and equipment. The gym has lots of equipment, unique features, and group classes available. Lots of amenities are available—the gym is full of exercise equipment that would otherwise not be accessible. The average gym-goer cannot buy machines and gym equipment for working out. Just the same way we enjoy buildings, recreations facilities, and other civil engineering structures, people go to gym to enjoy facilities and amenities and to have some good time relaxing, that they would otherwise not be in a position to purchase for themselves. Having a home gym may be realistic. However, the cost and space of fitting multiple gym machines into your home may not be within reach. Depending on whether it's a private gym, or franchise gym, you can have access to specialty equipment like kettlebells, boxing bags, pilates machines, total resistance exercise (TRX) stations, peloton bikes, powerlifting grade equipment, and much more.

Beyond state-of-the-art equipment, top-notch gyms offer an array of luxurious amenities that justify membership fees. These may include sports facilities: basketball courts, squash courts, and indoor tracks; specialized studios: mirrored dance studios and functional training areas; wellness amenities: swimming pools, saunas, and steam rooms; relaxation and recovery: massage chairs and tanning services; and convenience: on-site supplement and protein shops merchandise stores. These premium amenities enhance the overall fitness experience, providing members with a comprehensive wellness solution.

6.3.1.6 Shaping desirable habits and lifestyles

People go to the gym for lifestyle—the gym helps form positive habits and lifestyles. The gym can initiate positive habits and a healthy lifestyle! Going to the gym on a regular basis can create a routine that helps you stay fit, reduce stress, and sleep better. Consistency is the key idea here. Functional and practical goals for shaping habits and lifestyles include improved overall fitness, enhanced athletic performance, increased energy levels, better bone density, and injury rehabilitation and prevention.

Gym-goers get more in tune with their body, health, and nutrition. It is more likely that if you're making the right decisions for your health and wellness, you become more aware of other areas that need improvement. Going to the gym can have a ripple effect in your life. An important area that exercise can help improve is stress relief and sleep.

Committing to a gym routine requires self-discipline. You set goals, plan workouts, and stick to a schedule. This discipline often translates into other areas of life—in this regard, civil engineering! You might find yourself more focused on tasks, better at time management, and more diligent in achieving personal or professional goals as a civil engineer!

Make fitness a non-negotiable part of your routine, and you'll see the results. The more you stick to it, the more automatic it will become. Over time, this routine turns into a habit. Once the habit is established, it becomes a natural part of your lifestyle.

Going to the gym promotes regular exercise, which helps with the important factors of stress reduction and sleep. When stress is reduced and sleep improves, you're likely to feel more energized and focused, enabling you to tackle daily challenges more effectively. Exercise can help regulate your circadian rhythm, the body's internal clock that controls when you feel awake and sleepy. Additionally, exercise contributes to deeper and more restorative sleep, allowing you to wake up feeling more refreshed and energized.

Establishing a gym routine can act as a buffer against stress. Winding down with a workout can signal to your body that it's time to transition to a more relaxed state, priming you for better sleep. Recent research indicates that exercise decreases sleep complaints and insomnia in patients, according to John Hopkins Medicine [10].

6.3.1.7 Socialize and build community

People go to the gym to make friends as the gym allows you to socialize and build a community. The gym provides an opportunity to build relationships and socialize. Social and fun reasons are meeting new people and making friends. Gyms have a social environment. The gym brings people from all backgrounds together. There's an opportunity to form connections.

Though many people work out with headphones on and don't want to talk, friendly gestures, spotting someone's lift, and even cleaning equipment can go a long way to gain a good gym neighbor. If you get the chance to have a workout partner, it makes working out a fun experience.

Fun and enjoyment (dancing, swimming, etc.): Gyms offer more than just a place to work out. It's a social hub where you can have shared experiences with other gym-goers. The gym can create a sense of belonging, and the support can feel amazing. Of course, individuals may have unique motivations, and reasons may overlap. Gyms offer a structured environment for people to pursue their fitness goals, whatever they may be!

Group fitness classes and social events: It's a cool feeling to have a 'gym family.' The gym family represents those individuals who are always present when you're at the gym.

Supportive environment and community: When you see the same people over and over, you get comfortable with each other. You might never talk, but you see the other person pushing, so you push harder. It's a fitness camaraderie and competition that's hard to replicate elsewhere.

Working out with a partner can give you support and motivation, besides building a bond from sweating it out. A good way to meet people is with group classes. A challenging class can naturally lead to friendships.

6.3.2 Lessons for a Civil Engineer

The civil engineer can learn a lot from going to the gym! From tool box safety meetings, to socialization and networking, to health and safety lessons, a civil engineer can really benefit—comparing these activities to the gym.

From understanding point loading to watching how the anatomical systems are a source inspiration for marvelous civil engineering projects, there are several transferable skills and benefits that can enhance a civil engineer's professional development by virtue of visiting the gym!

The gym supports the engineer's physical and mental well-being: improved focus and concentration at work. The regular exercises are magical in the sense that they can boost mental clarity to tackle complex engineering problems. Going to the gym and exercising can be a great way to manage stress and, thus, be able to meet project deadlines and handle high-pressure roles or situations.

Learning time management, problem-solving skills, and attention to detail are just a few of the transferrable skills you can get from visiting the gym, as a civil engineer! Weightlifting, treadmilling, and stationary biking and other gym exercises do require problem-solving skills, such as adjusting the form or technique. Such a skill can be applied to engineering challenges. Have you seen how civil engineers, civil engineering surveyors, or surveyors approach equipment and tools they use in the field? Some gym exercises require enough attention to the proper form and technique. This can translate to proper attention to detail in civil engineering projects. Many exercises are timed in the gym, sessions last between 30 and 60 minutes and besides, balancing gym workouts with work and other responsibilities, can help the civil engineer develop effective time management skills.

The gym teaches the importance of consistency and patience. Progress in fitness doesn't happen overnight—it's a gradual process that requires dedication. This helps the civil engineer understand the importance of consistency and patience in their professional life.

Learning about safety in the gym cannot be avoided. It is one of the major benefits a civil engineer can really enjoy. See Section 6.3.3.1.

Networking opportunities: the gym can be a place to meet new people including potential professional contacts or mentors. You can build relationships when your regularly meet the same people at the gym.

Regular workout can develop discipline and motivation which you can later apply in your professional life.

Setting goals such as fitness goals can help you develop goal-setting skills, which can be applied to engineering projects and career development.

Unconventional benefits include understanding structural principles; these are outlined in Section 6.3.3.3. For example, weightlifting and other gym exercises help develop an intuitive understanding of structural principles, such as balance and stability. The gym can also support an appreciation of materials science, working with different types of equipment, and weights can make us appreciate the materials science and the properties of various materials.

6.3.3 Relationship between the Gym and Civil Engineering

The gym and civil engineering share a lot in common. In fact, advancing civil engineering can gain lots of insights from the gym, the gym-goer, and the equipment they interact with. This section highlights the **transferable skills and principles** between the gym and civil engineering, such as planning, execution, maintenance, safety, and teamwork.

In that regard, there are several lessons a civil engineer can learn from the gym. The gym can be a great source of inspiration for understanding civil engineering concepts, possibly conducting interdisciplinary research, and/or attaining lifelong learning skills. This is because the gym deals with loads, equipment, and structures which are the same items that civil engineers deal with at a massive scale. The gym compares with small-scale equipment and loads. You can basically learn about forces and moments in the gym.

The concept of safety can be well-understood when you visit the gym and you realize if it is not integrated into planning, and while exercising or working out, the gym-goers can be exposed to risks. The same thing civil engineers strive to do every day—identifying workplace hazard to eliminate risks, design engineering controls, and/or minimize risks. Also, when appraising projects, the designer ensures that a factor of safety is incorporated. As a civil engineer, take some time and visit the gym! It is all fun dealing with loads!

The civil engineering industry is vigilant on safety, environmental sustainability, and climate awareness. This rhymes well with the gym, especially on safety and environmental sustainability. Safety in civil engineering involves addressing workplace hazards through various organizational approaches. Some common hazards include working at height, working around moving vehicles or heavy equipment, and pandemic-related risks. To improve safety, civil engineers should use engineered safety concepts, wear reflective jackets, and ensure that projects meet established safety and quality standards. The same thing happens in the gym!

In terms of safety, the gym requires identifying potential risks to prevent injury of gym-goers, and civil engineering generally requires hazard mitigation. Establishing safety guidelines is essential by following a proper form for the gym and construction codes for the civil engineer. Emergency preparedness is important, that is, planning for unexpected events—in the gym, injury response; civil engineering deals with emergency response plans.

In civil engineering, statics and dynamics of bodies are essential topics that gym instructors practice daily in their classes, although they don't explicitly teach it to their students. Statics is an essential prerequisite for many branches of civil engineering, which address the various consequences of forces. Unlike static analysis, which deals with forces in equilibrium, dynamic analysis considers forces and motions that change with time. This type of analysis helps us predict and evaluate a structure's response when subjected to dynamic forces such as vibrations, impacts/shocks, seismic events, floods, or wind gusts.

Weight lifting and dealing with loads, generally, shed light on how stresses build up in structures. You basically feel where stress concentrates, accumulates, or builds up and compare this with a pattern when a civil engineering structure is loaded. For example, holding a dumbbell close to the chest and away from the chest, and stretching out just as the Vitruvian Man, you feel stresses in the arm differently. In this regard, as we learn about stresses and strains, gyms are a perfect illustration, following the body—doing varying exercises with loads such as dumbbells. Exercises for the back, legs, stomach, arms, chest, back (upper, lower, and spine), neck, shoulders, hips, knees, ankles, heart, and other organs are all different, and techniques vary widely.

Through these techniques, we basically feel sensational stresses differently, and we can compare these with civil engineering structures loaded differently. These stresses are felt by the musculoskeletal system. Musculoskeletal stress can build up due to compression (squishing forces), tension (stretching forces), shear (sliding forces), and torque (twisting forces). Musculoskeletal stress occurs when our muscles and other components of the musculoskeletal system experience tension, stiffness, or discomfort due to stressors. These forces have been illustrated in Chapter 4. The factors contributing to musculoskeletal stress include the weight and size of loads, posture and lifting techniques, frequency and duration of load-carrying, body position and alignment, and muscle strength and endurance.

Effects of prolonged musculoskeletal stress include fatigue, muscle strain, joint pain, injury (acute or chronic), and long-term damage (e.g., herniated disks and osteoarthritis). To mitigate musculoskeletal stress, use proper lifting techniques, maintain good posture, strengthen core and back muscles, take regular breaks, and use assistive devices (e.g., harnesses and wheelbarrows).

In civil engineering designs, therefore, we may learn from this to come up with safe, reliable, and sustainable designs. On the other hand, understanding this analogy also supports our approach to ergonomics, whose goal is primarily to minimize stress, discomfort, and injury by fitting tasks, tools, and environments to human capabilities and limitations.

A civil engineering class can be compared with a gym class. The gym and fitness centers have group classes that offer a variety of workouts such as yoga, spin, high-intensity interval training (HIIT), and dance, adding diversity to fitness routines and making workouts fun and engaging. They also have structured sessions where group classes follow a structured format led by an instructor, making it easier for individuals who prefer guidance during their workouts.

Forces induce loads in both civil engineering structures and gym-goers. The direction of causality is the same: forces come first, and loads follow as a result.

Both civil engineers and gym-goers plan and execute their plans. So, planning and design are vital for the gym, that is, workout routine for the gym and coming up with a project blueprint for a civil engineer. In the same way, civil engineers strategize effectively and sequence activities; for example, by construction phasing, gym-goers exercise order (do this first and end with that). The civil engineer in the present scenario must be adaptable, which means flexible and willing to adjust as circumstances may dictate, for example, adapting to site conditions. Gym-goers, on the other hand, modify workouts to suit the conditions as instructed by the instructor.

We need regular maintenance of civil engineering structures to prevent deterioration. Similarly, the gym-goers need consistent exercise to keep fit and attain their goals. Inspection, that is, periodic assessment and evaluation, is necessary for gym-goers (progress tracking) just the same way structural inspections are conducted for civil engineering structures.

There is also continuous improvement and optimization. Gym-goers always engage in new exercises, while civil engineering advances innovative materials from time to time.

The civil engineer is buried in collaboration with project teams the same way personal trainers in the gym interact and collaborate with their trainees or students. The gym requires clear communication to ensure success, that is, clear instructions, the same way civil engineers coordinate projects. Overcoming plateaus by gym-goers is essential just the same way resolving construction issues is. This collective process of problem-solving and troubleshooting happens in the gym and in a civil engineer's work.

These parallels highlight the transferable skills and principles between the gym and civil engineering, such as planning, design, execution, maintenance, safety, and teamwork.

6.3.3.1 Gym: safety tips/protocols

In Section 9.7, we share the importance of the following procedures. This same habit is good in the gym! The gym provides safety procedures to follow just like any civil engineering project. As a gym-goer, you should endeavor to consult a doctor before starting new exercises as it is a good practice to know whether your health condition is fit to carry on with the new exercise.

Working with a personal trainer or fitness coach is a good practice because they can offer customized guidance such as tailored workout plans addressing specific goals, needs, and limitations. They can also help track efficient progress by optimizing workout routines for better results. They can also help set clear goals, objectives, and strategies for achievement. This can be related to a trainee under the supervision and guidance of a qualified engineer.

As a gym-goer, setting realistic fitness goals is essential. Take time to listen to your body (rest when needed). This can be related to civil engineers setting realistic career goals and project design and/or development goals in their practice. Gym-goers follow gym rules and regulations, and so does the civil engineer.

6.3.3.1.1 General safety precautions

Typical precautions include warming up before exercising, learning proper exercise techniques and form, starting with lighter weights, and progressing gradually. Others include use of spotters or safety equipment (e.g., squat racks), staying hydrated and fueled, and reporting injuries or concerns to gym staff. These safety precautions can be related to safety precautions at construction sites.

6.3.3.1.2 Equipment safety

Gym-goers are instructed to inspect equipment before use (e.g., worn cables or cracks) and to use equipment for the intended purpose only. Prior to start, one should adjust the equipment to fit their body. Secure weights and bars properly and avoid overloading equipment, among other equipment safety protocols.

6.3.3.1.3 Personal safety

Gym-goers are usually instructed to be mindful of surroundings (e.g., people and equipment). Respect personal space and boundaries, inform staff of medical conditions, keep valuables secure (e.g., lockers), and avoid exercising alone, especially at night.

6.3.3.1.4 Hygiene and cleanliness

Gym-goers are required to wipe down equipment after use. They are required to shower and change clothes regularly and also wear clean, breathable clothing. They must use sanitary products (e.g., hand sanitizer). In case of areas that do not meet standard sanitation, they should report cleanliness concerns to staff. These hygiene requirements can be related to welfare standards at construction sites (not so different from the necessary hygiene at construction sites).

6.3.3.1.5 Gym etiquette

Just as the gym has etiquettes, so does civil engineering. We have professional bodies that regulate the civil engineer's conduct. In the gym, you are required to respect others' workouts, share equipment and space, reduce noise levels, refrain from using strong perfumes/colognes, and be considerate of others' cultural differences. In civil engineering, you are required to train others and mind your safety, health, and welfare and that of others.

6.3.3.1.6 Emergency procedures

As a good practice, gym-goers must know gym emergency contact information, familiarize themselves with gym evacuation routes, report injuries or incidents immediately, follow gym protocols for medical emergencies, and know basic first aid and CPR. These are not so different from emergency procedures at construction sites!

6.3.3.1.7 Gym staff responsibilities
These include the following:

- Provide Personal protective equipment (PPE) information and training
- Ensure availability of PPE
- Monitor PPE usage
- Maintain PPE equipment
- Enforce PPE policies

6.3.3.1.8 PPEs in the gym and civil engineering
The benefits of PPE in the gym include reduced risk of illness, for example, eye infections, injury prevention, enhanced performance, increased confidence, and compliance with regulations (e.g., occupational health). This compares exactly with the role of PPEs in civil engineering.

The gym PPE categories are as follows: head and face protection, eye protection, hearing protection, hand and finger protection, body protection (torso, arms, and legs), and foot protection. Other gym-related PPEs include gloves (for weightlifting and grip support), wrist wraps (support and stability), knee sleeves (support and protection), elbow sleeves (support and protection), mouthguards (contact sports and heavy lifting), headphones (noise cancellation and focus), eye protection (goggles and sports-specific), footwear (non-slip shoes and athletic shoes), orthotics (support and stability), and protective clothing (e.g., sleeves and pants). PPEs are used in the following activities:

- High-impact activities (e.g., boxing and HIIT)
- Heavy lifting or weight training
- Contact sports (e.g., basketball and soccer)
- Group fitness classes (e.g., spinning and aerobics)
- Using equipment with moving parts (e.g., treadmills and ellipticals)

6.3.3.1.9 Lessons for the civil engineer

All these gym safety tips share insights into civil engineering safety protocols! From engineering controls, to identifying hazards in the work place, following procedures, using PPEs, and sharing responsibility as an employer, staff, worker, or designer, civil engineers learn a lot.

6.3.3.2 Analogies

Civil engineering plan, design, construction, operation, and maintenance usually analogize or share exciting scenarios, principles, and goals with the gym. As illustrated earlier, civil engineers mimic nature and anatomical systems greatly to innovate practical solutions that advance humanity, and the gym supports them by generating profound ideas that have not been there before. You can learn several civil engineering insights from the gym. From safety to mechanics to structural forces and moments, you can learn a lot of civil engineering concepts in the gym, and during the design or planning of civil engineering projects, you may analogize and mimic the human anatomy, borrowing techniques of working out in the gym and the equipment and structures gym-goers deal with. As a civil engineer, there is much to learn from the gym that could nurture your innovative thinking.

6.3.3.3 Pedagogical illustrations

This section clarifies complex engineering concepts for students, apprentices, and graduates. In other words, it supports instructional goals as it engages the learner's attention, fostering understanding and retention. It encourages critical thinking and inquiry, overall. Gym-goers primarily interact with three key components to achieve their fitness goals and other goals, that is, loads, equipment, and structures.

- Loads (maybe about 40%–50% of the time)—these include weights (free weights and weight machines), resistance bands, and other forms of resistance to challenge muscles. Typical strength training for beginners might last 45–60 minutes. For more advanced lifters, the sessions can potentially be up to 90 minutes.
- Equipment (maybe about 30%–40%)—these include cardio machines (treadmills, stationary bikes, and ellipticals), strength training machines (leg press, chest press, rowing machines, etc.), functional training tools (kettlebells, medicine balls, and TRX), and free weights (dumbbells, barbells, and plates). Cardio sessions typically range from 30 to 60 minutes.

- Structures (maybe about 10%–20%)—these include the gym layout and design, flooring (rubber, turf, and hardwood), storage systems (weight racks and equipment racks), and amenities (showers, locker rooms, and seating areas). Warm-up and cool-down sessions, at the beginning and end of workout, may range from 5 to 10 minutes.

Note: The specific time spent on loads, equipment, and structures depends on the individual workout plan. Safety is a very important issue in the gym, just as civil engineers mind safety in design, construction, and operation and maintenance of structures. The gym instructor ensures that gym-goers are safe and instructions are issued to follow.

As a gym owner, the well-being and security of gym members and staff should be the top priority. Every day, numerous individuals visit fitness centers to achieve their fitness goals, which is why creating a safe and secure environment is of utmost importance.

6.3.3.3.1 Gym roller

The gym mobility or stretching roller is a perfect example of a roller support in civil engineering. It resists vertical forces but cannot resist a moment or horizontal forces. See Figure 6.17.

6.3.3.3.2 Pin-connected, fixed-connected, and simply supported gym machines

In structural engineering, a support refers to a point or surface that bears the weight or load of a structure. A connection refers to the physical link or joint between two or more

FIGURE 6.17 Michael is using a mobility roller to stretch.

FIGURE 6.18   Michael and Easther working with the Smith machine.

structural elements, such as beams and columns. In the gym, structures are connected/supported differently presenting opportunities for an engineer to learn fixed, pinned, and roller connections/supports.

See the Smith machine in Figure 6.18, simply supporting the weight plates on a rod. The Smith machine is not fixed on the floor! It gains its stability from the self-weight. It is pin-loaded and has a barbell to select the weight load for chest pressing. The **Smith machine** is a weight machine used for weight training. The barbell of this machine is fixed within steel rails, allowing for only strict vertical movement. Some Smith machines have the barbell counterbalanced. The machine can be used for a wide variety of exercises, including, but not limited to, squats, the bench press, the shoulder press, and deadlifts.

Generally, pin-loaded machines include leg press machine, chest press machine, pulldown machine, and seated row machine. These machines use a pin to select the weight load. Fixed-connected machines include cable crossover machine (arms are fixed to cables), leg extension machine (legs are fixed to the lever), and others. These machines have fixed movement patterns.

6.3.3.3.3 Working out on a treadmill

Let's see how a treadmill showcases **energy conservation** that civil engineers cherish in their designs, analysis, and evaluation of structures. Gym visits can help you shed some unwanted pounds or gain some healthy weight. Your body burns calories faster when you work out. Working out in a gym means burning calories. You can get a close estimate of the calories burned during exercise using the following formula [11]:

$$\text{Calories burned per minute} = \frac{3.5 \times \text{MET} \times W}{200} \quad (6.3)$$

where

MET: metabolic equivalent
W: body weight in kilograms

MET is the ratio of your working metabolic rate relative to your resting metabolic rate [9]. Metabolic rate refers to how quickly fuels (such as sugars) are broken down to keep the organism's cells running. There are general differences in the metabolic rate among species, and the environmental conditions and activity level of an individual organism will also affect its metabolic rate.

Your metabolic rate is the rate of energy used per unit of time, whether you are active or sitting still. It is a term that gives you an idea of the intensity level of a particular activity. MET is standardized so that the value can be used for all groups of people, irrespective of their age, sex, and genetics. The value makes it easier to compare different activities to each other.

One MET is your resting or basal metabolic rate (BMR). So, if there is an activity with **MET 5**, it simply means that you are spending five times more energy than you would spend while you are doing nothing. In equation (6.3), 1 MET equals 3.5 mL of oxygen consumed per kilogram of body weight per minute.

Your body will burn 3.5 mL of oxygen per kg of bodyweight per minute when sitting still, and this is equivalent to one MET. To translate this into calories burned, the formula is as follows: 1 METs = 3.5 × weight in kg ÷ 200 Taking an average man weighing 75 kg, sitting still for 1 minute: 1 MET = 3.5 × 75 kg ÷ 200 × 1 minute = **1.31 calories** burned per minute.

Example 6.1: Treadmill workout

Michael (Figure 6.19), a 28-year-old gym-goer and trainer, weighing 90 kg (approximately 198.42 pounds), runs on a treadmill for 30 minutes, at a treadmill speed

$$E_{chem} = E_{pot} + E_{kin} + E_{therm} \qquad (6.5)$$

where

E_{chem}:	Chemical energy
E_{pot}:	Potential energy
E_{kin}:	Kinetic energy
E_{therm}:	Thermal energy

Energy expenditure on a treadmill involves the body's metabolism, and a significant portion of the energy is dissipated as heat, but the total energy expenditure is not simply the sum of these three forms of energy. It could potentially be much more than this. While kinetic and potential energy play a role, equation (6.5) is a simplified version. Potential energy (E_{pot}), assuming a negligible change in height, is approximately 0.

Note: Primary factors to consider for calculating energy expenditure are body weight, speed, treadmill incline, and duration of exercise.

So,

$$E_{chem} = 543.39 \text{ calories}$$

$$KE\left(E_{kin}\right) = \frac{1}{2}mv^2 \qquad (6.6)$$

$$E_{kin} = \frac{1}{2} \times 90 \times 3.13^2$$
$$= 440.86 \text{ J } (105.368 \text{ calories}) \qquad (6.7)$$

Thus,

$$E_{therm} = 438.02 \text{ calories}$$

Michael starts running, converting potential energy (E_{pot}) into kinetic energy (E_{kin}). His body's stored energy (chemical energy) is released as kinetic energy (E_{kin}) to move his limbs and propel him forward. As Michael runs, his kinetic energy (E_{kin}) increases, while his potential energy (E_{pot}) decreases. When he starts running, his muscles contract, releasing stored chemical energy (E_{pot}). This energy is converted into motion, increasing his velocity (E_{kin}). As his speed accelerates, his E_{pot} decreases (less stored energy). Simultaneously, his E_{kin} increases (more energy of motion) and the body sweats, dissipating heat energy.

Michael's body produces heat energy (E_{therm}) through friction, muscle contractions, and other thermodynamic processes. Some of the kinetic energy (E_{kin}) is converted into heat energy (E_{therm}), which is dissipated into the environment.

Breakdown

- **80.61%** of energy expended as heat (E_{therm})
- **9.39%** of energy expended as external work (E_{kin})

6.3.3.3.3.1 Lessons for the civil engineer The treadmill cardio machine would indicate the calories burnt as 543.39. It is clear that these calories are burnt at 7 mph, at a given inclination. What happens when you vary the inclination? As Michael works out, he sweats and warms up. The body changes slowly as he continues the exercise.

Thus, from the KE equation, one can quickly determine the kinetic energy dissipated in the 30 minutes because you

FIGURE 6.19 Michael and Easther treadmilling.

of 7 mph (3.13 m/s) and a MET value of 11.5. Evaluate the calories burnt.

Solution

The formula would work as follows:

$$\text{Calories burnt} = \frac{11.5 \times 3.5 \times 90}{200}$$
$$= 18.113 \text{ calories per minute} \qquad (6.4)$$

Michael runs for 30 minutes; thus, he burns about 543.39 calories. To convert a measurement in calories to a measurement in joules, multiply the energy by the following conversion ratio:

4.184 joules/calorie
So,

18.113 calories per minute ≈ 75.785 joules per minute

Note: The calorie and its symbol, Cal, usually refer to the small unit, the large one being called kilocalorie (kcal). However, the kcal is not officially part of the International System of Units (SI) and is regarded as obsolete, having been replaced in many uses by the SI-derived unit of energy, the joule (J), or the kilojoule (kJ) for 1,000 joules. Calories are burned during exercise.

Let's consider a simplification of the principle of energy conservation. The intensity and duration of the exercise determine the amount of energy expended. Thus, equation (6.5) is a simplified version:

can visibly adjust the treadmill speed. Here, Michael runs at 7 mph. Remember, though, kinetic energy has no component of duration in the formula. However, the calories burnt have a component of time; see equation (6.4). Once KE is determined, the question remains as to where the remaining dissipated energy goes; PE is approximately 0 because the treadmill is on the floor.

The sweat indicates heat energy (thermal energy). However, there are other energies dissipated, e.g., sound. Thus, equation (6.5) is a simplified version. The question 'Where is the remaining dissipated energy after subtracting KE from the total calories displayed by the treadmill?' is essential. This kind of question is what engineers would phrase or pose when designing systems to properly account for all factors—account for all forces and moments, for example. No wonder, the universe is full of energy, and energy equations are versatile and applicable in all branches of civil engineering and other engineering fields.

You have the technology; you design a system, but you realize instability somewhere or spot a weakness somewhere. You realize that the system is not in equilibrium, so you need to account for the remaining factors to have things balanced.

In this particular case, KE is straightforward, the technology (treadmill machine) displays 543.39 calories, burnt in 30 minutes, but where is the remaining dissipated energy? Is it all heat energy or not? This treadmill illustration/exercise can be used to comprehend the second law of thermodynamics.

In Chapter 3, we saw the application of technology in scientific research, how science and engineering reciprocally advance each other, and how technology plays a pivotal role in the development of both engineering and science. Here, the treadmill exercise is a perfect basic example of how you can pose many questions to further research and development in the energy conservation context—see Section 3.3.3—and remember that research can never get exhausted unless problems cease to exist! For example, AI alone will likely bring about unprecedented chaos in the bid to tame it, especially when it eliminates the role of slow learners, coming from the functionalist perspectives! Therefore, many ongoing researches are carried out on how AI is developed, implemented, and integrated into the society—this serves as an example of how problems germinate from advancing technologies.

6.3.3.3.3.2 Real-world application   Michael's body converts chemical energy into mechanical energy while running on the treadmill, with some energy lost as heat. However, did you know that working out on a treadmill can generate clean electricity? The modern world has developed exercise treadmills, where ellipticals and cycles turn workouts into electricity, feeding it back into the building through an electrical socket. These systems showcase energy conservation by reducing reliance on fossil fuels and generating electricity, depending on the treadmill type.

Some treadmills convert user movement into electricity, while others promote energy efficiency through lower speeds and incline settings. This is an energy source burned naturally every day that isn't being properly harnessed: calories. There are a lot of gyms that are adopting green technology in a variety of ways.

There are some places where electricity is generated from walking on floors! In Japan, several public spaces, particularly in Tokyo, are utilizing piezoelectric technology to generate electricity from pedestrian traffic. These include railway stations, sidewalks, and even some stadiums. Specifically, Tokyo Metro has implemented this technology in its subway floors to convert the energy from footsteps into electricity, powering the stations and infrastructure.

The treadmill's maximum output is about 200 W/h. The average person uses about 28,000 W-h a day in the United States, for example [12]. The maximum treadmill workout, generating 200 W for an hour, would save 2.4 cents, assuming an electricity cost of $0.12 a kilowatt-hour, in addition to the power that would have been used by a motorized machine. On the treadmill, a 147-pound person running at an 8-minute, 20-second mile pace produces about 150 watts of mechanical power, which in 30 minutes is enough to power a typical 10-watt Wi-Fi router for about 7.5 hours. Considering energy losses due to generator efficiency, battery storage, and inversion of about 30 - 35%, the 30-minute run might actually give you closer to 4.5 to 5 hours of Wi-Fi. A 176-pound person lightly jogging for 20 minutes could power a 60-W lightbulb long enough to light the room while they're working out.

6.3.3.4 *Working out on a stationary bike*

Stationary bike—a cardio machine—is shown in Figure 6.20. This machine can be related to the generation of electricity. It mimics generation of electricity through potential ellipticals and cycles that can turn workouts into electricity.

FIGURE 6.20   Michael and Easther on stationary bikes.

6.3.3.5 Weightlifting

Through weightlifting and aerobics in the gym, a civil engineer can learn about point loading (e.g., a potbelly approximates to a point load; rather uneven loading)—on a person's body, particularly on their spine and lower back due to the concentrated weight and mass of the abdominal area. Generally, the body is built with equally distributed body mass. So, developing a pot belly is an indicator of underlying issues to resolve within the body. With the changes your body undergoes when you consistently go to the gym, you realize your body is not built to sit still, especially when the pot belly is cleared after consistent workouts! The body was meant to be loaded uniformly (uniform distribution of weights around the body!). You will realize why engineers love dealing with uniformly distributed loads.

Weightlifting and dumbbell exercises are perfect pedagogical illustrations for civil engineers in the making and training. Let's compare a human body to a civil engineering structure. Let's use dumbbells to illustrate structural forces and moments. In a gym, forces induce loads. When you contract your muscles (force), you induce a load (weight or resistance) to move or lift. Your muscular forces (pushes or pulls) cause the loads (weights or resistance) to be applied to your body or the machine. In both civil engineering and gym, forces induce loads. The direction of causality is the same: forces come first, and loads follow as a result.

Figure 6.21 is a perfect illustration of the impact of forces and moments on a structure: Michael is holding dumbbells in his outspread arms just like Leonardo da Vinci's Vitruvian Man. The weight of the dumbbell creates a force that makes him rotate his arm downward. The moment of this force around his shoulder joint (the pivot point) depends on two factors, that is, the force (F) due to the weight of the dumbbell and the distance (L) due to the length of the arm from the shoulder joint to the point where he is holding the dumbbell.

Assumption: Arm has a constant cross-sectional area. Although we have changes in body shapes and muscles, we assume uniformity. When the arms are stretched

FIGURE 6.21 Michael lifts 7.5 kg dumbbells with outspread arms.

FIGURE 6.22 Michael lifts 7.5 kg dumbbells close to the body, at ≈90° to the upper arm.

out completely (outspread arms), similar to the Vitruvian Man's (Section 2.5.1.1) exercise, the holder feels more muscle-fatigued/-stressed than when they hold the dumbbells close to the body, say at 90°, that is, when the fore arm holding the dumbbell is at a right angle to the upper arm (see Figure 6.22).

So,

$$\text{Moment} = \text{Force} \times \text{distance} \tag{6.8}$$

If the moment is taken at the shoulder, it would be as follows:
When the arm is outspread,

$$M = -\text{FL kNm} \tag{6.9}$$

When the fore arm is at a right angle with the upper arm (see Figure 6.22),

$$M = -\frac{\text{FL}}{2} \text{ kNm} \tag{6.10}$$

where L is the length of the arm in meters.

In this illustration, the moment is the rotational effect of the dumbbell's weight around Michael's shoulder joint. The greater the weight or the longer your arm, the greater the moment. Therefore, it is obvious that the holder feels more muscle-fatigued in the upper arm, the area that does most of the work! In this particular case, relating this to a cantilever beam point-loaded at the extreme end, the moment is always greatest at the fixed end and 0 at the extreme end. It would, therefore, be wise that the shear reinforcements are placed closer to each other at the fixed end and more spaced in other parts of the beam. Consider a cantilever[i] beam (Figure 6.23) point-loaded at the extreme end to explain this illustration. See Figure 6.24 for the shear force diagram (SFD) and bending moment diagram (BMD) for the point-loaded cantilever beam.

Reactions:

$$R_A = P$$

$$R_B = 0$$

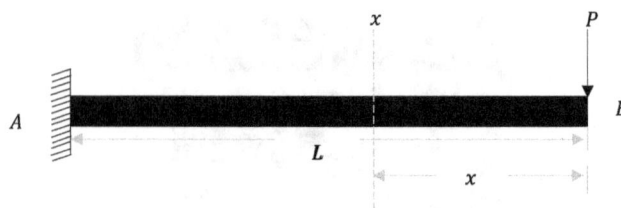

FIGURE 6.23 Cantilever beam point-loaded at the end.

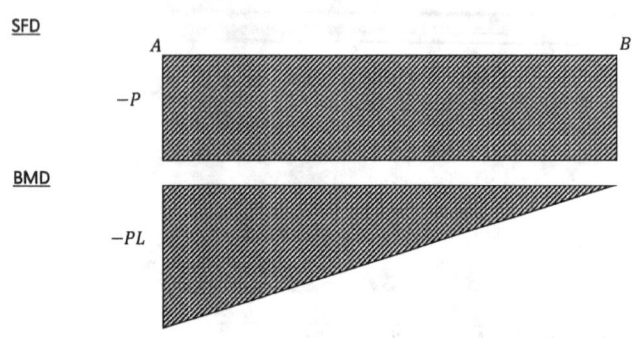

FIGURE 6.24 SFD and BMD for point-loaded cantilever beam (Figure 6.25).

The moment the force, P, creates at point x away from the force is

$$M_{x-x} = -Px$$

is negative because of the downward direction.

The moment the force, P, creates at point B is

$$M_{B(x=0)} = 0$$

The moment that the force, P, creates at point A is

$$M_{A(x=L)} = -PL$$

A moment refers to a measure of the turning or rotational effect of a force around a pivot point or axis.

Note:

Shear force at point x:

$$SF_{x-x} = -P$$

Shear force at A and B:

$$SF_A = SF_B = -P$$

Maximum deflection

Assumptions

- The beam is prismatic (i.e., has a constant cross-sectional area).
- The beam is subjected to a point load at the extreme end.
- The beam is cantilevered (i.e., fixed at one end and free at the other).
- The beam material is linearly elastic (i.e., follows Hooke's law). See Section 3.3.2.

Consider a cantilever beam with the following parameters: length, L; cross-sectional area, A; area moment of inertia, I; and modulus of elasticity, E. Using the Euler–Bernoulli beam theory, we can write the differential equation (6.11) for the deflection (y) of the beam as follows:

$$EI \frac{d^2 y}{dx^2} = M(x) \tag{6.11}$$

The Euler–Bernoulli beam theory (also known as engineer's beam theory or classical beam theory) is a simplification of the linear theory of elasticity, which provides a means of calculating the load-carrying and deflection characteristics of beams. According to the Euler–Bernoulli beam theory, the bending stiffness of a section of material is given by EI, where E is the elastic modulus of the material, I is the area moment of inertia, and $M(x)$ is the bending moment at a distance x from the free end. For a cantilever beam with a point load at the extreme end, the bending moment is as follows.

The moment the force, P, creates at point x is

$$M_{x-x} = M(x) = -Px$$

Substituting this expression into the differential equation (6.11), we get

$$EI \frac{d^2 y}{dx^2} = -Px \tag{6.12}$$

Integrating equation (6.12) twice with respect to x, we get

$$EI \frac{dy}{dx} = -\frac{Px^2}{2} + C \tag{6.13}$$

$$EIy(x) = -\frac{Px^3}{3} \tag{6.14}$$

$$y(x) = -\frac{Px^3}{3EI} \tag{6.15}$$

The maximum deflection occurs at the extreme end $(x = L)$, so we substitute $x = L$ into the equation (6.15). Thus, the deflection is

$$y(x) = -\frac{PL^3}{3EI} \tag{6.16}$$

If more than one point load and/or uniform load (q) are acting on a cantilever beam, the resulting maximum moment at the fixed end A and the resulting maximum deflection at end B can be calculated by summarizing the maximum moment in A and maximum deflection in B for each point and/or uniform load. This is analogized from the gym, in Figure 6.25, where Michael lifts a uniform wooden bar of constant stiffness, on top of his right hand.

The weight of the wooden bar creates a force that urges him to rotate his arm downward. The moment of this force around his shoulder joint (the pivot point) depends on two factors, that is, the force (F) due to the weight of the uniform wooden bar, whose length is equal to the arm, and the distance

FIGURE 6.25 Michael lifts a uniform wooden bar.

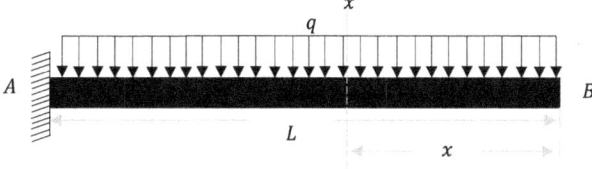

FIGURE 6.26 Uniformly loaded cantilever beam.

(L) due to the length of the arm from the shoulder joint to the point where the wooden bar ends as well. The issue here is that the holder feels that the force is different from when holding the dumbbell of equal weight. See the uniformly loaded cantilever beam in Figure 6.26.

The total force, q, along the beam is $q \times L$ acting midway the beam at point where $x = \dfrac{L}{2}$.

The moment the force, q, creates at point x is

$$M_{x-x} = -qLx \tag{6.17}$$

negative because of the downward direction.

The moment the force, q, creates at point B is

$$M_{B(x=0)} = 0$$

The moment that the force, q, creates at point A is

$$M_{A(x=L)} = -\frac{qL^2}{2} \tag{6.18}$$

Note:

Shear force at point x:

$$SF_{x-x} = -P \tag{6.19}$$

Shear force at A and B:

$$SF_A = SF_B = -P \tag{6.20}$$

6.3.3.6 Aerobics exercises—a gateway to stress analysis in structures

For the idea of a structure designed to imitate the position in Figure 6.27—Michael conducting sit-ups—the exercise meant to strengthen core muscles, particularly the abdominal muscles. Michael lifts his torso, curling up toward his knees. The primary body areas that could feel more stressed out are the abdominal muscles, that is, the rectus abdominis, obliques, and transverse abdominis muscles which are engaged to lift the torso. The lower back muscles, such as the erector spinae, may experience strain, especially if the proper form is not maintained. Michael definitely feels the body stretched out, and this helps show where the stresses are concentrated.

In Figure 6.27, you can tell where the stirrups should be concentrated just in case you are developing a structure that mimics this aerobic position or sit-up style. This can also be insightful for the level of shear reinforcement required! In that position (Figure 6.27), you feel discomfort or muscle fatigue more on the stomach (abdominis muscles) and the start of the legs (hip flexors)—the muscles in front of the hips may also feel engaged or strained, as secondary strained areas. The gym

FIGURE 6.27 Michael doing the sit-ups.

instructor would recommend maintaining proper form and engaging the core muscles effectively to minimize stress on the lower back. In case you are mimicking a structure to look like that—that is, the stable position where Michael lifts his torso, curling up toward his knees—you would need to get the feel of where stirrups and reinforcements will be concentrated in beams and columns and how they must be arranged and reasonably mimicking the gym instructor's advice of maintaining proper form. The sizes of stirrups and reinforcement depend on dead and live loads and the overall loading configuration.

The same thing happens when you do several other kinds of exercises. You learn several concepts to apply in structures. The physical pain, discomfort, or muscle fatigue you experience is a gateway to stress analysis in structures that try to mimic that position or aerobics styles. You've probably seen a number of architectural styles that mimic natural creatures or the environment—either in part or whole. Tai Chi is another form of exercise that could be mimicked—it may produce some interesting design ideas! Although we have changes in body shapes and muscles, we assume uniformity. This sit-up style (Figure 6.27) is commonly used for shedding belly fat. Following this analysis, which column is correct, A, B, or C in Figure 6.28?

As shown in Figure 6.28, adding more stirrups in the middle of the columns is over-design. Remember we are in an era where design optimization cannot be avoided to attain sustainability goals! Over-design or redundancy must be justified; economical use of materials and minimizing wastage are recommended. The spacing of ties is calculated based on the plastic moment capacity of the column, the induced moments from the slab strength, or earthquake loads. However, more reinforcement is needed at the column ends because that's where the moments are greatest and most cracking occurs. To minimize the risk of concrete spalling, engineers provide tighter confinement at the ends to enhance the ductility and overall performance. This is why shear ties are typically placed closer together at the ends. So, the correct answer is Fig. 6.28(a). As a rule of thumb, these stirrups are continuously added especially in areas where stress is greatest.

6.3.3.7 Stress concentration points

A stress concentration is a **localized area** in a structure at which the stress is significantly higher than in the surrounding material. Stress concentration occurs at joints, holes, bends, sharp corners, and even within the same material if it changes the cross-sectional area or shape. The increased stress lowers the resistance of the material to impact and fatigue loadings and is one of the more important factors contributing to structural failures. Localized increase in stress within a material or structure due to geometric features or discontinuities, such as sharp corners or holes, disrupts the smooth flow of stress, potentially leading to premature failure.

The fatigue you feel in your shoulder while holding a dumbbell can be compared to stress concentration in structures. See Figure 6.29. When a structure has a weak point or a sharp corner, stress tends to concentrate at that point, leading to potential failure. For instance, sharp corners or notches create localized increases in stress, whereas rounded edges (fillets) can help distribute the load more evenly.

In Figure 6.30a–j, the most fatigued points are shown mimicking the stress concentration points in structures where the moments are greatest and most cracking would occur if it were a structure. In beams and girders, the maximum moment often occurs at the midpoint or at supports, which can also be areas of high stress concentration. When a structure is subjected to bending or torsion, the resulting moments can cause stress concentrations at specific points, such as the extreme fibers of a beam or the surface of a shaft. The locations of maximum moments in a structure often coincide with stress concentration points. For example, in a simply supported beam, the maximum moment occurs at the midpoint, which can also be a stress concentration point.

In cantilever beam, the maximum moment occurs at the fixed end which can also be a stress concentration point. This is mimicked by Figure 6.21. High localized stresses can lead to premature failure, such as cracks or fractures, especially under cyclic or dynamic loads, so, if not accounted for in design and installation/construction, failures are inevitable. Therefore, choosing materials with higher strength and ductility can help resist stress concentrations. Structures with high stress concentrations may have a shorter fatigue life, meaning that they are more prone to failure under repeated loading, especially flexible pavements in roads. That's why we do traffic counts/assessments and establish the heaviest trucks meant to use the road.

In design and analysis of structures, we consider a stress concentration factor (SCF). The SCF is a measure of how much the stress is amplified at a point of stress concentration compared to the nominal stress in the material. A higher SCF indicates a greater risk of failure at that point.

Engineers can minimize stress concentrations by using rounded corners instead of sharp corners, avoiding sudden

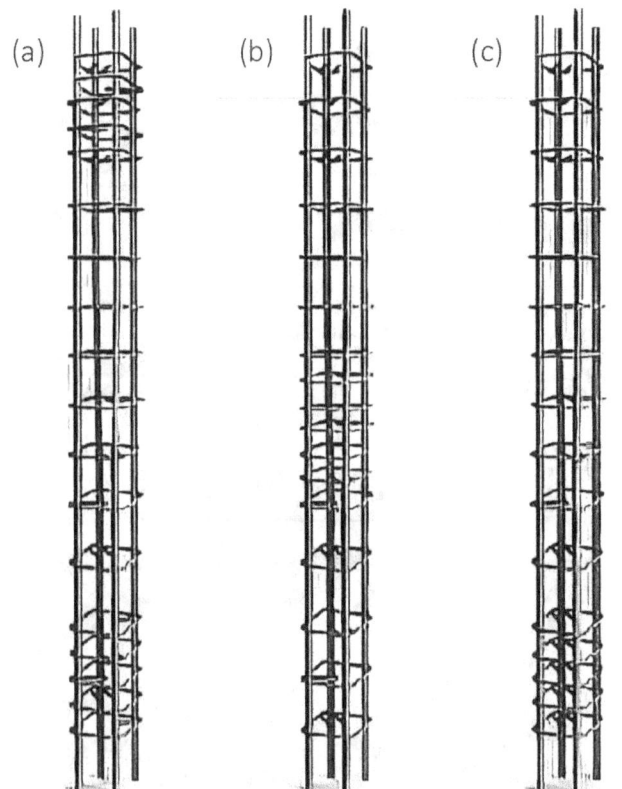

FIGURE 6.28 Column steel work and stirrups.

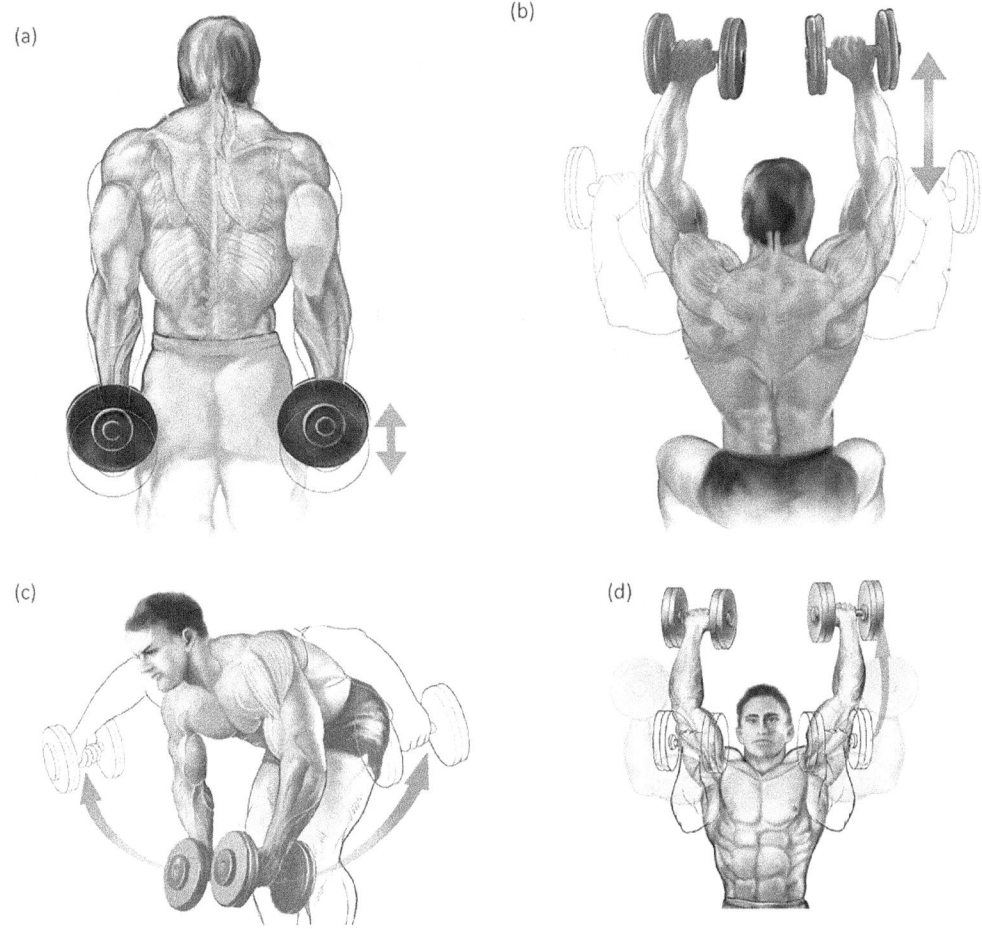

FIGURE 6.29 Exercising with dumbbells.

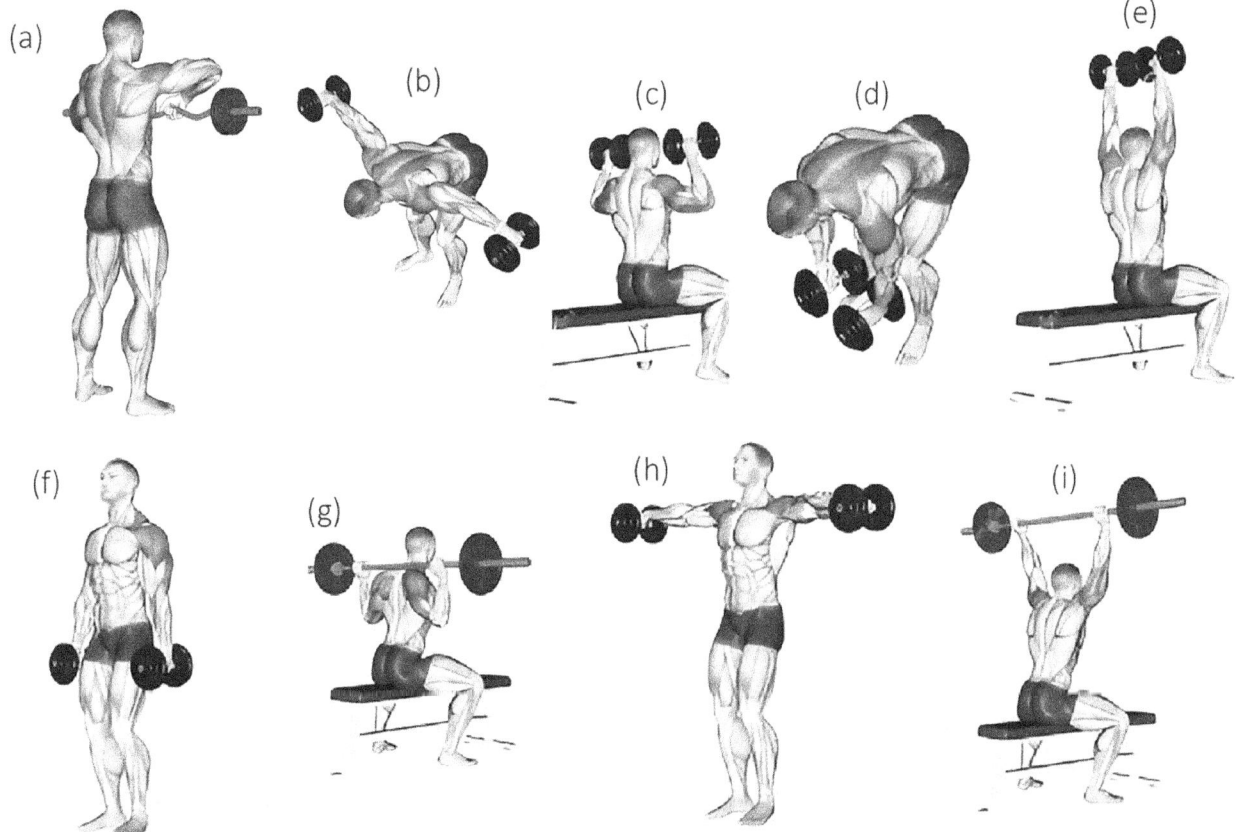

FIGURE 6.30 Demonstration of most fatigued areas when weightlifting dumbbells.

changes in the cross-sectional area, and adding fillets or radii at stress concentration points. In concrete structures, stirrups (transverse reinforcement) help resist shear stresses. In composite materials, adding reinforcement such as fibers or particles can enhance strength and reduce stress concentrations.

6.3.3.7.1 Stress concentration factors

Stress concentration factors are **numerical values** that quantify the increase in stress at specific points in a material, often due to geometric discontinuities such as notches, holes, or abrupt changes in the cross-section. These factors are crucial in materials like steel and metals because they help engineers predict where failures might occur under load and design components to withstand such stresses effectively. Stress concentration factors play a critical role in the design and safety of metal structures by allowing engineers to identify potential weak points where failures might occur. Stress concentration factors are crucial when evaluating fatigue failure in metallic materials as they highlight areas where cyclic loads may cause cracks to initiate and propagate over time.

Stress concentration factors can significantly affect the durability and performance of steel and metal structures, making it essential for engineers to account for them during design. The common geometric features like holes, grooves, and fillets can have varying stress concentration factors, influencing how stresses are distributed across a component [13].

The stress concentration factor is often determined using analytical methods or finite-element analysis to accurately predict stress distributions in complex geometries. Higher stress concentration factors can lead to premature failure in components subjected to dynamic or static loads, highlighting the importance of proper design practices. By using appropriate design techniques like adding fillets or changing cross-sections, engineers can reduce stress concentrations and improve the overall strength of metal components.

6.4 MIND MAPPING QUESTIONS

1. Identify civil engineering disciplines or innovative technologies that parallel the functions of the 12 human anatomical systems, extending beyond those mentioned in Section 6.2.1.
2. How can we mimic nature to develop sustainable infrastructure?

6.5 CONCLUSION

This chapter shares how civil engineers mimic the natural world, from mimicking anatomy, physiological processes, to the natural environment.

Previously, civil engineers only focused on harnessing the great sources of power in nature for the convenience of man. However, due to the growing demand to achieve sustainable development, they changed goalposts learning from the ecosystem. Today, civil engineers only design and develop systems that are in harmony with the natural world, which greatly minimize harm on the environment as much as applicable.

Civil engineers look at the 12 anatomical systems, the physiological processes, the natural world, and the ecosystem functioning to design and develop civil engineering structures and advance or extend new civil engineering materials—to refine materials, generally. Specifically, engineers have harnessed the circulatory and respiratory systems to advance pump technology throughout history! The combination of these two systems is powerful, drawing insights into how pump technology can be improved.

Scholars, students, and professionals in all fields of civil engineering design harness biomimicry and its potential to yield sustainable outcomes when practiced in a deep and thoughtful way. The design community, therefore, is an important leverage point for fueling a conversation about biomimicry because designers work at the nexus of values, attitudes, needs, and actions.

Civil engineers also think together with other fellow engineers and scientists through interdisciplinary approaches to improve the world. All this needs is a better understanding of the natural world to contribute to sustainable development.

Civil engineers go to the gym and work out as they learn from safety to mental health, physical health, engineering principles, socialization, networking, feeling good, and even career advancement. The gym provides a wealth of learning opportunities you wouldn't want to miss!

NOTE

1 Cantilever beams are the special types of beams that are constrained by only one given support. These types of objects would naturally deflect more due to having support at one end only.

REFERENCES

1. Kemp, M., et al. (1987). Inventions of Nature and the Nature of Inventions, in: Ed. P. Galluzzi, *Leonardo da Vinci Engineer and Architect*. Montreal Museum of Fine Arts, Montreal.
2. Vacuum Pump: What Is It? How Does It Work? Types of Pumps (iqsdirectory.com). Retrieved from: https://www.iqsdirectory.com/articles/vacuum-pump.html
3. Ren, J., & Xia, F. (2024). Brain-inspired Artificial Intelligence: A Comprehensive Review. Retrieved from: https://arxiv.org/html/2408.14811v1
4. Mechanical Boost. (n.d.). Diaphragm pump. Retrieved from: https://mechanicalboost.com/diaphragm-pum/
5. Laboratory pumps for biopharma processes | PSG. Retrieved from: https://www.psgdover.com/quattroflow
6. igus. (n.d.). Glossary: Ball and socket joints. Retrieved from: https://www.igus.co.uk/info/glossary-ball-an-socket-joints
7. Ssempeera, P. (2023). *Integrated Drainage Systems Planning and Design for Municipal Engineers* (1st ed.). CRC Press. https://doi.org/10.1201/9781003255550

8. Zhang, Y, Li, G, Liu, C, Guan, J, Zhang, Y, & Shi, Z. (2023). Comparing the Efficacy of Different Types of Exercise for the Treatment and Prevention of Depression in Youths: A Systematic Review and Network Meta-Analysis. *Front Psychiatry*, 14, 1199510. https://doi.org/10.3389/fpsyt.2023.1199510. Erratum in: *Front Psychiatry*, 14, 1304302. https://doi.org/10.3389/fpsyt.2023.1304302.

9. Digital health tracking tools help individuals lose weight, study finds. Retrieved from: https://med.stanford.edu/news/all-news/2021/02/digital-health-tracking-tools-help-individuals-lose-weight-study-finds.html

10. Exercising for Better Sleep | Johns Hopkins Medicine. Retrieved from: https://www.hopkinsmedicine.org/health/wellness-and-prevention/exercising-for-better-sleep

11. Bushman, B. (2017). *Complete Guide to Fitness and Health* (2nd ed.). American College of Sports Medicine. Human Kinetics, Champaign, IL.

12. Retrieved from: https://www.renewableenergyworld.com/energy-storage/battery/treadmills-that-generate-electricity-may-be-headed-for-your-gym/

13. Fiveable. (2024, July 31). Stress concentration factors – Intro to Civil Engineering. Retrieved from: https://library.fiveable.me/key-terms/introduction-civil-engineering/stress-concentration-factors

FURTHER READING

1. Ssempeera, P. (2023). *Integrated Drainage Systems Planning and Design for Municipal Engineers* (1st ed.). CRC Press. https://doi.org/10.1201/9781003255550

Equations and Their Formulation

7

7.1 INTRODUCTION

Throughout the course of history, engineering and mathematics have developed in parallel. All branches of engineering depend on mathematics for their description, and there has been a steady flow of ideas and problems from engineering that has stimulated and sometimes initiated branches of mathematics.

It is essential to note that mathematics is not about formulae and equations! Not at all; it is about ideas! Modern mathematics is a combination of exact formulas and approximate computations [1]. It is about studying formulae and equations and understanding what they mean, but not just calculations. It is about learning and understanding how to formulate equations. It is about formulating equations and never about mere calculations. Failing to do a calculation does not automatically translate to being poor at math. No, not at all! At least not all the time. It may mean you simply do not know the technique used to solve a mathematical problem. And once you get the technique at your fingertips, the rest is history!

Equation formulation is defined as the process of creating an equation by expressing a mathematical relationship between variables, constants, and operations. It is a process of identifying relevant variables (**independent** and **dependent**), defining relationships between them, and expressing relationships mathematically using equations. It involves translating a problem or situation into a symbolic representation that can be solved or analyzed mathematically. This process involves mathematical modeling, representing real-world phenomena or systems, and finally deriving equations that describe behavior, dynamics, or interactions. This process is crucial in all engineering disciplines.

Mathematics is the creative energy of science. Engineers make breakthroughs through mathematical formulations! Mathematics has never been important than today in the era of artificial intelligence (AI) and machine learning (ML). Lots of people learn about formulae and equations and grasp how to use them to bring a desired output. However, they fail to investigate how they are derived, theoretically and empirically. Thus, they fail to appreciate the flexibility in their applications—that is, when and how to modify them depending on unique circumstances and how to derive new techniques to solve them.

This skill is essential for today's engineers because the primary work of an engineer is to apply mathematical techniques to solve societal problems, particularly, as they innovate new solutions to engineering problems. Therefore, as an engineer, learn how equations are built, their differences, how they are modified following units, how they are grouped, and the variances in applications as this enables you to work smart, inspiring others through innovations.

The most versatile mathematical fields that civil engineers harness include probability and statistics, calculus, geometry, linear algebra, graph theory, optimization techniques, topology, numerical methods, and ML. This chapter provides a glimpse or basic knowledge of how to construct equations and possibly understand 'how' and 'when' to apply them and/or not to apply them. You should be able to understand the tricks engineers use to manipulate equations—by manipulating variables. You should also be able to figure out how you can advance new solutions to engineering problems through formulation of equations.

Generally, reading this chapter will help you differentiate between numerical, analytical, and empirical mathematical techniques that engineers use to solve engineering problems and how and when to use them; learn about the basic techniques engineers use to formulate and construct equations to model engineering problems; apply and interpret equations appropriately, recognizing their limitations; identify and exploit mathematical tricks and techniques used by engineers to manipulate equations; appreciate the role of logarithms and graphs in engineering; and appreciate the role of probability and statistics in engineering science.

Mathematics plays an important role in engineering design. It has been used as an engineer's tool to design and communicate engineering information throughout history. As a primary tool for engineers, it provides the analytical and problem-solving tools necessary for designing, analyzing, and optimizing systems, ensuring safety, efficiency, and performance. Engineers use mathematics to make informed decisions and drive technological advancements across various engineering disciplines—civil engineering being one of them.

It is correct to say that mathematical ideas influence engineering innovations, providing the theoretical foundations for new technologies and solutions, and engineering innovations, in turn, have advanced mathematical concepts throughout history, as engineers have developed new mathematical tools and techniques to solve real-world problems. This reciprocity between mathematics and engineering has driven progress in both fields throughout history. Examples of this reciprocity are as follows:

- Calculus, developed by mathematicians and physicists, is widely used in engineering to model and optimize systems.

DOI: 10.1201/9781003561620-7

- Mathematical modeling and simulation have enabled engineers to design and test complex systems, leading to innovations in fields like aerospace and biomedical engineering.
- Engineering applications have led to the development of new mathematical fields, such as topology and graph theory.

As engineering continues to evolve, it relies on mathematical ideas to innovate and advance, and in turn, engineering innovations lead to new mathematical discoveries and advancements. So, as a practicing engineer, looking through the whole field of mathematics and what happens with mathematicians around the globe to generate ideas is not such a bad idea! That's curiosity!

Therefore, mathematics and engineering are closely related. Engineering mathematics, as it is usually called, is an interdisciplinary subject that helps engineers deal with practical, theoretical, and other considerations outside of their specialization.

Engineering mathematics is a collection of mathematical concepts that find applications in several branches of engineering. Engineers extract from the whole field of mathematics what is needed for use in design, construction, operation, and maintenance of engineering systems. Thus, they are able to define and redefine the scope of engineering mathematics as it constantly evolves. Engineering itself involves using natural science, mathematics, and the engineering design process to solve technical problems and improve systems.

The ability to appropriately map practical applications onto analytical models is a key part of becoming a competent engineer and is essential for any engineering analysis. An intuitive understanding for how well analytical models approximates real-world structural behavior, for example, is something that comes with practice and experience. Intuition is a product of experience! See Chapter 9.

Please note that a formal training in engineering largely focuses on developing knowledge of **analytical models** and how to manipulate them. During the first 3–5 years as a practicing engineer, the intuition for relating real-world behavior to mathematical models is developed. Here, you largely learn more about **empirical** and **numerical** techniques. This is one of the reasons why it typically takes a minimum of 4–5 years of post-graduation experience before you can become a professional engineer.

Mathematics is a fascinating discipline. We have three primary mathematical methods or techniques that engineers use to solve engineering problems: **analytical, empirical, and numerical mathematical methods/techniques.** It is important to learn about these techniques.

7.2 MATHEMATICS: A PRIMARY TOOL FOR CIVIL ENGINEERS

Mathematics is the language engineers use to communicate. It is the foundation upon which innovative designs are built and the key to simulating and optimizing complex systems. And its reach extends far beyond engineering—it impacts technology, transportation, healthcare, finance, and so much more.

Mathematics is often referred to as the universal language of science, and for good reason. Mathematics is the foundation on which engineering principles, innovations, and technological advancements are built. It plays a crucial role in engineering, which, in turn, touches every aspect of our daily lives. Mathematics is a primary tool for engineers—it is very helpful in communication, modeling phenomena, interpreting data, driving innovation, and optimizing design, among many other roles.

7.2.1 Communication

Unlike other professions, engineers primarily communicate their work through mathematics. An engineer's report without some kind of calculation is definitely boring. Engineer's reports typically compare one thing against the other, relate one thing with another, and/or show how things measure up! Mathematics helps piece things together logically and communicate them effectively.

Mathematics helps engineers solve complex problems, think critically, and communicate precisely. Engineers know that mathematics is the backbone of engineering. It provides a precise and concise way to express theories and complex ideas, which allows engineers to communicate and collaborate effectively.

Mathematics enables engineers to communicate complex ideas effectively, which would otherwise be difficult or lengthy to share without mathematical illustrations. Mathematics as a universal language enables engineers to share and discuss their ideas, designs, and solutions logically.

An engineer's report, with mathematical illustrations, can be brief, time-saving, and can actually say much with very few words, characters, or text.

7.2.2 Modeling

Engineers use models to design systems. They actually model real-world problems, and mathematics helps them describe and analyze complex systems, making it easier to understand and predict their behavior.

Mathematical models and equations are the tools engineers use to design, analyze, and optimize structures, systems, and processes.

From calculating forces in bridges to simulating fluid flow in pipelines, mathematics enables engineers to predict and understand the behavior of physical phenomena accurately. Without mathematics, engineering would lack the precision and rigor to achieve safe and efficient solutions.

Without mathematics, the models cannot be synchronized. Mathematics provides the foundation for modeling and simulation, enabling modelers to create accurate, reliable, and predictive models that can be synchronized and used to inform decision-making. Mathematical modeling and simulation are crucial tools in engineering fields.

7.2.3 Prediction

Making predictions is possible with mathematical calculations. Engineers are interested in predicting how their designs, products, and services would render value to the clients, and this is possible with mathematical equations, calculations, illustrations, and/or analysis.

Mathematical models and calculations enable engineers to make predictions and optimize designs with considerable levels of precision and accuracy. Engineers can simulate and analyze various scenarios with mathematical equations and algorithms without expensive and time-consuming physical prototypes.

Mathematical models enable engineers to understand the behavior of systems, predict outcomes, and optimize designs before committing to physical construction. Using mathematical simulations, engineers can predict potential material failures.

7.2.4 Design Optimization

As outlined in Section 7.2.3, you cannot optimize designs without mathematical analysis, modeling, and evaluations. Mathematics allows engineers to evaluate different design options, optimize performance, and minimize costs. An optimal design refers to a design that weighs costs, efficiency, effectiveness, and environmental impact. That is primarily balancing competing factors, such as cost vs. performance or functionality vs. aesthetics—**trade-off analysis** (see Section 9.2.11.3).

Mathematics helps optimize structures' shape and dimensions, predict materials' performance, and improve energy efficiency.

An optimal design refers to a design that achieves the best possible outcome, given a set of constraints, requirements, and objectives. It basically meets or exceeds requirements; satisfies the necessary conditions, specifications, and standards; minimizes or optimizes constraints—effectively manages limitations, such as cost, time, resources, or environmental impact; maximizes performance and efficiency; and achieves the best possible outcome, whether it's related to functionality, usability, sustainability, or other key performance indicators (KPIs).

Engineers rely on mathematical concepts, such as calculus, algebra, statistics, and differential equations, to analyze and solve complex problems. These mathematical tools allow engineers to model real-world phenomena, design structures, predict behaviors, and optimize systems.

7.2.5 Interpreting Data

Mathematics enables engineers to interpret data and make sense out of it. Mathematics helps engineers extract insights from data, making informed decisions and improving their designs. Mathematics fosters problem-solving skills, logical reasoning, and analytical thinking skills that are valuable in various engineering disciplines. Engineers deal with large data sets, and they cannot make sense out of it without mathematical tools. Statistical analysis, for example, helps in interpreting data and making sense of trends and patterns.

7.2.6 Driving Innovation

The power of mathematics pushes the boundaries of engineering! Mathematics enables engineers to explore new ideas, simulate novel designs, and push the boundaries of what is possible.

Whether it's calculating forces on a structure, predicting the behavior of a fluid flow in pipelines, or designing a retaining wall, mathematics provides engineers with the necessary framework for innovation and problem-solving.

Through mathematical formulations, engineers can express physical laws, develop mathematical models—analytical, numerical, or empirical models—and simulate scenarios to ensure the reliability and efficiency of their designs.

Mathematics is the backbone of innovation and technological advancements across various civil engineering disciplines. And today, that information technology has dramatically changed the face of civil engineering: mathematics is accelerating interdisciplinary learning and hence fostering rapid progress in innovations.

You notice that, for instance, in computer science and information technology, mathematics underlies algorithms, cryptography, data analysis, and AI. See Chapters 9 and 10. These mathematical concepts enable the development of efficient algorithms for data processing, secure communication protocols, and ML algorithms, all relevant to engineers, especially civil engineers.

7.2.7 Breaking Down Complex Problems

Even if you are not an engineer, problem-solving skills are essential to success. Mathematics is a subject that can help you develop and nurture these skills as it teaches individuals to break down complex problems into manageable components, develop logical reasoning, and formulate strategies to find solutions. The problem-solving abilities learned through mathematics are transferable and applicable to various real-life scenarios, making it understandable.

7.3 PRIMARY MATHEMATICAL TECHNIQUES FOR ENGINEERS

Three primary mathematical methods or techniques every engineer must know and use to solve engineering problems are **analytical, empirical, and numerical mathematical methods/techniques**. It is quite important to differentiate between these three techniques because as an engineer from any discipline, you must understand what they mean, how they are derived, how they are used, and why they are used.

7.3.1 Analytical Methods

Using analytical methods is a problem-solving approach that involves using established mathematical rules and operations (such as algebra, calculus, and trigonometry) to manipulate equations and derive an exact, explicit solution for a given variable.

In applied mathematics, the phrase analytical methods refer to using manipulation of formulae and equations (using rules of algebra, trigonometry, calculus, etc.) to solve for a particular variable, giving yourself a straightforward expression to calculate its value. These methods provide **exact solutions** to a mathematical problem. They provide exact solutions and provide **exact, closed-form solutions** to mathematical problems. For example, an analytical method uses the fundamental theorem of calculus to find the exact, closed-form solution. They are based on theoretical foundations, theoretical frameworks and principles, axioms, and assumptions. They typically involve a symbolic manipulation of mathematical expressions, using techniques like algebra, calculus, and geometry.

Analytical methods have universal applicability—they can be applied to a wide range of problems, without relying on specific data or observations. Differential equations, which are one of the chief equations that engineers deal with, are well known for having analytical solutions, meaning that some differential equations used to model real-world scenarios can be solved by analytical methods. In differential calculus, an analytical solution is preferable to a numerical one because an analytical solution is a mathematical function, and so the numerical value of the dependent variable can be computed for any value of the independent variable.

7.3.1.1 Example [7.1]: Linear equation

Solving for x in a linear equation (7.1),

$$2x + 7 = 11 \tag{7.1}$$

We can use an analytical method to isolate x by subtracting 7 from both sides and then dividing by 2, resulting in the exact solution $x = 2$.

7.3.1.2 Example [7.2]: Solution by inspection for first-order differential equation

Find the solution for the differential equation (7.2):

$$\frac{dx}{dt} = -4x \tag{7.2}$$

Recall, by inspection,[1] that if $x(t) = e^{-4t}$, then

$$\frac{dx}{dt} = -4e^{-4t} = -4x \tag{7.3}$$

In other words, the function $x(t) = e^{-4t}$ is the analytical solution of the differential equation

$$\frac{dx}{dt} = -4x \tag{7.4}$$

7.3.1.3 Example [7.3]: Exact first-order differential equation

Solve an exact first-order differential equation (7.5)

$$\left(\ln\sin t - 2x^2\right)\frac{dx}{dt} + x\cot t + 4t = 0 \tag{7.5}$$

Solution

Since

$$\frac{\partial}{\partial t}\left(\ln\sin t - 2x^2\right) = \cot t \tag{7.6}$$

and

$$\frac{\partial}{\partial x}(x\cot t + 4t) = \cot t \tag{7.7}$$

That means an appropriate function $h(x, t)$ may exist, in such a way that the variable x is a function of t, and that by chain rule of differentiation,

$$\frac{dh}{dt} = \frac{\partial h}{\partial x}\frac{dx}{dt} + \frac{\partial h}{\partial t} \tag{7.8}$$

Now, if a first-order differential equation is of the form,

$$p(t, x)\frac{dx}{dt} + q(t, x) = 0 \tag{7.9}$$

And the function $h(x, t)$ can be found such that

$$\frac{\partial h}{\partial x} = p(t, x) \tag{7.10}$$

and

$$\frac{\partial h}{\partial t} = q(t, x) \tag{7.11}$$

Then, equation 7.9 is equivalent to the equation

$$\frac{dh}{dt} = 0 \tag{7.12}$$

And the solution must be $h(t, x) = C$.

Now,

$$\frac{\partial h}{\partial x} = \ln\sin t - 2x^2 \tag{7.13}$$

gives

$$h = x\ln\sin t - \frac{2}{3}x^3 + C_1(t) \tag{7.14}$$

and

$$\frac{\partial h}{\partial t} = x\cot t + 4t \tag{7.15}$$

gives

$$h = x\ln\sin t + 2t^2 + C_2(x) \tag{7.16}$$

where $C_1(t)$ and $C_2(x)$ are arbitrary functions of t and x, respectively. Note that these functions play the same role in the integration of partial differential relations as the arbitrary constants do in the integration of ordinary differential relations. Comparing the two results, we observe that

$$h(t, x) = x \ln \sin t - \frac{2}{3}x^3 + 2t^2 \qquad (7.17)$$

satisfies equations (7.10) and (7.11), and so, the solution of the differential equation is

$$x \ln \sin t - \frac{2}{3}x^3 + 2t^2 = C \qquad (7.18)$$

The solution here is not an explicit expression for $x(t)$ and t. Instead, it's a cubic polynomial in x, where the coefficients change based on t. If an initial condition had been given, say

$$x\left(\frac{\pi}{2}\right) = 3 \qquad (7.19)$$

To apply the initial condition, we would plug the given values into the implicit equation. This would allow us to solve for the constant of integration, C.

Thus, $x\left(\frac{\pi}{2}\right) = 3$, which gives

$$3 \ln \sin \frac{\pi}{2} - \frac{2}{3} \cdot 3^2 + 2 \times 0^2 = C \qquad (7.20)$$

or

$$0 - 6 + 0 = C \qquad (7.21)$$

So,

$$x^3 - x \ln \sin t - 2t^2 - 6 = 0 \qquad (7.22)$$

Conclusion

All examples, **7.1**, **7.2**, and **7.3** are solved using analytical methods.

7.3.2 Empirical Methods

Empirical methods in mathematics are techniques that use direct observation and experimentation to solve problems and answer questions. Graphs are usually plotted following experiments and observations and empirical equations derived.

Empirical methods involve collecting data through observation or experimentation. They are data-driven and based on observations, experiments, and data collection. In many cases, empirical methods collect data points along the curve and use statistical analysis to approximate the area.

Empirical methods use objective, quantitative observations. Empirical methods establish approximate relationships between variables, using statistical analysis and curve fitting. They are designed to be replicable so that other scientists can validate the findings. They have practical applications more often used in applied fields such as engineering, economics, and social sciences, where data are abundant.

Sometimes, they have limited generalizability and may not be universally applicable as the results are specific to the data and context.

Important: Geotechnical engineering is one of the civil engineering fields well known for heavily relying on empirical methods.

7.3.2.1 Example [7.4]: Manning's equation

The equation is used to solve **open-channel flow problems**. Consider Manning's formula which expresses flow velocity with channel geometry, that is,

$$V = \frac{1}{n}R^{2/3}S^{1/2} \qquad (7.23)$$

Check the dimensional consistency,

$$[R] = \frac{[A]}{[P]} = \frac{M^2}{M} = M$$

$$\left[R^{2/3}\right] = M^{2/3}$$

$$[n] = \text{dimensionless}$$

$$[S] = \text{dimensionless}$$

However,

$$[V] = M / T$$

Therefore, by the rule of thumb, dimensions on the left are not always necessarily the same as those on the right:

$$\text{LHS} \neq \text{RHS}$$

This means that the formula is empirically derived from curve fitting, observing phenomena, experimentation, and studying several equations that were combined to make an approximate equation for solving open-channel flow problems. To comply, the equation must be customized. In analytical methods, equations/formulae are predominantly dimensionally consistent and customizations are pretty straightforward. In this case, to make the formula (7.9) applicable, units must be observed. The formula in equation (7.23) has two versions: the metric and English system. Equation (7.23) is the metric version, and the velocity obtained is expressed in m/s. The English version stands as in equation (7.24):

$$V = \frac{1.49}{n}R^{2/3}S^{1/2} \qquad (7.24)$$

Note the 1.49 coefficient (conversion factor) introduced.

We note that from Manning's equation, the product of the cube root of the square of the hydraulic radius and square root of the slope of the channel and the inverse of the coefficient of the channel provides a flow velocity estimate. The question is, what about the units—are they considered? Therefore, it is an approximate equation obtained from empirical data collection—basically through observation, experimentation, and curve fitting.

7.3.3 Numerical Methods

As far as engineering mathematics is concerned, these are the mostly used techniques in engineering. These are a set of techniques used to approximate solutions to mathematical problems by performing calculations with numbers, particularly when exact solutions are difficult or impossible to find analytically, often relying on iterative processes and computer algorithms to achieve an accurate result; essentially, it's a way to solve complex problems using numerical approximations instead of exact formulas.

Numerical methods are techniques to approximate mathematical processes (examples of mathematical processes are integrals, differential equations, and nonlinear equations' evaluations).

To make it simpler: numerical methods fill the gap when analytical solutions are elusive, driving research and inspiring new analytical breakthroughs. They are basically open for research. Numerical methods are used as a 'stand-in' for unknown or elusive analytical solutions. Numerical methods are often used when analytical solutions are unknown, impractical, or too complex, and they can provide approximate solutions, guide research, and inspire new analytical techniques.

In the case of differential equations, where we may not be able to find an analytical solution in such a form—either because no suitable mathematical technique for finding the solution exists or because there are no analytical techniques—we often use a numerical solution.

To obtain a numerical solution to a differential equation, for example, boundary conditions must be defined. This interdependence highlights the value of considering the entire problem—including both the differential equation and its boundary conditions—as an integrated unit, rather than treating them separately.

The numerical method uses a numerical algorithm, such as the trapezoidal rule or Simpson's rule, to approximate the area using discrete, numerical representations. Numerical analysis finds application in all fields of engineering and the physical sciences. They approximate solutions to mathematical problems using numerical algorithms and computational techniques.

They have discrete representations, which means they represent continuous problems using discrete, numerical representations, such as grids, meshes, or finite elements. What about the finite-element method? They have computational efficiency, allowing for fast and accurate solutions to complex problems. They can be applied to a broad range of problems, including those that are difficult or impossible to solve analytically.

7.3.4 Example [7.5]: Numerical Method Example

Find an approximation to

$$\int_0^2 2x^2 dx \qquad (7.25)$$

using the trapezium rule, with four equally wide strips.

TABLE 7.1 Values of x_i and $f(x_i)$

x_i	0	0.5	1	1.5	2
$f(x_i)$	0	0.5	2	4.5	8

Solution

With four strips, we need five points. The points are 0, 0.5, 1, 1.5, and 2. Table 7.1 shows both x_i and $f(x_i)$:

By the given formula,

$$h = \frac{b-a}{n} = \frac{2-0}{4} = \frac{1}{2} \qquad (7.26)$$

That means our approximation to the integral is given by the following:

$$\int_0^2 2x^2 dx \approx h\left\{\frac{1}{2}\left[f(x_0)+f(x_n)\right]+\sum_{i=1}^{n-1}f(x_i)\right\}$$

$$= \frac{1}{2}\left\{\frac{1}{2}[0+8]+(0.5+2+4.5)\right\} = 5.5 \qquad (7.27)$$

Note: If we were to evaluate this integral '**analytically**,' we would obtain

$$\frac{16}{3}$$

which is close to 5.5, which shows this is a good approximation. In many cases, numerical methods are used where analytical solutions do not exist! These are very common in water resource engineering and geotechnics, among others.

7.4 FORMULAE AND EQUATIONS

An equation and a formula are related but distinct concepts in mathematics. What is most important is knowing that a formula, as it is sounded, is a procedure for calculating a specific value or outcome, often used to solve problems, make predictions, or model real-world phenomena.

7.4.1 Equation and Its Characteristics

An equation is a statement that expresses the equality of two mathematical expressions. An equation is analogous to a scale into which weights are placed; see Figure 7.1. When equal weights of something are placed into the two pans, the two weights cause the scale to be in balance and are said to be equal. If a quantity of grain is removed from one pan of the balance, an equal amount of grain must be removed from the other pan to keep the scale in balance. More generally, an equation remains in balance if the same operation is performed on its both sides.

An equation typically contains variables, constants, and mathematical operations. Solving an equation involves finding the value(s) of the variable(s) that make the equation true.

FIGURE 7.1 Weighing scale.

In the quadratic equation (7.28), a, b, and c are constants, x is a variable, while $+$, $-$, and $=$ are mathematical operations.

$$ax^2 + bx - c = 0 \tag{7.28}$$

$$ax^2 = c - bx \tag{7.29}$$

7.4.2 Formula and Its Characteristics

A formula is a standard equation (specialist equation) universally acceptable for obtaining a specific outcome, e.g., the area of a circle is given by the formula.

A formula, as it is sounded, is used for calculating a specific value or outcome, often used to solve problems, make predictions, or model real-world phenomena. In other words, it is a general rule or expression that describes a mathematical relationship or operation. It is often used to calculate a specific quantity or value and typically involves variables, constants, and mathematical operations. Consider formula (7.12) for calculating the area of a circle:

$$A = \pi r^2 \tag{7.30}$$

In the formula (7.12),

A: Area of a circle
π: Pie, an irrational number/constant equal to 3.142….
r: Radius of the circle

Consider formula (7.13) that describes the relationship between the volume and radius of a circle and height of cylinder, allowing you to calculate the volume for any given radius and height of a cylinder.

$$V = \pi r^2 h \tag{7.31}$$

In the formula,

V: Volume of a cylinder
π: Pie, an irrational number/constant equal to 3.142….
r: Radius of the circle
h: Height of the cylinder

7.4.3 Key Differences between Formula and Equation

See Table 7.2.
Important

- A formula can be considered a type of equation. Therefore, formulas are a subset of equations, and while there is some overlap, not all equations are formulas.
- All formulas are equations, but not all equations are formulas.
- A formula is an equation that has a specific purpose or application.

7.4.4 Theoretical Equations

Theoretical equations are mathematical expressions derived from underlying theories or principles, aiming to explain and predict phenomena. They are derived by concepts of mathematics, and a logical explanation is possible. The theoretical equation or formula can be derived and proved.

Deriving theoretical equations from a theory typically involves several stages: starting with a clear understanding of the underlying theory, identifying relevant assumptions, using mathematical techniques to manipulate the theory's core principles, and finally, formulating a new equation that encapsulates the relationships derived. Standard equations are built or formulated from scientific theories through a process that involves the several stages as follows.

TABLE 7.2 Key differences between a formula and an equation

NO.	EQUATION	FORMULA
Purpose	Equations aim to solve for unknown values. For example, the equation: $x + 2 = 6$ $x = 4$ Solves the unknown value, 4. • Solve for unknown values. • Describe a specific situation or problem. • Represent a mathematical statement.	Formulas provide a general rule for calculation. For example, area of a circle: $A = \pi r^2$ With a circle of 10 m, the area would be 314.159 m²
Structure	Equations **typically** have an equal sign (=).	Formulas often use equal sign (=) and other operators (e.g., =, +, −, ×, /).
Application	Equations are used to solve specific problems.	Formulas are used to describe general relationships or perform calculations.

7.4.4.1 Theoretical framework

A scientific theory provides a conceptual foundation, describing the underlying principles and relationships—that is, understand the theory. Therefore, begin by thoroughly understanding the fundamental principles and assumptions of the theory. Then, know the theory's domain of applicability and possible limitations. This is accompanied by identifying the key concepts and variables within the theory that are relevant to the problem you're trying to address.

7.4.4.2 Identify relevant assumptions

Simplifications and assumptions are made to facilitate mathematical treatment, such as linearization, idealizations, or neglecting certain factors. You need to clearly state the assumptions made within the theory because they can significantly impact the derived equations. If the theory involves models, then it is necessary to understand how these models simplify the real-world system.

7.4.4.3 Mathematical formulation

Theoretical concepts are translated into mathematical language, using variables, constants, and mathematical operations. Equations translate the theory's principles into mathematical equations. We employ appropriate mathematical techniques, for example, differential equations, calculus, and algebra, to manipulate these equations.

Then, we derive the equation. The necessary steps are performed to derive the desired equation, combining the initial equations and simplifying them as needed. Mathematical derivations are performed to obtain the desired equation, often involving techniques such as differentiation, integration, or algebraic manipulation.

Empirical validation can also be considered. The resulting equation is tested against experimental data, if possible, ensuring it accurately predicts real-world phenomena. Following the empirical validation, the equation may be refined or modified based on new experimental evidence, leading to a more accurate or comprehensive representation of the theory in a mathematical sense.

7.4.4.4 Formulating the equation

The final step is to present the derived equation in a clear and concise manner. Interpret the derived equation in the context of the original theory and the problem you are trying to solve. If possible, validate the derived equation by comparing it with experimental data or other existing theoretical predictions.

7.4.4.5 Examples of theoretical equations

Some civil engineering examples of theoretical equations include those derived from Newton's laws of motion, Hooke's law, laws of thermodynamics, **conservation equations**, and energy and momentum equations, among others. Therefore, scientific theories are transformed into mathematical equations, enabling precise predictions, simulations, and problem-solving across various civil engineering disciplines.

7.4.4.5.1 Force equation from Newton's second law of motion

Newton's second law of motion states that 'Force acting on a body is equal to its rate of change of momentum.' Thus,

$$F = ma \tag{7.32}$$

in which

F: Force acting on a body
m: Mass of the body
a: Acceleration of the body

In other words,

$$F = \frac{d\rho}{dt} \tag{7.33}$$

$$\rho = mv; (\Delta v / \Delta t = a) \tag{7.33}$$

$$\rho = \text{momentum}$$

The equation/formula (7.32) can be proved by theoretically.

7.4.4.5.2 Rational formula for peak flow discharge computation

This formula (7.34) expresses the rainfall intensity and catchment area as independent variables:

$$Q = 0.278\text{CiA} \tag{7.34}$$

See Section 7.6.9.5 for the derivation of equation (7.34).

7.4.5 Empirical Equations

Empirical equations are based on observed data without necessarily having a theoretical basis. An empirical equation is not based on first principles formed from the theories of science and physics. It is derived from curve fitting to observed data. Several equations that engineers use to solve engineering problems are empirically derived.

Empirical means computed directly from the data. The word empirical means the following: based on, concerned with, or verifiable by observation, or experience rather than theory or pure logic. Characteristics of an **empirical equation**:

• It is not derived by mathematics directly.
• It is derived from experimental data collection and data analysis.
• This equation does not have a mathematical proof.
• Usually, it is an equation of a graphical curve.

7.4.5.1 Examples of empirical equations

This section outlines some common empirical equations in civil engineering and a brief history of their derivation, as appropriate.

7.4.5.1.1 Manning's equation

Manning's equation is a famous equation in water resource engineering and hydraulics. Robert Manning expressed the velocity of flow with the channel's geometry—applicable in open-channel flows. Manning's equation is derived from the combination of several empirical and theoretical considerations.

The Manning equation is a widely used formula in water resource engineering and hydraulics. It can be used to compute the flow in an open channel, derive the capacity of a pipe, compute the friction losses in a channel, check the performance of an area–velocity flow meter, and has many more applications.

The famous equation (7.35) was developed by the Irish engineer Robert Manning (1816–97), and its first version appeared in 1891 in Manning's paper 'On the flow of water in open channels and pipes' (*Transactions of the Institution of Civil Engineers* of Ireland). This equation is expressed as follows (metric system):

$$V = \frac{1}{n} \cdot \left(\frac{A}{P}\right)^{2/3} \cdot S^{1/2} \tag{7.35}$$

in which

V: Velocity of flow
n: Manning roughness coefficient
A: Cross-sectional area
P: Wetted perimeter
S: Channel slope

Robert Manning launched his career in 1846, joining the Arterial Drainage Division of the Irish Office of Public Works. He quickly rose through the ranks, becoming an assistant engineer to Samuel Roberts later that year, and then district engineer from 1848 to 1855.

During this period, he developed a strong interest in hydraulics after reading *Traité d'Hydraulique*. Manning then worked for the Marquis of Downshire from 1855 to 1869, overseeing the construction of the Dundrum Bay Harbor in Ireland and designing a water supply system for Belfast. He returned to the Irish Office of Public Works in 1869 and eventually became chief engineer in 1874, a position he held until his retirement in 1891. Manning did not receive any education or formal training in fluid mechanics or engineering. His accounting background and logic influenced his work and drove him to reduce problems to their simplest form.

What did Manning do exactly?

Manning compared and evaluated seven best-known formulas of the time, that is, Du Buat (1786), Eyelwein (1814), Weisbach (1845), St. Venant (1851), Neville (1860), Darcy and Bazin (1865), and Ganguillet and Kutter (1869) [2]. He then calculated the velocity obtained from each formula for a given slope and for hydraulic radius varying from 0.25 to 30 m. Then, for each condition, he found the mean value of the seven velocities and developed a formula that best fitted the data. The first best-fit formula he developed was as follows:

$$V = 32\left[RS\left(1 + R^{1/3}\right)\right]^{1/2} \tag{7.36}$$

He then simplified this formula to

$$V = CR^x S^{1/2} \tag{7.37}$$

In 1885, Manning gave x the value of 2 / 3 and hence wrote his formula as follows:

$$V = CR^{2/3}S^{1/2} \tag{7.38}$$

In a letter to Flamant, Manning stated, 'The reciprocal of C corresponds closely with that of n, as determined by Ganguillet and Kutter; both C and n being constant for the same channel.' On 4 December 1889, Robert Manning, then 73 years old, initially presented his groundbreaking formula to the Institution of Civil Engineers (Ireland). Two years later, in 1891, Manning's formula was formally introduced in his seminal paper, 'On the flow of water in open channels and pipes,' which was published in the *Transactions of the Institution of Civil Engineers* (Ireland). Manning did not like his own equation (7.38) for two reasons:

- First, it was difficult in those days to determine the cube root of a number and then square it to arrive at a number to the 2 / 3 power. Arriving at a cube root was not a trivial concern at that time.
- In addition, the equation was dimensionally inconsistent, and so to obtain dimensional correctness, he developed the following equation:

$$V = C\left(gS\right)^{1/2}\left[R^{1/2} + \left(0.22 / m^{1/2}\right)\left(R - 0.15m\right)\right] \tag{7.39}$$

where m is the height of a column of mercury which balances the atmosphere and C is a dimensionless number 'which varies with the nature of the surface.'

He did not mention his first equation (7.38) in his 1891 paper but rather mentioned equation (7.39) and named it after him. It is interesting to see that others preferred the earlier version of the equation (7.38) he initially presented to the Institution of Civil Engineers (Ireland). And in some late 19th-century textbooks, the Manning formula was written as follows:

$$V = \frac{1}{n}R^{2/3}S^{1/2} \tag{7.40}$$

In his *Handbook of Hydraulics*, King [3] popularized Manning's monomial formula as we know it today, as well as to the acceptance that Manning's coefficient C should be the reciprocal of Kutter's n, hence leading to the widespread use of the formula. From then, n is referred to as Manning's friction factor, or Manning's constant.

It is interesting to see that an equation that was rejected by its author who was an accountant-turned-self-taught engineer as a result of the Irish famine regained widespread adoption.

Key interesting facts

- This empirical equation was developed by aggregating seven best formulas known at the time. See 'On the flow of water in open channels and pipes,' which was published in the *Transactions of the Institution of Civil Engineers* (Ireland) in 1891.
- Manning had mixed thoughts about the dimensional consistency of his initial equation, which led him to seek a more dimensionally consistent formula. So, dimensional consistency is essential in empirical equations. That was the primary reason that diverted him from his first equation trying to pursue

a better dimensionally consistent formula. However, others preferred his first equation which gained widespread adoption. Note,

- **Metric system**

$$V = \frac{1}{n} R^{2/3} S^{1/2} \qquad (7.41)$$

- **English system**

$$V = \frac{1.49}{n} R^{2/3} S^{1/2} \qquad (7.42)$$

- Manning's empirical equation aimed to optimize velocity predictions based on channel geometry, synthesizing the results from seven existing formulas. Its development relied on observational data, and the incorporation of cube roots and square roots was an empirical discovery.
- This derivation combines theoretical and empirical elements to arrive at Manning's equation, widely used for open-channel flow calculations. In fact, some equations, for example, Chezy's or Darcy–Weisbach, are semi-empirical. While these other equations, that is, Chezy's or Darcy–Weisbach, may be more theoretically sound or accurate in specific cases, Manning's equation offers a practical balance of simplicity, flexibility, and reliability, making it a popular choice for many hydraulic engineering applications.

7.4.5.1.2 Flexural strength

Flexural strength, also known as bending strength, is a measure of a material's ability to resist deformation and fracture when subjected to bending forces. See Section 5.5.3.6. It is an important mechanical property of materials, particularly in structural and civil engineering applications. Flexural strength is defined as the maximum stress a material can withstand without failing when subjected to a bending load. It is typically measured in units of force per unit area, such as pounds per square inch (psi) or pascals (Pa):

$$f_{cr} = 0.7 \left(f_{ck} \right)^{0.5} \qquad (7.43)$$

where

- f_{cr} is the flexural strength
- f_{ck} is the characteristic cube compressive strength of **concrete** in N/mm²

Note

$$E_c = 5{,}000 \left(f_{ck} \right)^{0.5} \qquad (7.44)$$

E_c is the short-term static modulus of elasticity in N/mm². Both these formulas are empirically derived!

Note

- The above equations are derived from experience and observed data.
- They are derived from graphs and experiments.
- They are approximate values.

7.4.5.1.3 Bearing capacity equations (i.e., Terzaghi's equation)

Bearing capacity q_f is defined as the pressure that would cause shear failure of the supporting soil immediately below and adjacent to a foundation. It is the capacity of soil to support the loads that are applied to the ground above. It depends primarily on the type of soil, its shear strength, and its density. It also depends on the depth of embedment of the load—the deeper it is founded, the greater the bearing capacity. The bearing capacity of a shallow foundation on an undrained material can be written in a generalized form as

$$q_f = s_c N_c c_u + \sigma_q \qquad (7.45)$$

in which

q_f: Bearing capacity
s_c: Shape factor
N_c: Bearing capacity factor
c_u: Undrained shear strength
σ_q: Total surcharge pressure

Equation (7.45) is derived from principles of effective stress. The importance of the forces transmitted through the soil skeleton from particle to particle was recognized by Terzaghi [4], who presented his principle of effective stress, an intuitive relationship based on experimental data and observations. The principle applies only to fully saturated soils and relates the following three states:

- The **total normal stress (σ)** on a plane within the soil mass, being the force per unit area transmitted in a normal direction across the plane, imagining the soil to be a solid (single-phase) material;
- The **pore water pressure (u),** being the pressure of the water filling the void space between the solid particles;
- The **effective normal stress (σ′)** on the plane, representing the stress transmitted through the soil skeleton only (i.e., due to interparticle forces).

The relationship is

$$\sigma = \sigma' + u \qquad (7.46)$$

Everything was built on Terzaghi's principle of effective stress; the pore water pressures contributing resistance within the soil mass, equation (7.46). The bearing capacity of drained soils is different from that of undrained soils. Through empirical observations and experimentation, scholars and practitioners identified that the **shape of footing** influences the bearing capacity and that surcharge loads influence the bearing capacity, and depth and load inclination need to be taken into consideration. Bearing capacity factors (N_u, N_γ, and N_c) are dimensionless coefficients used in the Terzaghi bearing capacity equation to calculate the ultimate bearing capacity q_f of a shallow foundation in drained conditions. Thus, the equation is semi-empirical, combining theoretical concepts with **empirical coefficients** derived from experimental data and observations.

Role of bearing capacity factors

1. They account for soil strength. (N_q, N_γ, and N_c) represent the influence of soil strength parameters (effective cohesion, c', and effective friction angle, $\varnothing'$) on the bearing capacity.
2. They incorporate foundation geometry. The factors consider the shape and size of the foundation, such as the width (B') and depth (D').
3. They represent the failure mechanism. Each factor corresponds to a specific failure mechanism:
   - N_c represents the cohesive component of soil strength.
   - N_q represents the surcharge component of soil strength.
   - N_γ represents the weight component of soil strength.

Values of bearing capacity factors

The values of (N_q, N_γ, and N_c) depend on the (**effective internal angle of shearing resistance**) effective friction angle ($\varnothing'$) of the soil. These values can be determined from tables, charts, or equations, such as:

- Terzaghi's original values (1925).
- Hansen's values (1970).
- Vesic's values (1973).

Engineers typically use pre-calculated tables or software to determine the bearing capacity factors for a specific soil type and foundation geometry. By incorporating bearing capacity factors into the Terzaghi equation (7.47), engineers can estimate the ultimate bearing capacity of a shallow foundation and ensure a safe and stable design.

Note: The engineer must distinguish between undrained conditions (short-term loading, where pore water pressures are present and design is carried out for total stresses on the basis of $\varnothing_u$ and c_u) and drained conditions (long-term loading, where pore water pressures have dissipated), and design is carried out for effective stresses on the basis of $\varnothing'$ and c'. Thus, the bearing capacity of a shallow foundation (strip footing) on a **drained material** can be written in a generalized form as

$$q_f = s_q N_q \sigma'_q + \frac{1}{2}\gamma B s_\gamma N_\gamma + s_c N_c c' \tag{7.47}$$

where

q_f: bearing capacity
s_q: shape factor
N_q: bearing capacity factor
σ'_q: effective surcharge pressure (foundation)
γ: unit weight
B: footing width
s_γ: shape factor (bearing capacity)
N_γ: bearing capacity factor
s_c: shape factor (bearing capacity)
N_c: bearing capacity factor
c': cohesion intercept

Geo-laboratories determine the shear strength parameters (cohesion and angle of internal resistance).

Example [7.6]: Shallow foundation bearing capacity determination for illustrative purposes

A footing 2.25×2.25 m is located at a depth of 1.6 m in sand for which $c' = 0$ and $\varnothing' = 38°$. Determine the bearing resistance:

a. If the water table is well below the foundation level, and
b. If the water table is at the surface.

The unit weight of the sand above the water table is 18 kN/m^3; the saturated unit weight is 20 kN / m^3.

Solution

For $\varnothing' = 38°$, the bearing capacity factors are $N_q = 49$ (Equation 7.82) and $N_\gamma = 75$ (Equation 7.84). The footing is square (B/L=1), so the shape factors are $s_q = 1.62$ (from $s_q = 1 + \frac{B}{L}\sin\varnothing'$) and $s_\gamma = 0.70$ from ($s_\gamma = 1 - 0.3\frac{B}{L}$). The values of s_q and s_γ are both conservatives (as $\varnothing' > 30°$). As $c' = 0$, in this case, there is no need to compute N_c and s_c. For the case (a), when the water table is well below the founding plane:

$$q_f = s_q N_q \sigma'_q + \frac{1}{2}\gamma B s_\gamma N_\gamma \tag{7.48}$$

$$q_f = (1.62 \times 49 \times 18 \times 1.6) + (0.5 \times 18 \times 2.25 \times 0.70 \times 75)$$

$$q_f = 3,349.26 \text{ kPa}$$

When the water table is at the surface, the ultimate bearing capacity is given by

$$q_f = s_q N_q \gamma' d + \frac{1}{2}\gamma' B s_\gamma N_\gamma \tag{7.49}$$

$$q_f = \left[1.62 \times 49 \times (20 - 9.81) \times 1.6\right] + \begin{bmatrix} 0.5 \times (20 - 9.81) \times \\ 2.25 \times 0.70 \times 75 \end{bmatrix}$$

$$q_f = 1,896.06 \text{ kPa}$$

Comparing the two results, it can be seen that changing the hydraulic conditions (pore pressures) within the ground has a significant effect on the bearing capacity.

Analysis

See Figure 7.2.

$$q_f = s_q N_q \sigma'_q + \frac{1}{2}\gamma B s_\gamma N_\gamma + s_c N_c c' \tag{7.50}$$

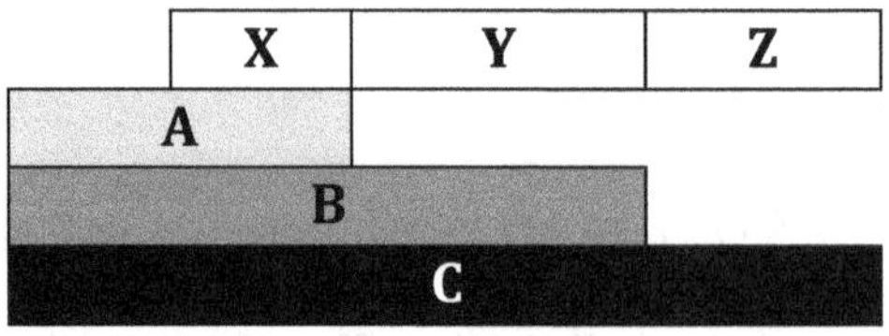

FIGURE 7.2 Bearing capacity equation.

- Z represents the cohesive component of soil strength.
- Y represents the weight component of soil strength.
- X represents the surcharge component of soil strength.

A drops Y and Z components, B drops the Z component, while C drops none. The refinement of this equation (7.31) underwent research and development (R&D) over the years, observing those three components! Thus, its application as illustrated in example **7.6** varies with different scenarios. We noted from example 7.6 that because $c' = 0$, there was no need to compute N_c and s_c. Surprisingly, equation (7.50) could still benefit from further refinement because it still may not account for complex soil behavior. The lesson to comprehend is that empirical formulas/equations are refined through empirical observations and experimentation—with advancements in technology, science, and engineering research. In many cases, refinements are triggered by failures in their applications at some point, thus triggering more research to understand the problems.

7.4.5.1.4 Design of highway intersections
The design of intersections is based on a number of empirical equations.

7.4.5.1.4.1 Average vehicle delays at the approach to a signalized intersection   Webster [5] derived the following equation (7.51) for estimating the average delay per vehicle at a signalized intersection:

$$d = \frac{c(1-\lambda)^2}{2(1-\lambda x)} + \frac{x^2}{2q(1-x)} - 0.65\left(\frac{c}{q^2}\right)^{\frac{1}{3}} x^{(2+5\lambda)} \quad (7.51)$$

where

d = average delay per vehicle (in seconds)
c = cycle length
λ = effective green time divided by cycle time
q = flow
s = saturation flow
$x = \dfrac{q}{\lambda s}$

Note

- The first term in equation (7.51) relates to the delay resulting from a uniform rate of vehicle arrival.
- The second term relates to the delay arising from the random nature of vehicle arrivals.
- The third term is an empirically derived correction factor, obtained from the simulation of the flow of vehicular traffic.

The above formula (7.51) can be simplified as follows:

$$d = cA + \frac{B}{q} - C \quad (7.52)$$

where

$$A = \frac{(1-\lambda)^2}{2(1-\lambda x)} \quad (7.53)$$

$$B = \frac{x^2}{2(1-x)} \quad (7.54)$$

C is the correction term, which can be taken as 10% of the sum of the first two terms. Equation (7.51) can thus be written in approximate form as follows:

$$d = 0.9 \times \left(cA + \frac{B}{q}\right) \quad (7.55)$$

7.4.5.1.4.2 Example [7.7]: Signalized junction average vehicle delay for illustrative purposes   This example presents a calculation of the average vehicle delay at the approach to a signalized junction. An approach has an effective green time of 65 seconds and an optimum cycle time of 100 seconds. The actual flow on the approach is 1,000 veh/h, with its saturation flow estimated at 1,750 veh/h. Calculate the average delay per vehicle using both the precise and approximate formulae.

Solution

$c = 100 \ s$

$\lambda = 0.65$

$q = 1,000 \ \text{veh/h} = 0.278 \ \text{veh/s}$

$s = 1,750 \ \text{veh/h} = 0.486 \ \text{veh/s}$

$$x = \frac{0.278}{(0.65 \times 0.486)} = 0.88 \quad (7.56)$$

$$A = \frac{(1-0.65)^2}{2(1-(0.65 \times 0.88))} = 0.14 \quad (7.57)$$

$$B = \frac{(0.88)^2}{2(1-0.88)} = 3.23 \quad (7.58)$$

$$C = 0.65 \times \left(\frac{100}{0.278^2}\right)^{1/3} (0.88)^{5.25} = 3.6 \quad (7.59)$$

Using the precise formula (7.51):

$$d = \frac{c(1-\lambda)^2}{2(1-\lambda x)} + \frac{x^2}{2q(1-x)} - 0.65\left(\frac{c}{q^2}\right)^{\frac{1}{3}} x^{(2+5\lambda)} \quad (7.60)$$

$$d = (100 \times 0.14) + (3.23 \div 0.278) - 3.6 = 22 \ \text{seconds} \quad (7.61)$$

Using the approximate formula (7.55):

$$d = 0.9 \times \left(cA + \frac{B}{q}\right) = 0.9 \times (14 + 12) = 23 \ \text{seconds} \quad (7.62)$$

Notes
   C, the correction term (see Section 7.6.8), was derived empirically from observation, experimental data, and curve fitting. It is usually taken as 10% of the sum of the first two terms. However, as an example, C *may* vary with advances in technology and engineering R&D. As a constantly evolving civil engineer, with the ability to study equations, you must challenge conventional knowledge as appropriate from time to time. That means being able to conduct market or industry research (see Section 7.7) through the design process is essential. The correction term adjusts for errors or inaccuracies in original equation—factors that were overlooked. Equation (5.1) is used for illustrative purposes.

In Chapter 3, the reasons for moving with latest technology were advanced. From the steam engine to AI and robots, the world has come a long way. If engineers from the 1900s were to return to the present day, they would likely face significant challenges in adapting to modern technologies. For instance, they would encounter unfamiliar tools such as AutoCAD. This highlights the importance for contemporary engineers to invest time in staying abreast of technological advancements and integrating new or existing technologies into their practice. Thus, they can be able to provide effective and up-to-date solutions to engineering problems. Failure to do so may result in difficulties in keeping pace with the evolving demands of the field.

Today, AI is prevalent. To stay ahead, it's essential to understand how AI learns from us, not just us learning from AI. Failing to adapt and leverage AI will make it challenging to collaborate with those who develop and utilize AI to benefit the humanity. Constantly evolving engineers understand how equations are built, used, and their limitations as these are part of the primary tools that open new gates to innovations. This example provides insights into how empirical equations get challenged from time to time, and it is the role of the practicing engineer to recognize drawbacks in their applications.

7.4.5.2 Curve fitting

Curves are fitted to communicate engineering information and to develop empirical formulas and equations. Curve fitting is the process of finding the best-fitting curve (or mathematical function) to a set of data points. This process aims to find a function that closely represents the relationship between the

data points, possibly with some constraints or considerations for noise or outliers. Throughout history, scientists and engineers have relied on experiments and observations to model real-world phenomena. In so doing, they develop graphs and charts (from data they collect) as a tool to ease their work.

The complex graphs are created to communicate engineering information. They highly base on empiricism! Graphs are used to develop approximate methods to civil engineering problems. And they are very essential in civil engineered design, analysis and evaluations, or predictions.

Some of the commonest graphs we know in water resource engineering or hydraulics are the Moody diagram and the nomograms. See Figure 7.3[2] for the Moody diagram. These graphs were developed from empirical observations. The developers plotted data collected to ease their work. Of course, using logarithms, as shown in Section 7.8 to ease the plots. Some of these graphs use logarithmic scales.

7.4.5.2.1 The Moody diagram
Let's use the Moody diagram in fluid mechanics to explain curve fitting. The **Moody chart** or **Moody diagram** (also **Stanton diagram**), Figure 7.3, is a graph in nondimensional form that relates the Darcy–Weisbach friction factor f_d, Reynolds number R_e, and surface roughness for fully developed flow in a circular pipe. It can be used to predict the pressure drop or flow rate down such a pipe.

The chart plots Darcy–Weisbach friction factor f_d against Reynolds number Re for a variety of relative roughness, the mean height of roughness of the pipe to the pipe diameter or $\frac{\epsilon}{d}$. The Moody diagram, also known as the Moody chart, was developed by Lewis F. Moody in 1944. It's a graphical

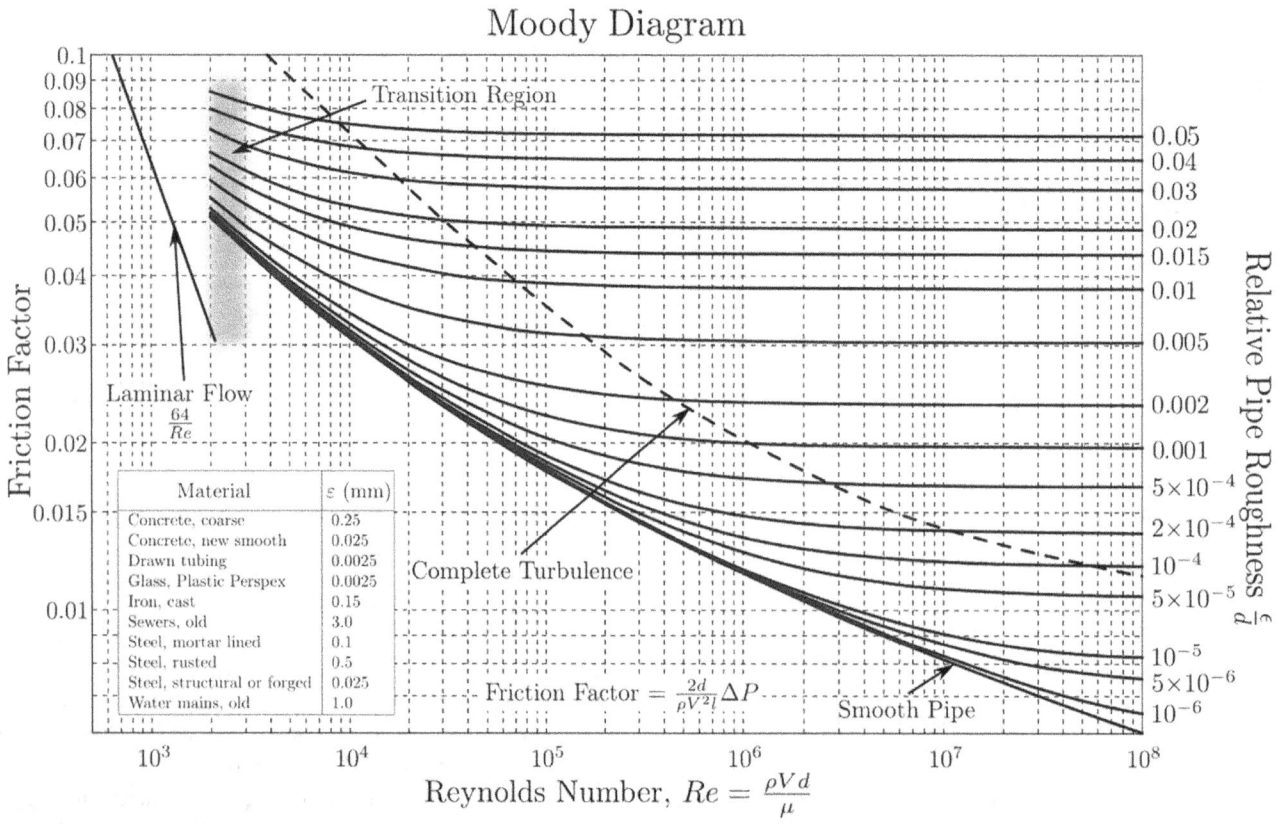

FIGURE 7.3 Moody diagram.

representation of the Darcy–Weisbach equation, used to determine the friction factor (f) in turbulent flow through pipes. It is very helpful in determining the flow resistance and pressure loss. The chart allows engineers to quickly determine the friction factor for a given flow condition, pipe material, and diameter, making it a valuable tool in fluid mechanics and pipe flow calculations.

The **Darcy–Weisbach equation** is an empirical equation that relates the head loss, or pressure loss, due to friction along a given length of the pipe to the average velocity of the fluid flow for an incompressible fluid. The equation is named after Henry Darcy and Julius Weisbach. Currently, there is no formula more accurate or universally applicable than the Darcy–Weisbach supplemented by the Moody diagram or **Colebrook equation**.

Darcy and Weisbach noted that the head loss in pipes was greatly influenced by the friction between the flowing fluid and the pipe material. And that there was a **correction factor** (friction factor) that varied with the pipe material, velocity of flow, viscosity, density of fluid, and pipe geometry and other flow conditions. However, explicit determination of friction factor f_d was not quickly established, until Moody and Colebrook presented the Moody diagram and Colebrook equation, respectively. Today, you can use either the Moody diagram or the Colebrook equation to determine f_d.

Darcy–Weisbach equation

$$\Delta H = f_d \frac{lV^2}{2gd} \tag{7.63}$$

or

$$\Delta P = f_d \frac{\rho lV^2}{2d} \tag{7.64}$$

where

ΔH: head loss (m)
f_D: Darcy friction factor (unitless)
l: length of the pipe (m)
d: diameter of the pipe (m)
V: average velocity of the fluid (m/s)
g: acceleration due to gravity m/s^2
ΔP: pressure drop (head loss) in the pipe (Pa)
ρ: density of the fluid (kg/m^3)

Colebrook equation

For the turbulent flow regime, the relationship between the friction factor f_d, the Reynolds number Re, and the relative roughness ϵ / d is more complex. One model for this relationship is the Colebrook equation (which is an implicit equation in f_d):

$$\frac{1}{\sqrt{f_d}} = -2.0 \log_{10} \left(\frac{\epsilon/d}{3.7} + \frac{2.51}{Re\sqrt{f_d}} \right) \tag{7.65}$$

Both the Colebrook equation and the Moody diagram are used to determine the **Darcy friction factor** (f) in the Darcy–Weisbach equation, which is essential for calculating head loss in pipes.

The Moody chart can be divided into two regimes of flow: laminar and turbulent. For the laminar flow regime

$Re <\sim 3,000$, roughness has no discernible effect, and the Darcy–Weisbach friction factor f_d was determined analytically by Poiseuille:

Laminar flow,

$$f_d = \frac{64}{Re} \tag{7.66}$$

Note

$$Re = \frac{\rho Vd}{\mu} \tag{7.67}$$

The Reynolds number (Re) is a dimensionless quantity that predicts fluid flow patterns in different situations by measuring the ratio between inertial and viscous forces within a fluid that is subjected to relative internal movement due to different fluid velocities. At low Reynolds numbers, flows tend to be dominated by laminar (sheet-like) flow, while at high Reynolds numbers, flows tend to be turbulent.

How was the Moody diagram plotted?

1. **Data collection:** Moody compiled experimental data from various sources, including his own research, on pipe flow experiments with different pipe materials, diameters, and flow conditions.
2. **Dimensionless parameters:** Moody identified the key dimensionless parameters that influence the friction factor, including the following:
   • Reynolds number (Re)
   • Relative roughness (ϵ/d)
3. **Plotting:** Moody plotted the friction factor (f) against the Reynolds number (R_e) for various relative roughness values (ϵ/d). He used a log–log plot, see Section 7.8.3; see the graph x-axis that runs from 10^3 to 10^8.
4. **Curve fitting:** He fitted smooth curves to the data points, creating a continuous relationship between f, Re, and ϵ/d.
5. **Chart development:** Moody developed the chart by plotting the curves for different relative roughness values, creating a comprehensive diagram.
6. **Refinement:** The chart has undergone refinements and updates over the years, incorporating new data and research, with the latest one as shown in Figure 7.3.

Notes

• Friction factor (f) on the y-axis.
• Reynolds number (Re) on the x-axis.
• Relative roughness (ϵ / d) as a parameter, represented by different curves.
• Friction factor is less than 1, implying that it is a fraction reducing the head or pressure in the pipe. See Section 7.6.

7.4.5.2.2 Nomographs and tabulations

Nomograms are graphical tools, also known as nomographs or alignment charts, used to perform mathematical calculations or estimate outcomes based on relationships between variables.

These graphical tools usually contain three parallel scales graduated for different variables so that when a straight line connects values of any two, the related value may be read directly from the third vertical line at the point intersected by the line.

7.4.5.3 Key takeaways

- Units are very important in empirical equations. This is because equations are built through experience and experimental observations, where logic refines intuition. Dimensional consistency may have to be forged, making the equation factual.
- Empirical equations are popular in geotechnical and structural engineering—geotechnical engineering is particularly reliant on empirical equations due to the inherent uncertainties and complexities of working with earth materials.
- Empirical equations are complemented with graphs and tabulations. Empirical equations are used hand in hand with graphs, nomograms, and tabulations to solve engineering problems.

7.4.6 Numerical Methods in Civil Engineering

As outlined in Section 7.3.3, these are a set of techniques used to approximate solutions to mathematical problems by performing calculations with numbers, particularly when exact solutions are difficult or impossible to find analytically, often relying on iterative processes and computer algorithms to achieve an accurate result; essentially, it's a way to solve complex problems using numerical approximations instead of exact formulas. Numerical methods usually refer to systematic and algorithmic approaches to solve mathematical problems, such as iterative methods (e.g., Newton–Raphson and bisection), approximation methods (e.g., Taylor series and asymptotic expansions), discretization methods (e.g., finite difference and finite element), and Euler's method.

Fact notes: Numerical methods are the commonest methods used throughout all civil engineering branches! Some differential equations, particularly, do not have analytical solutions, and the only way we can solve them is through numerical techniques.

7.4.6.1 Numerical method in water resource engineering

In hydraulics system designs, numerical solutions are possible and are used in most practical applications especially finite-difference and finite-element methods—discretization techniques. In hydraulics system designs, particularly water systems, different numerical schemes are used to solve Saint Venant equations. Equations (7.68) and (7.69) are together called the Saint Venant equations in the honor of Saint Venant who developed these equations in 1871. It is not possible to solve them together analytically, except in some very simplified cases.

Saint Venant proposed the mathematical model of a water flow in rivers, based on the laws of conservation of momentum and mass of fluid. Saint Venant shallow water equations are derived from the Navier–Stokes equations. Numerical

solutions are possible and are used in most practical applications especially finite-difference and finite-element methods. The governing hydrodynamic flow equations in 1-D, with a momentum coefficient equal to 1 and neglecting lateral inflows and wind shear, the continuity equation takes the form:

$$\frac{\partial A}{\partial t} + \frac{\partial Q}{\partial x} = 0 \tag{7.68}$$

The momentum equation (7.69), in its complete form, is complex to solve except in very simplified cases, where some components are dropped. In many cases, software programs are used setting initial boundary conditions. So, because of its complexity, numerical methods are used to solve these helpful equations in hydraulics:

$$\frac{\partial Q}{\partial t} + \frac{\partial \left(Q^2 / A \right)}{\partial x} + gA\frac{\partial Z}{\partial x} + \frac{gn}{AR^{4/3}} Q|Q| = 0 \tag{7.69}$$

Temporal   Spatial   Gravity   Friction

where

Q: discharge
t: time
A: cross-sectional area
n: friction coefficient
g: acceleration due to gravity
x: distance
R: hydraulic radius
Z: stage

Equations (7.68) and (7.69) are nonlinear analytical solutions which are both partial differential equations; a finite-difference method (discretization numerical method) is used to solve these two governing equations. Third- or higher-order accurate numerical methods are used to solve these equations. Several explicit and implicit finite-difference schemes are used to solve 1-D Saint Venant equations for solving open-channel flow problems. These include MacCormack discretization scheme, Lax-Wendroff discretization scheme, Preissmann implicit model, and the method of characteristics (MOC). See reference [7] (Chapter 6) for numerical methods used in solving hydraulic flow routing techniques.

7.4.7 On Dimensional Consistency

Dimensional consistency refers to the property of an equation or expression or formula where the dimensions of the LHS are identical to the dimensions of the RHS. In simple terms, dimensional consistency ensures that the units of measurement on both sides of an equation or expression are the same and that the dimensions of the physical quantities being described are properly matched.

Let's consider Newton's second law of motion, which states that the acceleration of an object depends upon two variables—the net force acting on the object and the mass of the object. The acceleration of the body is directly proportional to the net force acting on the body and inversely proportional to the mass of the body. This means that as the force acting upon an object is increased, the acceleration of the object is increased. Likewise, as the mass of an object is increased, the acceleration of the object is decreased. Newton's second law can be expressed mathematically as follows:

$$F = ma \tag{7.70}$$

where

F: force (measured in Newtons, N)
m: mass (measured in kilograms, kg)
a: acceleration (measured in meters per second squared, m/s^2)

In equation (7.70), the dimensions of the LHS (force, F) are $\left[\mathrm{MLT^{-2}}\right]$, which matches the dimensions of the RHS (mass × acceleration, $m \times a$), which is also $\left[\mathrm{MLT^{-2}}\right]$. **Note:** M stands for mass, L for length, and T for time.

Let's now look at more complex equations, especially empirical ones. How do we ensure dimensional consistency? Units are a crucial aspect of dimensional consistency. Dimensional consistency is ensured when the units of measurement on both sides of an equation or expression are compatible and cancel out correctly.

- The units on both sides of an equation must be compatible. For example, you can't add meters and seconds or mass and meters.
- When units are multiplied or divided, they must cancel out correctly. For example, meters per second (m/s) can be multiplied by seconds (s) to obtain meters (m).
- Some quantities, like angles or ratios, are dimensionless. These quantities can be used in equations without affecting the dimensional consistency.
- When converting between units, it's essential to ensure that the conversion factors are correct and dimensionally consistent.

This is all done to ensure that errors are avoided and that equations and models communicate physically meaningful and insightful information. Units come with conversion factors from one measurement system to another, and these are embedded in equations and formulae. The SI standard selects the following dimensions and corresponding dimension symbols: time (T), length (L), mass (M), electric current (I), amount of substance (N), luminous intensity (J), and absolute temperature (Θ).

In the realm of international business, do you know why fortunes can be made or lost in a day? This is due to a single, often overlooked factor: conversion rates. The adage 'time is money' takes on a profound significance in this context. The accurate conversion of currencies, units of measurement, and other relevant factors is important in international trade. A slight miscalculation or misunderstanding or time delay can result in substantial financial losses or gains. As an engineer, you ought to note the following while performing any calculation applying formulae or equations:

- Start with the questions
  - Am I in SI units or English units or other units?
  - Am I in local/regional standards or not?

- As a rule of thumb, start with units for empirical equations or formulae. There are always standard units for each empirical formulae/equation.

For example, let's use Manning's equation (7.71) to illustrate this important concept. Manning's equation is an empirical formula which expresses flow velocity with channel geometry. Manning's equation differs by a conversion factor of 1.49 in English system; in other words, multiply 1.49 with the SI system-derived output. Do you know how this comes about—specifically the 1.49 conversion factor? Let's see how the 1.49 conversion factor comes about. Let's consider an open U-drain (Figure 7.4) with 1 m height and 1 m width with the following attributes:

- Roughness factor, $n = 0.012$
- Hydraulic radius, $A/P = 1/3$
- Slope, $S = 0.002$

In the metric system,

$$V = \frac{1}{n} \cdot \left(\frac{A}{P}\right)^{2/3} \cdot S^{1/2} \tag{7.71}$$

$$V = \frac{1}{0.012} \cdot \left(\frac{1}{3}\right)^{2/3} \cdot 0.002^{1/2} = 1.79 \text{ m/s} \tag{7.72}$$

However, converting to feet, the dimensions of the U-drain (Figure 7.5) appear as follows:
From (7.71),

$$V = \frac{1}{0.012} \cdot \left(\frac{10.765}{9.843}\right)^{2/3} \cdot 0.002^{1/2} = 3.95 \tag{7.73}$$

However, $1.79 \times 3.281 = 5.873$ ft/s.

Therefore, to get the **conversion factor**, we assume it to be x, for which

$$3.95 \text{ of } x = 5.873$$

$$x = 1.487 \approx 1.49$$

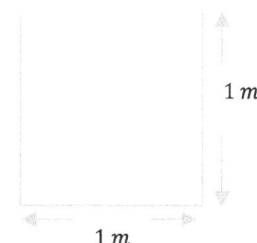

FIGURE 7.4 U-drain (metric system).

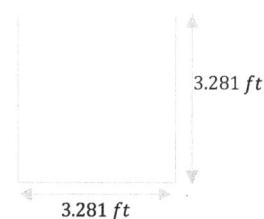

FIGURE 7.5 U-drain (English system).

Therefore, in English units, Manning's equation is

$$V = 1.49 \cdot \frac{1}{n} \cdot \left(\frac{A}{P}\right)^{2/3} \cdot S^{1/2} \tag{7.74}$$

Strict adherence to unit consistency is vital in empirical equations as these equations are derived from observational data, curve fitting, variable manipulations, or experimental results and are therefore approximate. Inaccurate unit handling can compromise the **validity** and **reliability** of the equation, leading to incorrect predictions or conclusions.

In conclusion, when working with empirical equations, it is crucial to exercise utmost care and attention to unit consistency. Don't mess with units! Again, don't mess around especially when you deal with empirical equations. Failure to do so can lead to inaccurate results, compromised design integrity, and potentially catastrophic consequences. In regions where both local and international systems of units are employed, the risk of unit-related errors is amplified. You can propose a flawed design if you don't take units very seriously. It is essential to ensure that all units are correctly matched and converted to avoid introducing errors into calculations. Thus, inadequate attention to unit consistency can result in incorrect design proposals, compromised safety and performance, and costly rework and revisions. To mitigate these risks, engineers/designers must prioritize unit consistency and carefully verify all calculations to ensure accuracy and reliability.

7.4.8 Grouping Equations

Engineers group equations to simplify work. They often organize and categorize equations into groups or sets to make their work more manageable and efficient. By grouping equations, engineers can reduce complexity, identify patterns and relationships, and hence simplify calculations. Engineers may group equations to form **dimensionless quantities**, such as Reynolds number or Mach number, which can help simplify complex problems. See the following example for Reynolds number equation, Darcy–Weisbach equation, and Colebrook equation.

Reynolds number

$$\text{Re} = \frac{\rho V d}{\mu} \tag{7.75}$$

Darcy–Weisbach equation

$$f_d = \Delta P \frac{2d}{\rho l V^2} \tag{7.76}$$

Colebrook equation

$$\frac{1}{\sqrt{f_d}} = -2.0 \log_{10}\left(\frac{\varepsilon/d}{3.7} + \frac{2.51}{Re\sqrt{f_d}}\right) \tag{7.77}$$

Therefore, Colebrook equation groups Reynolds number equation and Darcy–Weisbach equation! Grouping equations is one of the techniques engineers primarily use to solve complex engineering problems and come up with empirical equations, among other things. Section 7.4.5.1.1 shows how Robert Manning (1816–1897) grouped the Darcy and Weisbach equations to come up with an empirical equation. And others! Another typical example is how energy and momentum equations are grouped to solve problems in fluid mechanics and hydraulics—and particularly how the Froude number groups and ties the energy and momentum equations together.

By grouping and solving equations, engineers can develop innovative solutions and improve the performance of complex systems. Improve communication and collaboration and enhance problem-solving efficiency.

Engineers often group equations to solve complex engineering problems by identifying relevant variables and relationships. They also organize equations into categories (e.g., energy, mass, and momentum). Sometimes, engineers use substitution or elimination (or combination methods) to reduce the number of equations. They finally solve the resulting system of equations using numerical or analytical techniques. This process helps engineers to simplify complex problems, identify key parameters and relationships, develop mathematical models and simulations, optimize system performance and design, and troubleshoot and analyze problems.

7.4.8.1 How engineers group equations

Engineers identify the specific engineering problem or system to be analyzed. Then, gather/collect all relevant equations that describe the system's behavior based on their physical meaning and relationships. The grouping of equations, in many cases, takes into account **dimensional consistency**. Units are strictly followed especially in empirical equations.

Engineers collect equations that can potentially describe or relate to a phenomenon or system's behavior. For example, conservation laws (e.g., mass, energy, and momentum) are grouped together—common in water resource engineering (hydraulics) and constitutive equations (e.g., material properties) for material physical properties (such as stress and strain)—common in structural engineering. Also, other equation groups common in structural engineering include forces, moments, and/or torque, which can be referred to as balance equations. Another collection is the kinematic and dynamic equations for analyzing dynamic systems found in all civil engineering branches.

Some common examples of equation grouping in civil engineering include Navier–Stokes equations in fluid dynamics, energy and momentum equations in thermodynamics, and structural analysis equations in civil engineering.

Engineers can then simplify and reduce equations using mathematical techniques (e.g., substitution, elimination, and combination) to simplify and reduce the number of equations, eliminating redundant or unnecessary ones/equations/variables. Combination includes techniques used to simplify and solve systems of equations by combining two or more equations into a single equation.

Engineers may then use numerical or analytical methods to solve the resulting system of equations, often using computational tools like MATLAB, Python, MAPLE, MS Excel, or AI tools. Then, engineers examine and interpret the solutions to gain insights into the system's behavior, performance, and optimization opportunities, if possible. The results are validated and refined against experimental data, physical laws, or other verification methods and refine the model as needed.

Specific techniques engineers use in equation grouping include dimensional analysis where equations are identified and grouped based on their physical dimensions and units; see Section 7.4.8.1.1 (Buckingham's Pi theorem). Scaling analysis is also used to identify and group equations based on their relative importance and scaling behavior. See the role of coefficients, Section 7.6. Another technique involves linearization—that is, approximating nonlinear equations with linear ones to simplify the analysis. Decoupling is another technique used by engineers where coupled equations are separated into independent ones to simplify the solution process. This structured approach enables engineers to effectively group and solve equations to tackle complex engineering problems.

To solve mathematical/engineering problems, several of these techniques can be used—say combination methods, decoupling, and dimensional analysis may be used.

7.4.8.1.1 Buckingham Pi (π) theorem

Buckingham's Pi theorem states that if there are n variables in a problem and these variables contain m primary dimensions (e.g., M, L, T), the equation relating all the variables will have $(n-m)$ dimensionless groups [6]. Buckingham referred to these groups as π groups. The final equation obtained is in the following form:

$$\pi_1 = f\left(\pi_2, \pi_3, \ldots \ldots \pi_{n-m}\right) \tag{7.78}$$

Buckingham provides a framework for dimensional analysis, allowing engineers to identify the key variables and relationships in complex problems. The π groups must be independent of each other, and no one group should be formed by multiplying together powers of other groups. This method offers the advantage of being simpler than the method of solving simultaneous equations for obtaining the values of the indices (the exponent values of the variables). In this method of solving the equation, there are two conditions:

- Each of the fundamental dimensions must appear in at least one of the m variables.
- It must not be possible to form a dimensionless group from one of the variables within a recurring set. A recurring set is a group of variables forming a dimensionless group.

7.4.8.1.1.1 Worked example [7.8]: Buckingham Pi (π) theorem for illustrative purposes
Question: Establish the relationship of the effect of the variables d, L, ρ, μ, and v on pressure drop, (ΔP), in a pipeline.

Here,

d: diameter of the pipeline
L: length of the pipeline
ρ: density of the fluid
μ: viscosity of the fluid
v: velocity of the fluid

Solution

$$f\left(\Delta P, d, L, \rho, \mu, v\right) = 0 \tag{7.79}$$

No. of variables, $n = 6$, that is, ΔP, d, L, ρ, μ, and v.

No. of fundamental dimensions, $m = 3$, that is, $[M], [L], [T]$

According to Buckingham's theorem, the no. of dimensionless groups, $n - m = 6 - 3 = 3$.

The recurring set must contain three variables that cannot themselves be formed into a dimensionless group. In this case, there are two restrictions:

- Both L and d cannot be chosen as they can be formed into a dimensionless group, (L/d).
- ΔP, ρ, and v cannot be used since

$$\left(\frac{\Delta P}{\rho v^2}\right)$$

is dimensionless.

Thus, the variables d, v, and ρ are chosen as the recurring set. The dimensions of these variables are as follows:

$$d = [L]$$

$$v = \left[LT^{-1}\right]$$

$$\rho = \left[ML^{-3}\right]$$

Rewriting the dimensions in terms of the variables chosen:

$$[L] = d$$

$$[M] = \rho d^3$$

$$[T] = dv^{-1}$$

Thus, the dimensionless groups are formed by taking each of the remaining variables ΔP, L, and μ in turn.

ΔP has dimensions of

$$\left[\Delta P M^{-1} T^{-2}\right]$$

Therefore,

$$\Delta P M^{-1} T^{-2}$$

is dimensionless.

Substituting the dimensions in terms of variables,

$$\pi_1 = \Delta P \left(\rho d^3\right)^{-1} (d)\left(dv^{-1}\right)^2 \tag{7.80}$$

$$\pi_1 = \frac{\Delta P}{\rho v^2}$$

L has dimensions of $[L]$

$$L[L]^{-1}$$

is therefore dimensionless.

$$\pi_2 = \frac{L}{d}$$

μ has dimensions of

$$\left[ML^{-1}L^{-1}\right]$$

$\mu\left[\mathrm{M^{-1}LT}\right]$ is, therefore, dimensionless:

$$\pi_3 = \mu\left(\rho d^3\right)^{-1}(d)\left(dv^{-1}\right)^2 \tag{7.81}$$

$$\pi_3 = \frac{\mu}{dv\rho}$$

Thus,

$$f\left(\frac{\Delta P}{\rho v^2}, \frac{L}{d}, \frac{\mu}{dv\rho}\right)$$

Hence, the derived relationship is

$$\frac{\Delta P}{\rho v^2} = f\left(\frac{L}{d}, \frac{dv\rho}{\mu}\right)$$

Therefore, the Buckingham Pi (π) theorem provides a framework for dimensional analysis, allowing engineers to identify the key variables and relationships in complex engineering problems. This is very helpful in equation formulation and grouping.

7.4.8.1.2 Decoupling equations

In the context of mathematics and engineering, **decoupling equations** means transforming a system of interconnected equations into a set of independent equations that can be solved separately. This simplification makes the problem easier to analyze and solve. In several civil engineering branches, we have coupled systems of differential equations as governing equations. To simplify their analysis, we usually decouple them. A perfect example is the coupled continuity and momentum equations in water resource engineering, applicable especially in hydraulic structure designs. Decoupling the sets of differential equations often involves linear transformations and matrix operations. **Eigenvalues** and **eigenvectors** are applied in decoupling systems of differential equations. See reference [7] (page 222) for application of eigenvalues to decouple systems of differential equations.

7.4.8.1.2.1 Typical geotechnical example [7.5] for illustrative purposes   See Figure 7.13.

This graph plots equations (7.82)–(7.84) for N_q, N_c, and N_γ, respectively, in terms of $\varnothing'$. These equations are grouped by graphing. They are used to obtain the bearing capacity factors used in equation for calculating the bearing capacity for shallow foundations in drained conditions—full equation (7.86).

Bearing capacity factors:

$$N_q = \frac{\left(1+\sin\varnothing'\right)}{\left(1-\sin\varnothing'\right)}e^{\pi\tan\varnothing'} \tag{7.82}$$

Parameter N_c can be similarly derived for soil with non-zero c' to give

$$N_c = \frac{N_q - 1}{\tan\varnothing'} \tag{7.83}$$

$$N_\gamma = \left(N_q - 1\right)\tan\left(1.32\varnothing'\right) \tag{7.84}$$

In EC7, the following expression is provided:

$$N_\gamma = 2\left(N_q - 1\right)\tan\varnothing' \tag{7.85}$$

Values of N_q, N_c, and N_γ are plotted in terms of $\varnothing'$ in Figure 7.13.

The bearing capacity of a shallow foundation (strip footing) on a drained material (long-term loading) can be written in a generalized form as

$$q_f = s_q N_q \sigma_q' + \frac{1}{2}\gamma B s_\gamma N_\gamma + s_c N_c c' \tag{7.86}$$

Engineers group equations to solve engineering problems by following a structured approach. Therefore, Figure 7.13 groups these equations/formulas perfectly well to aid solve equation (7.86).

7.4.8.1.3 Combination methods

Combination methods are techniques used to simplify and solve systems of equations by combining two or more equations into a single equation. These are chief techniques engineers use to sort equations. They are used to simplify complex systems of equations, reduce the number of equations, isolate variables, and solve for unknowns. Thus, engineers and mathematicians can efficiently solve systems of equations and analyze complex problems in various fields by applying combination methods. Some common combination methods include the following:

1	Multiplication method	This is where we multiply two or more equations by necessary multiples to eliminate one or more variables.
2	Elimination method	This is where equations are added or subtracted to eliminate one or more variables, leaving a single equation with one variable.
3	Substitution method	This is where we solve one equation for a variable and substitute that expression into another equation.
4	Linear combination method	Take a linear combination of two or more equations to eliminate one or more variables.
5	Gaussian elimination	This is a systematic method of using elimination and substitution to solve a system of linear equations.
6	Cramer's rule	This refers to a method that uses determinants to solve a system of linear equations.

7.4.8.1.3.1 Worked example [7.9] for illustrative purposes   Use combination methods to group the pair of simultaneous equations

$$x + 3y = 6 \tag{7.87}$$

$$2x - y^2 = 2.5 \tag{7.88}$$

Solution

Express x in terms of y.
From equation (7.87)

$$x = 6 - 3y \tag{7.89}$$

Substitute x in equation (7.88):

$$2(6-3y)-y^2 = 2.5 \tag{7.90}$$

$$y^2 + 6y = 14.5 \tag{7.91}$$

Therefore, x is finally eliminated from the grouped equation (7.91).

7.5 BUILDING EQUATIONS IN CIVIL ENGINEERING

Equations are built in different ways. As outlined in Section 7.4, theoretical means are used to develop equations and in other ways that empirical derivation is relied on. Simply put: equations are built either: analytically/theoretically or empirically! And they can be solved either analytically or numerically!—engineers use these methods to develop and solve equations.

- An equation can be built through curve fitting—this is, of course, an empirical one.
- An equation can be built from theory: this a classical analytical equation. The perfect example is one built from Newton's laws of motion, which are foundational equations for civil engineering!
  - **Resolving forces**—this is one of the primary methods that civil engineers use to develop equations. By resolving forces to find equilibrium, engineers can develop a set of equations especially in structural engineering.
- An equation can be solved numerically—through approximations! These equations are also very common throughout the civil engineering field.

Note: A mathematical equation can be considered a formula when it is standardized, simplified, and specifically designed to provide a direct, unique solution to a well-defined problem or a set of problems, making it a reusable and applicable mathematical tool. Through curve fitting, approximate methods are used as tools to solve engineering problems.

7.5.1 How Newton's Discoveries Inspired Engineers and Scientists

Benchmarking on Sir Isaac Newton's discoveries of the laws of motion, several scientists and engineers learnt to group and construct meaningful equations, hence advancing science and engineering. Newton's laws provided a foundation for the development of classical mechanics, particularly, and inspired later scientists and engineers, such as Leonhard Euler, Joseph-Louis Lagrange, and William Rowan Hamilton, to develop more advanced mathematical frameworks for describing physical systems. It is important to note the work done by Newton to popularize the meaning of the 'product of variables.'

For instance, let's consider Newton's second law of motion, which states that the acceleration of an object depends upon two variables—the net force acting on the object and the mass of the object. The acceleration of the body is directly proportional to the net force acting on the body and inversely proportional to the mass of the body. This means that as the force acting upon an object is increased, the acceleration of the object is increased. Likewise, as the mass of an object is increased, the acceleration of the object is decreased.

In layman understanding, when you are standing, and someone throws an object at you, you feel the impact force differently when they vary a) the velocity at which the object is thrown at you and b) the mass of that object. That is Newton's second law of motion, which states that the rate of change of momentum is equal to the force acting on the object doing that work. You experience different impact forces (magnitude) when an object is thrown at you depending on two factors: how fast (velocity) the object is thrown at you and its mass (weight). This illustrates Newton's second law of motion:

Force = rate of change of momentum

In simpler terms:

- The heavier the object (more mass), the greater the force you'll feel.
- The faster the object is thrown (more velocity), the greater the force you'll feel.

Newton's second law shows that force, mass, and velocity are interconnected. The law is often mathematically expressed as follows:

$$F = ma \,(\text{force} = \text{mass} \times \text{acceleration}) \tag{7.92}$$

This fundamental principle helps us understand how forces affect motion in our daily lives, from catching a ball to designing safe transportation systems. Thus, the product of mass and acceleration is termed 'momentum':

$$\sum F = \Delta MV \tag{7.93}$$

Remember force is the product of mass and acceleration—acceleration being change of velocity. This is a product of mass and velocity. We encounter analogous identities in many engineering applications—in both scalar and vector spaces. Sir Isaac Newton opened doors for a better understanding of physical and engineering systems!

This kind of observation and experimentation is crucial in engineering—studying products of two independent variables, primarily. Although these concepts can be explained in both integral and differential calculus, the idea remains the same—a product of two variables.

In engineering, most formulae typically involve independent variables, which are quantities that can be changed or controlled without affecting each other. These variables are often represented by separate symbols or parameters in equations. However, some engineering formulas and/or equations may involve dependent variables, where changes in one variable affect others.

To further illustrate Newton's observation, imagine a stationary object stuck in a conduit transmitting an uncompressible fluid in such a way that the fluid just passes it with difficulty. Did you know that varying the velocity of the fluid would eject it out of the conduit? Similarly varying the quantity of flow would eject it out while conserving the momentum.

For example, if the velocity at which the fluid flows is 2 m/s, mass of the fluid flowing through the conduit 10 kgs, then momentum would be 20 kgm/s. The same momentum is achieved when the flow velocity is 10 m/s and fluid mass 2kgs, that is, 20 kgm/s, keeping other factors constant and assuming the initial velocity of the fluid is 0 m/s. Either way, the object is ejected out of the conduit considering the fact that the rapid momentum change required is 20 kgm/s.

Newton was able to identify that the product of two independent variables, mass and velocity, was essential in the study of dynamic systems. Whether you have a high velocity but small quantity or low velocity and big quantity, either way you achieve the goal of ejecting the object stuck in the conduit.

So, to calculate the impact force in this case, the rate of change of momentum is obtained. Remember, the object was stationary, implying at rest, implying 0 kgm/s momentum, but due to rapid change of momentum, it is ejected out, and if it takes 1 second, the force would be 20 kgm/s^2, which is 20 N. Some of the practical applications of the momentum equations/function in civil engineering include the following:

- **Transportation:** momentum affects vehicle stopping distances and crash safety.
- **Hydraulic design:** momentum is applied in design of hydraulic systems.
- **Mechanical systems:** momentum considerations ensure stable and efficient mechanical systems.

Ever since Newton proposed the universal laws of motion opening the platform to study sound classical mechanics, engineers learnt to investigate the product of (independent) variables, through innovative approaches, hence coming up with essential significant empirical and theoretical formulas. Of course, Sir Isaac Newton was not the first to make these observations, but he popularized them, opening gates for many scientists and engineers to understand products of variables much better—this became more solidified with his invention of calculus!

What is truly important is that he invented calculus—differential and integral, together with Gottfried Wilhelm Leibniz (1646–1716). Differential calculus is concerned with rates of change, while integral calculus refers to summing up infinitesimal changes!

Therefore, by manipulating variables (in equations), engineers and scientists study several phenomena to advance new knowledge in several spheres of engineering disciplines. Further examples in civil engineering that show how engineers have investigated products of variables include the **moment**, **Young's modulus**, and **flexural rigidity** in structural engineering. In structural engineering, primarily, the concepts of **stress and strain** have had far-reaching success, opening doors to the understanding of materials science much better.

7.5.1.1 Moment

Moment is force (F) times distance (x):

$$M = Fx \tag{7.94}$$

Imagine the applied force is constant, but the distance varies from 0.0 to 0.5 m. In this case,

$$\sum M = F \sum_{n=0}^{0.5} dx$$

$$M = F \int_{0}^{0.5} dx$$

7.5.1.2 Axial stress

This is the result of a force acting perpendicular to an area of a body, causing the extension or compression of the material. See Chapter 4:

$$\sigma = \frac{F}{A_0} \tag{7.95}$$

Alternatively, this can be written as follows:

$$\sigma = \frac{1}{A_0} \times F \tag{7.96}$$

7.5.1.3 Strain

The strain or relative deformation is the change in length, $L_n - L_0$, divided by the original length L_0:

$$\varepsilon = \frac{(L_n - L_0)}{L_0} \tag{7.97}$$

Alternatively, this can be written as

$$\varepsilon = \frac{1}{L_0} \times (L_n - L_0) \tag{7.98}$$

7.5.1.4 Young's modulus

Young's modulus, E, is numerical constant, named after the 18th-century English scientist, Thomas Young, that describes the elastic properties of a solid undergoing tension or compression in only one direction, as in the case of a metal rod that after being stretched or compressed lengthwise returns to its original length. Thomas Young investigated the product of two variables, that is, stress σ and the inverse of the strain $\frac{1}{\varepsilon}$, named as Young's modulus. Young's modulus is a measure of the ability of a material to withstand changes in length when under lengthwise tension or compression. Sometimes referred to as the modulus of elasticity, Young's modulus is equal to the longitudinal stress divided by the strain. Young's modulus is expressed mathematically as

Young's modulus, $E = \dfrac{\sigma}{\varepsilon} = \dfrac{1}{\varepsilon} \times \sigma = \dfrac{F/A_0}{(L_n - L_0)/L_0}$

$$= \frac{FL_0}{A(L_n - M_0)} \quad (7.99)$$

This is a specific form of Hooke's law of elasticity. See Section 3.3.2. The units of Young's modulus in the English system are pounds per square inch (psi) and in the metric system newtons per square meter (N/m^2). This inspired engineers to study the stiffness of materials, which primarily depends on Young's modulus and area moment of inertia.

The area moment of inertia, also called the second moment of area, is a parameter that defines how much resistance a shape (like the cross-section of a beam) has to experience bending because of its geometry. It *is a property of a two-dimensional plane shape*—a geometrical property of an area which reflects how its points are distributed with regard to an arbitrary axis.

The value of Young's modulus for aluminum is about 1.0×10^7 psi, or 7.0×10^{10} N/m^2. Young's modulus value for steel is about three times greater, which means that it takes three times as much force to stretch a steel **bar** the same amount as a similarly shaped aluminum bar. In the preceding statement, 'similarly shaped' is mentioned because if the materials are not of the same shape, the force to stretch the bars would be different either way.

7.5.1.5 Flexural rigidity

Flexural rigidity is the product of Young's modulus and moment of inertia. This value is used to describe the resistance of a beam or other structural member to bending:

$$\text{Flexural rigidity} = EI \quad (7.100)$$

where E is Young's modulus (material property; material stiffness) and I is the moment of inertia (geometrical property). Thus, the product of the material property and its geometrical

property communicates essential information, that is, flexural rigidity! It's a critical parameter in structural engineering and mechanics of materials. In essence, EI predicts how much a beam will resist deformation when subjected to external forces, such as bending moments or loads. This parameter is used in several equations for beam analysis.

7.5.1.6 Buckling load

The critical buckling load is calculated by the following formula (7.101):

$$P_{cr} = \frac{\pi^2 EI}{(KL)^2} \quad (7.101)$$

This is a product of flexural rigidity EI and $\dfrac{\pi^2}{(KL)^2}$.

7.5.1.7 Area moment of inertia

Also known as the second moment of area, it is a geometrical property of a two-dimensional area that describes how its area is distributed with respect to a given axis. It quantifies the resistance of a cross-section, such as a beam, to bending. It's a key parameter in structural engineering for analyzing beams and columns:

$$\bar{I}_x = \int y^2 dA \quad (7.102)$$

7.5.2 Fundamental Equations That Have Advanced Civil Engineering

Table 7.3 outlines the equations that changed the face of civil engineering throughout history. These equations and laws have had a profound impact on the development of civil engineering, enabling the design and construction of safer, more efficient, and more sustainable infrastructure.

TABLE 7.3 Fundamental equations that have advanced civil engineering

DISCOVERER	EQUATION		ROLE
Navier–Stokes equations (1845)	Continuity equation $$\nabla \cdot \vec{V} = 0$$ Momentum equation $$\rho \frac{D\vec{V}}{Dt} = -\nabla p + \rho \vec{g} + \mu \nabla^2 \vec{V}$$	(7.103) (7.104)	Describes fluid dynamics and behavior, fundamental to hydraulic engineering, pipe flow, and water resource management. They are a set of partial differential equations describing the motion of viscous fluid substances which were developed over a period of decades, with Claude–Louis Navier's work starting around 1822 and Sir George Stokes finalizing the equations in 1845. These equations are fundamental in fluid dynamics and are used to model various phenomena, including weather patterns, ocean currents, and fluid flow in engineering applications.
		A	B	C	D		
	A: Total derivative. B: Pressure gradient. C: Body force term. D: Diffusion term.		
Euler–Bernoulli beam equation (1750s)	See equation (6.11)		Describes the bending of beams under loads—instrumental in the design of bridges, buildings, and other civil engineering structures.

(Continued)

TABLE 7.3 (Continued) Fundamental equations that have advanced civil engineering

DISCOVERER	EQUATION	ROLE
Mohr–Coulomb failure criterion (1871)	$\tau = c + \sigma_n \tan(\varnothing)$ (7.105) where τ is the shear stress at failure, σ_n is the normal stress on the failure plane, $\varnothing$ is the angle of internal friction, $\tan(\varnothing)$ is the slope of the failure envelope, and c is often called the cohesion.	Predicts the failure of soils and rocks under different stress conditions—crucial for geotechnical engineering.
Darcy–Weisbach equation (1845)	See equation (7.63)	Relates the head loss in a pipe to the flow rate, pipe diameter, and friction factor. It is essential for hydraulic engineering and pipeline design.
Rankine's earth pressure theory (1857)	Active pressure $$P_a = \frac{1}{2} K_a \gamma h^2 \qquad (7.106)$$ where K_a is the active earth pressure coefficient, γ is the unit weight of the soil, and h is the height of the wall. For passive pressure, K_a is replaced with K_v	Calculates the lateral earth pressure on retaining walls. The equation is fundamental to geotechnical engineering and structural design.
Buckingham Pi (π) theorem (1914)	If there are n variables in a problem and these variables contain m primary dimensions (e.g., M, L, T), the equation relating all the variables will have $(n-m)$ dimensionless groups. Buckingham referred to these groups as π groups. The final equation obtained is in the form of $$\pi_1 = f\left(\pi_2, \pi_3, \ldots\ldots\pi_{n-m}\right)c \qquad (7.107)$$	Provides a framework for dimensional analysis, allowing engineers to identify the key variables and relationships in complex engineering problems.
Timoshenko beam equation (1921)	The governing equations are the following coupled system of ordinary differential equations: $$\frac{d^2}{dx^2}\left(EI\frac{d\varphi}{dx}\right) = q(x) \qquad (7.108)$$ $$\frac{d\omega}{dx} = \varphi - \frac{1}{kAG}\frac{d}{dx}\left(EI\frac{d\varphi}{dx}\right) \qquad (7.109)$$ where L is the length of the beam. A is the cross-section area. E is the elastic modulus. G is the shear modulus. I is the second moment of area. k is called the Timoshenko shear coefficient, depends on the geometry. Normally, $k = 5/6$ for a rectangular section. $q(x)$ is a distributed load (force per length). ω is the displacement of the mid-surface in the z-direction. φ is the angle of rotation of the normal to the mid-surface of the beam.	Extends the Euler–Bernoulli beam equation to include the effects of shear deformation and rotary inertia. It is extremely important for dynamic analysis and vibration studies.
Terzaghi's consolidation theory (1925)	See Section (7.4.5.1.3)	Describes the settlement of soils under load, making it crucial for geotechnical engineering, foundation design, and soil mechanics.
Hazen–William's head loss equation (1900s)	Metric units $$H = \frac{10.583 \times L \times Q^{1.85}}{C^{1.85} \times d^{4.87}} \qquad (7.110)$$ Imperial units $$H = \frac{4.72 \times L \times Q^{1.85}}{C^{1.85} \times d^{4.87}} \qquad (7.111)$$	Relates the head loss in a pipe to the flow rate, pipe diameter, and roughness. It is widely used in hydraulic engineering and water supply system design.

(Continued)

TABLE 7.3 (Continued)   Fundamental equations that have advanced civil engineering

DISCOVERER	EQUATION	ROLE
	where • H = head loss (m or ft) in the pipe • L = length of pipe (m or ft) • d = diameter of pipe (m or ft) • Q = flow rate in the pipe (m³/s or cfs) • C = Hazen–William's roughness coefficient The Hazen–Williams coefficient C varies from about 150 for smooth pipes to about 70 for very rough pipes.	
Newmark's seismic design equation (1950s)	$$U_{max} = \left(\frac{v_{max}^2}{2a_y}\right) \cdot \left(\frac{a_{max}}{a_y}\right)$$ (7.112) where U_{max}: upper-bound permanent displacement. a_{max}: maximum velocity of ground motion. a_{max}: peak horizontal ground acceleration. a_y: yield acceleration.	Provides a framework for seismic design, calculating the response of structures to earthquake loading, essential for earthquake engineering and structural design.
Newton's equations of motion (1687)	See Section 3.3.1.1.	Describes the relationship between a body and the forces acting upon it. It is a fundamental equation to understanding motion, forces, and energy.
Manning equation (1890)	See equation (7.71)	Relates the flow rate in an open channel to the channel's slope, roughness, and hydraulic radius—(channel geometry) and it is widely used in hydraulic engineering, drainage design, river engineering, and flood control systems,
Hooke's law (1676)	See Section 3.3.2	Describes the linear relationship between stress and strain in elastic materials, essential for understanding material behavior.
Energy equations	See Section 3.3.3	Describes the relationships between energy, work, and forces, fundamental to understanding thermodynamics, mechanics, and electrical engineering.
Young's modulus (1807)	See equation (7.99)	Describes the stiffness of a material, relating stress and strain in the linear elastic region, essential for understanding material behavior. It is very helpful in studying deformable bodies).
Saint Venant equations (1855)	See equations (7.68) and (7.69)	Fundamental in modeling one-dimensional, unsteady open-channel flow. See Chapter 6 of further reading 4. These equations are widely used in hydraulic engineering for designing structures, analyzing floods, and managing water resources systems.
Bernoulli's equation	See equation (3.35)	Describes the relationship between pressure, velocity, and elevation of a fluid in motion, playing an important role in designing and analyzing pipeline systems and other structures.

7.5.3 Questioning Mathematical Operators in Equations and Formulas

In building equations, questioning operators is essential. These operators are arithmetic operators, comparison operators, logical operators, exponential and root operators, and many other operators. However, civil engineers greatly rely on the primary operators, which are the arithmetic operators.

7.5.3.1 Arithmetic operators

Arithmetic operators are used to perform mathematical calculations. They are also called binary arithmetic operators.

OPERATOR	RESULT	EXAMPLE
(+)	Addition of two numbers	$2 + 3$
(−)	Subtraction of two numbers	$6 - 4$
(×)	Multiplication of two numbers	5×1
(÷)	Division of two numbers	$8 \div 6$
(mod or %)	Modulus (remainder)—divides two numbers and returns the remainder	10 mod 3 equals 1 because when you divide 10 by 3, you get a remainder of 1

7.5.3.2 Comparison operators

Comparison operators compare values and return a true or false value. They are also known as relational operators.

OPERATOR	RESULT	EXAMPLE
(=)	Equal to	$2 + 2 = 4$
(≠)	Not equal to	$3 \neq 4$
(>)	Greater than	$5 > 3$
(<)	Less than	$-15 < 4$
(≥)	Greater than or equal to	$A \geq B$
(≤)	Less than or equal to	$C \leq D$

7.5.3.3 Logical operators

Logical operators are used to determine if a condition is true or false. They are often used in conditional statements, such as IF statements—very helpful in programming and software development. These operators are very helpful in equations building for software development.

OPERATOR	RESULT	EXAMPLE
(∧ or& &)	And	$x = 7$ $y = 4$ x(9 & & y)2 Return True
(∨ or ‖)	Or	$x = 7$ $y = 4$ $x = 5 \parallel y = 5$ Return False
(~ or !)	Not	$x = 7$ $y = 4$!(x = y) return True

7.5.3.4 Exponential and root operators

OPERATOR	RESULT	EXAMPLE
(**or)	Exponentiation	3^2 or 3^2 (gives 9)
$(\sqrt{\ })$	Square root	$\sqrt{4}$ (square root of 4 is 2)
$(\sqrt[3]{\ })$	Cube root	$\sqrt[3]{8}$ (cube root of 8 is 2)

7.5.3.5 Example [7.10] for illustrative purposes

You notice that the primary operators, that is, the arithmetic operators, are the key operators. When building or analyzing equations, civil engineers look at how these operators

influence, magnify, or reduce the variables in equations or expressions. For example, in an equation (7.113):

$$xy + \frac{e^{-2x}}{y} - 4 = 0 \qquad (7.113)$$

The expression

$$xy + \frac{e^{-2x}}{y} - 4$$

has three independent terms, that is, xy, $\frac{e^{-2x}}{y}$, and −4. We may say that the first two terms are added and the third one subtracted from the result.

You need to question this way

- if I knew x, what would happen to the products xy and $\frac{e^{-2x}}{y}$ if y is a fraction or not?
  - What if x is irrational?
- if I knew y, what would happen to the products xy and $\frac{e^{-2x}}{y}$ if x is a fraction or not?
  - What if y is irrational?
- And you notice that the term −4 reduces the output of $xy + \frac{e^{-2x}}{y}$.
- And again, you notice that equation (7.113) involves a natural exponential function, e^{-2x}. The presence of the natural exponential function in the equation suggests that the relationship between the variables x and y involves exponential growth or decay. Simply put, as x increases, e^{-2x} decreases exponentially. As x approaches negative infinity, e^{-2x} approaches positive infinity.

Imagine we have a well-defined scale for x, as a fraction.

X	Y_1 (SERIES 1)	Y_2 (SERIES 2)
0.1	0.206	39.794
0.2	0.169	19.831
0.3	0.138	13.195
0.4	0.114	9.886
0.5	0.093	7.907

Let's assume graph y versus x, Figure 7.6a. We realize that to visibly fit y_1 and y_2 on the same graph, we may need a log scale. See Section 7.8.2. So, y_1 is plotted alone in Figure 7.6b for visibility. Figure 7.6c provides the graphs for y_1 and y_2 when the log scale is used—when we choose the logarithmic scale to the base of 10.

Now, imagine the expression $xy + \frac{e^{-2x}}{y} - 4$ is a formula for calculating a dependent variable T, and then

$$T = xy + \frac{e^{-2x}}{y} - 4 \qquad (7.114)$$

This means the dependent variable T heavily depends on the term xy and has an exponential growth with respect to x, while 4

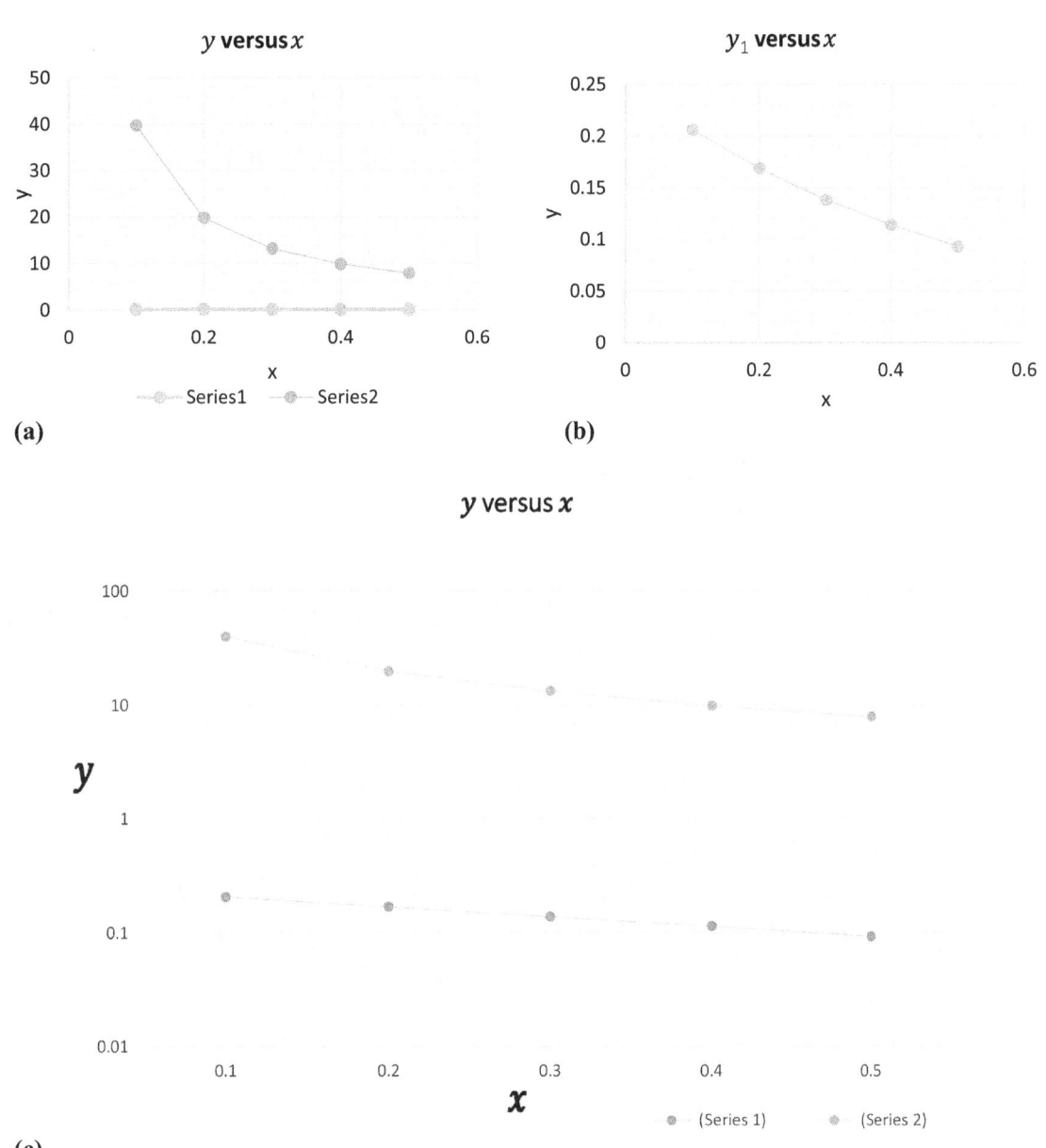

FIGURE 7.6 Graph y versus x (a)—left and (b)—right. (c) Graph of y_1 and y_2 with log scale to base 10.

is a correction factor of a sort because it reduces T. It is essential to realize that in the formulate equation (7.114), 4 may be a correction factor that refines an earlier version of the formula through research, while the coefficient 2 in e^{-2x} may vary in magnitude and sign as better understanding of T advances. Did you realize a drastic change in T when the correction factor is a fraction?

See Example 7.6 and equation (7.47) for the bearing capacity of a shallow foundation (strip footing) on a drained material. Each factor corresponds to a specific failure mechanism:

- N_c represents the cohesive component of soil strength.
- N_q represents the surcharge component of soil strength.
- N_γ represents the weight component of soil strength.

Graphs in Figure 7.6 are for $T = 0$. However, the fact that we have a well-defined scale, and we pick a point $x = 0.1$, and then we vary y from 0.206 to 0.25, then $\Delta y = 0.044$, and we obtain

T as 0.825. This kind of investigation would lead to set conditions as we formulate our equation for a better understanding of T. Another important attribute are the units as this may be an empirical formulation of T. 'What are the units for x, y, and T?' is the question to answer.

Note

- Understanding the order of operations is important within an expression/equation. The order of operations in mathematics refers to the specific sequence in which mathematical operations should be performed within an expression, typically remembered by the acronym 'PEMDAS,' that is, Parentheses, Exponents, Multiplication/Division (left to right), and Addition/Subtraction (left to right), meaning you solve operations inside parentheses first, then exponents, then multiplication and division from left to right, and finally addition and subtraction from left to right.

7.5.4 Building Equations through R&D

In Section 7.7.4, we outline the roles of dependent and independent variables. In research, a dependent variable is studied by varying the independent variables to study their effects on the dependent variable. Through this work, we are able to study and build equations. Theoretical equations may be derived from careful examination of a theory while empirical equations may be derived from observations, experimentation, and graph plotting. In both cases, we come up with equations and formulae that can be used to support the work of engineers.

7.6 ROLE OF COEFFICIENTS IN EQUATIONS

It is essential to note that engineers widely use coefficients in equations in their work to communicate engineering information, design, develop, operate, and maintain engineering systems. These coefficients can be derived empirically, analytically, or numerically throughout the engineering fields. They help generalize engineering problems and classify them.

Through research and experimentation, engineers are able to define/obtain coefficients in equations that they standardize to communicate engineering information.

And as a practicing engineer, you must be able to define and manipulate variables in equations, and coefficients can be helpful in furthering industry and engineering research by studying them. This is a fundamental skill!

Coefficients can represent physical properties or constants, scale or normalize variables, and indicate the strength or magnitude of a relationship. Through understanding and working with coefficients, engineers and researchers can develop and refine mathematical models, analyze and optimize complex systems, and make predictions and informed decisions.

In industry and engineering research, coefficients can be incredibly valuable in materials science—understanding material properties and behavior; fluid dynamics—modeling and optimizing fluid flow; structural analysis—predicting stress and strain on structures; and data analysis—identifying trends and patterns in complex data sets.

As a practicing engineer, being able to define and manipulate variables and understand the role of coefficients is essential for advancing industry and engineering research. Coefficients play a crucial role in equations, serving several purposes. It is important to note that a design can be messed up because of a wrong coefficient used somewhere in the equations/analysis, leading to expensive, technically unreliable design, or even a flawed design, overall! It is thus important to realize the power of coefficients in design work.

7.6.1 Scaling

Coefficients can scale the value of a variable, making it larger or smaller. A scaling coefficient is a parameter that measures how much one variable change relative to another variable. A good example is the Weibull distribution coefficient that scales the x-variable. α is the scale parameter; it scales the $x$variable. For example, the Weibull distribution coefficients have far-reaching success in the field of quality control, reliability studies, and management of the production process! See Section 7.6.9.7.1.

7.6.2 Proportionality

Coefficients can indicate the proportionality between variables, showing how much one variable change when another changes. Chapter 6 shares how civil engineers mimic the natural world. Have you realized how the size of the mouth is somewhat proportional to the body mass of an animal? The size of an animal's mouth is generally proportional to its body mass. Larger mammals, such as elephants and hippopotamuses (Figure 7.7), have bigger mouths than smaller mammals, like rodents and bats.

FIGURE 7.7 Hippopotamus.

Larger fish species tend to have larger mouths relative to their body size. The size of a reptile's mouth is often correlated with its body size. This is because the mouth size needs to be large enough to accommodate the amount of food required to sustain the animal's bodily functions and support its growth. To what extent is this true? Do snakes conform, for example? Did you ever realize that there are exceptions and variations across different species, depending on factors like feeding behavior, diet, and evolutionary adaptations? In an attempt to study this proportionality, a coefficient may be developed. Two variables to start with include the mouth size and body mass.

7.6.3 Weighting

Coefficients can help weight the importance of different variables in an equation, indicating their relative contributions.

7.6.4 Parameterization

Coefficients can be used as parameters to define a family of equations or functions, allowing for flexibility and generalization.

7.6.5 Simplification

Coefficients can simplify complex equations by combining like terms or factoring out common factors. A common factor is a number that divides two or more numbers exactly. For instance, 10 and 20 have a common factor of 2 because they are both even. The greatest common factor (GCF) is the largest number that divides two or more numbers exactly. The numbers 10 and 20 have a greatest common factor of 10. So, coefficients can factor out common factors.

7.6.6 Physical Significance

In some cases, coefficients have physical significance, representing quantities like velocity, acceleration, or friction. Coefficients are essential components of equations, allowing to scale and proportion variables, weigh their importance, parameterize and generalize equations, simplify complex expressions, and convey physical significance. We can thus modify the behavior of equations by adjusting coefficients, making equations more accurate, flexible, or applicable to various situations.

7.6.7 Conversion Factors

A conversion factor is a numerical value used to convert a quantity from one unit of measurement to another. In other words, it's a multiplier that helps you change the units of a measurement without changing its value. See the conversion factor in Manning's equation (7.71) and rational formula (7.116) for calculating peak flow discharges from catchments.

7.6.8 Correction Factors

A correction factor is a factor multiplied with the result of an equation to correct for a known amount of systemic error. Correction factors are applied when the conditions during measurement or calculation deviate from the standard conditions under which a value was initially determined.

Notes

a. A constant can be a coefficient in an equation. In algebra, a coefficient is a constant or variable that is multiplied by a variable or an expression. In other words, a coefficient is a factor that is attached to a variable or expression. For example, in the algebraic expression,

$$4x + 7$$

- The number 4 is a coefficient because it is a constant that is multiplied by the variable x.

b. In general, when a variable is used as a coefficient, it means that the magnitude of the effect of the variable being multiplied is not constant, but rather depends on the value of the coefficient variable. Using variables as coefficients adds flexibility and complexity to equations, allowing for more nuanced modeling of relationships between variables. See the Weibull distribution equation (7.123). For example, in the algebraic expression:

$$xy + 6$$

- The variable x is a coefficient because it is multiplied by the variable y. And it could also be true the other way round, but the order matters.

c. **Keynotes:** Constant, variable, coefficient.

7.6.9 Typical Examples of How Coefficients Are Used in Civil Engineering

Examples of applications of coefficients are as follows.

7.6.9.1 Coefficient of friction, μ

This is well known and applied throughout classical mechanics! The friction force, $F = \mu N$, where N is the force normal to the frictional plane. Table 7.4 provides the coefficient of sliding friction μ for different materials rubbing against each other. The coefficient of sliding friction μ scales the force normal to the frictional plane. The resulting force can never be the same on different materials.

To understand the effect of friction, take an example of how donning hand gloves alters the tactile experience, compared to bare hands, when handling rough bricks or operating machinery in a workshop. The sensation is distinct, with gloves dampening the texture and temperature of surfaces, while bare hands feel every grit and groove. See Chapter 6 for more details.

TABLE 7.4 Coefficient of sliding friction μ for different materials rubbing against each other

MATERIALS	COEFFICIENT OF SLIDING FRICTION μ
Metal on metal	0.15–0.6
Metal on wood	0.20–0.60
Wood on wood	0.25–0.50
Rubber on paving	0.70–0.90
Nylon on steel	0.30–0.50
PTFE on steel	0.05–0.20
Metal on ice	0.02
Masonry on masonry	0.60–0.70
Masonry on earth	0.50
Earth on earth	0.25–1.00

Role: In this case, the coefficient of sliding friction μ indicates the proportionality between variables, that is, the normal force and the friction force, showing how much the friction force changes when the normal force changes.

7.6.9.2 Fill load coefficients for pipeline backfill loads

In buried sewer and water pipeline construction, fill loads are calculated based on **fill load coefficients**. For normal ordinary trench construction, the coefficient C can be obtained from the formula (7.115) to use in Marston's equation (7.143):

$$C = \frac{1 - e^{-2\mu'KH/B}}{2\mu'K} \tag{7.115}$$

where

H: depth of fill above the pipe
B: width of the trench just below the top of the pipe
K: ratio of active lateral pressure to vertical pressure
μ': **coefficient of sliding friction** between the fill material and sides of the trench (see Section 7.6.9.1).

The coefficient of friction, C, is applied in Marston's equation to calculate fill loads acting on buried pipes. Marston's equation (7.142) is theoretical.

Role: Influenced by the trench geometry (depth to width ratio) and the friction between the backfill and trench walls, its calculation is essential for ensuring the structural integrity and long-term performance of the pipeline (influences/scales the fill loads).

7.6.9.3 Coefficient of linear thermal expansion

The coefficient of linear thermal expansion is a material property that measures how much a material expands when it's heated. It's represented by the symbol α. Through experimentation, engineers were able to define a constant for several materials:

$$\alpha = \frac{\Delta L}{L_0 \Delta T} \tag{7.116}$$

where

- α is the coefficient of linear thermal expansion per degree Celsius
- ΔL is the change in length of the material due to heating or cooling
- L_0 is the original length of the material
- ΔT is the change in temperature

Role: It is used to predict the dimensional changes of materials during heating or cooling.

7.6.9.4 Manning's roughness coefficient

Manning's roughness coefficient is used in Manning's equation to calculate flow in open channels. Manning's roughness coefficient, denoted by letter n, is a parameter used to calculate the resistance to flow in open channels. It's also known as the Manning coefficient. Coefficients for some commonly used surface materials are listed in Table 7.5.

Role: It represents the resistance to flow in open channels and is an important parameter in hydraulic calculations. It quantifies the frictional effect of the channel's surface material and shape on flow velocity and discharge. Higher n values indicate greater roughness and reduced flow.

7.6.9.5 Runoff coefficient in the rational formula

The rational formula used to compute peak discharge is a theoretical equation derived from the first principles of calculus, where it expresses area and rainfall intensity as independent variables. The rational formula for peak flow discharge computation (metric units) is

$$Q = 0.278CiA \tag{7.117}$$

where

Q: discharge $\left(\text{m}^3\right)$
C: runoff coefficient
i: runoff intensity $\left(\text{mm/hr}\right)$
A: area $\left(\text{km}^2\right)$

TABLE 7.5 Manning coefficients for select materials

TYPE OF PIPE	MANNING'S n	
	MIN.	MAX.
Glass, brass, or copper	0.009	0.013
Smooth cement surface	0.010	0.013
Wood stave	0.010	0.013
Vitrified sewer pipe	0.010	0.017
Cast iron	0.011	0.015
Precast concrete	0.011	0.015
Cement mortar surfaces	0.011	0.015
Common-clay drainage tile	0.011	0.017
Wrought iron	0.012	0.017
Brick with cement mortar	0.012	0.017
Riveted steel	0.017	0.020
Cement rubble surface	0.017	0.030
Corrugated metal storm drain	0.020	0.024

The rational method expresses the intensity of rainfall and area as independent variables. In Section 3.3.1, we saw how Newton studied the product of independent variables introducing calculus, together with Gottfried Wilhelm Leibniz (1646–1716). From that time onward, the world learnt how essential rates of change were.

The rational formula is the simplest method to determine peak discharge from drainage basin runoff. Did you know that you might begin with a rate of change or simply units to investigate variables that can contribute to that formula?

The famous rational formula calculates the peak discharge in m^3/s or ft^3/s. However, discharge is the rate of change of volume through a surface per unit time. Here, a rate of change is a differential equation that can be phrased as

$$\Delta Q = \frac{\Delta V}{\Delta t} \tag{7.118}$$

Determination of volume is pretty straightforward; the same applies to time. For a cylindrical container, the volume can be determined from the usual formula, $\pi r^2 h$. From this formula, several variables, onto which calculus can be applied, that is, r, h, t, V, and Q depending on how they model comes out, are available.

Using the rational formula, the quantity of rainfall that falls on an area can be determined from the rainfall intensity measured in mm/hr and area in km^2.

This gives

$$Q = i \times A \tag{7.119}$$

$$Q = \frac{mm}{hr} \times km^2 = \frac{(0.001\ m)}{60 \times 60\ s} \times (1{,}000\ m)^2$$

$$Q = 0.278\ m^3/s$$

Therefore, 0.278 is the conversion factor. See Section 7.6.7.

So,

$$Q = 0.278iA \tag{7.120}$$

It was noticed that not all rainfall is converted into runoff; some evaporates, while the remaining infiltrates into the ground, and this depends on the type and nature of the surface where rain falls. Therefore, a runoff coefficient was introduced into the equation to cater for the losses. Since the coefficient is a reduction coefficient (scales down), it is between 0 and 1. Please note that if the coefficient is catering for additional runoff, then it would be greater than 1. Therefore, its role is to determine how much rainfall is scaled down due to imperviousness.

The runoff coefficient (C) is a dimensionless value between 0 and 1 that represents the proportion of rainfall that becomes runoff, rather than being absorbed or evaporated. C is the runoff coefficient, an empirical coefficient representing rainfall–runoff correlation. The typical runoff coefficient values include the following: impervious surfaces (such as roofs and roads): 0.7–0.9, pervious surfaces (such as lawns and parks): 0.1–0.3, mixed urban areas: 0.3–0.6, and forested areas: 0.01–0.1. You can see that the runoff generated is greatest for highly impervious surfaces, implying that minimal water infiltrates into the ground.

The factors influencing runoff coefficient include land use (urban, rural, and agricultural), surface type (paved, unpaved, and vegetated), slope and topography, soil type and permeability, climate, and rainfall intensity. Therefore, once the rational technique was paused, engineering researchers furthered the quest for knowledge to understand the systems' runoff determination much better. Therefore, the rational formula is given as

Metric units

$$Q = 0.278CiA \tag{7.121}$$

or simply

$$Q = CiA \tag{7.122}$$

Role: The runoff coefficient scales the peak discharge received in an area—depending on the imperviousness levels of the area; the peak discharge is controlled. 0.278 is a conversion factor.

7.6.9.6 Froude number in open-channel flow

The Froude number ties the energy and momentum equation together, and it is very helpful in hydraulic system designs. It works as a coefficient in energy and momentum equations, thus scaling the sizes of the hydraulic structures:

- For supercritical flow, $F_r < 1$.
- For supercritical flow, $F_r > 1$.
- For critical flow, $F_r \approx 1$.

For detailed information, visit reference 7, Chapters 5 and 6.

Role: Froude number ties the energy and momentum equation in hydraulic structure system designs—in open-channel flow problems. It is a condition in design that program conditions must obey. Froude number is a dimensionless number that compares the inertial forces to the gravitational forces in a fluid flow.

See Section 7.4.8 (grouping equations).

7.6.9.7 Probability distribution functions' coefficients

A probability distribution is a mathematical function that describes the probability of different possible values of a variable. Probability distributions are often depicted using graphs or probability tables. Engineers highly rely on PDFs to model engineering problems. The coefficients in the PDFs communicate essential information that is normally relied on to improve engineering systems. Some of the PDFs engineers work with include the Weibull distribution, exponential distribution, normal distribution, log-normal distribution, gamma distribution, Gumbel distribution, extreme value distribution, Poisson distribution, and Log-Pearson III distribution. Most of these distributions rely heavily on coefficients to predict and/or provide useful information necessary for designing, developing, planning, evaluation, maintenance, and operation of engineering systems. They are very necessary especially in quality control and management.

7.6.9.7.1 *Weibull probability distribution function coefficients*

The Weibull probability distribution function provides a classic example of how coefficients can be used to communicate to management essential engineering information for reliability improvement—quality management. Ernst Hjalmar Waloddi Weibull, a Swedish engineer and statistician, introduced the Weibull distribution in 1951. For the purposes of illustrating how coefficients are so powerful in PDFs, consider the Weibull two-parameter distribution given by the following formula:

$$f(x;\alpha,\beta) = \begin{cases} \dfrac{\beta}{\alpha}\left(\dfrac{x}{\alpha}\right)^{\beta-1} e^{-(x/\alpha)^{\beta}} & x \geq 0, \\ 0 & x < 0, \end{cases} \tag{7.123}$$

where

> x is the random variable, in this case the monthly maximum rainfall received in a year.
> α is the scale parameter; it scales the x variable. The scale parameter determines the range of the distribution.
> β is the shape parameter.

α and β are called Weibull coefficients. The Weibull shape parameter, called β, indicates whether the failure rate is increasing, constant, or decreasing. $\beta < 1.0$ indicates a decreasing failure rate. $\beta = 1.0$ indicates a constant failure rate. $\beta > 1.0$ indicates an increasing failure rate. See example 7.20.

Weibull's work revolutionized reliability engineering, and his distribution has since been widely used in various fields, including reliability engineering, quality control, materials science, biology, and finance. Weibull's discovery of the Weibull distribution was a significant contribution to statistics and probability and has had a lasting impact on many fields.

Role: Management of manufacturing companies uses Weibull coefficients to make lasting reliability improvements in their products. Failure data are collected that are assumed to follow the Weibull distribution. The distribution is important in reliability studies and quality control because it studies and models failure rates.

7.6.9.8 Buckling coefficient

Euler's column buckling theory is a fundamental concept in structural engineering that explores the stability and load-bearing capacity of slender columns. Also known as column buckling, it refers to the failure of a structural element subjected to high compressive forces. The critical buckling load is calculated by the following formula:

$$P_{cr} = \frac{\pi^2 EI}{(KL)^2} \tag{7.124}$$

where EI is the flexural rigidity, L is the length of the column, and the buckling coefficient could be $K = 0.5$ (fixed support), $K = 0.7$ (pinned support), $K = 1$ (pinned support), and $K = 2$ (free). See Figure 7.8.

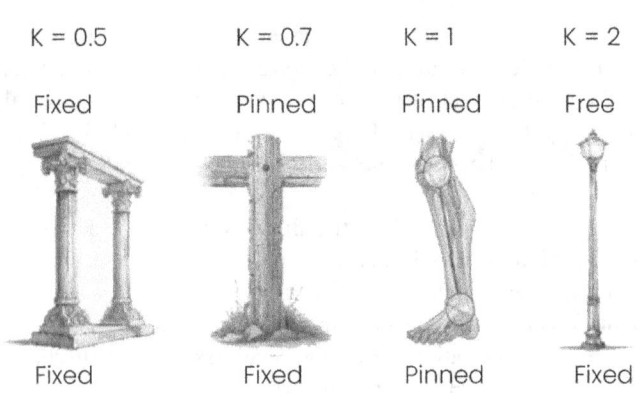

K = 0.5	K = 0.7	K = 1	K = 2
Fixed	Pinned	Pinned	Free
Fixed	Fixed	Pinned	Fixed

FIGURE 7.8 Buckling coefficients.

Note: Euler's differential equations are used to determine buckling coefficients, which is one of the applications of differential calculus in civil engineering science.

Role: The critical buckling coefficient changes its value in terms of loading and boundary condition. The buckling coefficient is a factor used to predict the critical load at which a slender structure, such as a column or plate, will buckle under compressive stress. It helps prevent structural failures and ensures the safety of buildings, bridges, and other structures.

7.6.9.9 Factor of safety

In engineering, a factor of safety (FOS), also known as safety factor, expresses how much stronger a system is than it needs to be for an intended load. It is the ratio of the maximum load a structure can support to the actual load applied. It is a measure of the safety margin or redundancy built into a structure and is typically expressed as a number greater than 1. A higher FOS indicates a more robust and reliable structure.

The FOS is a numerical value that expresses how much stronger a system or structure is than it needs to be for its specified maximum load. In the design of civil engineering structures, we have global FOS, local FOS, and material FOS.

- **Global FOS** applies to the entire structure or system.
- **Local FOS** applies to specific components or elements.
- **Material FOS** accounts for material uncertainties and variability.

FOS can be considered coefficients that are applied to various loads, stresses, or material properties to ensure safety and reliability in engineering design. Engineers ensure that their designs meet specific safety and reliability criteria by applying an FOS. FOSs are used to ensure that the designs meet a certain degree of **safety** and **reliability**, hence lowering the susceptibility to collapse, whatsoever.

$$\begin{aligned} \text{Factor of safety (FOS)} &= \frac{\text{Maximum strength}}{\text{Design load}} \\ &= \frac{\text{Ultimate stress}}{\text{Allowable stress}} \end{aligned} \tag{7.125}$$

7.6.9.9.1 Example [7.11] for illustrative purposes

Consider a bridge with a permissible stress of 1000 units and ultimate stress, that is, the point after which failure occurs, of 4000 units. The FOS, in this case, would be calculated as follows:

$$FOS = \frac{4,000}{1,000} = 4 \tag{7.126}$$

This implies that the bridge is four times stronger than it theoretically needs to bear the maximum load. If the bridge were to bear more weight than estimated, up to four times the initial assumption, it would not fail suddenly.

7.6.9.9.2 Role of factor of safety

In the realm of engineering, the FOS is a crucial concept. This safety buffer provides engineers with a sense of guarantee, ensuring that structures are robust enough to handle unforeseen stresses and loads without failure. Its role extends far beyond mere calculations and sheds light on the resilience of engineering designs against real-world conditions.

FOS plays a crucial role in engineering design, particularly in civil/structural and mechanical engineering. Its primary purpose is to ensure that a system, structure, or component can withstand various loads and stresses without failing. In civil engineering and building design, engineers employ an FOS to consider potential consequences of structural failure. FOS also takes into account the variability of loads, like wind and seismic activities. Key roles of the FOS include the following:

a. **Risk reduction:** FOS helps reduce the risk of failure by providing a buffer against uncertainties, variability, errors, and unexpected loads. FOS helps prevent catastrophic failures by providing a margin of safety.

b. **Reliability:** FOS ensures that a design can withstand various loads and stresses, providing a reliable and safe structure. It accounts for uncertainties in material properties, loading conditions, calculations, and other factors that may affect the design, thus enhancing reliability. FOS contributes to the overall reliability of the system or structure. FOS accounts for unpredictable variables, such as design inaccuracies, untested load conditions, and material inconsistencies. It provides a cushion against the unforeseeable.

c. **Economic considerations:** A higher FOS might engender extra costs due to increased material usage. Thus, determining the optimal FOS balances safety and financial feasibility. It helps balance safety with cost considerations, avoiding overdesign. By applying an appropriate FOS, designers can optimize material usage, reducing waste and costs.

d. **Legal implications:** In many territories, FOS is not optional. Various jurisdictions have established minimum FOS values for different structural elements and materials, rendering adherence a legal obligation. Engineers have a responsibility to ensure the safety and reliability of the products they design. FOS is not just a technical requirement but also a legal and ethical necessity in many industries.

e. **Code compliance:** Many design codes and standards require a minimum FOS to ensure public safety and structural integrity. FOS often meets regulatory requirements and industry standards.

f. **Design optimization:** FOS can be used to optimize designs, balancing factors like cost, weight, and performance. FOS also ensures that the system operates within a safe and efficient range. FOS accounts for variations in loads, such as wind, earthquakes, or other external forces. FOS introduces conservatism in design, ensuring that the system is more likely to withstand extreme conditions.

7.6.9.9.3 Factor of Safety—Key takeaways

- FOS is a crucial engineering concept that calculates the ratio of the actual strength of a structure or material (ultimate stress) to the maximum stress it should bear in service (allowable stress).
- Low value of FOS indicates a high level of risk, while a high FOS shows a cautious design approach. Legal and regulatory standards often require minimum FOS values for specific applications.
- FOS can be calculated using the formula FOS = Ultimate Stress/Allowable Stress.
- FOS helps manage uncertainties in design and provides a margin for errors and unforeseen conditions, thus ensuring safety in structures and materials used in engineering.
- The value of FOS varies for different materials like steel, concrete, wood, and aluminum, depending on their usage and the expected environmental conditions.

7.6.9.9.4 Typical examples of FOS in civil engineering

In civil engineering, FOSs are used to ensure that structures can withstand various loads and stresses without failure. Typical FOSs are shown in Table 7.6.

7.6.9.10 Obtaining coefficients in equations

Two approaches are used—that is, analytical/statistical approaches and laboratory methods.

7.6.9.10.1 Analytical and statistical approaches

See Weibull analysis example (7.20)—obtaining β and α. In the analytical and statistical approaches, coefficients are obtained by graphs generated from plotting data.

7.6.9.10.2 Laboratory

In Sections 2.5.2 and 7.4.5.1.3, we saw how Terzaghi opened doors for engineers to understand soil mechanics and geotechnical engineering much better. In the design of civil engineering structures, geotechnical engineers determine most of the coefficients in the laboratory by collecting field samples and testing them in the laboratory. In other cases, engineers do *in situ* testing of materials to come up with coefficients or factors necessary for building equations that support them to evaluate the feasibility of their designs or proposals. As an example, the following **parameters or coefficients** are determined in the geotechnical laboratory for use in geotechnical engineering such as foundation design. Geo-labs typically determine the following parameters:

TABLE 7.6 Typical values of FOS

NO.	DISCIPLINE/STRUCTURE	FOS	REFERENCE
1	Structural design	1.5–2.0 (e.g., building columns, beams, and foundations)	EC 2, ACI 318, and IS 456
2	Soil mechanics	2.0–3.0 (e.g., retaining walls, slopes, and foundations)	EC 7 (EN 1997)
3	Geotechnical engineering	2.5–3.5 (e.g., tunnels, excavations, and underground structures)	EC 7
4	Bridge design	1.2–1.5 (e.g., beams, girders, and suspension systems)	BS 5400 and EC 3 (EN 1993)
5	Highway design	1.1–1.5 (e.g., pavement thickness and slope stability)	EC7, AASHTO 93, and FHWA pavement thickness design guidelines
6	Water resources	1.2–1.5 (e.g., dam design, levees, and flood control structures)	EC 7 (EN 1997)
7	Foundation design	2.0–3.0 (e.g., shallow and deep foundations)	EC 7
8	Retaining wall design	1.5–2.0 (e.g., gravity walls and cantilever walls)	EC 7
9	Pavement design	1.1–1.5 (e.g., asphalt, concrete, and composite pavements)	EC7, AASHTO 93, and FHWA pavement thickness design guidelines
10	Hydraulic structures	1.2–1.5 (e.g., culverts, storm drains, and water supply systems)	BS EN 1916

1. **Shear strength parameters:**
   a. Cohesion (c)
   b. Angle of internal friction ($\varnothing$)
2. **Unit weight** (γ): The weight of the soil per unit volume.
3. **Water content** (w): The percentage of water in the soil.
4. **Porosity** (n): The ratio of the volume of voids to the total volume of the soil.
5. **Permeability** (k): The ability of the soil to transmit water.
6. **Atterberg limits**
   a. **Liquid limit (LL):** The water content at which the soil behaves like a liquid.
   b. **Plastic limit (PL):** The water content at which the soil behaves like a plastic.
   c. **Shrinkage limit (SL):** The water content at which the soil stops shrinking.
7. **Compaction characteristics**
   a. **Maximum dry density (MDD):** The maximum density achieved by compacting the soil.
   b. **Optimum moisture content (OMC):** The moisture content at which the soil can be compacted to its maximum density.
8. **Soil classification**
   These geo-laboratory tests provide essential data for various geotechnical applications, ensuring safe and stable designs for infrastructure projects.

7.7 FORMATIVE RESEARCH AND DEVELOPMENT: THE CIVIL ENGINEER'S EFFECTIVE DESIGN GUIDE

Formative research is used to gather insights and data during the initial stages of developing a project, program, or an intervention. Formative development in civil engineering, on the other hand, involves shaping and refining a project through design and construction to achieve its full potential. It is the process of designing, testing, and refining a product or service, often involving iterative feedback and improvement. Formative R&D aim to understand the problem; gather information about the issue, audience, or context to inform project or program design and development; and use findings to shape the design, necessary content, or intervention.

7.7.1 Role of Formative Research to a Civil Engineer

Engineers identify issues notes them to design effective project outputs and outcomes. As an example, in some cases, project budgets are often estimated by benchmarking costs from similar, previously completed projects—identified as case studies. This approach helps establish a reference point for budgeting and cost planning. From these case studies, engineers may draw study propositions, in both the commercial perspective and technical or engineering aspect, and either reject or confirm the propositions from the findings, leading to continuous improvement in the budgeting or planning/design process.

Formative research helps create project outputs that are more likely to achieve desired project outcomes, ultimately leading to more effective and successful projects. A perfect civil engineering example is a new bridge development connecting point A with B:

- **Output:** a new bridge is built to connect region A to region B previously disconnected—people and vehicles finding difficulty crossing form one point to another.
- **Outcome:** reduced traffic congestion, improved commute times, and increased economic activity.

Formative research helps engineers focus on user-centered design and development—prioritizing the needs and preferences of the end-users throughout the design process—seeing things at both output and outcome levels!

In civil engineering projects, formative research may overlap with 'appropriate research' or 'market research' or 'baseline surveys'—basically research intended to provide practical insights into the design, development, analysis, and evaluation of civil engineering projects. These types of researches can overlap with formative research, although they serve different purposes, in the classical sense. For example, market research focuses on market dynamics, while formative research focuses on user-centered design and development. Baseline surveys provide civil engineers with essential data on existing site conditions and socioeconomic status of beneficiary communities, serving as a reference point for project design, implementation, and monitoring. Engineers can understand the site's topography, geology, and environmental factors, informing design decisions and ensuring projects are tailored to specific site needs, leading to better outcomes and outputs and more sustainable infrastructure. However, when the research/survey provides valuable user insights that inform design and development decisions, it can be formative.

Civil engineers gather insights and collect various forms of data. This systematic investigative approach generates vital insights into target audiences, their needs, and contextual factors that shape behavior. Civil engineering is constantly evolving, and user comfort is a primary goal in an effective project design and development. Therefore, civil engineers always design functional and reliable systems from a well-informed point of view. They mind about people's attitudes, behaviors, and beliefs in their user-centered design philosophy.

While formative research may not always be strictly necessary, it can be highly beneficial in ensuring that civil engineering designs meet user needs and are effective, efficient, reliable, and sustainable. In many cases, organizations or companies that overlook formative research risk developing projects, programs, or interventions based on assumptions rather than evidence, potentially leading to reduced program or project effectiveness and resource wastage. Engineers strictly follow evidence as the right approach to design and develop reliable infrastructure systems. Formative research can help address these common issues in civil engineering design through an effective needs assessment integrated with feasibility studies.

Remember we design infrastructure for people, and we intend to solve community problems, from water distribution systems, to roads and highways, and hospitals, among many other infrastructure systems. All these infrastructure projects need appropriate research prior to designing and developing effective projects or program projects.

For example, in a community health program, formative research might reveal new barriers or emerging needs, prompting timely adjustments to the intervention strategy. To develop effective healthcare programs, healthcare interventions are necessary through conducting formative research. Public health awareness campaigns, water supply and sanitation projects, wastewater projects, roads and highways projects all need proper formative research.

Unlike summative research, which evaluates the outcomes after a project or program's completion, formative research focuses on understanding the needs, behaviors, and contexts of the target audience to tailor designs or interventions more effectively.

The primary purpose of formative research is to ensure that a final program or project is well suited to the target audience's characteristics and needs. It is dangerous to finish a project and realize that it doesn't serve the intended purpose.

Formative research is valuable for both programs and projects. It's a research process used to gather insights and data during the early stages of development to inform decision-making and improve outcomes. Essentially, it's a way to gather information and make adjustments throughout the development process, whether it's for a larger program or a smaller project.

In projects, formative research is used to understand user needs, gather feedback, and identify areas for improvement during the development process. This is particularly helpful in areas such as user experience (UX) research, where formative research can guide the design of products and services. For example, building a new road needs proper formative research to understand the social and economic factors that may impact road development at both outcome and output levels.

In the design and implementation of programs, formative research is used to understand the needs of the target audience, identify potential barriers, and ensure the program is culturally and geographically appropriate. For example, formative research might be used to determine the most effective ways to reach a specific population, understand their behaviors and attitudes, and identify the best strategies for promoting a desired behavior change. Also, developing a comprehensive transportation system, which might include multiple projects (e.g., building roads and improving public transit), needs a proper formative research. On the other hand, implementing a new traffic management system to reduce congestion could be an intervention that may require formative research.

7.7.2 Key Components of Formative Research

The key components of formative research include needs assessment, audience segmentation, pretesting, and continuous feedback. Engineers usually conduct needs assessment, pretesting, and continuous feedback in their work. This process helps ensure that the project addresses real and pressing issues. The assessment may be done using surveys, focus group discussions, evaluation of existing data, and many more. By understanding the specific needs and priorities of the target audience, researchers can tailor project to be more relevant and impactful.

Key steps in the formative research process include defining objectives, designing the research plan, data collection, data analysis, and applying findings. Formative research needs to be well defined, considering the context, selecting appropriate methods, analyzing data effectively, and translating findings into practical applications.

During the data analysis stage, mathematics is applied. Qualitative data analysis involves identifying themes, patterns, and trends from textual or visual data, while quantitative data analysis involves statistical analysis to determine the prevalence and relationships between variables. Tools and software such as SPSS or Excel for quantitative data and NVivo for qualitative data can be used to assist in this process.

7.7.3 Formative Research Methods

Although formative research often focuses on qualitative insights, quantitative methods are used to complement the findings.

- **Qualitative methods:** These include focus groups discussions, in-depth interviews, key informant interviews, observations, and case studies.
- **Quantitative methods:** These are surveys and structured observations.
- **Mixed methods:** This includes combining qualitative and quantitative approaches. For example, a study might use qualitative focus groups and key informant interviews to explore perceptions and then design a quantitative survey to measure the prevalence of these perceptions in the larger population.
- **Sequential and concurrent designs:** This approach allows for the initial qualitative findings to inform the design of the subsequent quantitative phase, or vice versa. Concurrent designs involve collecting both qualitative and quantitative data simultaneously and integrating the findings during the analysis phase. This enables provision of a comprehensive view of the research problem. By employing a variety of formative research methods, researchers can gather rich, multifaceted data that provides a thorough understanding of the target population, context, and issues at hand.

In formative research, engineers may not typically formulate equations in the classical sense. Instead, they may develop conceptual frameworks (Figure 7.9), identify relationships between variables, and create models (qualitative or quantitative). A conceptual framework is a visual or theoretical representation of the relationships between variables in a study or research project. A conceptual framework illustrates the expected relationship between variables. It defines the relevant objectives for the research process and maps out how they come together to draw coherent conclusions.

Notes

- **Dependent variables:** These are the variables that are being measured or observed in response to changes made to the independent variables. They are also known as **outcome** or response variables. The expected effect is the **dependent** variable (the response, or outcome variable).

- **Independent variables:** These are the variables that are manipulated or changed by the researcher to observe their effect. They are also known as predictor or explanatory variables. The expected cause is the **independent** variable (the predictor, or explanatory variable).
- **Moderating variables:** Affect the strength or direction of the relationship between independent and dependent variables.
- **Mediating variables:** Explain the underlying mechanism or process by which the independent variable affects the dependent variable.
- **Control variables** are factors that researchers keep constant in a study to ensure that any changes or effects observed are due to the independent variable (the variable being manipulated) and not other external factors.
- **Extraneous variables** are variables that are not of interest to the research study but can affect the outcome of the study.

When equations are involved, engineers might use statistical models (e.g., regression analysis) to analyze the data, develop mathematical representations of systems or processes, and apply existing mathematical frameworks to analyze data. The goal is to understand complex systems or phenomena, identify patterns or trends, and inform design decisions. These research methods developed by social scientists are an excellent tool for empirical scientists and engineers because of their radical mechanisms of quickly breaking through barriers and use results to improve systems. In doing so, graph plotting can be accelerated from which empirical equations and formulas can be derived—and that is the primary role of an engineer.

7.7.4 The Role of Dependent and Independent Variables

In Figure 7.9, we saw how independent variables inform the dependent variables. Engineers apply this concept throughout their work, from formative R&D to various studies and projects, to drive innovation and improvement. Independent variables are manipulated or changed by the researcher to observe their effect on the dependent variable—variables that are being measured or observed in response to changes made to the independent variable. As an example, engineers deal with differential equations. A differential equation is a mathematical equation that relates some function with its derivative.

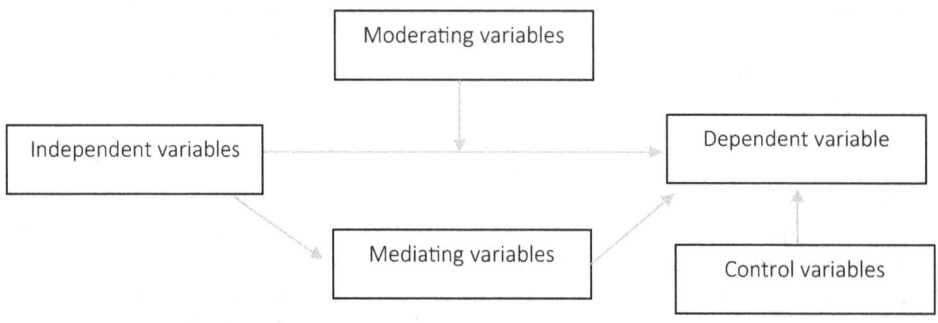

FIGURE 7.9 Typical conceptual framework.

The variables with respect to which differentiation occurs are called independent variables, while those that are differentiated are dependent variables. Ordinary differential equations (ODEs) involve a single independent variable, whereas partial differential equations (PDEs) involve multiple independent variables. A single ODE typically features one independent variable and one dependent variable. Just as algebraic equations can form systems that require simultaneous solutions, ODEs can also form **coupled systems** with one independent variable and multiple dependent variables.

Dependent variable is an outcome or response variable, that is, the variable being measured or observed in an experiment or study. It is affected by the independent variable—the changes in the dependent variable are influenced by the independent variable. The independent variable is a cause or predictor variable being manipulated or changed in an experiment or study. It influences the dependent variable as the changes in the independent variable directly affect the dependent variable.

In the PDE (7.127)

$$\frac{\partial f}{\partial x} + \frac{\partial f}{\partial y} = x^2 + 0.5y \tag{7.127}$$

The independent variables are x and y, and the dependent variable is f.

In the ODE (7.128)

$$\frac{d^2 f}{dx^2} - x\frac{df}{dx} = \sin 3x \tag{7.128}$$

The independent variable is x, and the dependent variable is f. These concepts are important in civil engineering modeling—how the changes in independent variables, extraneous variables, mediating, moderating, and control variables impact the dependent variable in research. A perfect civil engineering example is the rational formula, which expresses the rainfall intensity and area as independent variables, informing the peak discharge of a catchment area:

$$Q = CiA \tag{7.129}$$

$$\frac{\partial Q}{\partial A} = Ci$$

7.8 LOGARITHMS AND ITS APPLICATIONS IN ENGINEERING SCIENCE

The logarithm is the inverse function to exponentiation. That means that the logarithm of a number x to the base b is the exponent to which b must be raised to produce x. For example, since $1000 = 10^3$, the logarithm base 1010 of 1000 is 3, or $\log_{10}(1000) = 3$. The logarithm of 1000 to base 10 is 3. Logarithms have numerous applications in engineering science.

7.8.1 Simplify Complex Calculations

This involves applying logarithmic properties to rewrite and simplify a complex equation. Engineers usually rewrite equations and apply logarithmic properties. Consider example 7.12. By using logarithms, you can quickly simplify complex exponential equations and solve for the variable. This technique is useful in engineering where exponential relationships are common. So, engineers heavily rely on logarithms where exponential relations are involved in models they develop.

Example [7.12]: Using logarithms to solve equations
   Solve

$$4^x = 8x \tag{7.130}$$

Solution
   By inspection, we notice that

$$4^2 = 16$$

And that

$$8 \times 2 = 16$$

This would definitely mean $x = 2$, but how do we solve the equation manipulating the LHS and RHS?

To solve the equation $4^x = 8x$ logarithms are used to isolate the variable x.

Step 1: Take the logarithm of both sides

$$\log 4^x = \log 8x$$

Step 2: Apply the logarithmic properties

- $\log(a^b) = b\log(a)$
- $\log(ab) = \log(a) + \log(b)$

Thus, the equation becomes

$$x\log(4) = \log(8) + \log(x) \tag{7.131}$$

Step 3: Isolate the term with x

$$x\log(4) - \log(x) = \log(8)$$

Step 4: Factor out x

$$x = \frac{\log(8)}{\left(\log(4) - \dfrac{1}{\log(x)}\right)} \tag{7.132}$$

Note that we can also express the equation as follows:

$$2^{2x} = 2^3 x \tag{7.133}$$

In terms of the base 2, the equation can be expressed as follows:

$$2x = 3 + \log_2 x \qquad (7.134)$$

Factoring out x gives

$$x = \frac{3}{\left(2 - \dfrac{1}{\log_2 x}\right)} \qquad (7.135)$$

Note that equations (7.131) and (7.134) are implicit solutions, meaning x cannot be solved explicitly. Therefore, x can be approximated using numerical methods or graphing. Generating graph plots for the two equations—see Table 7.7.

We then plot the graphs of

- $y = x \log 4$
- $y = \log 8 + \log x$
- $y = 2x$
- $y = 3 + \log_2 x$

In that case, we clearly see that the role of logarithms and graphical methods cannot be overemphasized in engineering science. Logarithms are a fundamental tool in engineering science, enabling engineers to tackle complex problems and analyze various phenomena across different disciplines. This helps engineers model real-world phenomena, analyze and visualize data, and make accurate predictions and designs. The linearized relationships on a graph plot can help identify patterns, such as straight lines, curves, or deviations from expected behavior. See Figures 7.10 and 7.11.

TABLE 7.7 Log–log values for equations (7.86) and (7.88)

x	$y = x\ LOG\ 4$	$y = x\ LOG8 + LOG\ x$	$y = 2x$	$y = 3 + LOG_2\ x$
1	0.60206	0.90309	2	3
1.2	0.722472	0.982271	2.4	3.263034
1.4	0.842884	1.049218	2.8	3.485427
1.6	0.963296	1.10721	3.2	3.678072
1.8	1.083708	1.158362	3.6	3.847997
2	1.20412	1.20412	4	4
2.2	1.324532	1.245513	4.4	4.137504
2.4	1.444944	1.283301	4.8	4.263034
2.6	1.565356	1.318063	5.2	4.378512
2.8	1.685768	1.350248	5.6	4.485427
3	1.80618	1.380211	6	4.584963
3.2	1.926592	1.40824	6.4	4.678072
3.4	2.047004	1.434569	6.8	4.765535
3.6	2.167416	1.459392	7.2	4.847997
3.8	2.287828	1.482874	7.6	4.925999
4	2.40824	1.50515	8	5
4.2	2.528652	1.526339	8.4	5.070389
4.4	2.649064	1.546543	8.8	5.137504
4.6	2.769476	1.565848	9.2	5.201634
4.8	2.889888	1.584331	9.6	5.263034
5	3.0103	1.60206	10	5.321928

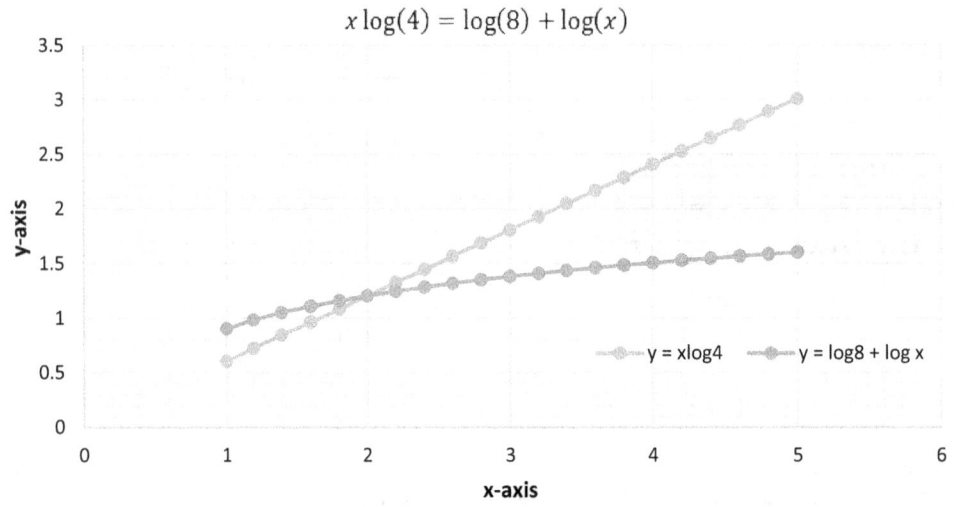

FIGURE 7.10 Plot of solution for $x \log(4) = \log(8) + \log(x)$.

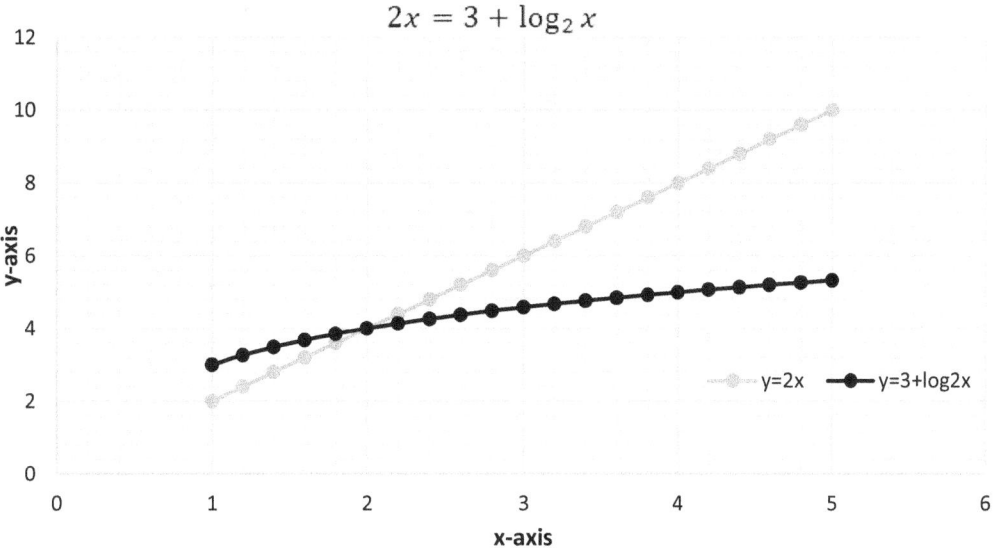

FIGURE 7.11   Plot of solution for $2x = 3 + \log_2 x$

7.8.2 Logarithmic Scales

A logarithmic scale is a way to display numerical data that spans a large range of values. It's often used in graphs and charts to show how quickly values are changing. A logarithmic scale is a method for graphing and analyzing a large range of values in a compact form. Logarithmic scales are useful when the data you are displaying are much less or much more than the rest of the data or when the percentage differences between values are important [8].

Unlike commonly used linear functions that show an increase or decrease along equivalent or equally spaced-out increments, log scales are exponential, which means increasing quickly by large numbers. This is the primary technique engineers use to generate complex graphs.

On a logarithmic scale, numerical values are represented on an axis, where the scale is proportional to the logarithm of the values. In other words, each tick mark on the axis represents a multiplicative change (e.g., $10\times$, $100\times$, and $1000\times$) rather than an additive change (e.g., $+10$, $+20$, and $+30$).

7.8.2.1 Example 7.13 for illustrative purposes

What is the log scale between 10 and 100?

The logarithm of 10 is 1.0, and the logarithm of 100 is 2.0, so the logarithm of the midpoint is 1.5. What value has a logarithm of 1.5? The answer is $10^{1.5}$, which is 31.62. So, the value half-way between 10 and 100 on a logarithmic axis is 31.62. See Figure 7.12.

Notes

- The multiplicative change (factor) is 1.122.
- On the logarithmic scale, mid-way between 10 and 100 is 31.62 instead of 60.
- Adding more grid lines (tick marks) across the scale would show an uneven distribution. For example, tick marks between 79.79 and 100 would appear closer together due to the logarithmic scaling, while

the grid lines (tick marks) between 10 and 19.95 would appear farther apart. This uneven distribution is due to the logarithmic scale. In a logarithmic scale, the same physical distance on the graph represents a much larger change in value at the higher end of the scale compared to the lower end due to multiplicative change.

Suppose we wanted a plot for Figures 7.6c, 7.10, and 7.11 on the same graph, the graph would look like in Figure 7.13a:

- $y = x\log 4$ (series 3)
- $y = \log 8 + \log x$ (series 4)
- $y = 2x$ (series 5)
- $y = 3 + \log_2 x$ (series 6)
- $xy + \dfrac{e^{-2x}}{y} - 4 = 0$ (series 1 and 2)

7.8.2.2 Examples of logarithmic scales in civil engineering

Bearing capacity factors plotted for shallow foundations bearing capacity determination under drained conditions. This graph in Figure 7.13 utilizes the logarithmic scale on the y-axis that denotes bearing capacity factors. See equations (7.82), (7.83), and (7.84). The graph is a plot of bearing capacity factors against the internal angle of shearing resistance (internal angle of friction).

7.8.3 Log–Log Plots

A log–log plot is a scatterplot that uses logarithmic scales on both the x-axis and y-axis. This type of plot is useful for visualizing two variables when the true relationship between them follows a power law.

A log–log plot, also known as a log–log graph or double logarithmic plot, is a type of graph where both the x-axis (horizontal) and y-axis (vertical) are scaled logarithmically. In a

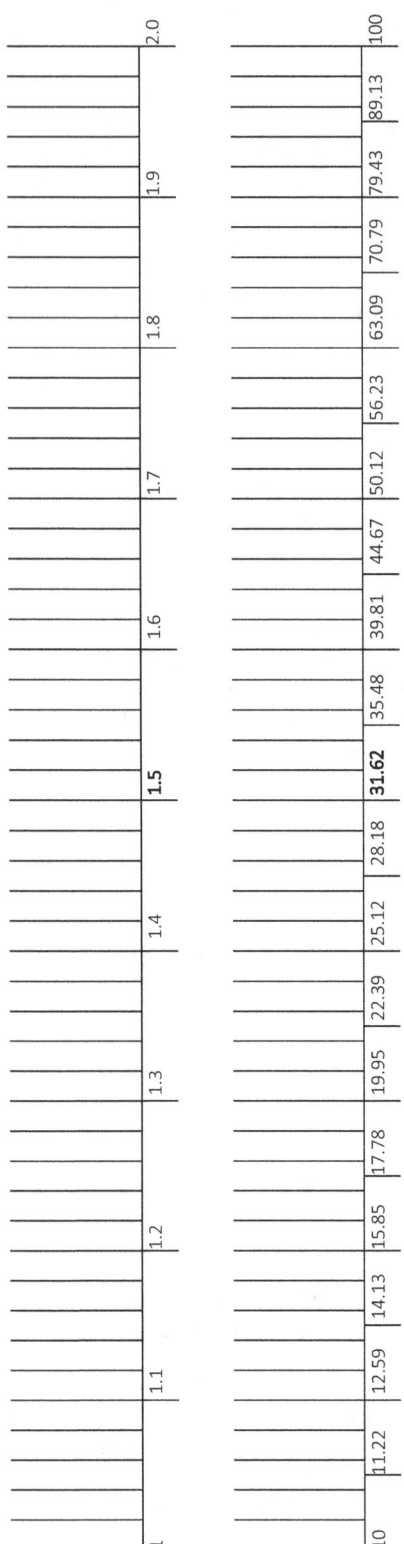

FIGURE 7.12   Logarithmic scale between 10 and 100.

replaced with their logarithms (usually $\log(x)$) and the y-axis values are replaced with their logarithms (usually $\log(y)$). Logarithms are a fundamental tool in engineering science, enabling engineers to tackle complex problems and analyze various phenomena across different disciplines.

7.8.4 Logarithms Help Deal with Large Numbers

Logarithms provide a powerful tool for working with large numbers, making them an essential part of many scientific and engineering applications! Logarithms help in data analysis and visualization where logarithmic scales help display large data-sets with varying orders of magnitude. Logarithms reduce the scale of large numbers, making them more manageable.

For example:

$$\log_{10}(1,000,000,000) = 9$$

Dealing with 9 which is the logarithm of 1 billion to base 10 is much easier. A logarithm is a technique of expressing numbers; it is utilized in all sorts of calculations in engineering and science. It is a system of evaluating multiplication, division, powers, and roots by appropriately converting them to addition and subtraction. Some practical examples where large numbers are scaled include the following:

- Decibel (dB) scale for measuring sound pressure and power.
- Richter scale for measuring earthquake magnitude.
- pH scale for measuring acidity and basicity.
- Logarithmic spirals in design and nature.

7.8.5 Common Logarithm

A common logarithm is a logarithm with a base of 10, often denoted as 'log' without a base. It represents the exponent to which 10 must be raised to obtain a given number. For example, $\log(100) = 2$ because $10^2 = 100$. Logarithms, particularly log 10 (common logarithm) and ln (natural logarithm), are commonly used due to their historical significance: Log10 was used for centuries in astronomy and mathematics, while ln was introduced by John Napier in the 17th century.

The widespread use of log10 and ln stems from their historical roots, ease of use, and versatility in various fields, making them essential tools in mathematics and science. Ease of calculation—before calculators, logarithms simplified complex calculations, especially multiplication and division. The logarithm is one of the many shortcuts of science and engineering, to capture a series of data in one single name. The metric system uses base 10 to simplify conversions and calculations.

7.8.6 Binary Logarithm

This is also called base two logarithm. The binary logarithm $(\log_2 n)$ is the power to which the number 2 must be raised to obtain the value n. That is, for any real number x.

log–log plot, the x-axis is scaled logarithmically (typically with a base-10 or natural logarithm). The y-axis is also scaled logarithmically (typically with the same base as the x-axis). This means that both the horizontal and vertical axes are compressed or expanded logarithmically, allowing for a more efficient representation of data that spans many orders of magnitude.

When interpreting a log–log plot, look for straight lines that indicate power-law relationships, curves that indicate nonlinear relationships, and clusters or patterns that indicate correlations or trends. In a log–log plot, the x-axis values are

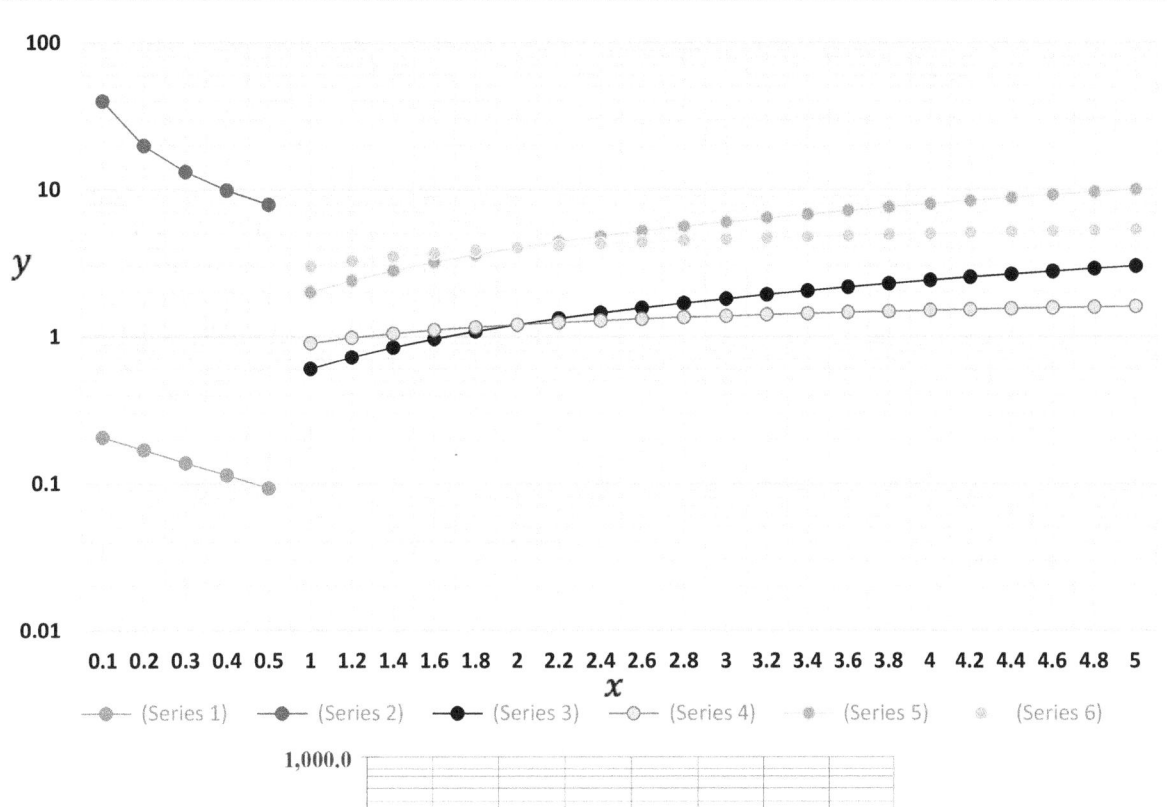

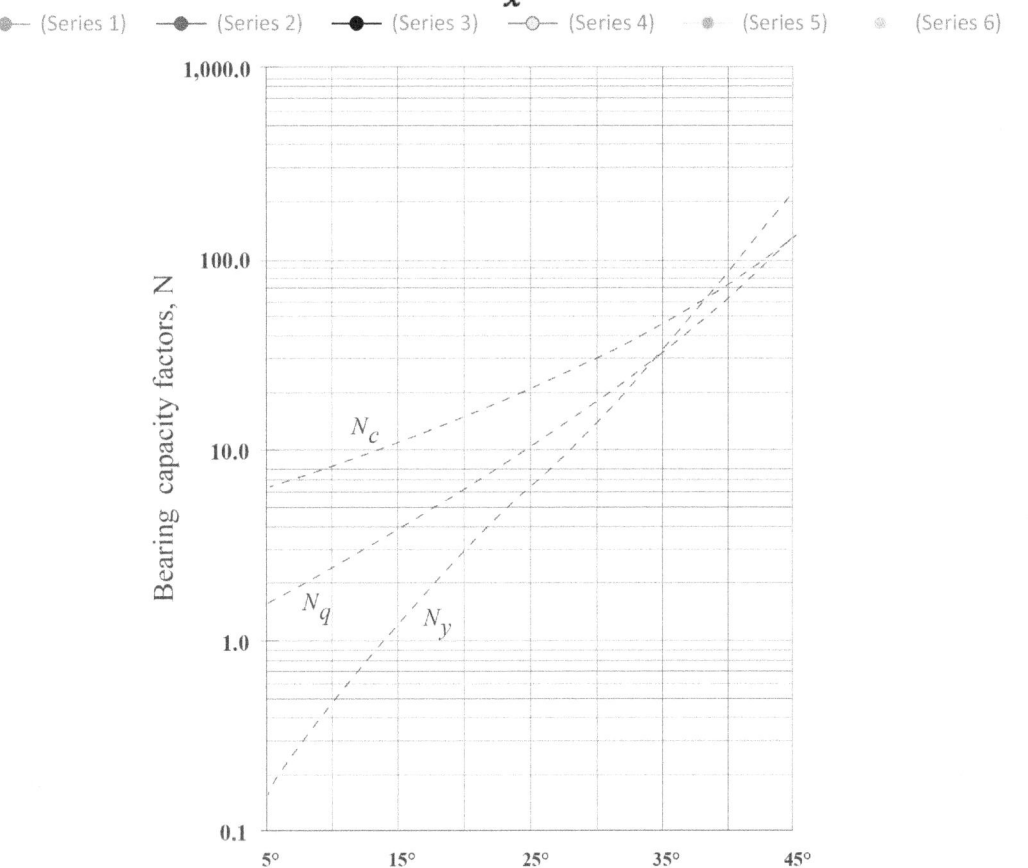

FIGURE 7.13 (a) Combined chart for Figures 7.6a, 7.6b, 7.10, and 7.11. (b) Bearing capacity factors for shallow foundations under drained conditions.

It is applicable in computer science. The compiler uses the binary system. Binary logarithms can be used to calculate the length of the representation of a number in the binary numeral system or the number of bits needed to encode a message in information theory. In computer science, they count the number of steps needed for binary search and related algorithms.

7.8.7 Natural Logarithm

The natural logarithm $\ln x$ is the logarithm to the base e, where e is a mathematical constant approximately equal to 2.71828. It's called 'natural' because it's based on the exponential function e^x, which is fundamental in mathematics. So, the natural logarithm is the inverse of the exponential function e^x.

Exponential function	The exponential function with base e is defined as function $f(x) = ex$.
Natural logarithm	The natural logarithm, denoted as $\ln(x)$, is the inverse function of the exponential function with base e.
Inverse property	Two functions are inverses if applying one function, and then the other results in the original input value.

Suppose, we take the natural logarithm of an exponential function $\ln(e^x)$, the result is x:

$$\ln(e^x) = x \qquad (7.136)$$

Similarly, if we apply the exponential function to the natural logarithm $e^{(\ln(x))}$, the result is x:

$$e^{(\ln(x))} = x \qquad (7.137)$$

So,

$$\ln(e^x) = e^{(\ln(x))} \qquad (7.138)$$

$$\ln(x) = \frac{1}{e^x} \qquad (7.139)$$

Simply put, e is defined as the $\lim\limits_{n\to\infty}\left(1+\dfrac{1}{n}\right)^n$.

In other words, the limit of $\left(1+\dfrac{1}{n}\right)^n$ as n approaches infinity is $e = 2.71828$.

The natural logarithm is called natural because it is based on the intrinsic properties of the exponential function with base e, rather than being defined an arbitrary choice of base. In other words, the natural logarithm answers the question 'To what power must e be raised to obtain a given value?'

For example,

$$\ln(10) \approx 2.3026 \text{ means that } e^{2.3026} = 10$$

7.8.7.1 Evolution of e

1. The ancient Greeks knew about the concept of **exponential growth**.
2. John Napier introduced logarithms, which led to the discovery of e in the 17th century.
3. Jacob Bernoulli studied the limit of $\left(1+\dfrac{1}{n}\right)^n$ which approaches e around 1683.

$$\text{The } \lim\limits_{n\to\infty}\left(1+\frac{1}{n}\right)^n \text{ approaches } e = 2.71828.$$

4. Leonhard Euler introduced the notation e to describe the properties of $\lim\limits_{n\to\infty}\left(1+\dfrac{1}{n}\right)^n$ and developed its properties in the 18th century. Euler, along with other mathematicians such as Bernoulli and Lagrange, further explored $e's$ properties.
5. $e's$ importance in calculus, number theory, and probability theory became clear in the 19th century.

Because of its unique properties, it was inevitable to enter the field of probability and statistics through the exponential function. Things growing exponentially must have excited mathematicians, coining the exponential function. The function $f(x) = e^x$. The exponential function is the unique differentiable function that equals its derivative and takes the value 1 for the value 0 of its variable.

6. e's role in modern mathematics, science, and engineering solidified in the 20th century.

e's unique properties and widespread appearance in mathematics and science have made it a fundamental constant, earning it the nickname 'Euler's number.'

7.8.7.2 Properties of natural logarithm

- Base—e (approximately $=2.71828$).
- Inverse of the exponential function (e^x). Therefore,

$$\ln(x) = \frac{1}{e^x}$$

$$\ln(e^x) = x$$

$$e^{\ln y} = y$$

for every real number x and every positive real number y.

- Derivative:
  - $\dfrac{d}{dx}(e^x) = e^x$
  - $\dfrac{d}{dx}(\ln x) = \dfrac{1}{x}$
- Integral: $\displaystyle\int e^{-x}dx = -\frac{1}{e^{-x}}+C, \; C \in \mathbb{R}$
- Integral: $\displaystyle\int \frac{1}{x}dx = \ln(|x|)+C, \; C \in \mathbb{R}$
- $\lim\limits_{h\to 0}\left(\dfrac{e^h-1}{h}\right) \approx 1$

- **Irrationality:** e is an irrational number, approximately equal to 2.71828, which means it can't be expressed as a simple fraction, but rather a non-repeating, non-terminating decimal.
- **Transcendence:** e is a transcendental number, meaning it's not the root of any polynomial equation with rational coefficients.
- **Base of the natural logarithm:** e is the base of the natural logarithm, making it a fundamental constant in calculus and probability theory.
- **Euler's number:** e is named after Leonhard Euler (1707–1783), who introduced it in the 18th century.

7.8.7.3 Application of e in constructing equations

This section illustrates a basic example of how the natural logarithm is utilized to formulate equations. We will use the following property of the natural logarithm:

$$\int \frac{1}{x}\,dx = \ln x + C \qquad (7.140)$$

First let's prove this property,

Recall that

$$\frac{d}{dx}\left(e^x\right) = e^x; \text{ this is used in the proof}$$

Let

$$y = \ln x$$

So,

$$x = e^y$$

Now,

$$\frac{dx}{dy} = \frac{d}{dy}\left(e^y\right) = e^y = x$$

So,

$$\frac{dx}{dy} = x$$

$$\therefore$$

$$\frac{dy}{dx} = \frac{1}{x}$$

$$\Rightarrow$$

$$dy = \frac{1}{x}\,dx$$

But,

$$\int dy = y + C$$

So,

$$y + C = \int \frac{1}{x}\,dx$$

But,

$$y = \ln(x)$$

$$\therefore$$

$$\int \frac{1}{x}\,dx = \ln x + C$$

This property is utilized in developing several techniques in civil engineering—a perfect example is developing the formula to calculate the soil's permeability coefficient, in the falling head test.

Falling head test

The falling head test, also known as the variable head test, is a method used to determine the permeability of soils,

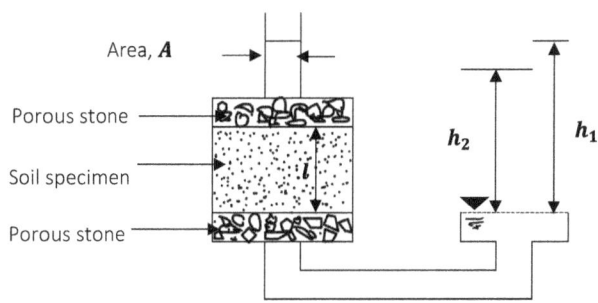

FIGURE 7.14　Falling head test.

particularly those with low permeability, such as clays and silts. The purpose of the falling head test is to measure the rate at which water flows through a soil sample under a decreasing hydraulic head, allowing engineers to calculate the soil's permeability coefficient. See Figure 7.14.

$$Q_{\text{Out}} = v \cdot A \qquad (7.141)$$

$$Q_{\text{Out}} = k\left(\frac{h}{L}\right) \cdot A$$

$$t = \int_{h_1}^{h_2} \frac{A_s dh}{Q_{\text{In}} - Q_{\text{Out}}}$$

$$t = \int_{h_1}^{h_2} \frac{a\,dh}{0 - k\left(\frac{h}{L}\right) \cdot A}$$

$$t = -\frac{aL}{kA}\int_{h_1}^{h_2} \frac{dh}{h}$$

At this stage, the property of the natural logarithm is introduced. By inspection, we see that

$$\int_{h_1}^{h_2} \frac{dh}{h}$$

is analogous to the property (7.139)

$$\int \frac{1}{x}\,dx = \ln x + C$$

Therefore,

$$t = -\frac{aL}{kA}\left(\ln h_2 - \ln h_1\right)$$

$$t = \frac{aL}{kA}\ln\left(\frac{h_1}{h_2}\right)$$

$$k = \frac{aL}{At}\ln\left(\frac{h_1}{h_2}\right)$$

Now, you can see that, the equation can be expressed as follows:

$$\frac{h_1}{h_2} = e^{\left(\frac{Atk}{aL}\right)}$$

or as

$$h_1 = h_2 e^{\left(\frac{Atk}{aL}\right)} \qquad (7.142)$$

Imagine we assume that

$$y = \frac{Atk}{aL}$$

Then,

$$h_1 = h_2 e^{(y)}$$

Conclusion

That is how engineers construct and group equations; in this case, the expression $\frac{Atk}{aL}$ is bundled as y. See Section 7.4.8 as well.

Marston's equation

Marston's equation for calculating fill loads on a buried pipeline is given by the formula

$$W = CwB^2 \qquad (7.143)$$

where

> W: load on the pipe per unit length
> C: coefficient that depends on the depth of the trench, characteristic of construction, and fill material.
> B: width of the trench just below the top of the pipe
> w: weight of the fill material per unit volume

For normal ordinary trench construction, C can be calculated from the following:

$$C = \frac{1 - e^{-2\mu'KH/B}}{2\mu'K} \qquad (7.144)$$

where

> H: depth of fill above the pipe
> B: width of the trench just below the top of the pipe
> K: ratio of active lateral pressure to vertical pressure
> μ': coefficient of sliding friction between the fill material and sides of the trench.

See full derivation in reference [5], pages 275–277, to appreciate how e finds application in the equation (7.143).

Proof for $\frac{d}{dx}(e^x) = e^x$

$$\frac{d}{dx}(e^x) = e^x \qquad (7.145)$$

Using the definition of a derivative:

$$f'(x) = \lim_{h \to 0}\left(\frac{f(x+h) - f(x)}{h}\right)$$

$$f'(x) = \lim_{h \to 0}\left(\frac{e^{(x+h)} - e^x}{h}\right)$$

$$f'(x) = \lim_{h \to 0}\left(\frac{e^x \cdot e^h - e^x}{h}\right)$$

$$f'(x) = e^x \lim_{h \to 0}\left(\frac{e^h - 1}{h}\right)$$

Note that

$$\lim_{h \to 0}\left(\frac{e^h - 1}{h}\right) \approx 1$$

Therefore,

$$f'(x) = e^x \cdot 1 = e^x$$

and

$$\frac{d}{dx}(e^x) = e^x$$

7.8.7.4 General applications of the natural logarithm

The natural logarithm is a fundamental tool in mathematics, science, and engineering and is widely used due to its unique properties and versatility. The natural logarithm, which is the logarithm to the base e, is used in various mathematical contexts, such as the following:

1. **Calculus**: integration, differentiation, and optimization problems.
2. **Number theory**: prime numbers, modular arithmetic, and cryptography.
3. **Probability theory**: random variables, distributions, and stochastic processes.
4. **Engineering**: e.g., geotechnical engineering.

7.8.7.4.1 Basic examples of e in use

e is a fundamental constant that shows up in many areas of mathematics, science, and engineering, often describing processes that involve growth, change, or randomness.

Compound interest

The earliest occurrence of the exponential function was in Jacob Bernoulli's study of compound interests in 1683. If you deposit $1 into a bank account with a 100% annual interest rate, compounded continuously, you'll have approximately e ($2.71828) after 1 year.

Compound interest formula

$$A = P\left(1 + \frac{r}{n}\right)^{nt} \qquad (7.146)$$

where

> A: amount
> P: principal
> r: rate of interest (%)
> n: number of times interest is compounded per year
> t: time (in years)
> $\left(1 + \frac{r}{n}\right)^{nt}$: growth factor

Population growth

If a population grows at a rate proportional to its current size, the growth factor will approach e over time. This is very much applied in population growth rates in water supply system designs:

$$P_n = P(1+r)^n \qquad (7.147)$$

where

P_n: population after n years
P: present population
r: annual growth rate (%)
$(1+r)^n$: growth factor

Probability

e appears in the study of random events, like the number of trials needed to get a specific outcome. An exponential function is a mathematical function used to calculate the exponential growth or decay of a given set of data. The exponential function takes the form

$$f(x) = a^x \qquad (7.148)$$

where the input variable x occurs as an exponent. Exponential function with base e is given as

$$f(x) = \frac{1}{a} e^{-\left(\frac{x}{a}\right)} \qquad (7.149)$$

where a is a constant and $0 \le x \le \infty$. Because of e's unique properties, it garnered applications in several continuous probability distributions, such as the exponential distribution, Weibull, normal, log-normal, gamma, Gumbel, extreme value, and Log-Pearson III, thus forming the backbone of distributions that are essential in quality control and management—statistical process control and reliability analysis and risk analysis which is basically decision-making under uncertainty.

Hydrology is almost entirely probability distributions—for example, modeling water flow, flood frequencies, and droughts. Another interesting field where probability and statistics are extensively applied is traffic engineering. These fields are pre-occupied by random events!

Thus, probability distributions help us model uncertainty, that is, to describe random phenomena. We can also analyze data to understand patterns and trends and to make predictions and forecast future events. See Section 7.9.2 for more information about the role of probability in civil engineering.

Modeling decay

e is used to describe the rate of decay in exponential functions, such as radioactive decay or chemical reactions. The formula for exponential decay is often expressed as

$$N(t) = N_o e^{-\lambda t} \qquad (7.150)$$

where $N(t)$ is the quantity remaining at time t, N_0 is the initial quantity, that is, the quantity at time $t = 0$, λ is the decay rate, and t is the time. This is the same approach used to study population growth or decline.

7.9 VERSATILE MATHEMATICAL FIELDS FOR CIVIL ENGINEERS

This section outlines civil engineers' mostly applied areas of mathematics. The most versatile mathematical fields that civil engineers harness include probability and statistics, calculus, geometry, linear algebra, graph theory, optimization techniques, topology, numerical methods, and ML.

7.9.1 Calculus

Invented by Sir Isaac Newton, calculus is the branch of mathematics that deals with the finding and properties of derivatives and integrals of functions, by methods originally based on the summation of infinitesimal differences. The two types of calculus are differential calculus and integral calculus. These include differential equations, integration, and optimization.

In civil engineering, calculus plays an important role in multiple areas, such as the design of structures, the calculation of load and stress factors, soil mechanics, and fluid dynamics. These fields require precise mathematical models to ensure safe and efficient infrastructure development.

In soil mechanics, differential equations which are a form of calculus prove invaluable because they help in forecasting the behavior of soils under varying loads and pressures, providing engineers with insights into the ground stability and safety for construction.

Calculus is helpful in infrastructure design—the load-bearing capacity, weight distribution, and stress factors must be calculated accurately for structural systems such as bridges and buildings. These important factors are evaluated using integral calculus because it gives an understanding of the accumulated values of various elements.

Similarly, fluid dynamics, which involves the motion of gases and liquids, rely on calculus for investigating characteristics such as velocity, pressure, and viscosity. In that case, engineers utilize this information for designing systems such as water supply networks, wastewater systems, and flood control systems (see further reading 4).

7.9.1.1 Differential calculus

Differential calculus is a branch of calculus that focuses on the study of rates of change of functions, using concepts such as derivatives and differentials to analyze how quantities change with respect to other quantities. A differential equation is a mathematical equation that relates some function with its derivatives. The process of finding the differential coefficient of a function is called the differentiation.

In engineering applications, the functions usually represent physical quantities, the derivatives represent their rates of change, and the equation defines a relationship between the two. They are solved by finding an expression for the function that does not involve derivatives—a concept that is the cornerstone for how engineers formulate equations. See Section 7.7.4 for the role of independent and dependent variables.

Differential equations are used to model processes that involve the rates of change of the variables and are widely used in engineering. Thus, for a single-variable calculus, the derivative of a function $y = f(x)$ can be defined as

$$f'^{(x)} = \frac{dy}{dx} = \lim_{\delta x \to 0}\left(\frac{\delta y}{\delta x}\right) = \lim_{\delta x \to 0}\left[\frac{f(x + \delta x) - f(x)}{\delta x}\right] \tag{7.151}$$

This means that y is differentiated with respect to x. Differential equations (ODE and PDE) are used for modeling dynamic systems.

7.9.1.1.1 Balancing volume in water reservoir design, A

In water reservoir design for water supply schemes, balancing volume/storage refers to the process of ensuring that the volume of water stored in the reservoir is balanced with the volume of water inflow and outflow. This is crucial to maintain a stable water level, prevent overflow or underflow, and ensure efficient water management. The balancing volume can account for about 20–80% of the storage volume.

The balancing storage requirement in a water supply system is caused by the cyclical variations in water demand over a period of time. There are several reasons why the rate at which a city's water reservoir tank is filled might change over time, and these include

- Multiple inlets.
- Fluctuations in the pumping—efficiency. The pump's capacity to fill the tank might change over time due to maintenance, upgrades, or equipment failures.
- Fluctuations in power, that is,
  - Solar-powered
  - Hybrid
  - Hydroelectric power (HEP)
- Head losses in pipelines.
- **Changes in pipe diameter and friction**: Changes in pipe diameter, material, or condition can affect the flow rate and pressure, altering the rate at which the tank is filled.
- **Variations in water supply**: The rate at which water is pumped into the tank might change due to fluctuations in the water supply source, such as a river, borehole, or reservoir.
- **Elevation and gravity**: If the tank is located at a higher elevation than the pumping station, gravity might affect the flow rate, causing it to slow down over time.
- **Valve operations**: Manual or automated valve adjustments might alter the flow rate into the tank.
- **Leaks or bursts**: Undetected leaks or pipe bursts can reduce the effective flow rate into the tank.
- **Seasonal changes**: Changes in temperature, precipitation, or humidity might affect the water treatment process, influencing the rate at which the tank is filled.
- **Maintenance and repairs**: Scheduled maintenance or unexpected repairs might temporarily reduce the flow rate into the tank.
- **Upgrades or expansions**: Modifications to the pumping station, pipes, or tank might alter the filling rate.

These factors can contribute to changes in the rate at which a city's water tank is filled over time. These factors can influence the development of a standard equation for filling the tank, which is a function of time—single-variable calculus. However, another equation may be developed depending on the findings involving other factors, leading to multivariable calculus. Consider a water tank being filled at a rate that is changing over time. The volume of water in the tank (V) at any time t is given by the equation:

$$V(t) = 3t^3 - 5t^2 + 3t + 1 \tag{7.152}$$

where

$V(t)$: volume of water in the tank (m^3) at time t
t: time (hours)

Find the rate at which the water level is rising when $t = 2$ hours.

Note: A water tank is can be filled at a rate that is changing over time, for example, when you have two inlets from different pumping stations. The fluctuations in discharges can occur due to pumps' efficiencies, pipe diameter, frictional losses, etc. Therefore, $V(t) = 3t^3 - 5t^2 + 3t + 1$ can be an empirical equation developed from observation of how the filling of the tank varies with time.

Solution

To find the rate at which the water level is rising, we need to find the derivative of $V(t)$ with respect to time (t). This will give us the rate of change of volume with respect to time.

Using the power rule of differentiation:

$$\frac{dV}{dt} = 9t^2 - 10t + 3 \tag{7.153}$$

$$\frac{dV}{dt} = \frac{\text{rate of change of volume of water in}}{\text{the tank with respect to time}}$$

Now, we need to find the rate at which the water level is rising when $t = 2$ hours :

$$\Rightarrow \frac{dV}{dt} = 19 \text{ m}^3/\text{h} \tag{7.154}$$

So, the water level is rising at a rate of 19 m³/h when $t = 2$ hours. This example illustrates how differential calculus can be used to find rates of change and slopes of curves, which is essential in engineering.

7.9.1.1.1.1 Worked example [7.14]: Tank balancing volume determination

If a community is estimated to consume 456 m^3 of water per day or a given period of time (say a week), it doesn't necessarily translate to a 456 m^3 tank to be constructed for the area. We can use differential calculus to model/determine the balancing volume. The balancing volume in water reservoir design involves calculating the following:

1. **Inflow volume:** the volume of water entering the reservoir from sources such as rivers, streams, underground aquifers, or rainfall.
2. **Outflow volume:** the volume of water released from the reservoir through outlets, such as pipes or gates.
3. **Storage volume:** the volume of water stored in the reservoir.

4. **Evaporation and seepage losses:** the volume of water lost due to evaporation and seepage through the reservoir walls or floor.

The balancing volume is calculated using the following equation:

Balancing volume (BV) = inflow volume (V_i)

$-$outflow volume $(V_o)-$evaporation and seepage losses (E)

$$(7.86)$$

The goal is to maintain a balanced volume, where the inflow volume is equal to the outflow volume, plus any losses due to evaporation and seepage. This ensures that the water level in the reservoir remains stable, and the reservoir operates within its design parameters. For example, if the inflow volume is 456 cubic meters per day, the outflow volume is 400 cubic meters per day, and the evaporation and seepage losses are negligible, the balancing volume would be

Balancing volume $= 456\ \mathrm{m}^3/\mathrm{day} - 400\ \mathrm{m}^3/\mathrm{day} - 0\ \mathrm{m}^3/\mathrm{day}$

$$= 56\ \mathrm{m}^3/\mathrm{day} \qquad (7.155)$$

Note

$$Q = Q_{\mathrm{in}} - Q_{\mathrm{out}} \qquad (7.156)$$

$$dQ = \frac{dV}{dt} \qquad (7.157)$$

This means that the reservoir needs to store 56 m³ of water per day to maintain a balanced volume. We don't need to construct a reservoir of 456 m³. To determine the total volume of the tank, we need to consider the number of days the tank needs to supply water.

Let's assume the tank needs to supply water for 1 or 3 days.

1 day

Balancing volume = balancing volume × number of days.

= 56 m³/day × 1 day

= 56 m³

3 days

Balancing volume = balancing volume × number of days

= 168 m³/day × 3 days

= 168 m³

So, this can account for about 20%–80% of the storage volume, considering other factors/special water needs—such as firefighting.

So, the tank needs to have a balancing volume of 168 m³ to store enough water to meet the balancing volume requirements for 3 days. To determine the size of the tank, we need to calculate the total volume of water that needs to be stored. And to determine the tank's dimensions, we need to consider factors such as the tank's shape (e.g., rectangular face (cuboidal), circular cross-section (cylindrical)), the desired height and width, and any constraints such as space limitations or construction requirements. For example, let's assume a cuboidal tank with a length (L), width (W), and height (H). The volume of the tank can be calculated as follows:

$$\mathrm{Volume} = \mathrm{L} \times \mathrm{W} \times \mathrm{H} \qquad (7.158)$$

Using the total volume calculated earlier (168 m³) and assuming it accounts for 80% of the total reservoir volume, the reservoir volume would be 210 m³. Thus, we can determine the tank's dimensions—and given the fact that the height of the community water reservoir tank is sometimes restricted to maybe 5 m in some manuals/jurisdictions, especially for tanks raised on towers for reasons to do with pressure optimization—a cuboidal tank may have the following dimensions: 6.1 m (W), 4.88 m (H), and 7.05 m (L)—and depending on the standards followed, for example, if HDG pressed steel panels are used, the standard sizes may have to be followed as per manufacturer details—so, for practical purposes, we may develop an ≈218 m³ tank, considering panel sizes of 1.22 m by 1.22 m, manufactured according to BS 1564:1975, with tank dimensions of 7.32 m (L), 4.88 m (H), and 6.1 m (W), thus garnering a safety factor of 1.04. See Section 7.6.9.9. Note that these dimensions are just an example and may vary depending on specific design requirements and constraints.

7.9.1.1.1.2 Required storage capacity from available data   To illustrate balancing volume further, consider the data in the Table 7.8 for a gravity flow scheme of a small rural community.

TABLE 7.8   Data for a gravity flow scheme of small rural community

TIME PERIODS	SUPPLY	DEMAND	DIFFERENCE
Schedule 1			
6 a.m.–8 a.m. (2 hours, 30%)	32,400	54,000	−21,600 (water withdrawn)
8 a.m.–4 p.m. (8 hours, 40%)	129,600	72,000	+57,600 (tank overflows)
4 p.m.–6 p.m. (2 hours, 30%)	32,400	54,000	−21,600 (water withdrawn)
Largest deficiency			**21,600 L**
Schedule 2			
5 a.m.–7 a.m. (2 hours, 10%)	32,400	18,000	+14,400 (tank overflows)
7 a.m.–11 a.m. (4 hours, 25%)	64,800	45,000	+19,800 (tank overflows)
11 a.m.–1 p.m. (2 hours, 35%)	32,400	63,000	−30,600 (water withdrawn)
1 p.m.–5 p.m. (4 hours, 20%)	64,800	36,000	+28,800 (tank refilling)
5 p.m.–7 p.m. (2 hours, 10%)	32,400	18,000	+14,400 (tank overflows)
Largest deficiency			**30,600 L**

In this gravity flow water scheme development, the projected population of a village is 4,000 persons, with no other special water needs. The safe yield of the source is 4.5 LPS (16.2 m3/hour), and five tap stands are to be built. Since the source is not large enough to supply more than two of the taps stands by itself, a reservoir tank is required. Using two demand schedules, the following water demands are calculated.

For this example, the required storage capacity is determined by schedule 1, at 21,600 L (21.6 m³). For practical design, consider this to be 22,000 L (22 m³)—which is the largest deficiency. The lesson to learn here is that this information can be helpful in formulating **balance volume equation** or an equation for the volume of water in the tank (V) at any time t. Another lesson is that water reservoir tank designs are based on the perceived or predicted water deficiencies, and rates of change can be helpful to model these behaviors.

7.9.1.2 Integral calculus

The basic principle of integration is to add up all the tiny contributions that make up a larger quantity. Integrals work by summing up an infinite number of small elements to find the whole. In other words, we can say that integral calculus is defined as the study of the area beneath a curve. For a single variable, specifically, the **definite integral** is mathematically expressed as follows:

$$\int_a^b f(x)\,dx = \lim_{\substack{n\to\infty \\ all\ \Delta x_i \to 0}} \sum_{i=1}^{n} f(\tilde{x}_i)\Delta x_i \tag{7.159}$$

where $a = x_0 < x_1 < x_2 < \cdots < x_{n-1} < x_n = b$, $\Delta x_i = x_i - x_{i-1}$, and $x_{i-1} \le \tilde{x}_i \le x_i$. Geometrically, we can interpret this integral as the area between the graph $y = f(x)$, the x-axis and the lines $x = a$ and $x = b$ as illustrated in Figure 7.15.

More generalized, integral calculus is a branch of calculus concerned with the theory and applications of integrals. While differential calculus focuses on rates of change, such as slopes of tangent lines and velocities, integral calculus deals with total size or value, such as lengths, areas, and volumes. Integral calculus is very helpful in optimization. More advanced integration includes both vector and scalar point functions.

The idea of integration can be extended to multiple dimensions. Examples are line integrals, double integrals, triple (volume) integrals, and surface integrals. Each one lets you add infinitely many infinitely small values, where those values

might come from points on a curve, points in an area, or points on a surface. These are all very powerful tools, relevant to almost all real-world applications of calculus, especially in engineering and associated fields. They are applied especially when integrating multivariable functions.

Line integrals, surface integrals, volume integrals, and double integrals are different types of integrals used in multivariable calculus[3], each integrating over different dimensional regions. Line integrals integrate along a curve (1D), surface integrals integrate over a surface (2D), and volume integrals integrate over a volume (3D). These concepts are greatly applied in the design of engineering systems. **Note:** Line integrals are generalizations of simple integrals, surface integrals are generalizations of double integrals, and volume integrals are a specific application of triple integrals. In terms of dimensions, line integrals integrate over curves (1D), surface integrals over surfaces (2D), and volume integrals over volumes (3D).

Notes

- A simple integral (also called definite integral) is a special case of various types of integrals, and line integrals are one of the generalizations of simple integrals.
- A surface integral is generalization of double integral—integration over a 2D surface in 3D space. Double integrals are a special case of surface integrals where the surface is flat and can be used to find volume under a surface—surface integrals integrate over a general surface.
- Volume integrals are a specific application of triple integrals.
- A multiple integral is any type of integral.

7.9.1.2.1 Line integrals

Line integrals refer to integration along a curve. Line integrals calculate the integral of a function along a specific curve or path in 1D, 2D, or 3D space. For example, it is used for calculating work done by a force along a path, circulation of a fluid, or flux along a curve. This is represented by the notation:

$$\int_C f(x, y)\,dx \tag{7.160}$$

Or, for a vector field $F(r)$, the line integral takes the notation:

$$\int_C F \cdot dr \tag{7.161}$$

The letter under the integral sign indicates that the integral is evaluated along the curve (or path) C. This path is not restricted to two dimensions and may be in as many dimensions as we please. It is normal to omit the points a and b since they are usually implicit in the specification of C.

Imagine

$$\int_a^b f(x, y)\,dx \tag{7.162}$$

where

$$y = g(x)$$

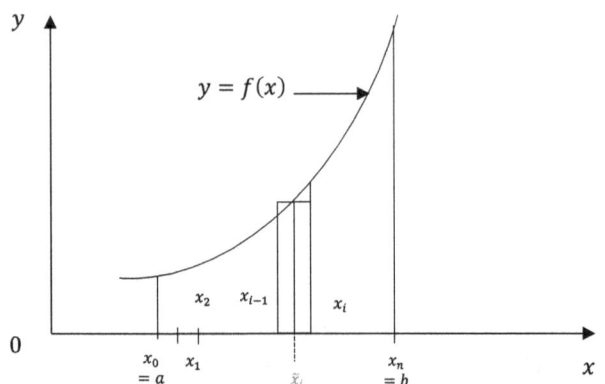

FIGURE 7.15   Definite integral as an area.

Here, we substitute for y in terms of x in the integrand and then perform the integration:

$$\int_a^b f(x, g(x)) \tag{7.163}$$

The value of the integral will depend on the function $y = g(x)$—interpreted as evaluating the integral $\int_a^b f(x, y) dx$ along the curve. It is important to note that the integral is not represented in this case by the area under the curve. This type of integral is called a line integral.

7.9.1.2.1.1 Worked example [7.15]: Line integral   Evaluate

$$\int_C xy dx$$

from A (1,0) to B (0,1) along the curve C (Figure 7.16) that is the portion of $x^2 + y^2 = 1$ in the first quadrant.

Solution

The curve C is the first quadrant of the circle, as shown in Figure 7.16. On the curve, $y = \sqrt{(1-x^2)}$.

So,

$$\int_C xy dx = \int_1^0 x\sqrt{(1-x^2)} dx = -\frac{1}{3} \tag{7.164}$$

7.9.1.2.2 Double integrals

Double integrals calculate the integral of a function over a 2D region, often a flat surface, such as an area in the xy-plane. Double integrals are applicable in finding the area of a region, the volume under a surface, or the average value of a function over a 2D region. This is represented by the notation:

$$\iint_R f(x, y) dA = \iint_R f(x, y) dx dy = \lim_{n \to \infty} \sum_{i=1}^n f(\tilde{x}_i, \tilde{y}_i) \Delta x_i \Delta y_i \tag{7.165}$$

It is possible to evaluate the integrals of the type $\iint_R f(x, y) dx dy$ as repeated single integrals in x and y, and they are usually called double integrals.

In Figure 7.17, we defined the definite integral for a function $y = f(x)$ of one variable.

Now, consider $z = f(x, y)$ and the region R of the (x, y) plane, as shown in Figure 7.17, and thus building on the concepts for definite integral for a function $y = f(x)$ of one variable, the double integral is defined as

$$\iint_R f(x, y) dA = \iint_R (x, y) dx dy = \lim_{\substack{n \to \infty \\ all \ \Delta A_i \to 0}} \sum_{i=1}^n f(\tilde{x}_i, \tilde{y}_i) \Delta A_i \tag{7.166}$$

where $\Delta A_i (i = 1, \cdots, n)$ is a partition of R into n elements of area ΔA_i and $(\tilde{x}_i, \tilde{y}_i)$ is a point in ΔA_i.

Now, $z = f(x, y)$ represents a surface, and so $f(\tilde{x}_i, \tilde{y}_i) \Delta A_i = \tilde{z}_i \Delta A_i$ is the volume $z = 0$ and $z = \tilde{z}_i$ on the base ΔA_i.

The integral $\iint_R f(x, y) dA$ is the limit of the sum of all such volumes, and so it is the volume under the surface $z = f(x, y)$ above the region R. The partition of R into elementary areas can be achieved using grid lines parallel to the x and y axes, as shown in Figure 7.18. Then, $\Delta A_i = \Delta x_i \Delta y_i$; thus, we can write

$$\iint_R f(x, y) dA = \iint_R f(x, y) dx dy = \lim_{n \to \infty} \sum_{i=1}^n f(\tilde{x}_i, \tilde{y}_i) \Delta x_i \Delta y_i \tag{7.167}$$

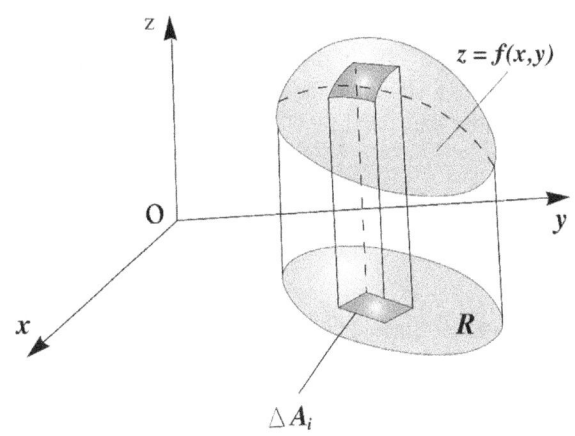

FIGURE 7.17   Volume as an integral.

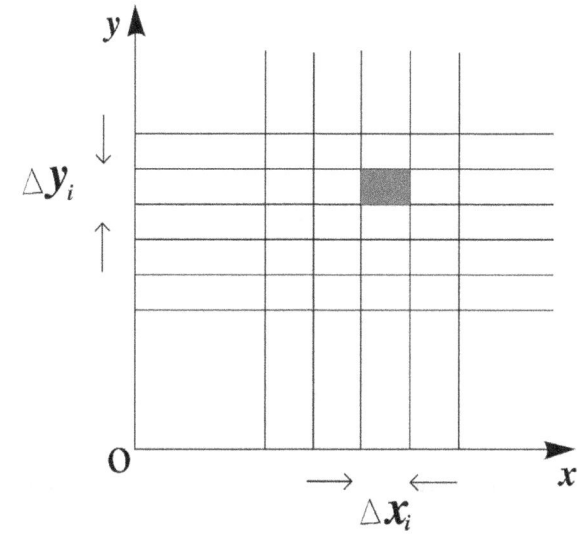

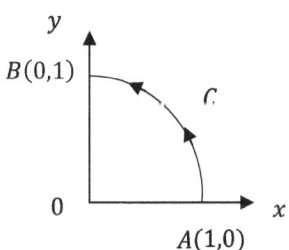

FIGURE 7.16   Portion of a circle.

FIGURE 7.18   Grid for the partition of R (rectangular Cartesian).

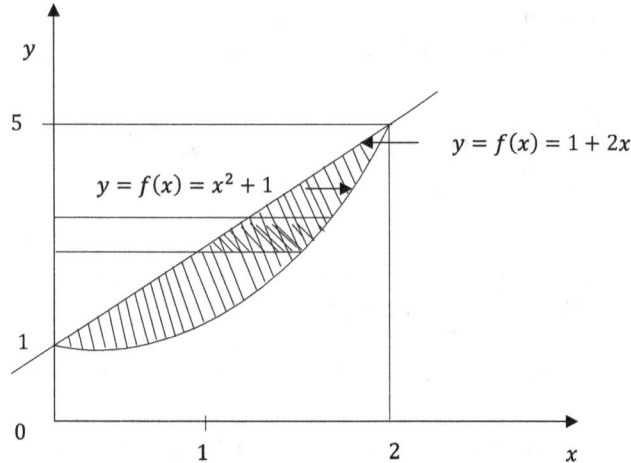

FIGURE 7.19   Domain of integration.

7.9.1.2.2.1 Worked example [7.16]: Double integrals   Evaluate

$$\iint_R (x+2y)^{-1/2} dA \qquad (7.168)$$

Over the region, $y - 2x \leq 1$ and $y \geq x^2 + 1$.

Solution

The bonding curves intersect where $2x + 1 = x^2 + 1$, which gives $x = 0$ (with $y = 1$) and $x = 2$ (with $y = 5$). The region R is shown in Figure 7.19. In this example, we choose to take x first because the formula for the boundary is easier to deal with $y = x^2 + 1$ rather than $x = (y-1)^{1/2}$. Thus, we obtain

$$\iint_R (y+2x)^{-1/2} dA = \int_0^2 \int_{x^2+1}^{2x+1} (y+2x)^{-1/2} dydx \qquad (7.169)$$

$$\iint_R (y+2x)^{-1/2} dA = \int_0^2 \left[2(y+2x)^{1/2} \right]_{y=x^2+1}^{y=2x+1} dx \qquad (7.170)$$

$$= \int_0^2 \left[2(4x+1)^{1/2} - 2(x+1) \right] dx \qquad (7.171)$$

$$= \left[\frac{1}{3}(4x+1)^{3/2} - x^2 - 2x \right]_0^2 = \frac{2}{3} \qquad (7.172)$$

7.9.1.2.3 Triple integrals

A triple integral is a mathematical operation that involves integrating a function of three variables (x, y, z) over a three-dimensional region. Triple integral and volume integral are used interchangeably, but technically, all volume integrals are triple integrals, but not all triple integrals are volume integrals.

A volume integral is a specific application of triple integrals, where the function being integrated is typically a density function or a constant, and the result represents the volume, mass, or other physical quantity of a 3D object. Triple integrals are applied to calculate the volume of a 3D object, the mass of a 3D object, or the average value of a function over a 3D region. This is represented by the notation:

$$\iiint_T f(x,y,z) dV = \lim_{\substack{n \to \infty \\ \text{all } \Delta V_i \to 0}} \sum_{i=1}^{n} f(\tilde{x}_i, \tilde{y}_i, \tilde{z}_i) \Delta V_i \qquad (7.173)$$

The idea presented under double integrals (Section 7.9.1.2.2) can be extended to define the integral of a function of three variables through a region T of three-dimensional space by limit

$$\iiint_T f(x,y,z) dV = \lim_{\substack{n \to \infty \\ \text{all } \Delta V_i \to 0}} \sum_{i=1}^{n} f(\tilde{x}_i, \tilde{y}_i, \tilde{z}_i) \Delta V_i \qquad (7.174)$$

where $\Delta V_i (i = 1, \cdots, n)$ is a partition of T into n elements of volume ΔV_i and $(\tilde{x}_i, \tilde{y}_i, \tilde{z}_i)$ is a point in ΔV_i, as illustrated in Figure 7.20.

In terms of rectangular Cartesian coordinates, the triple integral can, as illustrated in Figure 7.21, be written as

$$\iiint_T f(x,y,z) dV = \int_a^b dx \int_{g_1(x)}^{g_2(x)} dy \int_{h_1(x,y)}^{h_2(x,y)} f(x,y,z) dz \qquad (7.175)$$

Note that there are six different orders in which the integration in equation (7.175) can be carried out.

7.9.1.2.3.1 Worked example [7.17]: Triple integrals   Find the volume and the coordinates of the centroid of the tetrahedron defined by $x \geq 0$, $y \geq 0$, $z \geq 0$, and $x + y + z \leq 1$.

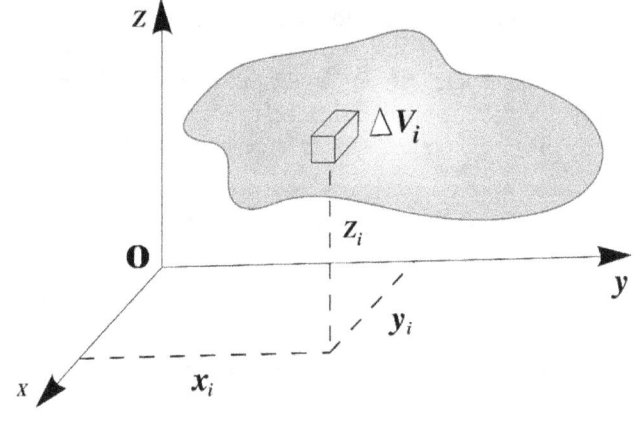

FIGURE 7.20   Partition of the region T into volume elements ΔV_i

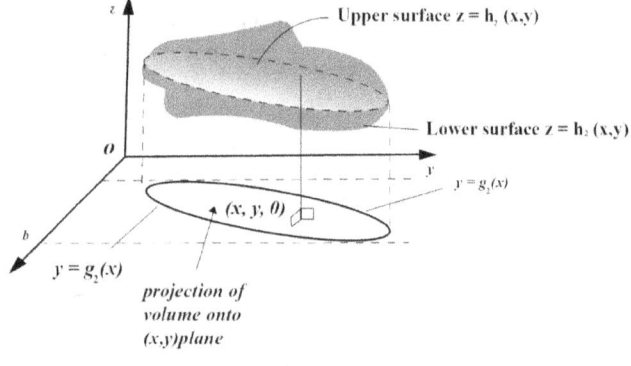

FIGURE 7.21   Cartesian coordinate system.

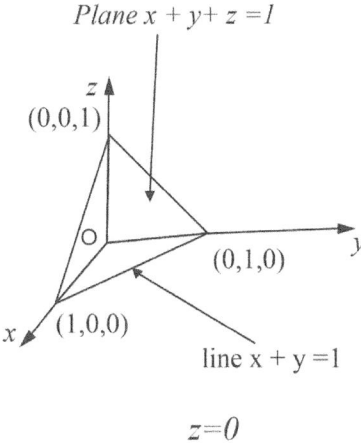

Plane x + y + z = 1

$z=0$

FIGURE 7.22 Tetrahedron.

Solution

The volume of the tetrahedron shown in Figure 7.22 is given by

$$V = \iiint\limits_{\text{tetrahedron}} dx\,dy\,dz = \int\limits_{x=0}^{x=1} dx \int\limits_{y=0}^{y=1-x} dy \int\limits_{z=0}^{z=1-x-y} dz \qquad (7.176)$$

$$= \int_0^1 dx \int_0^{1-x} (1-x-y)\,dy = \int_0^1 \frac{1}{2}(1-x)^2\,dx = \frac{1}{6} \qquad (7.177)$$

Let the coordinates of the centroid be $(\bar{x}, \bar{y}, \bar{z})$; then, taking moments about the line $x = 0$, $z = \bar{z}$,

$$\bar{x}V = \iiint\limits_{\text{tetrahedron}} x\,dV = \iiint\limits_{\text{tetrahedron}} x\,dx\,dy\,dz \qquad (7.178)$$

$$\bar{x}V = \int_0^1 dx \int_0^{1-x} dy \int_0^{1-x-y} x\,dz = \int_0^1 \frac{1}{2}x(1-x)^2\,dx = \frac{1}{24} \qquad (7.179)$$

Hence,

$$\bar{x} = \frac{1}{4}$$

By symmetry

$$\bar{y} = \bar{z} = \frac{1}{4}$$

7.9.1.2.4 Balancing volume in water reservoir design, B
Integral calculus is used to find the accumulation of a quantity such as water in a reservoir tank over a defined interval. Suppose we have a water reservoir tank with an inflow rate $Q_{\text{in}(t)}$ that varies with time according to the function (7.101):

$$Q_{\text{in}(t)} = 100 + 20\sin\left(\frac{\pi t}{12}\right) \qquad (7.180)$$

where

- $Q_{\text{in}(t)}=$ Inflow rate $\left(\text{m}^3/\text{h}\right)$ at time t.
- $t =$ Time (hours).

Suppose the outflow rate $(Q_{\text{out}(t)})$ is constant at $80\text{m}^3/\text{h}$. We want to find the balancing volume (BV) required to stabilize the water level in the tank over a 24-hour period. We can use the **definite integral** formula:

$$BV = \int \left(Q_{\text{in}(t)} - Q_{\text{out}(t)}\right) dt$$

So,

$$BV = \int \left(\left(100 + 20\sin\left(\frac{\pi t}{12}\right)\right) - 80\right) dt \qquad (7.181)$$

From 0 to 24:

$$BV = \int_0^{24} \left(\left(100 + 20\sin\left(\frac{\pi t}{12}\right)\right) - 80\right) dt \qquad (7.182)$$

Recall that

$$\int \sin\left(\frac{\pi t}{12}\right) dt = -\frac{12}{\pi}\cos\left(\frac{\pi t}{12}\right) + C \qquad (7.183)$$

$\therefore$

$$BV = 20(24-0) + 20\int_0^{24} \left(\sin\left(\frac{\pi t}{12}\right)\right) dt \qquad (7.184)$$

$$= 480 + 20\left[-\frac{12}{\pi}\cos\left(\frac{\pi t}{12}\right)\right]_0^{24}$$

$$BV = 480 - 20\left(\frac{12}{\pi}\cos(2\pi) - \frac{12}{\pi}\cos 0\right) \qquad (7.185)$$

$$BV = 480 - 20\left(\frac{12}{\pi} - \frac{12}{\pi}\right) = 480 \text{ m}^3 \qquad (7.186)$$

So, the balancing volume required to stabilize the water level in the tank over a 24-hour period is approximately 480 m³. This example illustrates how integral calculus can be used to determine the balancing volume in a water reservoir tank with varying inflow rates. We integrated the function $(Q_{\text{in}(t)} - Q_{\text{out}(t)})$ over the time interval $(0, 24)$ to find the total accumulation of water, which is the balancing volume.

Important: The issue is how this equation (7.180) is developed. Water may fluctuate in the reservoir due to several factors, which may include

- Multiple inlets.
- Fluctuations in the pumping—efficiency.
- Fluctuations in power.
  - Solar-powered
  - Hybrid
  - HEP-powered.
- Head losses in pipelines.

As stated earlier, in Section 7.4.5, equation (7.181) can be developed empirically or graphically from observations. As we know, the balancing storage requirement in a water supply system is caused by the **cyclical variations** in water demand and supply over a period of time. Pay attention to the constants 100 and 20 in the equation!

7.9.1.2.5 Worked example [7.18]: Hydrostatic force acting on one side of the submerged vertical plate

Calculus can be applied in determination of hydrostatic forces acting across an area in contact with a fluid. For example, in dam design, the hydrostatic forces influence dam stability and water pressure. In water tank design, the forces affect tank walls and foundation, and in lock gates, the hydrostatic forces impact gate design. By applying calculus, determine the magnitude of the hydrostatic force acting on one side of the submerged vertical plate shown below and determine the location of the center of pressure. See Figure 7.23.

Solution

See Figure 7.24.

$$\int dF = \int PdA$$

$$F = \int \delta h dA = \delta \int_0^6 hxdh \tag{7.187}$$

When $x = 0$, $h = 0$ and when $x = 10$, $h = 6$

$\therefore$

$$x = \frac{10}{6}h \tag{7.188}$$

Substituting $x = \dfrac{10}{6}h$ in equation (7.187)

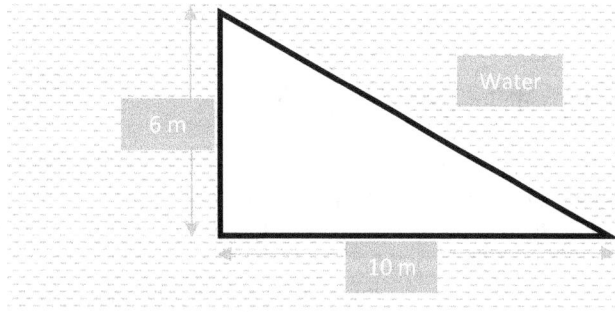

FIGURE 7.23   Submerged vertical plate.

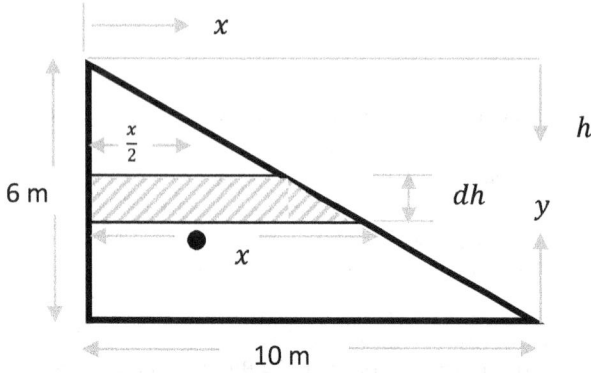

FIGURE 7.24   Center of pressure analysis sketch.

$$F = \delta \int_0^6 h\left(\frac{10}{6}h\right)dh = \delta \int_0^6 \frac{10}{6}h^2 dh = \delta\left[\frac{10}{6}\cdot\frac{h^3}{3}\right]_0^6 \tag{7.189}$$

$$F = 9,810\left(\frac{10}{6}\cdot\frac{6^3}{3}\right) = 1,177.2 \text{ kN} \tag{7.190}$$

The magnitude of the hydrostatic force acting on one side of the submerged vertical plate is 1,177.2 kN.

To determine the location of the center of pressure (cp),

$$y_{cp}F = \int ydF \tag{7.191}$$

Using,

$$y_{cp} + h_{cp} = 6 \tag{7.192}$$

Therefore,

$$y_{cp} = 6 - h_{cp} \tag{7.193}$$

$$\therefore h_{cp} = \frac{\int dM}{\int dF} = \frac{\int hdF}{\int dF} \tag{7.194}$$

$$h_{cp} = \frac{\delta \int_0^6 \frac{10}{6}h^3 dh}{1177200} \tag{7.195}$$

$$h_{cp} = \delta\frac{\left[\frac{10}{6}\cdot\frac{h^4}{4}\right]_0^6}{1,177,200} = 9,810 \times \frac{\frac{10}{6} \times \frac{6^4}{4}}{1,177,200} = 4.5 \text{ m} \tag{7.196}$$

$$\therefore h_{cp} = 6 - 4.5 = 1.5 \text{ m} \tag{7.197}$$

$$x_{cp} = \frac{\int dM}{\int dF} = \frac{\int \frac{x}{2}dF}{F} = \frac{\frac{1}{2}\int x\delta\left(\frac{10}{6}\right)h^2 dh}{F} \tag{7.198}$$

$$x_{cp} = \frac{\frac{1}{2}\delta \int \frac{10}{6}h\cdot\left(\frac{10}{6}\right)h^2 dh}{F} \tag{7.199}$$

$$\therefore x_{cp} = \frac{\delta \int_0^6 \frac{5}{6} \times \left(\frac{10}{6}\right)h^3 dh}{F} = 3.75 \text{ m} \tag{7.200}$$

Note

- Most of the hydraulics and fluid mechanics problems that involve calculus revolve around the integral of forces over an area in contact with the fluid.

7.9.1.2.6 Area moment of inertia

The area moment of inertia, also called the second moment of area, defines how much resistance a shape (like the cross-section of a beam) has to experience bending because of its geometry. Understanding the area moment of inertia is essential

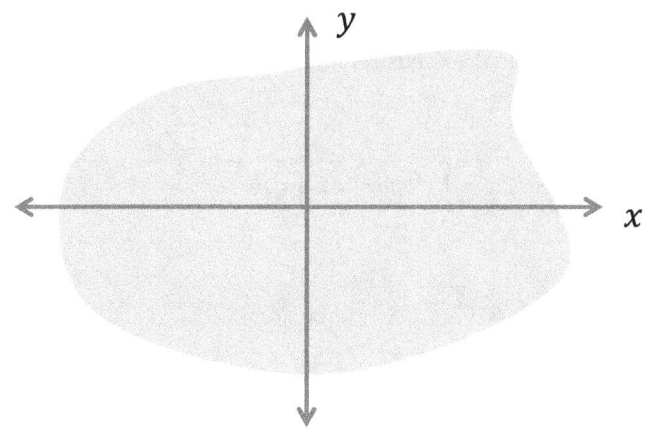

FIGURE 7.25 Cross-section of the material.

for designing and analyzing structural components, ensuring safety, efficiency, and optimal performance. It is important in beam design as it helps determine the beam's resistance to bending and deflection. Through the moment of inertia, we can calculate the stresses, loads, and stability of the structure. Material usage can be optimized by minimizing the weight of materials used while maintaining their strength requirements. Calculus can be used to determine the area moment of inertia. For any arbitrary cross-section like the one shown in the Figure 7.25, the area moment of inertia can be calculated using equations (7.201) and (7.202) [9].

Thus, by definition, the moment of inertia of an area is defined as follows:

$$\bar{I}_x = \int y^2 dA \tag{7.201}$$

$$\bar{I}_y = \int x^2 dA \tag{7.202}$$

The equations above can be used to calculate for any arbitrary shape. The x and y subscripts indicate that the area moment of

inertia is for bending about the x and y and axes, respectively. The SI units of the area moment of inertia are cm^4 or mm^4.

7.9.1.2.6.1 Worked example [7.19]: Area moment of inertia
Let's use calculus to determine the centroidal moments of inertia $\bar{I}_x$ and $\bar{I}_y$ of a rectangular area with base b and height h parallel to the x and y centroidal axes, respectively (Figure 7.26).

Solution

Using double integration
Here, we choose the second-order differentiation area element dA as shown below.

By definition,

$$\bar{I}_x = \int y^2 dA = \int_{-\frac{h}{2}}^{\frac{h}{2}} \int_{-\frac{b}{2}}^{\frac{b}{2}} y^2 dxdy = \int_{-\frac{h}{2}}^{\frac{h}{2}} by^2 dy \tag{7.203}$$

$$\bar{I}_x = b\left[\frac{y^3}{3}\right]_{-\frac{h}{2}}^{\frac{h}{2}} \tag{7.204}$$

$$\therefore \bar{I}_x = b\left[\frac{y^3}{3}\right]_{-\frac{h}{2}}^{\frac{h}{2}} = b\left[\frac{h^3}{24} + \frac{h^3}{24}\right] = \frac{1}{12}bh^3 \tag{7.205}$$

Similarly,

$$\bar{I}_y = \int x^2 dA = \int_{-\frac{h}{2}}^{\frac{h}{2}} \int_{-\frac{b}{2}}^{\frac{b}{2}} x^2 dxdy = \int_{-\frac{h}{2}}^{\frac{h}{2}} b^2 dy = \frac{1}{12}b^3h \tag{7.206}$$

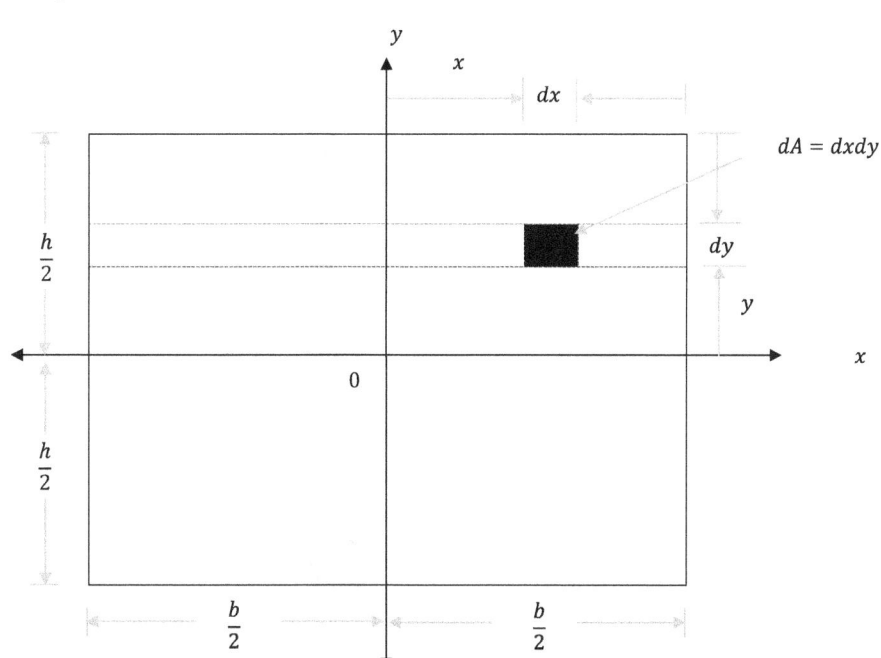

FIGURE 7.26 Determination of the area moment of inertia, double integration.

Using single integration

For determining $\bar{I}_x$, we choose dA to be a first-order areal strip parallel to the x-axis, as shown in Figure 7.27.

By definition,

$$\bar{I}_x = \int y^2 dA = \int_{-\frac{h}{2}}^{\frac{h}{2}} y^2 (bdy) = b \int_{-\frac{h}{2}}^{\frac{h}{2}} y^2 dy \qquad (7.207)$$

$$\bar{I}_x = \frac{1}{12} bh^3 \qquad (7.208)$$

For determining $\bar{I}_y$, we choose dA to be a first-order areal strip parallel to the y-axis (Figure 7.28).

Solution

$$\bar{I}_y = \int x^2 dA = \int_{-\frac{b}{2}}^{\frac{b}{2}} x^2 (hdx) = h \int_{-\frac{b}{2}}^{\frac{b}{2}} x^2 dx \qquad (7.209)$$

$$\bar{I}_y = \frac{1}{12} b^3 h \qquad (7.210)$$

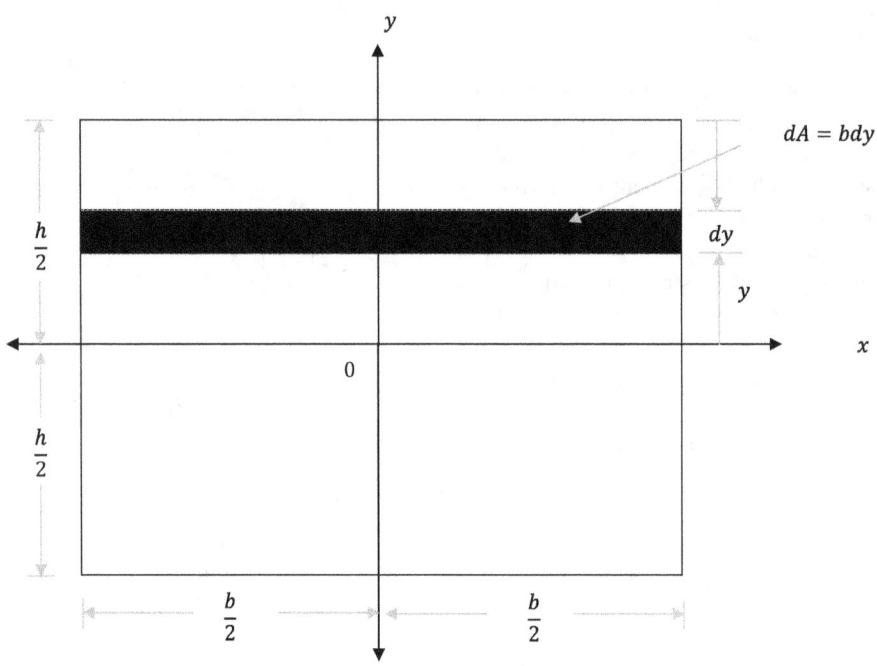

FIGURE 7.27   Determination of the area moment of inertia, I_x—single integration.

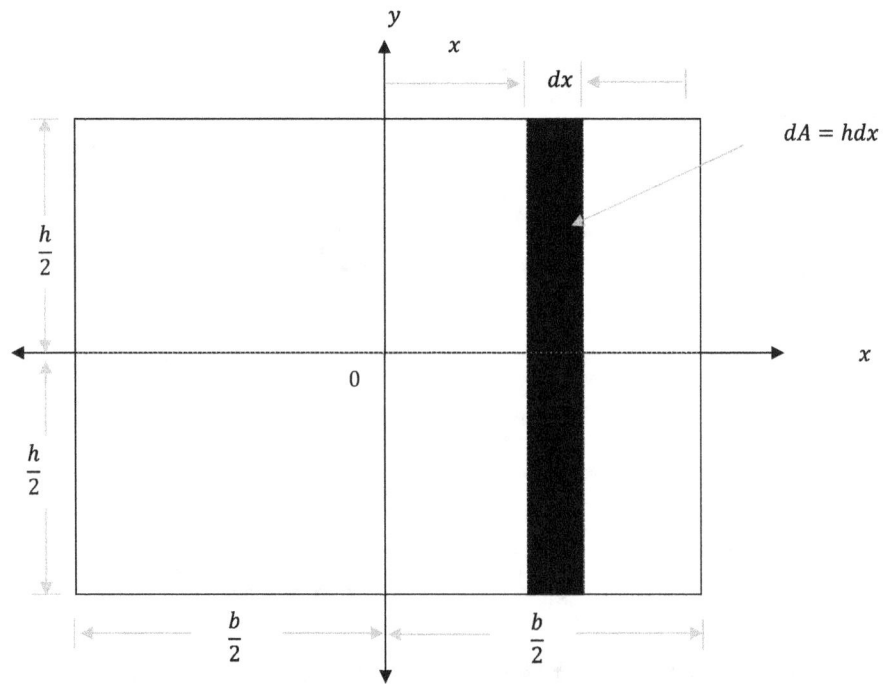

FIGURE 7.28   Determination of the area moment of inertia, I_y—single integration.

7.9.1.2.7 *Laplace and Fourier transforms*

Laplace and Fourier transforms are powerful mathematical tools in civil engineering used to analyze and solve complex problems related to structural behavior and system dynamics. Laplace transforms simplify the analysis of **transient responses and systems** with initial conditions, while Fourier transforms excel in **frequency domain analysis**, particularly for understanding periodic signals and vibrations.

Both transforms are essential in analyzing **structural dynamics**—particularly, the dynamic behavior of structures, which includes bridges, buildings, and dams, under various loading conditions. They play an important role in designing and analyzing control systems for infrastructure projects, such as vibration control systems for tall buildings (e.g., skyscrapers) or bridge dampers. Both transforms can be applied in **geotechnical engineering** to analyze soil behavior and wave propagation in soil—which is important for foundation design and earthquake engineering.

Specifically, the Laplace transform converts differential equations that describe the dynamic behavior of structures, into **algebraic equations**, making them easier to solve. On the other hand, the **Fourier transform** helps in understanding how structures respond to different frequencies of excitation, which is essential for designing structures to withstand dynamic loads such as earthquakes, wind, and/or traffic.

7.9.1.2.8 *Conclusion*

Ever since Newton invented calculus, whose basic application was to apply it as a tool for summing up areas and volumes, forces, and moments, several scholars and practitioners have expanded it to find useful applications beyond that. There is practically no engineering without calculus. It's the fundamental language that allows engineers to precisely articulate and manipulate the underlying phenomena they are dealing with. You can use calculus to represent or approximate any physical shape or phenomenon—a critical capability in any type of engineering.

In civil engineering, calculus plays an important role in multiple areas, such as the design of structures, the calculation of load and stress factors, soil mechanics, and fluid dynamics. These fields require precise mathematical models to ensure safe and efficient infrastructure development.

Whether you are designing a beam, column, road pavement, a hydraulic structure, or dynamic system, you may apply calculus to estimate the impact force, pressure, or stress over an area or volume. However, in dynamic systems, the story is a little different. The development of Laplace and Fourier transforms as powerful mathematical tools in civil engineering used to analyze and solve complex problems related to structural behavior and system dynamics benchmarked on calculus.

Engineers are able to figure out important details such as how much material or energy they need for a project by using calculus. This is essential for designing and checking their work. Calculus also helps engineers make things better and safer while reducing costs by providing a solid way to make decisions based on numbers.

7.9.2 Probability and Statistics

In civil engineering, probability and statistics are primarily applicable in uncertainty analysis, risk assessment, and data analysis. Probability refers to the mathematical study of the likelihood of an event occurring, essentially measuring the 'chance' of something happening, while statistics is a branch of mathematics that focuses on collecting, analyzing, interpreting, and presenting data to draw conclusions about a population based on a sample; it heavily relies on probability theory to make inferences about data.

Probability is all about chance, whereas statistics is more about how we handle various data using different techniques. It helps represent complicated data in a very easy and understandable way. Therefore, engineers rely greatly on probability density functions (PDFs) to model engineering problems. In fact, hydrologic studies, a discipline civil engineers employ in a number of civil engineering branches, are almost entirely made up of probability and statistics.

7.9.2.1 *The average dilemma*

A saying that 'You are an average of the 5 people you spend most of your time with' may be correct in a literary sense but invalid in the engineer's mind. You may live in undeveloped settings with atheists but think differently and be a believer. Associating with drunkards may not automatically make you one! That's why democracy will always be an ideal because it is unscientific.

Average is a measure of central tendency, just like the median and mode! Relying solely on measures of central tendency (such as mean, median, or mode) can be misleading when designing engineering systems because they can either lead to overdesign or underdesign or a flawed design. Therefore, averaging is not engineering!

In the aspect of overdesign, relying on averages can result in overly conservative designs, leading to increased costs and reduced efficiency. On the other hand, averaging can also lead to underdesign, resulting in inadequate performance or even failures as it fails to account for extreme values and variability in the design, etc. The following are the primary reasons relying on the measures of central tendency may create underdesign or overdesign.

- **Failure to cater for variability:** Central tendency measures fail to account for the spread or dispersion of data, which is essential in engineering design. The date may look alike when it is actually different. Relying on the measure of central tendency misleads!
- **Hides extremes:** By focusing on the 'average' or 'typical' value, central tendency measures can mask critical extreme values or outliers that might be catastrophic in engineering applications.
- **Assumes normality:** Many engineering applications don't follow a normal (Gaussian) distribution, making central tendency measures less representative. Engineering data rarely follow the normal distribution!
- **Overlooks skewness:** Central tendency measures can be skewed by asymmetric distributions, leading to inaccurate representations.
- **Fails to consider context:** Central tendency measures do not account for specific design requirements, operating conditions, or environmental factors.

- **Does not ensure robustness:** Relying solely on central tendency measures can lead to designs that are not robust or resilient.

Averaging or using any other measure of central tendency is not always logical engineering practice, if not, not engineering at all? Its appropriateness depends on the specific problem, data, and requirements. In conclusion, using measures of central tendency, such as averaging, can be a logical and useful approach in engineering, but it's essential to consider the context, data, and requirements carefully. Blindly applying averages without proper analysis or consideration of limitations can lead to illogical or misleading results. Therefore, in engineering design, it is essential to consider the entire distribution of data, including variability, extremes, and context, to ensure safe, reliable, and efficient systems. In that regard, the underlying distributions, uncertainties, and relationships between variables are considered to make informed decisions. Instead of relying solely on averages and mode, engineers use more sophisticated probability statistical methods/techniques, such as **probability distributions**, confidence intervals, reliability analysis, regression analysis, sensitivity analysis, and Monte Carlo simulations.

These methods provide a more comprehensive understanding of the problem, enabling engineers to make informed decisions and design robust, efficient, and safe solutions. So, as a rule of thumb, never take average from data collected to design civil engineering structures. For example, the strength of five concrete cubes at 28 days is 20, 16, 21, 28, and 18 MPa. Therefore, saying the concrete strength achieved in the construction project is roughly 20.6 MPa is false! Engineers cater for variability and uncertainty, the primary reason why the average is discouraged. That variability in those cube samples communicates a lot of information. Honestly, the data reveal a serious underlying problem that needs to be investigated. When you obtain such data, you may have to investigate the underlying issue—curve fitting may be one of the techniques to adopt, alongside the applicable probability distributions that model failures! Some of the chief distributions common in civil engineering are listed in Table 7.9.

7.9.2.2 Quality control distributions

The application of probability and statistical distributions appears in all civil engineering branches. Its role in quality control and management of the design and construction of civil engineering structures is fundamental. In quality control, several distributions are applicable, but some common ones include normal distribution, binomial distribution, Weibull distribution, and Poisson distribution. Probability distributions help us deal with large numbers and hence come up with quality control strategies by proportionating items.

The normal distribution is widely used in quality control due to the central limit theorem (many processes tend toward normality) and statistical process control (SPC) (control charts often assume normality). However, the choice of distribution depends on the specific problem and data characteristics. Quality control professionals apply probability distribution functions to better understand and manage process variability, ultimately leading to improved product quality and reliability. The famous distribution used in quality control includes the following:

- **Normal distribution (Gaussian distribution):** model continuous data with a symmetric, bell-shaped curve.
- **Binomial distribution:** models binary data, like pass/fail or yes/no outcomes.
- **Poisson distribution:** models discrete data, often used for count data, such as defects or errors.
- **Exponential distribution:** models continuous data with a **constant failure rate,** often used for reliability analysis.
- **Weibull distribution:** models continuous data with **varying failure rates**, often used for reliability and survival analysis.
- **Uniform distribution:** models continuous data with equal probabilities across a range.
- **Gamma distribution:** models continuous data with varying shapes, often used for reliability and quality control.
- **Beta distribution:** models continuous data between 0 and 1, often used for proportions or percentages.

These PDFs help quality control professionals to model process variability, analyze and predict defect rates, set control limits, monitor process stability, improve process capability, and make informed decisions. A perfect example of how PDFs control quality and improve reliability is illustrated in example 7.20. α and β are called **Weibull coefficients**. The Weibull shape parameter, called β, indicates whether the failure rate is

TABLE 7.9 Chief distributions common in civil engineering

DISTRIBUTION	SIGNIFICANCE
Exponential distribution	This models time between events in a Poisson process.
Weibull distribution	This is used in reliability analysis and failure rate modeling.
Normal distribution	A fundamental distribution in statistics, describing many natural phenomena. Often used to model continuous data, such as measurements.
Log-normal distribution	It models variables that are skewed and non-negative.
Gamma distribution	Used in reliability analysis, queuing theory, and insurance risk modeling.
Gumbel distribution	It models extreme events, such as floods and earthquakes.
Extreme Value distribution	Used to model rare events.
Log-Pearson III distribution	Used in hydrology to model flood frequencies.
Binomial distribution	Used for pass/fail or defective/non-defective data.
Poisson distribution	Models count data, such as defects per unit.

increasing, constant, or decreasing. $\beta < 1.0$ indicates a decreasing failure rate. $\beta = 1.0$ indicates a constant failure rate. $\beta > 1.0$ indicates an increasing failure rate.

The management uses these coefficients to make lasting reliability improvements. Failure data are collected that are assumed to follow the Weibull distribution, and based on the coefficients obtained, the management can take decisions to improve the product batch or manufacturing processes, etc. The distribution is important in **reliability studies** and quality control because it studies and models failure rates.

Note that reliability studies play a crucial role in defining how long a structure is expected to perform safely and effectively. **Reliability-based design** is especially used for bridges, buildings, tunnels, and other infrastructure projects—to inform the design life—which is the intended duration during which a structure should meet safety and serviceability requirements under expected loads and environmental conditions.

7.9.2.3 Example [7.20]: Weibull PDF application in reliability studies

X Construction Co. Ltd (XCCL) operates and manages the water supply system for municipal council **A**. It collects monthly revenues from the community for and on behalf of the municipal council. XCCL is entitled to receive 79% of the monthly collections. This means the higher the collections, the higher the income XCCL receives at the end of the month. Statistics show that the collections have never gone below $3,000,000 per month since when the water supply system was commissioned. XCCL bided for a contract to operate a typical water supply system of municipal council B in which XCCL is still entitled to receive 79% of the monthly collections. The Weibull distribution can help assess which station is more optimally reliable as far as XCCL objectives are concerned. The sum of company's monthly expenditures and overheads on municipal A's water supply scheme is $5,265,000. This includes materials/chemicals (e.g., aluminum sulfate), labor costs, equipment, salaries, electricity, and rents.

The desired reliability at $5,265,000 gross income receivable in a month for operating and managing the water supply on behalf of municipal council A is 0.79. In other words,

the income XCCL receives every month is 79% of the total monthly collections. Therefore, XCCL to receive $5,265,000, it should have collected $6,664,557. The reliability goal is expressed mathematically as $R(5,265,000) \geq 0.79$. The data for revenue collections for the two municipals (A&B) for 10 months of 2024 are presented in Table 7.10.

Assumption: since systems are typical, we assume the same/constant overheads and expenditure.

The data in Table 7.10 don't clearly indicate whether either water supply system meets the desired reliability goal. Both water supply systems had a monthly collection less than $6,664,557. Yet clearly, the average monthly collections for both water supply systems exceeded $6,664,557. A comparison of sample averages using Student's t-test reveals no statistical difference between the average collections of A and average collections of B. But as a simple measure of central tendency, the sample average gives no information about the spread or shape of the distribution of failure times to attain a collection of $6,664,557. Could the two water supply systems average monthly collections be the same, but their reliability be quite different? We can check this with the help of the **Weibull distribution**.

The Weibull cumulative distribution function (CDF) can be transformed so that it appears in the form $y = mx + c$, in the form of an equation for a straight line (Figures 7.29 and 7.30):

TABLE 7.10 The 2024, 10 months revenue, from municipals A and B

MONTH	MUNICIPAL A ($)	MUNICIPAL B ($)
January	5,250,150.00	6,306,700.00
February	6,234,550.00	8,600,780.00
March	5,737,050.00	7,500,000.00
April	7,845,150.00	6,873,000.00
May	7,337,900.00	9,300,350.00
June	8,178,200.00	7,821,400.00
July	11,696,550.00	8,262,000.00
August	9,174,350.00	5,450,000.00
September	7,427,800.00	5,608,250.00
October	11,074,350.00	7,790,700.00

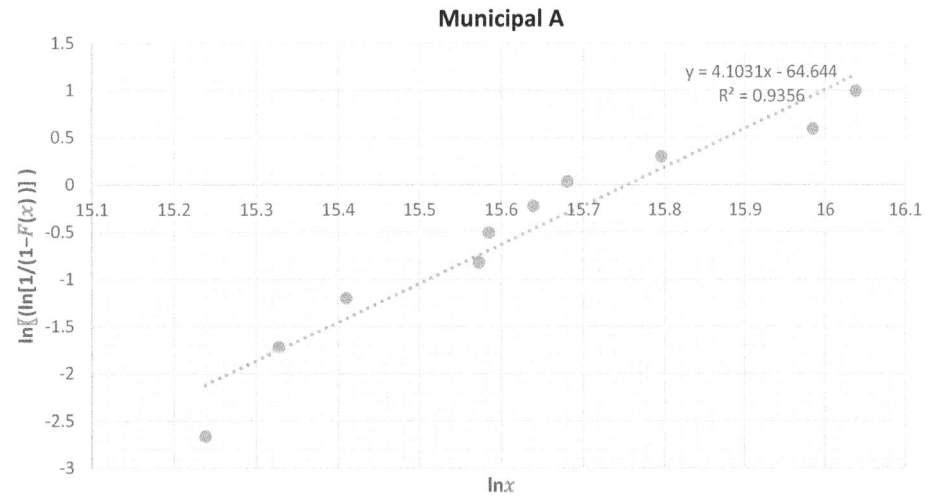

Municipal A

$y = 4.1031x - 64.644$
$R^2 = 0.9356$

FIGURE 7.29 Plot of Municipal A's results.

$$F(x) = 1 - e^{-\left(\frac{x}{\alpha}\right)^{\beta}} \qquad (7.211)$$

$$1 - F(x) = e^{-\left(\frac{x}{\alpha}\right)^{\beta}} \qquad (7.212)$$

Taking the natural logarithm on both sides

$$\ln\left(\frac{1}{1 - F(x)}\right) = \left(\frac{x}{\alpha}\right)^{\beta} \qquad (7.213)$$

which can be expressed in the form: $y = mx + c$

$$\ln\left(\ln\left[\frac{1}{1 - F(x)}\right]\right) = \beta \ln x - \beta \ln \alpha \qquad (7.214)$$

Comparing this equation (7.215) with the simple equation for a line, we see that the left side of the equation corresponds to y, $\ln x$ corresponds to x, β corresponds to m, and $-\beta \ln x$ corresponds to c. Thus, when we perform the linear regression, the estimate for the Weibull β parameter comes directly from the

slope of the line (Tables 7.11 and 7.12; Figures 7.29 and 7.30). The estimate for this must be calculated as follows:

Municipal A

Municipal A, graph plot for

$$\ln\left(\ln\left[\frac{1}{1 - F(x)}\right]\right) = \beta \ln x - \beta \ln \alpha \qquad (7.215)$$

Municipal B

Municipal B, graph plot for

$$\ln\left(\ln\left[\frac{1}{1 - F(x)}\right]\right) = \beta \ln x - \beta \ln \alpha \qquad (7.216)$$

Interpretation of the results

The Weibull shape parameter, called β, indicates whether the failure rate is increasing, constant, or decreasing, i.e., failure to collect \$6,664,557 and above. $\beta < 1.0$ indicates a decreasing failure rate. $\beta = 1.0$ indicates a constant failure rate. $\beta > 1.0$ indicates an increasing failure rate. With all this, still it does not reveal whether municipal A or B meet the reliability goal

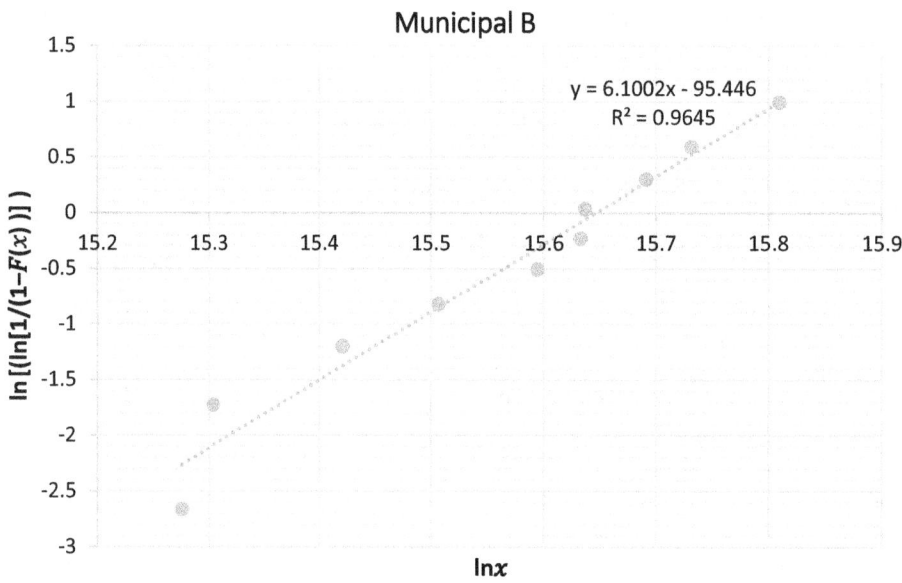

FIGURE 7.30 Plot of Municipal B's results.

TABLE 7.11 Analysis table for municipal A

A COLLECTION ($)	RANK	MEDIAN RANKS (MR)	$\frac{1}{(1-mr)}$	$\ln\left(\frac{1}{1-mr}\right)$	$\ln\left(\ln\left(\frac{1}{1-mr}\right)\right)$	IN (A)
4,147,618.50	1	0.067	1.072	0.070	−2.664	15.238
4,532,269.50	2	0.163	1.195	0.178	−1.723	15.327
4,925,294.50	3	0.260	1.351	0.301	−1.202	15.410
5,796,941.00	4	0.356	1.552	0.440	−0.822	15.573
5,867,962.00	5	0.452	1.825	0.601	−0.509	15.585
6,197,668.50	6	0.548	2.213	0.794	−0.230	15.640
6,460,778.00	7	0.644	2.811	1.033	0.033	15.681
7,247,736.50	8	0.740	3.852	1.349	0.299	15.796
8,748,736.50	9	0.837	6.118	1.811	0.594	15.984
9,240,274.50	10	0.933	14.857	2.698	0.993	16.039

TABLE 7.12   Analysis table for municipal B

A COLLECTION ($)	RANK	MEDIAN RANKS (MR)	$\frac{1}{(1-mr)}$	$\ln\left(\frac{1}{1-mr}\right)$	$\ln\left(\ln\left(\frac{1}{1-mr}\right)\right)$	IN (A)
4,305,500.00	1	0.067	1.072	0.070	−2.664	15.275
4,430,517.50	2	0.163	1.195	0.178	−1.723	15.304
4,982,293.00	3	0.260	1.351	0.301	−1.202	15.421
5,429,670.00	4	0.356	1.552	0.440	−0.822	15.507
5,925,000.00	5	0.452	1.825	0.601	−0.509	15.595
6,154,653.00	6	0.548	2.213	0.794	−0.230	15.633
6,178,906.00	7	0.644	2.811	1.033	0.033	15.637
6,526,980.00	8	0.740	3.852	1.349	0.299	15.691
6,794,616.20	9	0.837	6.118	1.811	0.594	15.732
7,347,276.50	10	0.933	14.857	2.698	0.993	15.810

of $R(5,265,000) \geq 0.79$. For this, we need to know the formula for reliability, assuming a Weibull distribution:

$$R(x) = e^{-\left(\frac{x}{\alpha}\right)^{\beta}} \tag{7.217}$$

where x is collection, which has to be attained to achieve the reliability goal

$$R(5,265,000)_A = e^{-\left(\frac{5,265,000}{6,954,434}\right)^{4.1}} = 0.727 \tag{7.218}$$

$$R(5,265,000)_B = e^{-\left(\frac{5,265,000}{6,238,654}\right)^{6.1}} = 0.701 \tag{7.219}$$

Both collections from the two water supply systems don't meet the reliability goal, but A's WSS is more reliable than B's WSS. So, relying on averages isn't logical! The reliability of the collections of systems could be affected by people who fail to settle their water bills in time, illegal connections, or unreliable power supply, etc. Municipal A's WSS seems more reliable than B's WSS maybe due to the population density of A could be bigger than that of B—settling bills promptly—of course with modern prepaid technological systems, fewer breakdowns of the A's WSS, multiple uses of the A's WSS, efficiency of workers, high yield from the source, good maintenance of the A's system, and politics involved could also sabotage the activities.

This can serve as a benchmark or warranty to a contractor or private operator willing to bid for a contract to manage these water supply systems. The reliability goal can help the contractor assess the rate of return on investment (ROI)—it can act as a source of reference—why these systems are not meeting the reliability goal? Is it constant breakdowns, poor maintenance, inadequate system design, embezzlement of funds, efficiency of workers, illegal connections, political intervention, unreliable power supply, or what? These questions can be answered in the bid to improve systems' reliability.

7.9.3 Other Versatile Mathematical Fields for Civil Engineers

- **Geometry:** This includes trigonometry, coordinate systems, and spatial analysis.

- **Graph theory:** A branch of mathematics that studies the properties and relationships of graphs, which are abstract structures consisting of vertices (or nodes) and edges (or links) that connect them. This includes network analysis, topology, and optimization.

- **Topology:** Study of shapes and spaces, focusing on properties that are preserved under continuous deformations, such as stretching, twisting, crumpling, and bending; that is, without closing holes, opening holes, tearing, gluing, or passing through itself—this includes network topology, spatial analysis, and geometric computing. Topology plays a big role in understanding and designing new materials with exotic properties, hence driving advancements in materials science and technology.

- **Optimization techniques:** This includes linear and nonlinear programming, dynamic programming—finding the best solution (minimum or maximum) to a problem by systematically selecting values from a set of alternatives, often subject to constraints.

- **Numerical methods:** The commonest techniques civil engineers use to model scenarios—real-world applications. These include discretization methods—finite-element analysis, finite-difference methods, and approximation techniques. See Section 7.4.6.

- **Statistics and machine learning:** This refers to data-driven modeling, pattern recognition, and predictive analytics.

- **Linear algebra:** This includes vector calculus (builds upon linear algebra), matrix operations, and **eigenvalue analysis**. Eigenvalue analysis is a powerful tool for solving matrices and differential equations, providing several insights into system behavior and stability. Specifically, eigenvalues and eigenvectors help diagonalize matrices—helping solve complex systems of **differential equations**, particularly in **linear algebra** and dynamical systems. Eigenvalues and eigenvectors are crucial in decoupling systems of differential equations. See Section 7.4.8.1.2.

7.10 MATHEMATICAL AND COMPUTATIONAL TOOLS FOR CIVIL ENGINEERS

7.10.1 MATLAB

MATLAB, in full Matrix Laboratory, is a high-level programming language specifically designed for numerical computation, data analysis, and visualization. MATLAB is a programming and numeric computing platform used by millions of engineers and scientists to analyze data, develop algorithms, and create models. Developed by MathWorks, Inc., MATLAB is widely used in various fields, including engineering: aerospace, biomedical, chemical, civil, electrical, and mechanical engineering.

MATLAB's key features include

- **Matrix operations:** Efficient manipulation of matrices and arrays.
- **High-level syntax:** Easy-to-read and write code, with a focus on the numerical computation.
- **Built-in functions:** Extensive libraries for tasks like linear algebra, optimization, and signal processing.
- **Visualization tools:** Powerful plotting and visualization capabilities.
- **Toolboxes and add-ons:** Optional packages for specialized tasks, such as image processing, control systems, and computer vision.

7.10.2 Simulink

Simulink is a MATLAB-based graphical programming environment for modeling, simulating and analyzing multidomain dynamical systems. Its primary interface is a graphical block diagramming tool and a customizable set of block libraries.

7.10.3 MAPLE

MAPLE is a mathematical tool. It's a computer algebra system (CAS) specifically designed to assist with mathematical tasks, such as algebra, calculus, geometry, and numerical analysis.

7.11 LOGIC AND THE ENGINEER

By the time you progress to this stage in the chapter, you must have seen how logic builds up equations and formulae and that mathematics is a fundamental tool for engineers. If not, you have to revisit the chapter again to appreciate how mathematics plays as an important tool for an engineer, relying on logic to formulate new equations for use in solving engineering problems. Developing the ability to see things without conscious effort means you have graduated or nearly graduating as an engineer.

That means your experience is marked by logical thought. Critical thinking is at the heart of engineering education.

Logic is the study of how to use reason to draw valid conclusions from the given information. It's a system of reasoning that's used to evaluate arguments and make inferences. In its basic sense, logic works through arguments—a set of premises and conclusions. In the context of logical reasoning, the conclusion is only true when the premise is true. The engineer heavily relies on logic to quicken the decision-making process and designs. The primary technique relied on is the fact that things must jigsaw at some point—and must be orderly. Engineers rely on limits, validity, utility, and reliability.

For example, while at a construction site, the engineer may check the depth of the excavation for a strip foundation. If it is 1 m and not 1.5 m (as specified on the working drawings), it communicates lots of information working the design backward. In other words, the required depth directly impacts the design/ specifications and must be achieved. If there any change orders or variations, they must be backed with evidence to support the logic why a change has been made. Other parameters that are commonly checked at site include formwork, steelwork, tools on site (safety precautions), concrete mixes, batching, and materials inspection. Specifically, everything around concrete production is checked prior to concrete placement to ensure everything is okay to pour/place concrete. From batching, to mixing, and placement, everything has the potential to impact the final concrete quality.

The logic is as follows: when the excavated depth for a strip foundation is reduced from 1.5 to 1 m, the design is tampered. Depth is a factor in determining the adequate/reliable foundation for the structure—a factor that is considered in bearing capacity determination of a strip footing. At this construction stage, we are not calculating, but we are inspecting to ensure everything is implemented as designed. The checklist is a tool recommended to evaluate whether works are implemented as required. From those inspection activities, intuition, a product of experience is activated, and logic refines intuition. Experience gives you insights from what you've done before on how to approach this particular work. See the basic logic examples below.

7.11.1 Logic Example [7.21]

Find height of the table, Figure 7.31.

Solution

If the height of the cat is x and that of the tortoise is y and that of the table z.

Then,

$$x + z - y = 180 \tag{7.220}$$

$$y + z - x = 140 \tag{7.221}$$

Adding these equations:

$$2z = 320$$

$$z = 160 \text{ cm}$$

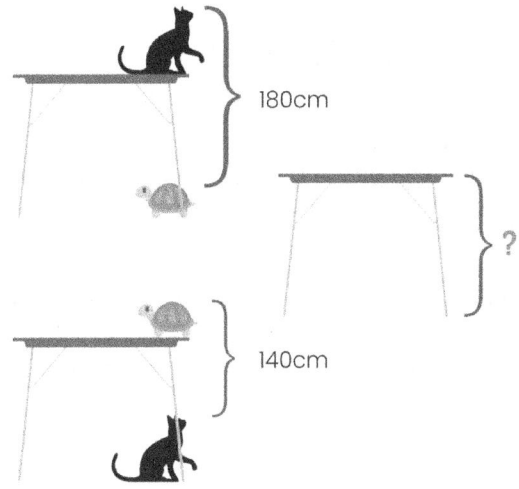

FIGURE 7.31   Table.

7	8	9
4	6	8
1	?	7

FIGURE 7.32   Puzzle to locate missing number.

In this example, we have used three variables and two equations, and we were able to find the height of the table, defying the odds of simultaneous equations but relying on logical reasoning! (Figure 7.32).

7.11.2 Logic Example [7.22]

Find the missing number.

7	8	9
4	6	8
1	?	7

?=4. Here is how

- 7+4 = 11; 11 − 10 = 1.
- 9+8 = 17; 17 − 10 = 7
- *So*, 8+6 = 14; 14 − 10 = 4.

7.11.3 Logic Example [7.23]

Find the missing number in Figure 7.33.
   Solution
   ?=25. We were able to recognize a pattern, where 5 (centered) is multiplied by the numbers to the left.

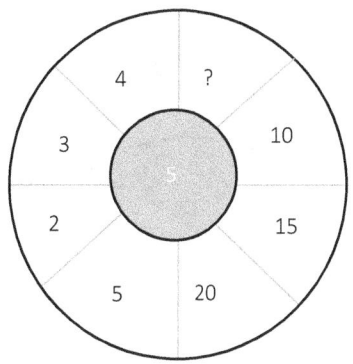

FIGURE 7.33   Circle to locate missing number.

Find the missing number.

2	0	5	8	3
0	7	4	4	5
1	8	3	9	1
1	5	4	7	?

FIGURE 7.34   Puzzle box to locate missing number.

7.11.4 Logic Example [7.24]

Find the missing number.

2	0	5	8	3
0	7	4	4	5
1	8	3	9	1
1	5	4	7	?

   ?=3. The recognized pattern is that the last row contains averages of the three numbers in the respective columns (Figure 7.34).

7.11.5 Logic Example [7.25]

Find the missing number in Figure 7.35.
   ?=17. The pattern suggests that each number in the circle equals the sum of the numbers (opposite half of the circle) in the main circle that the connecting line crosses.

7.11.6 Analysis

Realizing, recognizing, and inspecting patterns do not necessarily follow approved/standard mathematical principles or techniques! This concept is very important in design, analysis, and evaluation of civil engineering systems or structures. In the

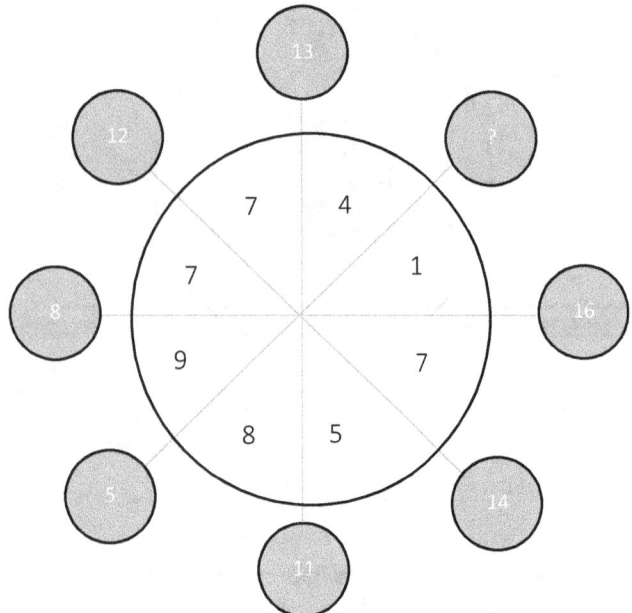

FIGURE 7.35 Diagram to locate missing number.

design and construction of civil engineering structures, analysis and evaluation are primarily based on logical identification of patterns in the information and data available. Engineers try to make sense of the available information borrowing knowledge of either standard techniques or logical steps. Playing with numerical, analytical, and empirical methods, engineers are able to propose breakthrough solutions. Solutions that make things tie logically.

7.12 MIND MAPPING QUESTIONS

1. Explore and map out coefficients used in various civil engineering disciplines.
2. Explore civil engineering equations that may need refinement, considering areas such as structures, hydraulics, soil mechanics, and geotechnics.
3. Create a mind map to visually organize and connect ideas on the challenges and limitations of using modal average in civil engineering.
4. How best can we advance logarithms in engineering education?
5. Find x.

$$x^2 + \left(\frac{3x}{x-3} \right)^2 = 16 \qquad (7.222)$$

6. Explore why failures matter.
7. Create a mind map to visually organize and connect ideas on applying the probability and statistics to advance civil engineering knowledge in the context of climate change.

7.13 CONCLUSION

Engineers use mathematics to model and analyze complex systems, optimize designs and processes, predict behavior and performance, make informed decisions under uncertainty, and develop new materials and technologies. Several mathematical techniques are applied across all branches of civil engineering, including structural analysis and design, transportation engineering, water resources and hydraulic engineering, geotechnical engineering, construction management, environmental engineering, and surveying and geomatics, among others.

Mathematics plays an important role in engineering design. It has been used as an engineer's tool to design and communicate engineering information throughout history. As a primary tool for engineers, it provides the analytical and problem-solving tools necessary for designing, analyzing, and optimizing systems, ensuring safety, efficiency, and performance. Engineers use mathematics to make informed decisions and drive technological advancements across various engineering disciplines—civil engineering being one of them.

Calculus and probability and statistics are the commonest mathematical fields for civil engineers. Other fields come next. Generally, it is a must for an engineer to appreciate the role of calculus, probability, and statistics, alongside the FOSs, coefficients, correction factors, capacity factors, and logarithms, among other topics. In addition, equations' grouping is a primary technique engineers use to unlock the impossible!

NOTES

1 The inspection technique is very helpful and powerful in understanding equations or crafting equations for engineering applications. It is not only helpful in solving equations but also helpful in design. Through intuition, the engineer develops an understanding of many equations, and the method of inspection is a gateway to understanding how equations can be manipulated to advance new solutions to new or existing engineering problems. Engineers craft problems and reduce complexity through the inspection technique so that they can fit constituent parts together to ease their work.

2 By Original diagram: S Beck and R Collins, University of Sheffield (Donebythesecondlaw at English Wikipedia) Conversion to SVG: Marc.derumaux - File:Moody_diagram.jpg, CC BY-SA 4.0, https://commons.wikimedia.org/w/index.php?curid=52681200

3 Multivariable calculus extends single-variable calculus by exploring functions with multiple input variables. It deals with differentiation and integration of these functions, including concepts like partial derivatives, multiple integrals, and vector calculus.

REFERENCES

1. Calculus. *Gilbert Strang*. Massachusetts Institute of Technology, Wellesley-Cambridge Press, Wellesley MA.
2. History of the Manning formula. Retrieved from: https://ponce.sdsu.edu/manning_history.html
3. King, H.W. (1918). Handbook of Hydraulics for the Solution of Hydraulics Problems (1st ed.). McGraw-Hill Book Co. Inc., New York
4. Terzaghi, K. (1943). *Theoretical Soil Mechanics.* John Wiley & Sons, New York.
5. Webster, F. V. (1958). Traffic Signal Settings. Road Research Technical Paper No. 39. The Stationery Office, London.
6. Coppi, P. (n.d.). Astro 520 (course materials). Retrieved from: https://www.astro.yale.edu/coppi/astro520/buckingham_pi/Buckinghamforlect1.pdf. Accessed 3/July/2025.
7. Ssempeera, P. (2023). *Integrated Drainage Systems Planning and Design for Municipal Engineers* (1st ed.). CRC Press. https://doi.org/10.1201/9781003255550
8. Logarithmic scale. Retrieved from: https://www.ibm.com/docs/en/cognos-analytics/11.1.0?topic=visualizations-logarithmic-scale#tasklogscalevis__steps__1. Last Updated 29/February/2024.
9. https://efficientengineer.com/area-moment-of-inertia/

FURTHER READING

1. James, G. (n.d.). *Modern Engineering Mathematics* (4th ed. with MyMathLab). Prentice Hall, England.
2. James, G. (n.d.). *Advanced Engineering Mathematics* (4th ed. with MyMathLab). Prentice Hall, England.
3. Yang, X.-S. (2017). *Mathematics for Civil Engineers: An Introduction* (1st ed.). Liverpool University Press, UK.
4. Ssempeera, P. (2023). *Integrated Drainage Systems Planning and Design for Municipal Engineers* (1st ed.). CRC Press. https://doi.org/10.1201/9781003255550

Common Civil Engineering Tools

8

8.1 INTRODUCTION

A tool is any gadget, system, or product used to aid any work or assignment. So, a civil engineering tool refers to a software, instrument, or device used to design, analyze, construct, or manage infrastructure projects and systems. Civil engineering applies several tools to derive solutions to engineering problems in the design, construction, operation, and maintenance of civil engineering structures.

These tools range from software applications, computers (hardware), equipment and machinery, drawings, to documents. Specifically, civil engineering tools include the following: software such as AutoCAD, Revit, Civil 3D, STAAD, and ETABS; instruments such as Total stations, GPS receivers, levels, and theodolites; and devices such as concrete testers, soil compaction meters, and asphalt density gauges.

Heavy tools used in construction include equipment and machinery such as cranes, excavators, bulldozers, and concrete mixers. See the detailed list of tools used in construction in Section 9.6.

Drawings such as hand-drawn sketches, computer-aided design (CAD) drawings, technical illustrations, diagrams and schematics, and maps and plans are tools.

Tools enhance the accuracy, efficiency, and productivity in civil engineering tasks, enabling engineers to create safe, sustainable, and innovative infrastructure solutions. Collectively, tools enable engineers to design and model buildings, bridges, roads, and other structures; analyze and simulate stress, strain, and other physical forces; plan and manage construction projects and timelines; conduct site surveys and inspections; test and measure materials and soil properties; and visualize and communicate complex data and designs, among other roles.

A camera used to take photographs during the study of a civil engineering problem can be referred to as a tool (instrument). Similarly, survey equipment that civil engineering surveyors or surveyors use to collect data are all tools that enable civil engineering studies and construction to be carried out. A civil engineer's computer in office is a tool, and the software installed on it is also a tool.

These tools are used to communicate engineering information. For example, a drawing is a primary tool civil engineers use to communicate engineering information. Engineering drawings differ and are explained in Section 8.6.

In academia, particularly, free-body diagrams, shear force diagrams, and bending moment diagrams are drawings that are used to communicate engineering information to students. On the other hand, these common tools are widely used by professionals, as well, to design engineering systems and communicate their designs.

Engineers build the technological tools through a combination of research, experience, intuition, creativity, innovation, and ingenuity. And they use them to advance civil engineering, science, and technology at the same time.

Chapter 6 showed how civil engineers mimic the natural world to engineer systems that support humanity. The human anatomy and nature are excellent tools that engineers use to communicate their concepts, proposals, and designs in the modern day.

8.2 STRUCTURAL ENGINEERING—A GATEWAY TO OTHER BRANCHES

In Section 1.6.5, we saw that structural engineering is supreme among all branches of civil engineering. Other branches primarily benchmark on structures. We discuss water resources engineering, geotechnical engineering, highway or transportation engineering, public health engineering, etc., all functions embedding within them structural components that carry loads. It is in this regard that structural engineers find themselves conducting interdisciplinary approaches in designing and developing structures. Therefore, this section introduces support and connection types in structural engineering that you need to know as a civil engineer.

These **supports** and **connections** carry loads imposed on the structures—dead and live loads. It is, therefore, good for all civil engineers to learn the basics of structural engineering because the primary concepts of stress and strain analysis that structural engineers master help understand many concepts in other branches of civil engineering, for example, geotechnical engineering, highway engineering (e.g., pavement designs), and pipeline analysis in water resources (hydraulic engineering)—for example, anchor blocks used in pipeline construction, the stress build-up on walls in pipelines, are well modeled in the structural engineering context.

Structural engineering's support and connection types aid further learning; for example, free-body diagrams, shear force, and moment diagrams are excellent tools that designers and planners use to communicate engineering information. FBDs are employed across multiple disciplines. Therefore, understanding support types is crucial for accurate calculations and structural analysis.

DOI: 10.1201/9781003561620-8

8.3 INTRODUCTION TO SUPPORT AND CONNECTION TYPES

We saw in Chapter 4 that civil engineering structures are built to withstand forces of various types. In doing so, structural systems that civil engineers develop are able to perform by transmitting loads to the ground through several support systems.

In load transfer to the ground, structural systems transfer or distribute loads from one structural member to another through supports and/or connections. Therefore, before the load is transferred to the ground, structural systems transfer their loading through a series of elements or structural members connected to each other or supported by others. Therefore, to accomplish this, engineers join system elements at their intersections. That means, each connection is designed to properly transfer or support a specific type of load or loading condition.

Structural integrity is achieved by carefully designing the intersections of elements. Each connection is tailored to transfer specific loads or withstand particular loading conditions. To analyze a structure, it's crucial to understand the forces that can be resisted and transferred at each support level. However, support and connection behavior can be complex, making detailed design a lengthy process.

The real behavior of support or connection system can be unique, and that means, if we are to consider various conditions, the design of each support or connection would be tedious. It is essential to note that the conditions at each of the supports greatly influence the behavior of the elements which make up each structural system. Despite this complexity, connections significantly impact the behavior of structural elements. Effective connection design ensures structural stability and performance. Different materials have unique connection methods. For example,

- Structural steel (e.g., reservoir tank towers) has either welded or bolted connections.
- Precast reinforced concrete has various mechanical connections.
- Cast-in-place concrete has monolithic connections (poured together). **Note: monolithic connections** are connections between concrete elements that are poured together at once.
- Timber is sometimes connected through nails, bolts, glue, or some other engineered connectors.

It is important to note that regardless of the material, connections must be designed with a specific rigidity in mind. The connections range from rigid (stiff or fixed), maintaining relative angles, pinned (hinged), allowing relative rotation, to partially rigid, offering controlled flexibility.

Rigid, stiff, or fixed connections lie at one extreme limit of this spectrum, and hinged or pinned connections bound the other. The stiff connection maintains the relative angle between the connected members, while the hinged connection allows a relative rotation. There are also connections in steel and reinforced concrete structural systems in which a partial rigidity is a desired design feature.

Key notes (supports versus connections)

- Supports primarily provide reaction forces to maintain equilibrium, while connections facilitate load transfer between structural members.
- Supports are designed to resist loads and provide stability, while connections are designed to transfer/distribute loads and stresses between members.
- Supports are typically located at the base or ends of structural members, while connections can be located anywhere along the length of a member.

8.4 THE THREE CONNECTION OR SUPPORT TYPES

Civil engineers use three primary types of theoretically idealized connections and support systems to analyze structures, that is, pin, roller, and fixed supports. These three common types of **supports** join a built structure to its foundation:

- **Roller support** which allows rotation and translation along the supporting surface.
- **Pinned support** which resists vertical and horizontal forces, allows rotation in one direction, and prevents translation.
- **Fixed support** which resists all types of forces and moments, with no rotation or translation.
- A fourth, less common type is the **simple support**, idealized as a frictionless surface.

All of these supports can be located anywhere along a structural element. These supports or connections can be found at midpoints, at the ends, or at any other intermediate points. This type of support connection determines the type of load that the support can resist. The support type also has a great effect on the load-bearing capacity of each element, and therefore the system.

To simplify structural analysis, supports are often idealized, similar to assuming massless, frictionless pulleys in physics. This approach ignores factors like friction and mass, enabling focused examination of specific issues. However, it's crucial to recognize that graphical representations of supports are abstracted from reality. Effort should be made to search out and compare the reality with the graphical and/or numerical model. It is often very easy to forget that the assumed idealization can be strikingly different from reality! Figure 8.1 indicates the forces and/or moments which are 'available' or active at each type of support. It is expected that these representative forces and moments, if properly calculated, will bring about equilibrium in each structural element or member.

8.4.1 Roller Supports

Roller supports allow for both rotation and translation, enabling unrestricted movement along the supporting surface. Translation, in this context, means movement or displacement without rotation. These supports can be positioned on horizontal, vertical, or inclined surfaces at any angle. The resulting

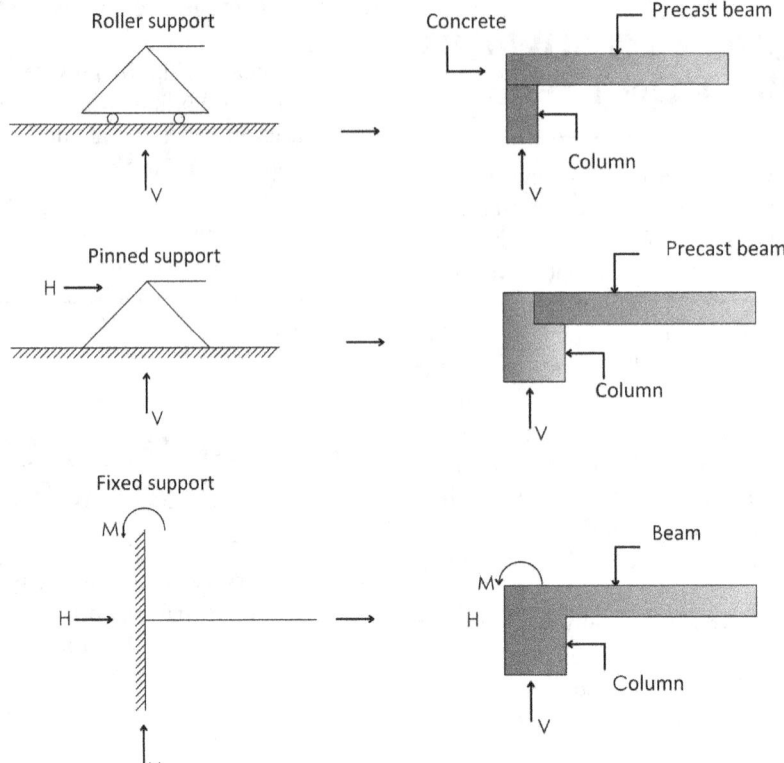

FIGURE 8.1 Support systems illustrated.

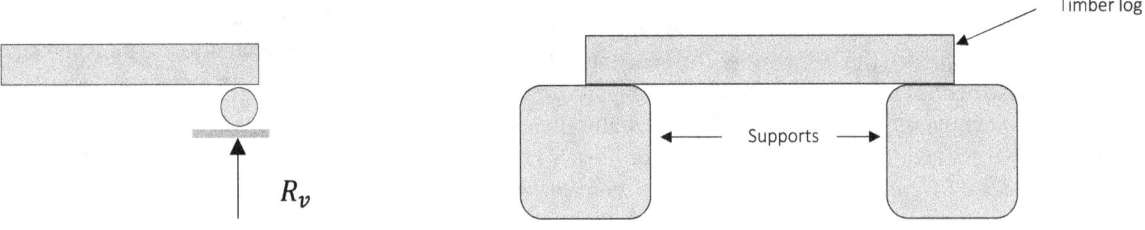

FIGURE 8.2 Roller.

FIGURE 8.3 Simply supported beam.

reaction force is a single force that acts perpendicular to the surface and away from the surface.

In real-world application, roller supports are commonly located at one end of long bridges. This allows the bridge structure to expand and contract with temperature changes. Roller supports are provided for thermal expansion. The expansion forces could fracture the supports at the banks if the bridge structure was 'locked' in place. Roller supports can also take the form of rubber bearings, rockers, or a set of gears designed to allow few lateral movements.

A roller support (Figure 8.2) cannot resist lateral forces, making it ineffective against sideways loads. To illustrate, imagine a person on roller skates or sliding gates—they remain stable under vertical loads, but any lateral force such as a push, wind gust, or earthquake can cause them to roll away. This highlights why roller supports, alone, are insufficient for most structures, which often experience various lateral loads. Consequently, buildings require complementary support systems to ensure stability. See the roller in the gym section used for mimicking the natural world in Figure 6.17.

8.4.1.1 Simple supports

Simple support and roller support share similarities, in that they primarily resist vertical forces. Idealized simple supports are often considered frictionless surface supports, resulting in a single perpendicular reaction force. However, they differ significantly in their ability to resist lateral loads. Unlike roller supports, simple supports cannot resist lateral loads of any magnitude. In practice, gravity and friction provide some minimal frictional resistance to moderate lateral loading, but this is not reliable.

Consider a timber log bridging a gap (see Figure 8.3). It remains stationary until disturbed, such as by a kick or movement. At that point, the plank shifts since the simple connection cannot resist lateral forces. This illustrates the limitations of simple supports. A simple support can be found as a type of support for long bridges or roof span. Simple supports are often found in zones of frequent seismic activity but also commonly used as crossovers for open drainage channels.

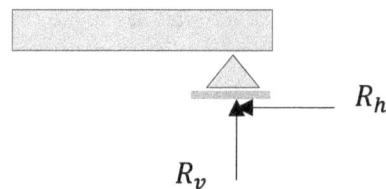

FIGURE 8.4 Pinned support.

8.4.2 Pinned Supports

A pinned support (Figure 8.4) resists both vertical and horizontal forces, but not moments (torques). It allows rotation in one direction while preventing translation in all directions. In reality, connections may resist small moments, but are idealized as pinned connections for simplicity. Some characteristics of pinned supports include the following: they resist vertical and horizontal forces, allow rotation in one direction, prevent translation in all directions, may resist rotation in other directions (e.g., hinged doors), and require additional support for structural stability—one support is not enough. The examples of pinned supports include hinged doors, knees (allowing rotation in one direction), and idealized connections in structural design.

The knee (Figure 6.15) can be idealized as a connection that allows rotation in only one direction and provides resistance to lateral movement. See also gym and civil engineer (Section 6.3). A pinned support can resist both vertical and horizontal forces, but not a moment. It allows the structural member to rotate, but not to translate in any direction. Many connections are assumed to be pinned connections, even though they might resist less moment in reality.

8.4.2.1 Pinned connections

Pinned connections, typical in trusses, offer design flexibility and versatility, allowing for articulated or hidden installations and expressive or subtle visual integration. A popular alternative to roller supports, pinned connections provide effective structural solutions.

We encounter pinned connections in our daily lives, often without realizing it. A simple example is a hinged door. When you push it open, the hinge functions as a pinned connection, enabling rotation around a fixed axis while restraining movement in two perpendicular directions (vertical and horizontal). The door hinge prevents vertical and horizontal translation. As a matter of fact, if a sufficient moment is not generated to create rotation, the door will not move at all.

Ever noticed how some doors open smoothly, while others require a harder push? The answer lies in the moment required to overcome the door's resistance, a calculation that reveals the subtle differences in door design. Ever noticed that the force required to unscrew or tighten tire nuts depends on the moment (torque) applied, which is calculated as follows:

$$\text{Torque } (T) = \text{Force } (F) \times \text{Distance } (d) \qquad (8.1)$$

where

- T = Torque (measured in Nm or lb-ft)
- F = Force (measured in N or lb)
- d = Distance (measured in m or ft).

8.4.3 Fixed Supports

Fixed supports can resist vertical and horizontal forces, as well as a moment. Since they restrain both rotation and translation, they are also known as rigid supports. This means that a structure only needs one fixed support in order to be stable. All three equations of equilibrium can be satisfied.

A column footing in a building is a good example of this kind of support. The column is embedded in the foundation, preventing movement and translation in any direction. Anchor bolts securing a steel beam to a concrete foundation are a perfect example of a rigid support. Foundation piles for bridges or high-rise buildings are also a perfect example of a rigid support. The representation of fixed supports always includes two forces (horizontal and vertical) and a moment.

8.4.3.1 Fixed connections

Fixed connections are very common. Its examples include welded connections, monolithic connections, and rigid frames.
Key notes

- Steel structures of many sizes are composed of elements welded together.
- A cast-in-place concrete structure is automatically monolithic, and it includes a series of rigid connections with the proper placement of the reinforcing steel. **Note:** A monolithic connection is a type of structural connection where two or more elements are joined together to form a single, integral unit, without any visible gaps or interfaces. The connection is rigid and continuous, with no relative movement between the connected elements. Typically, monolithic connections are connections between concrete elements added at once.
- Fixed connections demand greater attention during construction and are often the source of building failures.
- A connection refers to the joint or link between two or more structural elements, such as beams, columns, or slabs.
- A support refers to a structural element or system that provides reaction forces to a structure, holding it in place and resisting external loads.

8.4.4 Key Takeaways

In reality, actual supports deviate from idealized models and sometimes remain abstract in academia. However, these models—fixed, pinned, and roller supports—serve as helpful approximations for design and analysis. For instance, a foundation designed to prevent rotation at a point is typically modeled as a fixed support or built-in support. Therefore, the model developed will not allow rotation, and the structure's simulated behavior will reflect the absolute fixity. However, this does not mean that the actual foundation will have absolute rotational fixity. This must be comprehended.

Alternatively, if engineering judgment suggests the foundation will offer minimal resistance to rotation, we may idealize it as a pin or roller support, assuming zero rotational restraint. However, this simplification may not accurately reflect the foundation's actual behavior once constructed.

The essential understanding is that although three support models are well defined, placing too much trust in the output of analytical models does not warrant accuracy or desired reliability and stability. We must recognize that actual foundations rarely adhere perfectly to this idealized behavior. Therefore, the ability to map real-world behavior onto analytical or mathematical models is what sets a competent engineer from a junior. The transition from theory to practical application is where engineering truly manifests.

It is important that engineers understand the forces and moments that structures impose on their supports or foundations in order to design foundations that will not experience unacceptable settlements. Therefore, the type of restraint or support a particular foundation or support type will provide to the structure must be understood.

8.4.4.1 Key differences

See Table 8.1.

8.4.5 Reaction Forces and Statistical Determinacy and Indeterminacy

After the supports for a structure are idealized and modeled, then the magnitude and direction of the various reactions developed at the supports are determined. This enables us to define the amount of reaction forces available in each support. If the structure is subject to externally applied forces, the reactions must be developed at supports to ensure the structure remains in a state of static equilibrium. Therefore, support reactions are developed directly in response to the loads applied to the structure (and any self-weight of the structure).

To identify support reactions, we assume static equilibrium and apply the three equations of equilibrium:

$$\sum F_x = 0 \tag{8.2}$$

$$\sum F_y = 0 \tag{8.3}$$

$$\sum M = 0 \tag{8.4}$$

These equations enable us to solve up to three unknown support reactions. In these equations, we consider orthogonal components of forces experienced by an object separately [1].

This means that a force at an arbitrary angle can be resolved into two orthogonal components (which means they are at right angles to each other)—to a selected frame of reference. We could equally condense this and propose that the sum of all forces of an object must equal to zero, that is,

$$\sum F = 0 \tag{8.5}$$

So, if any of these conditions is not satisfied, the object will not be in a state of static equilibrium and will undergo a change in its velocity (i.e., it will experience an acceleration). Therefore, these requirements are both necessary and sufficient conditions for equilibrium [2]. To analyze static equilibrium, we use three fundamental equations to determine up to three unknown forces or moments in a system. These equations enable us to solve for forces in the x-direction (F_x), forces in the y-direction (F_y), and moments (M). With these three equations, we can resolve up to three unknowns, ensuring the system is in balance.

An object is said to be in a state of static equilibrium if the sum of all forces and moments acting on the object are zero and the object is at rest. This is a direct result of Sir Isaac Newton's first law of motion, which states that a body remains at rest or moving with a constant velocity unless an unbalanced force acts upon it.

Reaction forces are as a result of Newton's third law of motion, which states that to every action, there is an equal and opposite reaction. From this, we note that equilibrium can also apply to moving objects provided they are moving at a constant velocity. Such objects are said to be in a state of dynamic equilibrium. For now, we will just concern ourselves with the objects in a state of static equilibrium.

We consider forces and moments in a 2D plane and later elaborate in this chapter and in Chapter 9. An object is in a state of static equilibrium if it satisfies the three conditions/equations of equilibrium. That is equations (8.2–8.4).

It is important, though, to note that these equations apply to statically determinate structures. A statically determinate structure is a stable structure that can be analyzed using only the three equations of equilibrium to determine all of its unknown reactive forces. In summary, the equations of equilibrium state that the sum of forces and moments acting on a structure must be 0.

It is important to note again that statically determinate structures are only adopted in scenarios where simple structures are deemed necessary. For example, determinate structures are suitable for simple, small-scale structures, such as beams and frames—beam bridges utilize simple determinate structures. See Figure 8.3. Also, where temporary structures are desired, determinate structures are often used, like

TABLE 8.1 Key differences between fixed, roller, and pinned supports

SUPPORT TYPE	ROTATION	TRANSLATION	RESISTS	DEGREES OF FREEDOM
Fixed	No	No	All loads	0
Pinned	Yes	No	Vertical and horizontal	1
• Hinged	Yes	Limited	Vertical and horizontal Moment	2
Roller	Yes	Yes	Vertical	2
• Simple support	Yes	Yes	Vertical	2

scaffolding work and temporary bridges. Foot bridges may adopt statically determinate structures as well. Additionally, cases of low-load conditions such as small residential buildings may also require determinate structures.

A statically indeterminate structure is a stable structure that cannot be analyzed using only equations of equilibrium. This is because the number of unknown forces in the structure is greater than the number of equations available. To analyze a statically indeterminate structure, additional compatibility conditions are required. The degree of indeterminacy is the difference between the number of unknown reactions and the number of equations of equilibrium. Some examples of statically indeterminate structures include fixed beams, continuous beams, and two-hinged arches.

Again, it is important to note that statically indeterminate structures are the most common structures developed by civil engineers because they offer numerous advantages. These include the following:

- **Redundancy:** this is the primary benefit of indeterminate structures. Indeterminate structures have redundant members, which provide additional support and ensure the structure remains stable even if one member fails.
- Member forces in the determinate structure are not affected by the stiffness of the members, which means that all member forces and moments can be calculated using statics, in contrast to indeterminate structures.
- **Improved stability:** indeterminate structures exhibit greater stability under various loading conditions, including earthquakes and wind loads.
- **Distribution of forces:** indeterminate structures distribute forces more efficiently, reducing stress concentrations in structural members and minimizing the risk of structural failure. Some examples of statically indeterminate structures include cantilever bridges, domes, and multistory buildings.
- **Reduced deflections:** indeterminate structures tend to have reduced deflections, which provides a more rigid and comfortable structure.
- **Better resistance to dynamic loads:** indeterminate structures are more effective at resisting dynamic loads, such as vibrations and impact.
- **Aesthetics and architecture:** indeterminate structures offer more design flexibility, enabling architects to create complex and visually appealing structures.
- **Economic benefits:** while initial construction costs may be higher, indeterminate structures can be more economical in the long run due to reduced maintenance and repair needs.

Because of the above advantages, indeterminate structures are preferred over determinate structures. Thus, the limitations of statically determinate structures include limited redundancy, and because of this, they are more vulnerable to structural failure if a member is damaged. Higher stress concentrations in structural members for determinate structures increase the risk of structural failure. In terms of aesthetics and architecture, determinate structures offer less design flexibility, not

enabling architects to create complex and visually appealing structures—this simplified design approach has limited options, thus limiting architectural creativity and flexibility.

In summary, statically indeterminate structures offer numerous benefits, including redundancy, improved stability, and reduced deflections, making them a popular choice for complex and large-scale structures. However, statically determinate structures remain suitable for simpler, low-load applications.

The fact that as indeterminate structures cannot be solved by equilibrium equations, several theorists have developed techniques to address the structural analysis of these structures. These techniques include the force method (flexibility method), displacement method (stiffness method), and energy techniques/methods (variational principles). The choice of method depends on structure complexity, loading conditions, and desired outcome (e.g., internal forces or deflections).

In Section 3.3.3.3, we saw that energy equations/techniques have advanced civil engineering technology throughout history. These techniques/equations have had far-reaching success in all branches of civil engineering. In structural engineering, particularly, energy equations, also known as energy methods or variational principles, can be used to solve statically determinate and indeterminate structures.

It is important to note that energy methods offer advantages for solving indeterminate structures such as reduced computations as they often require fewer calculations. They also improve accuracy and offer flexibility because they can handle complex loading conditions. It is, therefore, essential to recognize the role of energy equations across multiple fields.

8.5 PRIMARY CIVIL STRUCTURAL ANALYSIS/GRAPHICAL TOOLS—FUNDAMENTALS

These are the essential/fundamental core graphical tools that civil and structural engineers use in the analysis, design, and communication of engineering designs and concepts. They are primarily graphical tools, which include free-body diagrams (FBDs), shear force diagrams (SFDs), bending moment diagrams (BMDs), and influence lines. FBDs help calculate reaction forces using equilibrium equations. Equilibrium equations are essential in analyzing statically determinate structures.

8.5.1 Free-Body Diagram

An FBD, also referred to as a force diagram, is a graphical representation of an object or system, showing the external forces acting on it and sometimes internal forces and moments (but this is optional). Every equilibrium problem begins by drawing and labeling an FBD! Therefore, an FBD shows the forces imposed on an object.

One of the first steps in analyzing a structure is to sketch out its FBD, identifying all of the forces that must be considered in the analysis. FBDs are used to evaluate the forces acting on a structure in equilibrium.

An FBD is used to visualize the applied forces, moments, and resulting reactions on a free body in a given condition. It depicts a body or connected bodies with all the applied forces and moments and reactions, which act on the body. An FBD is an essential tool for applying Newton's law of motion. It helps ensure that all forces that are applied are accounted for, including their proper directions.

The body may consist of multiple internal members (such as a truss) or be a compact body (such as a beam). A series of FBDs and other diagrams may be necessary to solve complex problems. Sometimes, in order to calculate the resultant force graphically, the applied forces are arranged as the edges of a polygon of forces or force polygon.

The types of FBDs include a 2D FBD (planar), 3D FBD (spatial), and a dynamic FBD, which includes inertial forces. An FBD simplifies complex problems as it clarifies force interactions and helps visualize force interactions much better. It also enhances problem-solving because it helps identify critical forces and moments, thus enabling engineers to analyze equilibrium conditions. An FBD is essential for engineering design and analysis because it helps determine unknown forces or moments. FBDs are helpful in analyzing structures. Therefore, sketching out an FBD is the first step in any civil structural analysis.

Notes

- A 2D planar structure is a type of structural system that lies in a single plane ($x - y$ plane), has no significant thickness in the third dimension ($z -$ axis), and

resists loads applied in the plane (in-plane loads). See Figure 8.5. So, a 2D FBD is a graphical representation of the external forces and moments acting on a 2D object or structure.
- A dynamic FBD includes inertial forces. See Figure 8.6.
- A 3D FBD is a graphical representation of the external forces and moments acting on a 3D object or structure. See Figure 8.11.

The key components of an FBD include object or system boundary and external forces (arrows). The external forces indicate the applied forces, that is, loads or weights, and reaction forces, indicate supports or friction.

In very limited circumstances, an FBD may indicate internal forces and moments. For example, normal forces, shear forces, bending moments, and torques can be shown by an FBD, but not always. So, the common symbols portrayed by an FBD are for external forces, moment, couple, support reaction, and weight.

In drawing an FBD diagram, you define the object or system and identify external forces acting on the object/system. Draw the object's boundary, and then represent forces as arrows, and label forces and moments.

An FBD has wide applications, especially in learning the mechanics of materials, statics and dynamics of bodies, structural analysis, machine design, robotics design and manufacturing, and aerospace engineering, among others. Some

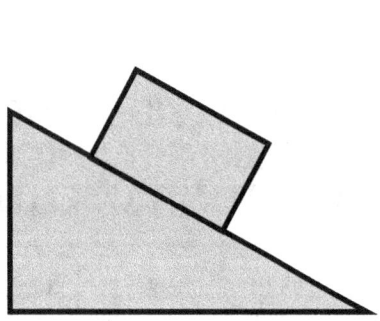
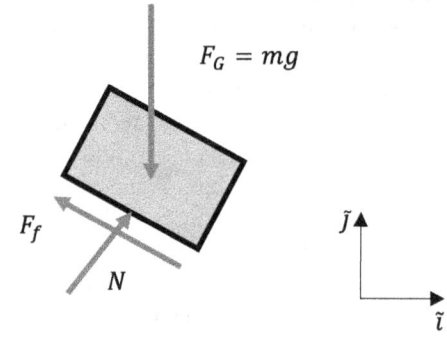

FIGURE 8.5 Block on a ramp and corresponding FBD of the block.

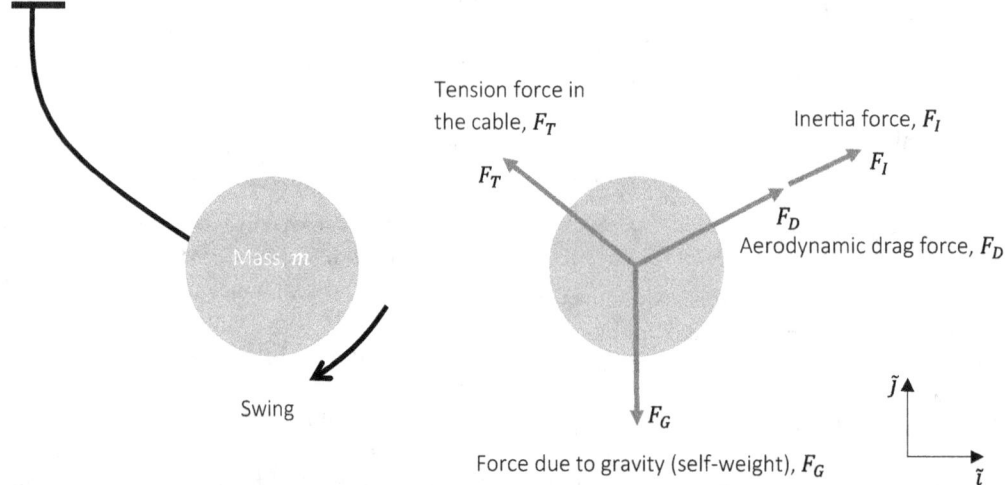

FIGURE 8.6 A dynamic FBD of a swinging pendulum.

of the software tools used to draw FBDs include Autodesk, SolidWorks, MATLAB, and Ansys. It is important to note that FBDs are used in various branches of civil engineering beyond structural engineering. Other typical applications in other branches include the following:

- Geotechnical engineering, that is, soil mechanics, foundation design, and slope stability analysis. For example, geotechnical engineers use FBDs to analyze soil–structure interaction, foundation settlement, and stability of slopes.
- Transportation engineering, that is, vehicle dynamics, pavement design, and traffic flow analysis. For example, transportation or highway engineers use FBDs in pavement design and analysis—they analyze vehicle loads, pavement stresses, and deflections with use of FBDs.
- Water resource engineering, that is, fluid mechanics, hydraulic structures, and dam design, among others. For example, water resource engineers use FBDs to study and analyze hydraulic structures—they analyze flow, pressure, and forces on dams, gates, and valves.
- Environmental engineering, that is, fluid flow, pipe networks, and water treatment plant design. For example, environmental engineers use FBDs to analyze pipe networks—fluid flow, pressure drops, and pipe stresses.
- Construction engineering management, that is, crane load calculations, rigging, and lifting operations. For example, construction managers use FBDs to calculate crane operations, that is, crane loads, stability, and rigging forces.

8.5.2 Steps/Procedure for Drawing FBDs

1. Identify the object/particle. Decide on the particle or object whose equilibrium is to be analyzed [3].
2. Isolate the object. Imagine this particle is 'cut' completely free (separated) from the body, structure, and/or its environment. That is
   a. In two dimensions, think of a closed line that completely encircles the particle.
   b. In three dimensions, think of a closed surface that completely surrounds the particle.
3. Sketch the particle (i.e., draw a point).
4. Add the forces:
   a. Sketch the external forces that are applied to the particle by the environment (e.g., weight).
   b. Wherever the cut passes through a structural member, sketch the forces that occur at that location. Forces at cut points (where the object was separated from other structures).
   c. Wherever the cut passes through a support, sketch the reaction forces that occur at that location.
5. Define the coordinate system to be used. Add pertinent dimensions and angles to the FBD to fully define the locations and orientations of all forces.

8.5.3 2D FBDs

These are representations of two-dimensional reactions of a planar structure.

8.5.3.1 Block on a ramp

A simple example of an FBD of a block of mass m resting on a ramp is shown in Figure 8.5. The influence of gravity is represented by the gravitational force $F_G = mg$, while the reaction force imposed by the supporting surface is represented by N. The friction force resisting the block from sliding is represented by F_f. From this, we can conclude that there are three forces acting on the block:

- The gravitational force, $F_G = mg$
- The friction force, F_f
- The reaction force, N

This is a trivial example, but FBDs are so helpful when analyzing complex systems. Note that in this example, the weight of the isolated body (block) is non-negligible. However, in the analysis of most complex systems, the mass of the isolated body is always neglected.

8.5.3.2 Swinging pendulum

Consider a slightly more complex example of a pendulum swinging through the air of mass, m (Figure 8.6)—a perfect example of a dynamic FBD. In this case, we identify the following:

- The force due to gravity, F_G
- The tension force in the suspension cable, F_T
- The drag force due to the presence of air, F_D
- The inertia force present due to the fact that the mass has an acceleration, F_I

8.5.3.3 Simply supported beam

The beam in Figure 8.7 is subjected to two point loads, 40 and 35 kN. The beam has a pin support at its left end and a roller support at its right end. This support configuration (pin and roller) is very common, and a structure with this support configuration is often said to be simply supported. See FBD in Figure 8.8.

FBD

FBD of the simply supported beam showing all forces. We have three unknown reactions. We can consider these as potential reactions since we don't know yet whether any of them evaluate to 0. Since there are three reactions to identify, we can use the three equilibrium equations identified earlier. The FBD is shown in Figure 8.8.

8.5.3.4 FBDs in other civil engineering branches

2D planar FBDs are also commonly applied in other branches of civil engineering, especially soil mechanics and geotechnical engineering. The study of stress and stain analysis in pavement and soil–structure interaction models uses FBDs. The steps are the same as outlined in Section 8.5.2. Isolating

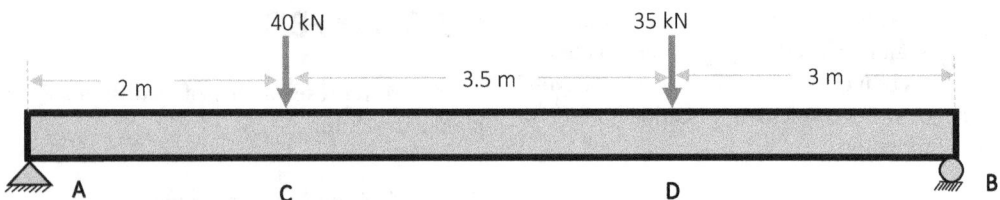

FIGURE 8.7   Simply supported beam.

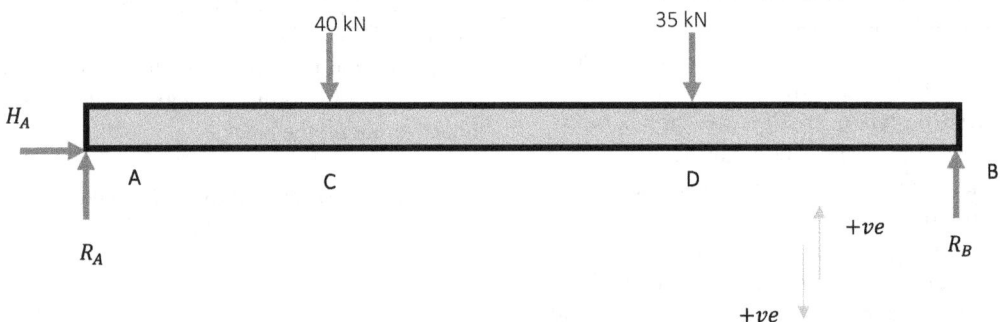

FIGURE 8.8   FBD for the simply supported beam.

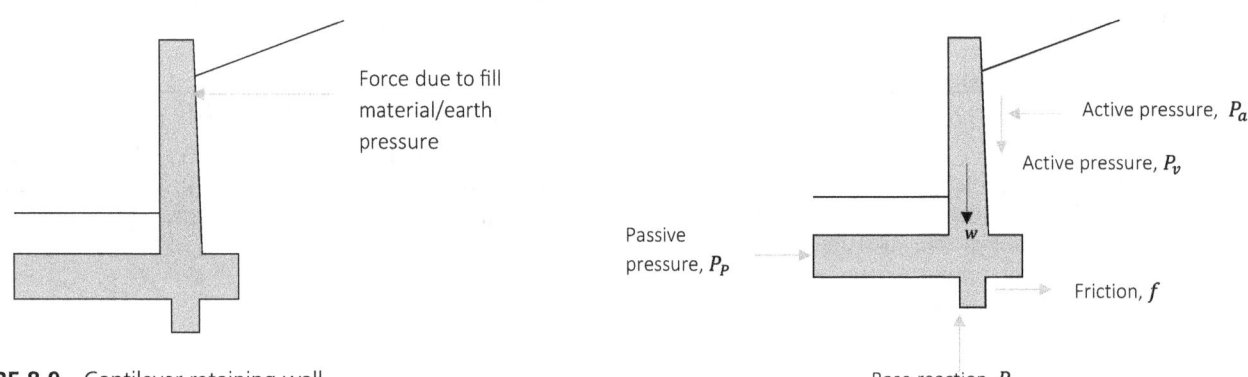

FIGURE 8.9   Cantilever retaining wall.

FIGURE 8.10   FBD for a cantilever retaining wall.

a body—this concept of drawing an FBD applies in all fields of civil engineering. For example, when designing a retaining wall (Figure 8.9), an FBD diagram can be drawn. You can isolate the body (wall) and identify forces that act on the wall in 2D (planar)—see Figure 8.9.

Cantilever retaining wall in 2D (planar) FBD

where w is the weight of the wall; sometimes, the weight of the wall is neglected in the analysis. When designing a retaining wall, engineers typically consider the loading exerted by the backfill soil on the wall. This loading is usually measured in kilonewtons per meter (kN/m), representing the force per unit length of the wall (in the z-axis). The reason is that designers assume a unit width of the retaining wall (usually 1 m) to simplify calculations, and they assume homogeneity of the retained soil mass (Figure 8.10).

An FBD is drawn for this unit width, isolating the forces acting on the wall. The backfill soil is assumed to be homogeneous and uniform across the retaining wall, meaning its properties (density, friction angle, etc.) are consistent. The design loads calculated for the unit width are then extrapolated to the entire wall length, assuming uniform conditions.

The key benefits of this approach include simplified calculations as it allows for standardized design procedures and enables easy scalability for different wall lengths. Common loads/forces considered in retaining wall design are as follows: earth pressure (active or at rest), surcharge loads (traffic, buildings, etc.), water pressure or pore water pressure (if applicable), and self-weight of the wall. Design standards and guidelines typically adopted in retaining wall design include BS 8002, Eurocode 7 (EC7), American Society of Civil Engineers (ASCE) 7-16, ACI 318, IS 456:2000 (India), and Australian Standard AS 4678-2009.

8.5.3.5 3D FBD

These are representations of three-dimensional reactions of a 3D structure. This is applicable in civil engineering 3D frames, bridges, towers, etc. It considers three-dimensional reactions. The main added complexity with three-dimensional objects is that there are more possible ways the object can move and also more possible ways to restrain it. As before, FBDs should show the reactions supplied by the constraints, not the constraints themselves.

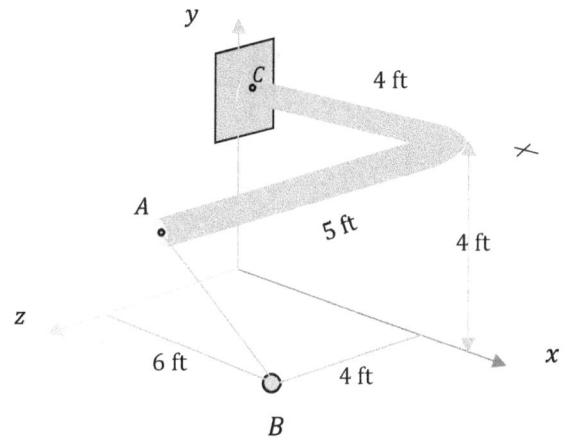

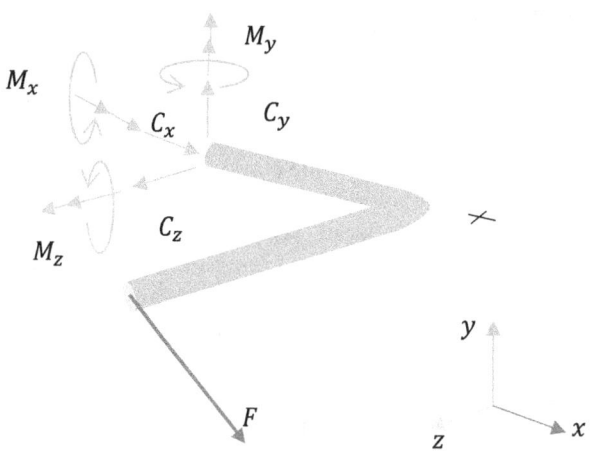

FIGURE 8.11 Bent bar.

8.5.3.5.1 3D Bar

A good example where we can apply 3D FBD analysis is on a door handle. Force applied on different doors differs from one door to another.

Example [8.1] for illustrative purposes

The bent bar shown in Figure 8.11 is held in a horizontal plane by a fixed connection at C, while cable AB exerts a 500 lb force on point A. Given $A = (4, 4, 5)$, $B = (6, 0, 4)$, and $C = (0, 4, 0)$, find the reaction force C and concentrated moment M with components M_x, M_y, and M_z required to hold the bar in this position under this condition.

Solution

1. Draw an FBD
    Begin by drawing an FBD.
2. Determine the force acting at point A in Cartesian form. The force of the cable acts from A to B. This direction is described by the displacement vector from A to B:

$$r_{AB} = (2i - 4j - 1k) \text{ ft} \tag{8.6}$$

or the corresponding unit vector

$$\lambda_{AB} = \frac{r_{AB}}{|r_{AB}|} \tag{8.7}$$

$$\lambda_{AB} = \frac{2i - 4j - 1k}{\sqrt{(2)^2 + (-4)^2 + (-1)^2}} = \frac{2i - 4j - 1k}{\sqrt{21}} \tag{8.8}$$

Multiplying the unit vector by the cable tension gives the force acting on A, as a three-dimensional Cartesian force vector:

$$F = \lambda_{AB} T = \left(\frac{2i - 4j - 1k}{\sqrt{21}}\right) \times 500 \text{ lb} \tag{8.9}$$

$$F = (2i - 4j - 1k)\left(\frac{500}{\sqrt{21}}\right) \text{ lb} \tag{8.10}$$

$$F = (218i - 4j - 109k) \text{ lb} \tag{8.11}$$

3. **Determine the moment about C**
    The moment about point C is found with the cross-product (4.3.4), where the moment arm is the displacement vector from C to A:

$$r_{CA} = (4i + 0j + 5k) \text{ ft} \tag{8.12}$$

$$M_C = r_{CA} \times F = \begin{matrix} i & j & k \\ 4 & 0 & 5 \\ 2 & -4 & -1 \end{matrix} \left(\frac{500}{\sqrt{21}}\right) \tag{8.13}$$

$$M_C = (2182i + 1528j - 1746k) \text{ ft.lb} \tag{8.14}$$

4. **Apply the equations of equilibrium**

$$\sum F = 0 \begin{cases} \sum F_x = 0 : C_x + F_x = 0 \\ \quad C_x = -218 \text{ lb} \\ \sum T_y = 0 : C_y - F_y = 0 \\ \quad C_y = +436 \text{ lb} \\ \sum T_z = 0 : C_z - F_z = 0 \\ \quad C_z = +109 \text{ lb} \end{cases} \tag{8.15}$$

$$\sum M = 0 \begin{cases} \sum M_x = 0 : M_x + M_{C_x} = 0 \\ \quad M_x = -2180 \text{ ft.lb} \\ \sum M_y = 0 : M_y + M_{C_y} = 0 \\ \quad M_y = -1530 \text{ ft.lb} \\ \sum MM_z = 0 : M_z + M_{C_z} = 0 \\ \quad M_z = +1750 \text{ ft.lb} \end{cases} \tag{8.16}$$

The resulting vector equations for the reaction force C and reaction moment M are as follows:

$$C = (-218i + 436j + 109k) \text{ lb}$$

$$M = (-2180i - 1530j + 1750k) \text{ ft.lb}$$

8.5.4 Shear Force Diagram

A SFD plays a crucial role in structural analysis, particularly in beam design. Its primary purpose is to visualize and quantify the internal shear forces within a beam under various loads. A shear force is defined as the algebraic sum of all the vertical forces, either to the left- or to the right-hand side of the section. The shear force at a section is the sum of all external vertical forces acting on the portion of the structure to one side of the section, including reactions, loads, and applied forces.

8.5.4.1 Function of SFD

The SFD is used to determine the location and magnitude of the maximum shear force which is crucial for designing beam cross-sections. This process helps engineers economize costs by minimizing material usage while ensuring safety, thus optimizing the beam design, overall.

The SFD is used to locate points where shear forces are greatest, and this helps designers to reinforce these sections. The process of shear reinforcement basically follows a rigorous assessment of the shear forces acting on the structure. Shear reinforcement design typically involves calculating shear forces acting on the structure using SFDs, assessing the structure's shear capacity, comparing applied shear forces to the structure's shear capacity, and designing reinforcement (e.g., stirrups and rebars) to resist excess shear forces.

Key standards and codes include ACI 318 (American Concrete Institute), Eurocode 2 (EN 1992-1-1), and BS 8110-1:1997—Section 3—Structural use of concrete—Code of practice for design and construction; Ultimate limit state— Shear and torsion; and BS 8110-2:1985—Section 5—Design charts for shear. Do you know where stirrups are typically or commonly concentrated in beams and columns of structures?

The SFD helps analyze the load distribution by visualizing how loads are transferred from the beam to structural supports. This also enables verification of the structural integrity of the structure by confirming whether the beam can withstand applied loads without failing due to shear forces.

8.5.4.2 Steps to draw SFD

Step 1: Draw the beam's FBD. Let's consider Figure 8.7.
Step 2: Calculate reactions at supports. From Figure 8.7, and using the equilibrium equations:

$$\sum F = 0$$

$$\sum F_x = 0$$

$$\sum F_y = 0$$

$$R_A + R_B = 75 \tag{8.17}$$

$$\sum M = 0$$

$$(40 \times 2) + (35 \times 5.5) = 8.5 R_B \tag{8.18}$$

$$R_B = 32.06 \text{ kN} \tag{8.19}$$

Step 3: Determine shear force at critical sections. The rules considered include

- Following the algebraic sum. We consider magnitude and direction (positive or negative). Another rule is that vertical forces are

considered only—SFDs focus on vertical forces only. Mathematically, the shear force can be represented as follows:

$$SF = \sum F_y = 0$$

- Do one side of the section: Choose either the left or right side of the section; the result will be the same.

Step 4: Plot shear force values along the beam length. See Figure (SFD for Figure 8.7).
Step 5: Connect points to form the SFD.
   Note: An SFD will always close. If not, then you must be going wrong somewhere.

Explanation

The **internal shear force** of 40.94 kN acts through the beam for a 2-meter stretch, that is, between support A and the 40 kN point load. Between the 40 kN point load and the 35 kN point load, with a span of 3.5 m, an internal shear force of 2.95 kN is resisted by the beam. These are positive forces because they push upward.

Beyond point D, the shear force of magnitude 32.06 kN acts downward (negative). This information (figures) influences the beam design, the reinforcement type and size, and spacing of bars and stirrups, among other things, with the detailing necessary. Now, imagine an SFD is presented to you without a FBD and/or beam with support sketches, and you can quickly visualize where maximum or major shear forces are concentrated that need attention (and their direction).

- This sheds light on why beams and columns of constant stiffness are widely adopted alongside uniformly distributed loads other than point loads. It is all about the proper distribution of forces and moments throughout the structure. Your goal must be to create a constancy in the structure, ensuring that forces are balanced, stresses are distributed evenly, deformations are minimized, and equilibrium is maintained.
- Note that the moment increases with distance from the loaded end, so the magnitude of the maximum value of M compared with V increases as the beam length increases. This is true of most beams, so shear effects are usually more important in beams with small length-to-height ratios.

8.5.5 Bending Moment Diagram

A BMD is a graph that shows how bending moments change along a beam's length.

8.5.5.1 Function of BMD

A BMD is used to analyze and design beams and to help determine the following:

- **Maximum loads**: where the maximum loads occur along the beam.

- **Design optimization**: how to optimize the design to prevent failures.
- **Weight and cost reduction**: how to reduce the overall weight and cost of the structure.
- **Beam deflection**: how to determine the deflection of a beam.
- **Beam member type, size, and material**: how to determine the type, size, and material of a beam member to support a given set of loads without failure.

BM diagrams are drawn on the compression side of members. Bending moments and shear forces depend on two factors: how the beam is loaded, and how it's supported.

8.5.5.2 Steps to draw BMD

The BMD follows the SFD. Let's consider Figure 8.7.

1. Draw the SFD. See Figure 8.12.
2. Start at one end of the beam and move along the length.
3. Plot the shear force at each point, using the following rules:
   - +ve shear force: plot above the beam.
   - −ve shear force: plot below the beam.
4. Connect the points to form the SFD.
   Start at one end of the beam, and move the location of the imaginary cut to the other end, applying the equilibrium equations and calculating the shear forces and bending moments as you move along the beam. Doing this along the full length of the beam will yield the shear force and BMD.
5. **Draw the BMD**
   Steps:
   - Calculate the bending moment at different points along the beam by considering the sum of moments of forces acting on one side of the section.
   - Plot these values to create the BMD. A point load will cause a linearly varying bending moment, a uniformly distributed load (UDL) will cause a parabolic change, and a uniformly varying load (UVL) will cause a cubic curve.

- The area under the SFD represents the change in the bending moment. Areas above the x-axis increase the bending moment, and areas below decrease it.
- Use the calculated values of bending moment at different points to draw the BMD.
- The shape of the BMD depends on the type of loading (point load, UDL, and UVL) and the beam's support conditions.

Explanation

Looking at the BMD, you notice that the maximum/large bending moments occurred at points where we have the concentrated loads (point loads). However, depending on the loading configuration, maximum bending moments can occur at supports (e.g., fixed ends or simply supported beams) and changes in loading patterns. The minimum/least bending moments often occur at points of zero shear force (where the SFD crosses the x-axis) or at inflection points (where the bending moment changes sign). The exact locations depend on the specific beam configuration, loading, and boundary conditions (Figure 8.13).

8.6 ESSENTIAL DRAWINGS FOR CIVIL ENGINEERS

It is important to know that drawings are a means of communication for the civil engineer, documentation, and problem-solving. They help civil engineers to visualize and convey complex ideas and designs; plan and detail infrastructure projects; communicate with stakeholders, interest groups, contractors, and other fellow engineers (mechanical, electrical, and building services engineers), among others; identify and solve design problems; and be able to record and document project progress and changes

Civil engineers must be able to create, read, and interpret drawings to communicate with architects, contractors, and other stakeholders. We ensure that accurate construction and implementation are done, identify potential design flaws or errors, and finally be able to meet project requirements and regulations.

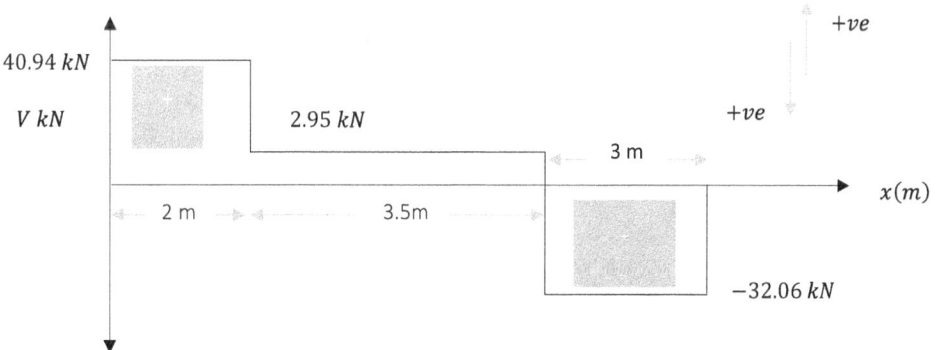

FIGURE 8.12   SFD for Figure 8.7.

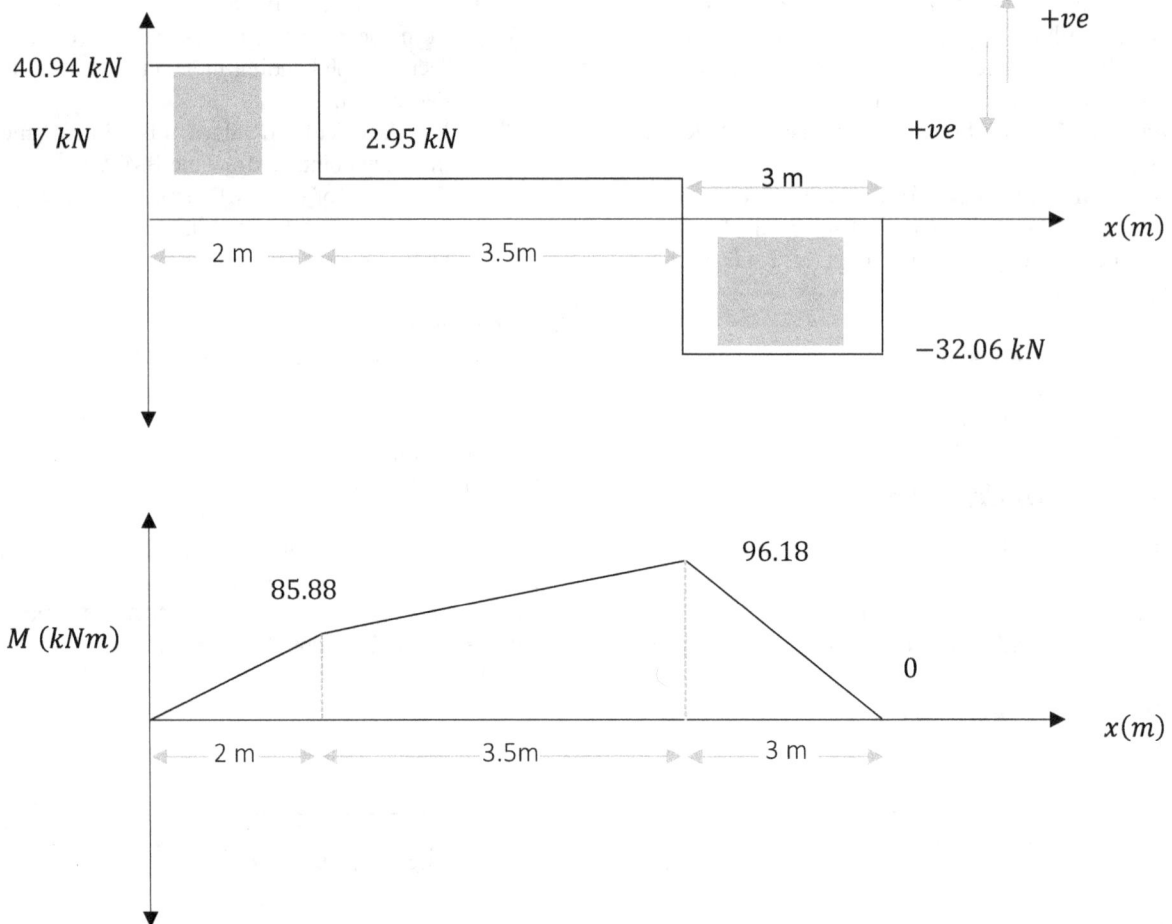

FIGURE 8.13 Bending moment diagram for Figure 8.6.

Proficiency in CAD software, such as AutoCAD or Revit, is essential for creating and managing these drawings. A civil engineer must be proficient in creating and/or interpreting various types of drawings, including plans, elevations, sections, detail, isometric, orthographic, flowchart, survey, site plan, structural, architectural, mechanical, electrical, and shop drawings. Civil engineers need to be proficient in creating and interpreting various types of drawings to effectively communicate their designs and plans. Some of the most common types of drawings are outlined henceforth.

8.6.1 Design Drawings

Design drawings are visual representations used to communicate and develop design ideas, especially in the early stages of a project. They serve as a way for designers and clients to visualize and refine concepts, often focusing on aesthetics and architectural elements. These drawings can also be used to communicate design intent to builders and contractors. They include concept designs and sketches. See Figure 8.14 for a building sketch.

8.6.2 Architectural Impressions

Architectural impressions (Figure 8.15) are visual representations of buildings, either existing or proposed, created to showcase design concepts and details before construction or to document existing structures accurately. These can range from simple sketches to detailed 3D models and photo-realistic images.

8.6.3 Site Layout (Plan) Drawings

A site layout (plan) (Figure 8.16) is typically a large-scale drawing that shows the full extent of the site for an existing or proposed development. They generally depict the layout of a site, including buildings, roads, and utilities. Depending on the size and complexity of the project, site plans are likely to be at a scale of 1:500 or 1:200. Site plans, along with location plans, may be necessary for planning applications. In most cases, site plans will be drawn up following a series of desk-based studies and site investigations. However, for very small projects, larger scales may be used, and for large projects, smaller scales, or even several drawings may be required, perhaps pulled together on one very small-scale plan.

FIGURE 8.14 Building sketch.

FIGURE 8.15 Architectural impression.

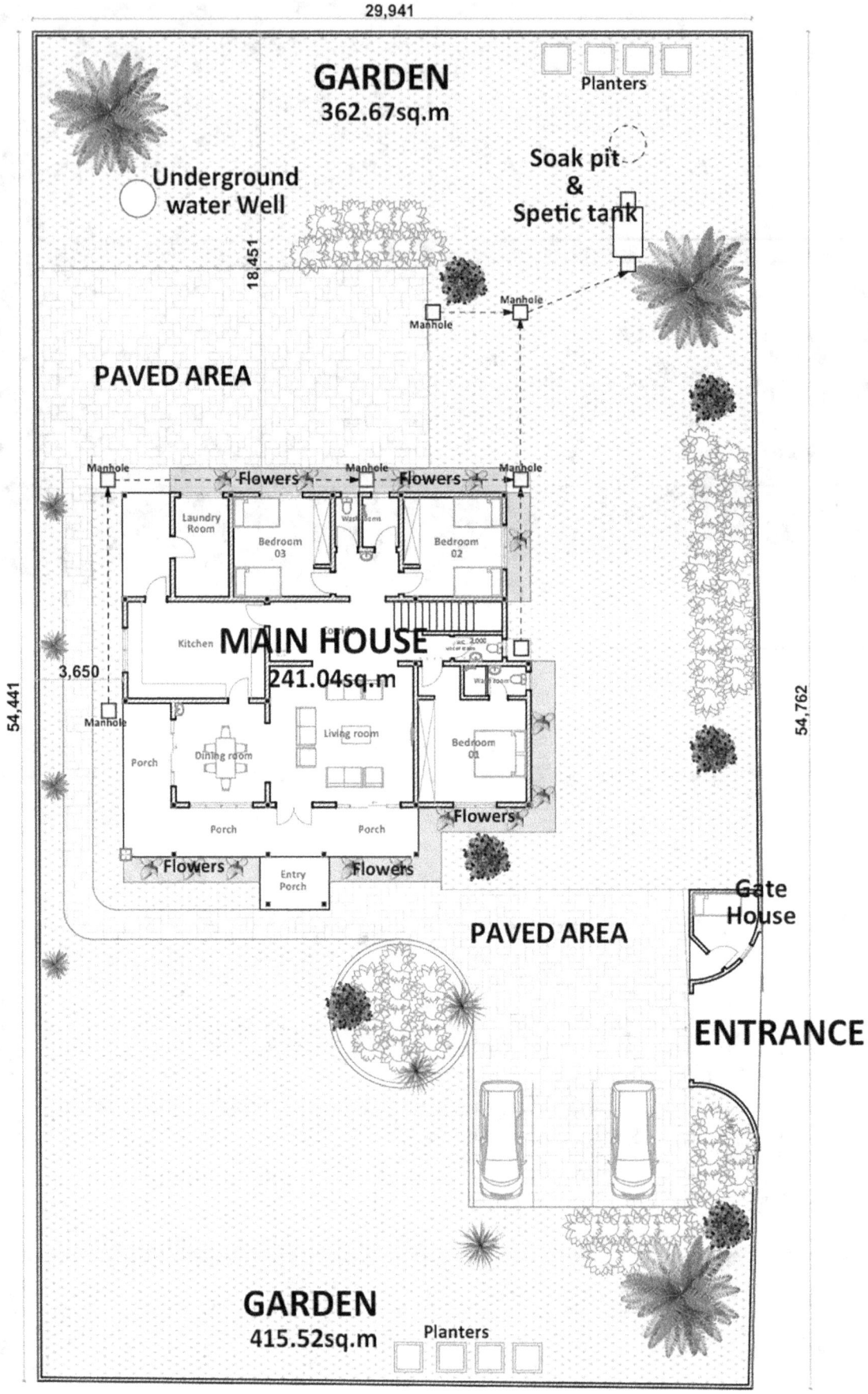

FIGURE 8.16 Site layout.

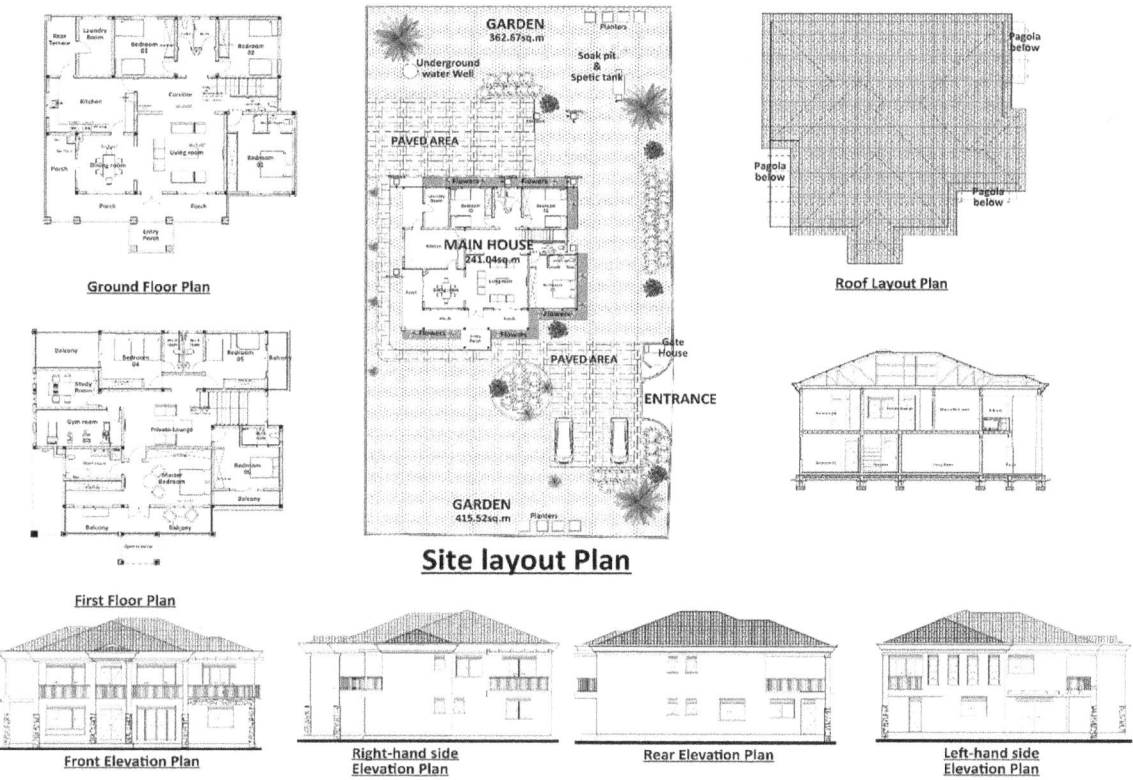

FIGURE 8.17 Architectural drawings.

8.6.4 Architectural Drawings

Architectural drawings (Figure 8.17) are technical illustrations that provide detailed information about a building or structure, encompassing its design, layout, and construction details. They are used by architects, engineers, and contractors to communicate design intent, guide construction, and ensure compliance with building codes. These drawings serve as a visual representation of the building plan, including elements like floor plans, elevations, sections, and site plans. Architectural drawings emphasize the aesthetic and functional aspects of a building. Architectural drawings are typically created using architectural CAD software.

8.6.5 Structural Drawings

Structural drawings are detailed engineering documents that outline the design and specifications of a building's structural components. Figure 8.18 shows a typical staircase structural drawing. They focus on the load-bearing elements such as foundations, beams, columns, slabs, and load-bearing walls, detailing their size, location, and materials. These drawings act as blueprints for contractors and builders, ensuring a structure is built safely and accurately.

8.6.6 Section Drawings

A 'section drawing,' 'section,' or 'sectional drawing' (Figure 8.19) shows a view of a structure as though it had been sliced in half or cut along another imaginary plane. It shows a vertical cut transecting, typically along a primary axis, an object or building. It is a cut-through view of a structure to show internal details. For buildings, this can be useful as it gives a view through the spaces and surrounding structures (typically across a vertical plane) that can reveal the relationships between the different parts of the buildings that might not be apparent on plan drawings.

8.6.7 Plan Drawings

This shows the layout of a structure or site from a top-down view. Plan drawings can be floor plans that illustrate the layout of each floor, including rooms, doors, windows, and dimensions. See Figure 8.20.

8.6.8 Elevation Drawings

An elevation drawing is a first angle projection that shows all parts of the building as seen from a particular direction, with the perspective flattened. They depict the exterior of a structure from the side or front. See Figure 8.21.

8.6.9 Isometric Drawings

Isometric drawings are a way to represent a 3D object on a 2D surface, where all three dimensions are drawn at a full scale, with angles between the axes typically being 120°.

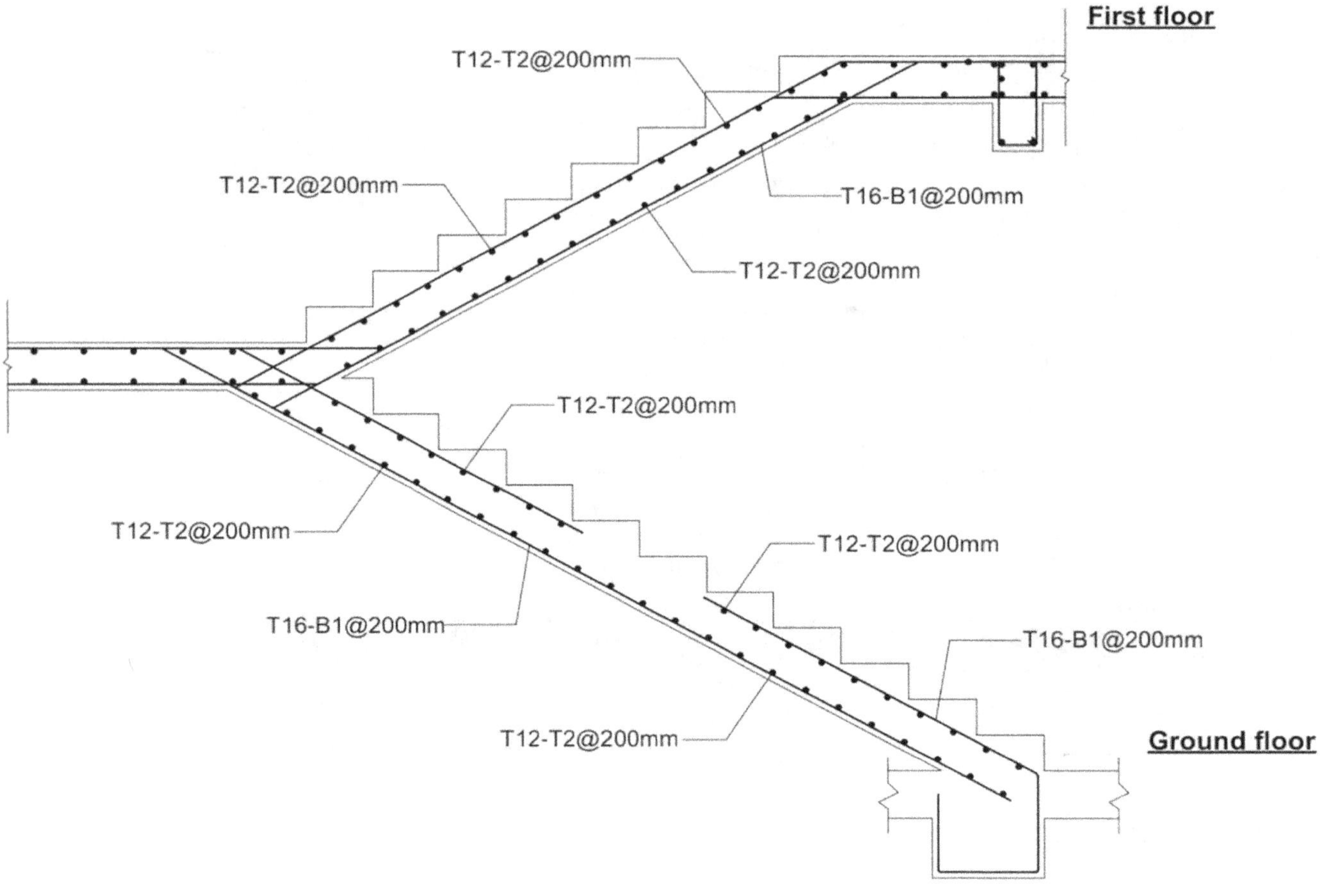

First floor

T12-T2@200mm

T12-T2@200mm

T16-B1@200mm

T12-T2@200mm

T12-T2@200mm

T12-T2@200mm

T12-T2@200mm

T16-B1@200mm

T16-B1@200mm

T12-T2@200mm

Ground floor

FIGURE 8.18   Structural drawing for staircase.

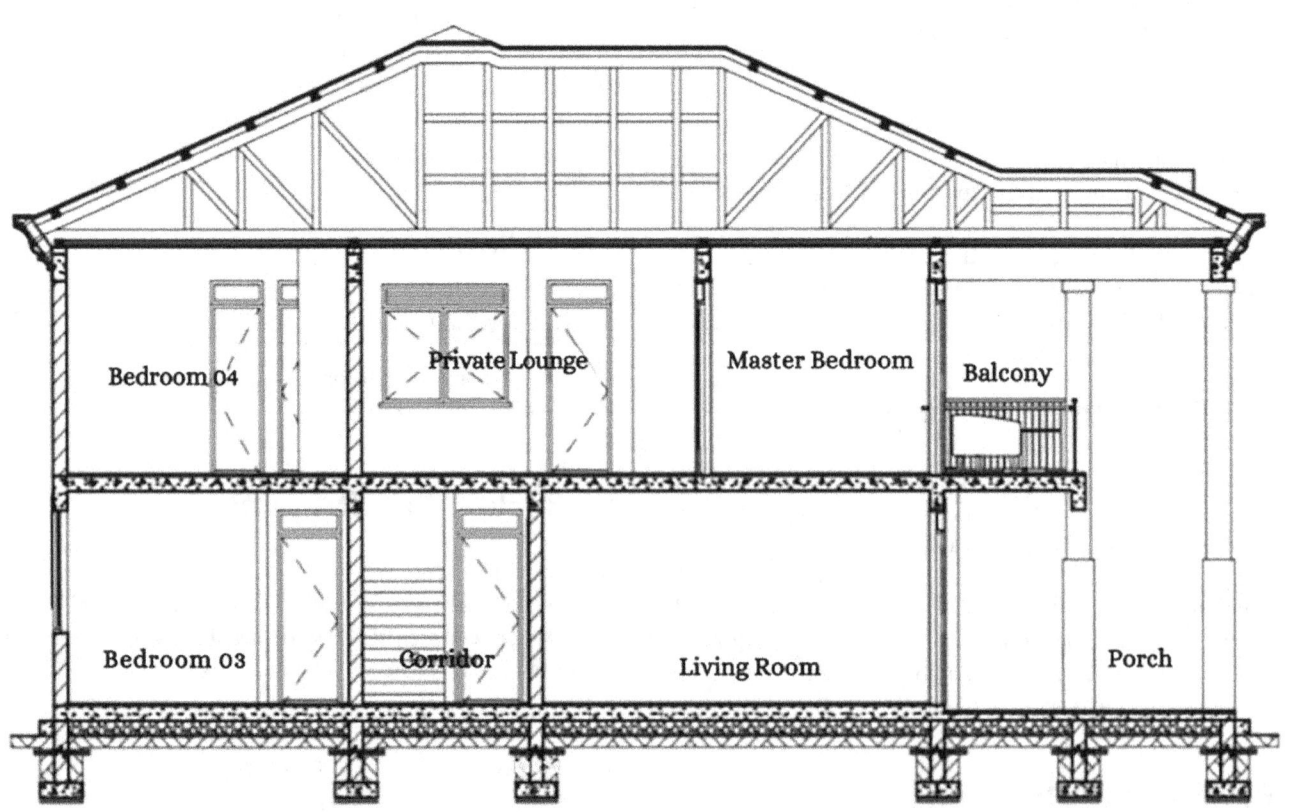

Bedroom 04

Private Lounge

Master Bedroom

Balcony

Bedroom 03

Corridor

Living Room

Porch

FIGURE 8.19   Two-story house section drawing for project in Figure 8.17.

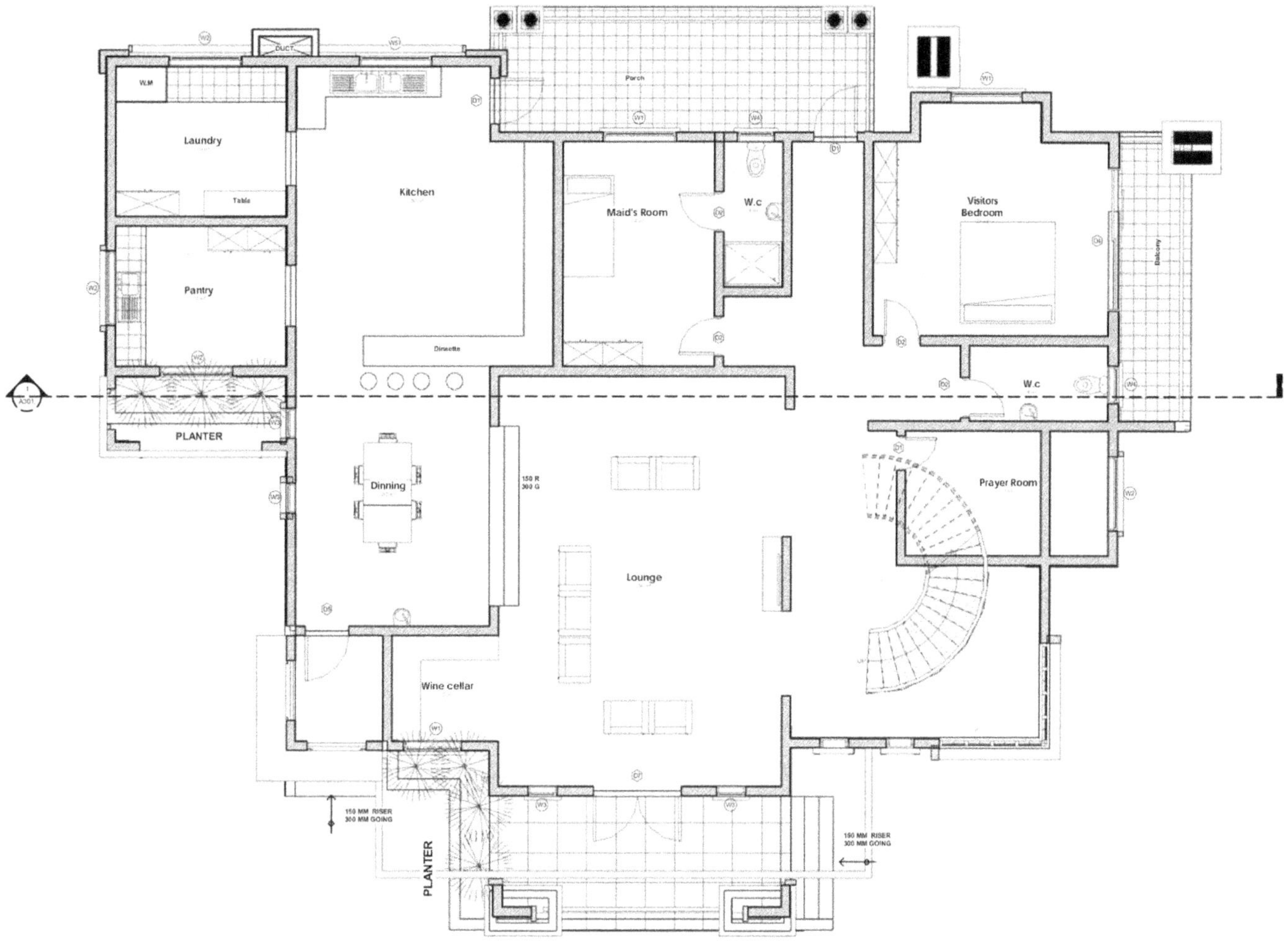

FIGURE 8.20   Plan drawing for Figure 8.15.

FIGURE 8.21   Elevation drawing for plan drawing (Figure 8.20).

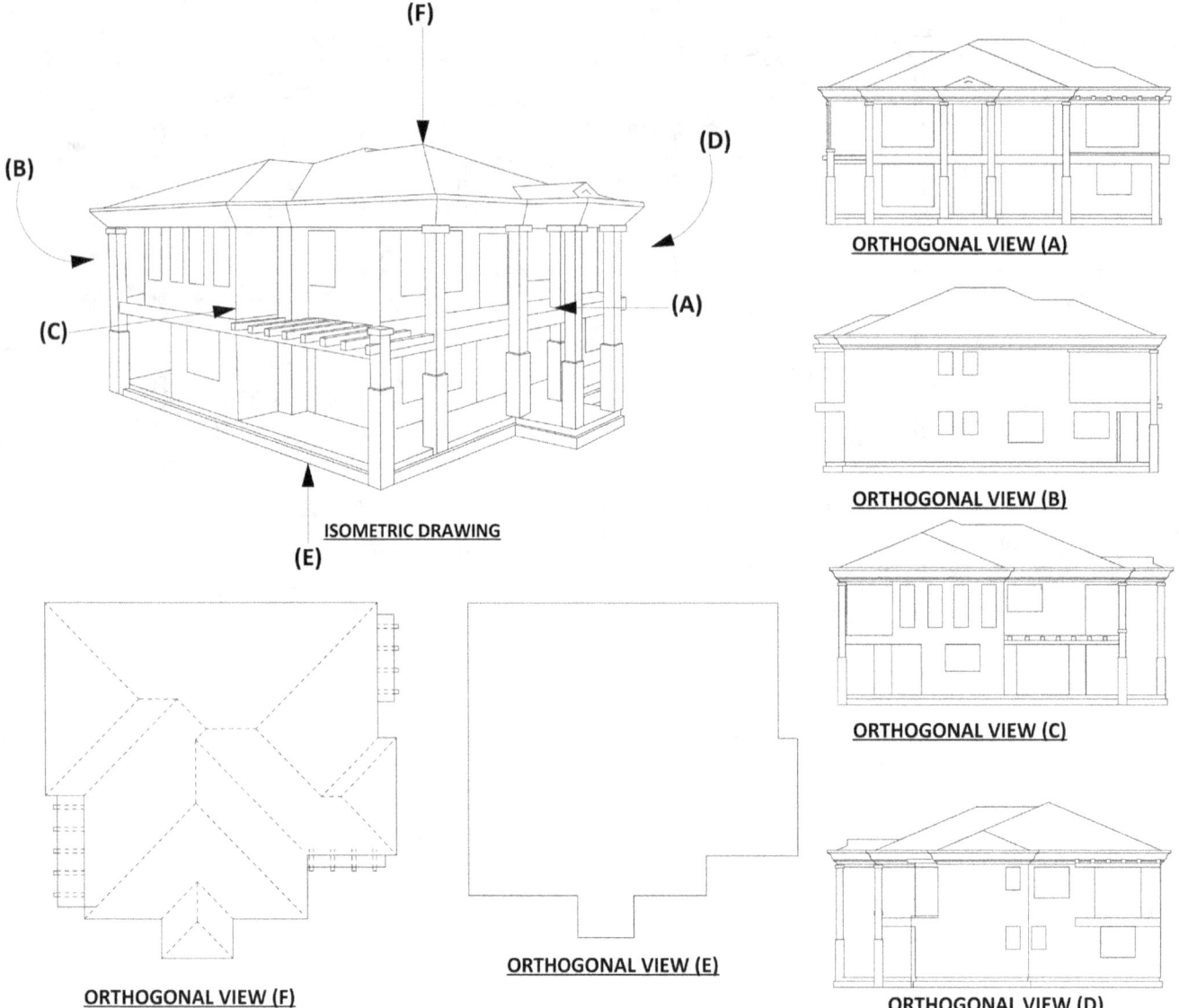

FIGURE 8.22   Orthographic drawings for project in Figure 8.17.

This technique gives the illusion of depth and space, making it useful for visualizing objects in a clear and understandable way. See Figure 8.22.

8.6.10 Orthographic Drawings

These are 2D representations of structures or systems from multiple angles. Orthographic drawings, also known as orthographic projections, are a way to represent three-dimensional objects in two dimensions using multiple two-dimensional views. These views, typically front, top, and side views, are drawn as if the object is projected onto a series of planes, each representing a different perspective. This method is commonly used in engineering and design to accurately depict objects from various angles. See Figure 8.22.

8.6.11 Detail Drawings

These are the enlarged views of specific components, such as connections, joints, and finishes. They are made up of smaller cross-section drawings and close-up views of parts of the building.

8.6.12 Mechanical Drawings

Mechanical drawings typically illustrate **heating, ventilation, and air conditioning (HVAC)** systems. A mechanical drawing is a technical drawing that displays information about various mechanical systems, such as HVAC. Mechanical drawings are often combined with electrical and plumbing drawings to create Mechanical, Electrical, and Plumbing (MEP) drawings.

8.6.13 Electrical Drawings

Electrical drawings, also called electrical plans or wiring diagrams, are technical drawings that provide visual representations of circuits or electrical systems. They are like a layout on paper before physically installing the required machines.

8.6.14 Plumbing Drawings

A **plumbing drawing** is a type of technical drawing that provides a visual representation and information of a plumbing system. It is used to convey the engineering design to plumbers or other workers who will use them to help install the plumbing system. It is used to show clearly the location of fixtures, sanitaryware, pipework, valves, and so on and illustrates how fresh water is to be supplied into a building and waste water removed.

8.6.15 Shop Drawings

Shop drawings are detailed, scaled drawings created by contractors, fabricators, or suppliers to illustrate the fabrication and installation of specific components or systems for a construction project. They are typically based on the design drawings and specifications provided by the architect or engineer. Shop drawings provide detailed, scaled instructions for fabricating and constructing components, ensuring accuracy, quality, and compliance with project specifications. They facilitate coordination and communication among stakeholders, reducing errors and discrepancies, and ultimately enhancing the efficiency and quality of the construction or manufacturing process. Shop drawings provide manufacturers and contractors with precise information for fabricating and constructing components, such as steel structures, precast concrete, or complex systems.

8.6.16 Construction Drawings

Construction drawings, also known as working drawings, are a set of detailed, scaled drawings that provide a comprehensive visual representation of the project's design and construction requirements. They are used to guide the construction process and ensure that the project is built according to the design intent. Construction drawings are typically created by architects, engineers, or designers using CAD software and are submitted to contractors, builders, and regulatory authorities as part of the construction documentation package. These are typically sets of drawings that may include site plans, floor plans, elevations, sections, detail drawings, schedules, and notes and legends. Construction drawings serve several purposes which include communicating design intent, guide construction, and ensure accuracy, and meet regulatory requirements.

8.6.17 As-Built Drawings

As-built drawings: Compare the current condition of a building to its original plans. As-built drawings, also known as red-line drawings or record drawings, are revised sets of drawings created after a construction project is complete. They document the actual construction as it was built, reflecting any changes or deviations from the original design plans. These drawings are crucial for various reasons, including verifying contractor performance, managing future renovations, and for legal purposes.

8.6.18 Excavation Drawings

These drawings provide details about trenches, pits, shafts, tunnels, and other soil removal techniques.

8.6.19 Site Drawings

These show information about grading, landscaping, building arrangement, topography, and other details.

8.6.20 Flow Charts and Diagrams

These illustrate processes, systems, or workflows. In civil engineering, flow charts and diagrams are used to visually represent and document processes, systems, or procedures. They illustrate the sequence of steps, decision points, and relationships within a project or process, aiding in communication, analysis, and planning.

8.6.21 Survey Drawings

These show property boundaries, topography, and existing conditions. Survey drawings are detailed, scaled maps prepared by surveyors that depict the features and boundaries of a property, including things like property lines, buildings, and other relevant details. They are used to record and delineate the physical characteristics of land and are often required for real estate transactions, construction projects, and other legal or development purposes.

8.6.22 Geotechnical Drawings

These show soil conditions, foundation designs, retaining walls, and other subsurface details.

8.6.23 Hydraulic and Hydrologic Drawings

Hydraulic and hydrologic drawings depict water flow, drainage, and flood defense or control systems. They are usually presented as a network of pipes and channels, together with cross-sections of the hydraulic structures.

8.6.24 Bar Bending Schedule

A bar bending schedule (BBS) is a detailed document used in reinforced concrete construction to specify the bending and

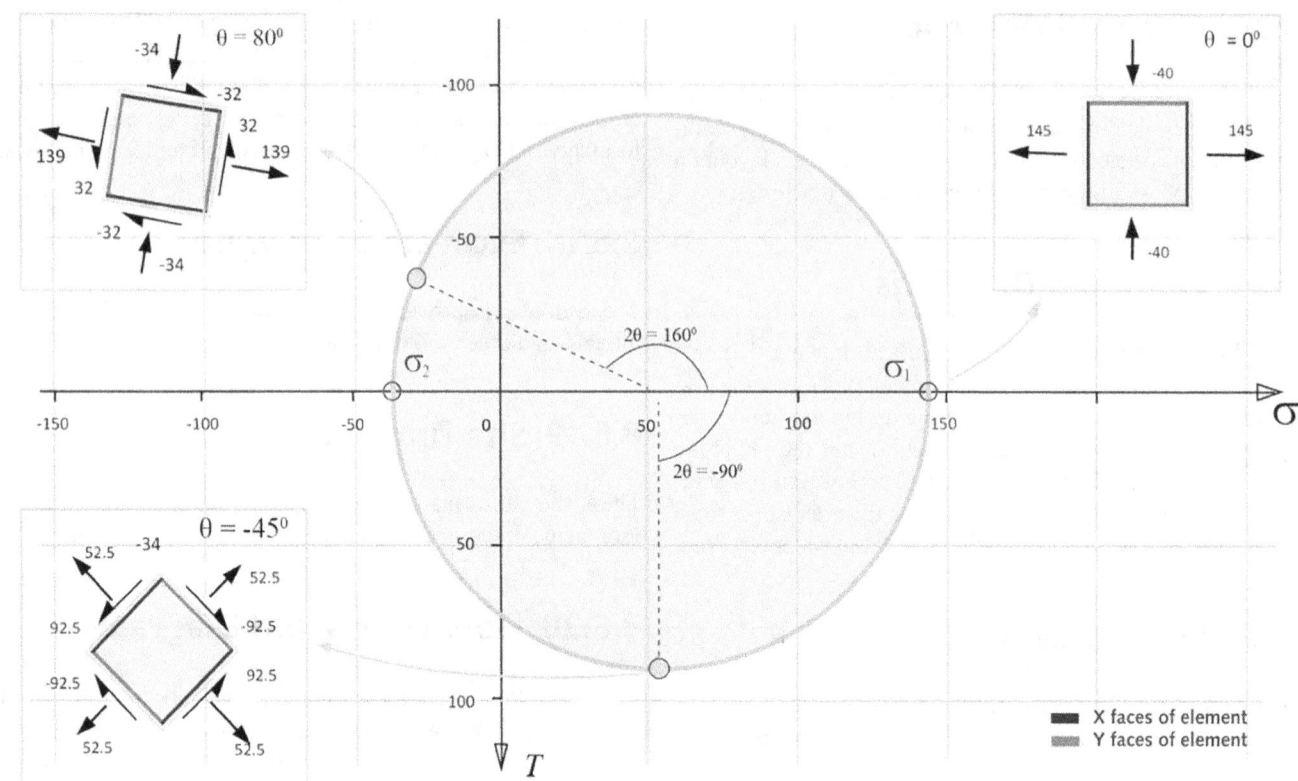

FIGURE 8.23 Mohr's circle.

shaping of steel reinforcement bars (rebars) for a particular project. The BBS provides a systematic and organized way to communicate the required bending and shaping of rebars to the fabrication team, ensuring that the steel reinforcement is accurately prepared and installed on-site. By using a BBS, construction teams can ensure that the steel reinforcement is accurately fabricated and installed, which is critical for the structural integrity and safety of the building. **Note:** A BBS is not a drawing itself, but rather a tabular document that provides detailed information about the reinforcement bars (rebars) required for a construction project. A typical BBS includes the information of bar mark, bar size, length, bending shape, bending dimensions, cutting requirements, and other special requirements.

8.7 SPECIAL DRAWINGS AND CHARTS

8.7.1 Structural/Geotechnical Engineering: Mohr's Circle

Named after the German engineer and mathematician, August Ferdinand Möhr, Mohr's circle is a graphical representation of the relationship between stress and strain on a material, commonly used in mechanics, geotechnical engineering, mechanical engineering, structural analysis, and materials science. It is a circle drawn on a graph that has normal stress on the horizontal axis and shear stress on the vertical axis. It's a useful tool for visualizing and analyzing the behavior of materials under different types of loading conditions. Mohr's circle is a powerful graphical method used to visualize and analyze the stress state at a single point within a body. The circle is constructed by plotting the normal stress (σ) against the shear stress (τ) for a material, and it provides a convenient way to determine the principal stresses, principal strains, and the maximum shear stress.

It helps determine the normal and shear stress components for different orientations of the stress element graphically, instead of using the stress transformation equations. An example for a plane stress case (i.e., two-dimensional stress) is shown in Figure 8.23. Each point on the circle defines the normal and shear stress components for a certain orientation of the stress element. Figure 8.23 shows the stress elements for three different points on Mohr's circle, corresponding to three different orientations of the stress element.

Mohr's circle can be used to

- Determine the maximum shear and normal stresses at a single point.
- Determine the principal stresses and the orientation of the principal plane at a single point (more about principal stresses later).
- Develop a more intuitive and complete understanding of the stress state at a single point.

8.7.2 Water Resources/ Hydraulics: Nomograms and Moody Diagram

Nomograms are graphical tools, also known as nomographs or alignment charts, used to perform mathematical calculations or estimate outcomes based on relationships between variables.

These graphical tools usually contain three parallel scales graduated for different variables so that when a straight line connects values of any two, the related value may be read directly from the third vertical line at the point intersected by the line.

The Moody diagram (Figure 7.3) is a graphical representation used in fluid dynamics to determine the friction factor (f) for fluid flow in pipes and to estimate head loss due to friction in pipes. It's a graphical representation of the Darcy–Weisbach equation, used to determine the friction factor (f) in turbulent flow through pipes.

8.7.3 Hydrology: IDF Curves

Intensity–duration frequency (IDF) curves are used to assess storm events, classify climates, and design stormwater and urban drainage systems for a particular geographic area. As rainfall events increase in frequency (in other words, how often they occur) and intensity (how severe the storms are when they do occur), IDF curves have to be updated periodically. IDF curves play a critical role in ensuring that stormwater management systems are designed to handle rainfall events of varying intensities and durations, reducing the risk of flooding and infrastructure damage.

8.8 PEDAGOGICAL TOOLS IN CIVIL ENGINEERING

Pedagogical tools are resources used to teach and facilitate learning. They are designed to help students understand a problem or lesson and are essential for acquiring new knowledge and skills. Pedagogical tools can also be used to accommodate students with special needs and to be inclusive of students from diverse backgrounds. For example, instructors or tutors or lecturers can provide texts at appropriate learning levels and activities for different learning styles. These pedagogical tools help engineering students develop practical skills, work on projects, and engage in collaborative learning. The pedagogical tools commonly used in engineering education include software, artificial intelligence (AI), machine learning (ML), and generative AI.

8.8.1 Software

This refers to a set of instructions that enable a computer or electronic device to perform specific tasks or functions. Software is a vital pedagogical tool utilized to enhance civil engineering education, supporting lectures, tutorials, and training programs. These specialized software programs are specifically designed and tailored to cater to the unique needs of civil engineers, facilitating interactive learning, simulation, and analysis. For example,

- **Simulation software:** These include tools such as MATLAB, Simulink, and Ansys that allow students to model and simulate real-world engineering problems.

- **CAD/CAE software:** These are programs such as Autodesk Inventor, AutoCAD, SolidWorks, and Catia that enable students to design and analyze engineering systems.
- **Virtual labs and remote experimentation platforms:** Tools such as LabVIEW, National Instruments' ELVIS, and remote access labs allow students to conduct experiments and gather data remotely.
- **Programming languages and development environments:** Tools such as Python, C++, and Java.

8.8.2 Artificial Intelligence

AI refers to the broad field of research and development aimed at creating machines that can perform tasks that typically require human intelligence. It encompasses various approaches, including rule-based systems, expert systems, and ML. AI is the broadest term, encompassing various approaches. AI focuses on reasoning and problem-solving, knowledge representation, planning and decision-making, natural language processing (NLP), and computer vision. AI can significantly aid civil engineering design, serving as a pedagogical tool for an instructor, in various ways:

- AI algorithms can effectively optimize design parameters, such as structural shapes, materials, and systems, to minimize costs, weight, and environmental impact.
- AI-powered simulations can analyze complex systems, such as structural behavior, fluid dynamics, and thermal performance, to predict and optimize design outcomes.
- AI can generate multiple design options based on performance criteria, allowing engineers to explore and evaluate various design solutions—generative design.
- AI can automate routine design tasks, such as drafting, detailing, and documentation, freeing engineers to focus on higher-level creative work.

8.8.2.1 Machine learning

ML is a subset of AI that enables machines to learn from data without being explicitly programmed. ML focuses on learning from data, for example, a recommendation system that suggests products based on user behavior (uses supervised learning). Neural networks are a fundamental component of ML, enabling machines to learn from data. ML focuses on developing algorithms that can learn patterns and relationships in data, make predictions or decisions based on data, improve performance over time, and adapt to new data or environments. ML is divided into supervised learning (e.g., classification and regression), unsupervised learning (e.g., clustering and dimensionality reduction), and reinforcement learning (e.g., game playing and robotics).

8.8.2.2 Generative AI

Generative AI can help you create content and be more creative, productive, and knowledgeable. Generative AI is a specific type of AI focused on generating new content. Generative

AI uses ML techniques (see Section 3.2.6). It is a powerful tool for civil engineers because it can help provide innovative insights on the desired 2D or 3D models for civil engineering structures. An example of generative AI is a system that generates new product designs or images which uses generative adversarial networks (GANs) or variational autoencoders (VAEs).

8.8.2.3 Danger of overly relying on AI tools

As a pedagogical tool, AI comes with drawbacks. It must be supplemented by human teaching rather than replacing it. AI tools generate new insights and content, but still limited by the data they are trained on and may not always understand the context of human knowledge; they basically quickly share the available knowledge as they can understand based on training. AI learns from you (the user). Sometimes, you have no choice but to be able to beat AI; you must judge correctly. For example, AI has no emotions, lacks human intuition, and may struggle to replicate human creativity, critical thinking, and common sense.

In technical reports and articles, it is good to always share statements/facts/calculations based on reliable sources (reliable research articles, standards, codes of practice, books, government guidelines/documents, etc.), but not shared by AI tools.

All engineers, educators, practicing engineers, researchers, student, etc. should be alert, especially the younger generation, who prepare research articles and technical reports. Some companies are reducing staff, citing AI's ability to manage most tasks. However, in engineering research and practice, we must not overly depend on AI. It is crucial to apply our fundamental knowledge and rigorously verify details using trusted sources.

This also applies to the use of commercial software in the analysis and design of structures, especially when the outcomes are not unique, as they often depend on user input and the selection of parameters.

As a note of caution, while the AI engine is constantly learning, we cannot guarantee that these AI-generated answers will always be without error. The user must use their own judgment on the suitability of the answer and review the outcomes/answer is based on.

You need proper judgment to use any AI or software. AI tools are constantly learning and improving. On the other hand, effective knowledge building requires human–AI collaboration. Humans provide context, domain expertise, and critical thinking, while AI contributes computational power, scalability, and pattern recognition.

Knowledge building refers to the process of creating, refining, and organizing knowledge through social and cognitive processes. It involves identifying gaps in existing knowledge or literature, generating new ideas and connections, elaborating and refining concepts, integrating new information into existing knowledge structures, and sharing and negotiating knowledge with others. Knowledge building is essential in various domains, including education, research and development (R&D), innovation, problem-solving, and decision-making.

If AI defeats you in thinking not in computational efficiency or power, you may not be good as you think! AI cannot defeat the human sense of inquiry! What we remain with as humans, particularly, engineers and scientists, is the ability to conduct scientific inquiry—to formulate hypotheses—because we have the natural talents developed through intuition and instincts. Thus, it is important to acknowledge that one of humanity's primary strengths in controlling AI lies in the capacity for scientific inquiry, critical thinking, and ethical decision-making.

While AI is a powerful tool, the human sense of inquiry remains unbeatable due to its unique blend of creativity, emotional intelligence, and curiosity.

The engineering profession is likely to shift away from charging hourly rates for work to charging based on the value added as generative AI technology advances. Although AI can enhance engineering solutions and reduce time, human oversight is still necessary for professional liability and adjustments. In the future, it may no longer be feasible to charge clients per hour for engineering services due to advancements in technology.

8.8.3 Drawings

Drawings can be used as pedagogical tools. See Section 8.6 for the different types of drawings a civil engineer should know.

- Sketches and conceptual drawings.
- Other drawings.

8.8.4 FBDs

FBDs can be effective pedagogical tools in civil engineering education because of the visual representation. See Section 8.5.1. FBDs can help students visualize complex forces and interactions much better. The software tools for creating FBDs include Autodesk, STAAD, SAP2000, ETABS, FreeBodyDiagram (online tool), and Graphical Analysis Software (GAS).

8.8.5 Graphs and Charts

These are standard visual tools to communicate information and data. They are very helpful in project management, reporting, project proposals, research, and engineering design. These include bar charts used to compare categorical data, line graphs used to show trends over time, scatter plots which illustrate relationships between variables, and pie charts used to display proportional data. These graphs and charts are used to communicate data to illustrate complex information, identify trends and patterns in data, and compare data, that is, facilitate comparison of different data sets.

8.8.6 Roller, Hinge, and Fixed Support Idealization

These support idealizations are used in various structural analysis applications, including beams, frames, and trusses. They serve as pedagogical tools to educate civil engineering students.

8.8.7 Computers and Associated Hardware/Software

Computers and associated software/hardware are used in various educational settings, including classrooms, online courses, and distance learning programs. Computers and associated hardware and software can enhance teaching and learning in various ways including interactive learning, engaging students with interactive simulations, games, and multimedia content.

As part of the outcomes, computers can transform the learning experience by increasing student engagement, thus making learning more interactive and enjoyable. They can improve accessibility by providing equal access to education for students with disabilities. They can enhance teaching effectiveness through supporting teachers in delivering high-quality instruction.

Through educational/specialized software, learning can be more interactive and enjoyable. Computers/software can support virtual experiments and simulations. Computers and associated software can support online learning platforms by leverage learning management systems (LMS) to deliver course content and track student progress. In the same way, multimedia resources incorporate videos, podcasts, and interactive multimedia to enhance student engagement.

8.8.8 Scale Models

A scale model is a smaller or miniature version, a physical representation of a real-world structure or system, like a building, bridge, or even a city, that is built to a specific ratio relative to the full-scale object. These models are used to study the structure's behavior, test its design, and communicate its features in a manageable and cost-effective way. A scale model of a building can be used for testing various aspects such as statics, dynamics, acoustics, lighting, aerodynamics, and thermodynamics for energy efficiency. A scale model is geometrically similar to the object.

8.8.9 The Gym Equipment

See Chapter 6, Section 6.3.3.3.

8.8.10 Visual Aids

Visual aids are visual materials, such as pictures, charts, and diagrams, that help people understand and remember information generally shared in an oral presentation. These are items of illustrative matter, such as a film, slide, or model, designed to supplement written or spoken information so that it can be understood more easily. The five common types of visual aids are photographs, infographics, diagrams, videos, and data charts and graphs, such as pie charts and bar charts.

A visual aid is typically used to help an audience understand and remember information in a presentation. It can also be used to maintain an audience's attention, inspire listeners to action, clarify the organization of a presentation, or make a presentation more persuasive.

8.8.11 Conclusion

These pedagogical tools help engineering students develop practical skills, work on projects, and engage in collaborative learning.

8.9 SOFTWARE TOOLS FOR CIVIL ENGINEERS

Civil engineering software solutions are specifically tailored to support practitioners across diverse disciplines—civil engineering branches, encompassing structural analysis, geotechnical engineering, transportation infrastructure, water resources management, and environmental sustainability. These cutting-edge tools streamline workflows, enabling users to seamlessly transition from conceptual design phases to intricate analysis, comprehensive documentation, and integrated project management.

Civil engineering software encompasses a range of tools to help civil engineers during the design and construction processes. These tools for civil engineers can help in every stage of your project including drafting and documenting, designing, and visualizing and analyzing.

Civil engineers use software tools to design, construct, operate, and maintain civil engineering structures. These tools enable engineers to be more efficient. The work that would otherwise be done in several days is reduced to hours.

Software also enhances design or construction precision. Several branches of civil engineering have customized software programs built by civil engineers for the civil engineering industry. The software is developed integrating civil engineering principles and techniques to aid the work of a civil engineer. The aim of the software is to increase the productivity of the engineer, improve efficiency, and reduce the design and construction time and the overall cost of design, construction, operation, and maintenance.

Civil engineering software development is a vital field that supports the complex and evolving needs of civil engineers. By creating specialized tools that enhance design, analysis, and project management, software developers contribute significantly to the efficiency, safety, and sustainability of infrastructure projects. As technology continues to advance, the demand for innovative software solutions will likely grow, presenting opportunities for developers to address emerging challenges in the industry. The software is used for designing and drafting projects and managing/operating projects.

8.9.1 Types of Civil Engineering Software

Civil engineering software can be broadly categorized into several types, each serving specific functions within the engineering process.

8.9.1.1 Geotechnical engineering software

Geotechnical software aids in analyzing soil properties and behavior for foundation design, slope stability, and excavation. Features may include soil analysis and classification, slope stability and retaining wall analysis, and groundwater flow modeling. Examples of geotechnical engineering software are PLAXIS, GeoStudio, OpenGround, and DeepEX.

8.9.1.2 Structural analysis software

This software helps engineers analyze the strength, stability, and performance of structures under various loads. Key features may include finite-element analysis (FEA), load and resistance factor design (LRFD), modal analysis for vibrations, and seismic and wind load analysis. Examples are SAP2000, ETABS, and STAAD.Pro, among others.

8.9.1.3 Design and drafting software

These applications are used for creating detailed drawings and designs of structures, including plans, sections, and elevations. They often include features like 2D and 3D modeling, parametric design tools, automated dimensioning and annotations, integration with Building Information Modeling (BIM), and BIM software for everyone. Examples are AutoCAD, Revit, and MicroStation, among others.

8.9.1.4 Transportation engineering software

This software supports the design and analysis of transportation systems, including roads, highways, and traffic management. Key features can include traffic simulation and modeling, highway design and alignment analysis, and pavement design and analysis. Examples are Verkehr In Städten-SIMulationsmodell (VISSIM), SYNCHRO, and Civil 3D, among others.

8.9.1.5 Water resource engineering software

The tools for hydrological modeling, hydraulic design, and water quality analysis fall into this category. Features often include stormwater management modeling, flood risk analysis, and water distribution system modeling. Examples are HEC-RAS, SWMM (Storm Water Management Model), Environmental Protection Agency Network Evaluation Tool (EPANET), and AutoDesk Storm and Sanitary Analysis.

8.9.1.6 Project management software

These applications help manage civil engineering projects, focusing on scheduling, budgeting, and resource allocation. Features may include Gantt charts and project timelines, resource management and allocation, and cost estimation and tracking. Examples are Microsoft Project, Primavera P6, and Procore, among others.

8.9.2 Key Features of Civil Engineering Software

When developing civil engineering software, several key features should be considered to ensure usability and effectiveness [4].

8.9.2.1 User interface (UI) and user experience (UX)

- **Customizable workspaces**: Users should be able to tailor their interface to suit their workflow.
- **Intuitive design**: The software should have a user-friendly interface that allows engineers to navigate easily and perform tasks efficiently.

8.9.2.2 Integration capabilities

- **Data interoperability**: The software should support data import/export from other tools and systems (e.g., GIS, CAD, and BIM).
- **API support**: Providing APIs for integration with third-party applications can enhance functionality.

8.9.2.3 Collaboration tools

- **Multi-user access**: This allows multiple users to work on a project simultaneously and share data in real time.
- **Cloud-based solutions**: Cloud hosting can facilitate collaboration and access from various locations.

8.9.2.4 Data management and reporting

- **Data storage**: Efficient management of large datasets related to designs, analyses, and project documentation.
- **Automated reporting**: Generating reports, drawings, and documentation automatically based on inputs and analyses.

8.9.2.5 Performance and analysis tools

- **Simulation and modeling**: Advanced tools for running simulations and analyses (e.g., structural loads and environmental impact).
- **Visualization tools**: 3D rendering and visualization capabilities to help clients and stakeholders understand designs better.

8.9.3 Development Process for Software Tools for Civil Engineers

Civil engineers use software applications to design and manage projects. Civil engineering software development involves creating specialized applications and tools that aid civil engineers in designing, analyzing, and managing infrastructure projects. This niche within software development focuses on addressing the unique challenges faced by civil engineers, such as structural analysis, project management, and compliance with regulatory standards. The following is a detailed exploration of the various aspects of civil engineering software development, key features, development process, and current trends.

Civil engineering software development entails crafting specialized solutions that empower civil engineers to design, analyze, and manage complex infrastructure projects. This specialized domain within software development concentrates on alleviating the distinctive challenges faced by civil engineers, including structural analysis, project management, and adherence to regulatory standards.

8.9.3.1 Requirements gathering/ analysis phase

- Identify the needs and challenges of civil engineers.
- Conduct surveys, interviews, and focus groups to gather requirements.
- Define the scope, goals, and deliverables of the software tool.
- Understanding user needs: conduct interviews and surveys with civil engineers to identify pain points and desired features.
- Market research: analyze existing software solutions to determine gaps in functionality or usability.

8.9.3.2 Planning and design phase

- Create a detailed design document outlining the software's features, functionality, and UI.
- Develop a project timeline, budget, and resource allocation plan.
- Define the software's architecture and technical specifications.
- Architecture design: define the software architecture, including databases, UI design, and integration points.
- Prototyping: create wireframes or prototypes to visualize the application's layout and workflow.

8.9.3.3 Development

Write the software code in a programming language (e.g., Python, C++, and Java). Programming language theory is the subfield of computer science that studies the design, implementation, analysis, characterization, and classification of programming languages. Examples of programming languages are Python, Ruby, Java, JavaScript, C, C++, C#, SQL, HTML, and CSS, among others.

- Develop the user interface and user experience (UI/ UX) components.
- Integrate databases, APIs, and other third-party tools as needed.
- An API is a software intermediary that allows different applications to communicate with each other and share data. APIs are a set of protocols and definitions that determine how one application can

access the data or functions offered by another software program.
- Conduct unit testing and debugging.
- **Programming**: writing code using appropriate programming languages and frameworks (e.g., Python, C#, and Java).
- **Database development**: setting up databases to manage and store project data efficiently.

8.9.3.4 Testing and quality assurance

- Conduct various types of testing (e.g., functional, performance, and usability).
- Perform validation and verification of the software's accuracy and reliability.
- Identify and fix bugs and errors.
- Conduct user acceptance testing (UAT) with civil engineers and other stakeholders. Involving real users to validate that the software meets their needs and expectations.
- **Unit testing**: testing individual components for functionality and reliability.
- **Integration testing**: ensuring that various parts of the software work together seamlessly.

8.9.3.5 Documentation and deployment

- Create user manuals, guides, and tutorials.
- Develop training materials and conduct workshops or webinars.
- Deploy the software to the production environment.
- Configure the software for different platforms (e.g., Windows, macOS, and Linux).

8.9.3.6 Deployment, maintenance, and support

- Monitor the software's performance and fix issues as they arise.
- Release updates and new versions to add features and improve performance.
- Provide technical support to users through various channels (e.g., email, phone, and forum).
- Gather feedback and iterate on the software's development.
  - **Deployment**: Releasing the software for public use, ensuring it is easily installable or accessible via the cloud.
  - **Ongoing support**: Providing customer support, bug fixes, and regular updates based on user feedback.

8.9.3.7 Continuous improvement

- Conduct regular review and assessment of the software's effectiveness.
- Gather feedback from users and stakeholders.
- Identify areas for improvement, and prioritize new features and enhancements.
- Refine the software's development process based on lessons learned.

TABLE 8.2   Software customization versus adaptation

FEATURE	CUSTOMIZATION	ADAPTATION
Core code	Modified	Not modified
Purpose	Tailoring to specific needs	Adapting to different environments/users
Implementation	Requires coding or development	Often done through configuration or plugins
Example	Adding a new feature to a software application	Making a web application responsive to devices

Note that this is a general outline, and the specific steps and phases may vary depending on the project's size, complexity, and methodology (e.g., Agile and Waterfall).

8.9.4 Software Customization and Adaptation Techniques

Software customization involves modifying existing software or creating new software to meet specific needs, while adaptation focuses on adjusting software to fit new environments or user requirements without altering the core code. See Table 8.2 for comparisons.

- Customization is the process of tailoring software to meet unique business or user requirements, often involving changes to the code or functionality.
- Adaptation involves making changes to software to ensure it functions correctly in different environments or with different user configurations, without altering the core code.

Software customization is essential especially for regions that rely on programs built elsewhere. This is due to the variances in standards and codes of practice that need to be followed. In Chapter 3, a highlight of units' conversion was provided just the same way as in Chapter 7. In that regard, software needs to be used with caution and experience. Customizing a civil engineering software is crucial to produce accurate and usable results. If not, the entire project can be misled.

Customization in this regard means checking and verifying the units, techniques, and equations the program relies on, coefficients, and constants, among other parameters—everything must be configured and localized as appropriate.

Please note that a software helps you quickly design or manage a civil engineering project but does not substitute your expertise and engineering judgment. A software is a tool just like any other that enables you to perform certain functions more efficiently and effectively. That means having a basic knowledge and understanding of its development is crucial.

Having a knowledge of its shortcomings, procedural steps necessary for using it, and customization to meet regional conditions is essential. The biggest mistake is to assume that a software can be used by anyone to design, construct, operate, or maintain a system. Civil engineering software systems need to be used with prior knowledge and skill of the subject matter. Sound knowledge of the civil engineering principles of the subject matter is deemed necessary.

While using a software, relate it to discussion with another potentially competent engineer. And that means to err is human! The output of a civil engineering software must be checked against conventional wisdom, employing professional judgment and attempt to compare it to hand-made (manual) calculations or designs. What does the manual calculations say? From experience and engineering judgment, would this work?

When working with software and AI, it's crucial to verify the output to ensure accuracy. Without this step, you risk producing nonsensical results. To get the most out of these tools, learn how to tailor them to your needs and understand the development process behind them. You have to learn how to use and customize software. This knowledge will enhance your skills and help you use the software effectively.

It is essential to note that a civil engineering software can only be used perfectly by a person experienced on the subject matter. That means you must have sound knowledge and understanding of the subject matter to critic and check results against manually produced results. An attempt to employ professional judgment in all ways applicable to the situation at hand is necessary.

Software customization and adaptation techniques allow developers to tailor software to meet specific user needs, industries, and/or applications. These techniques enable developers to adapt software to meet specific needs, improving user experience, efficiency, and productivity. A user has to learn steps required to customize the software platform to their needs. Some common software customization techniques for developers include the following:

- **Configuration:** Modifying software settings, options, or parameters to suit user preferences.
- **Localization:** Adapting software for different languages, cultures, or regions.
- **Integration:** Connecting software with other systems, services, or tools via APIs, SDKs, or middleware.
- **Application programming interfaces (APIs):** Leveraging APIs to integrate software with other systems or services.
- **Custom programming:** Writing new code to add features, functionality, or integrations.

8.10 MIND MAPPING QUESTIONS

1. What innovative tools do highway engineers use for road design and maintenance?
2. How do structural engineers leverage technology for building design, analysis, and evaluation?
3. What cutting-edge tools are water resource engineers utilizing for watershed management?
4. What interactive or pedagogical tools can enhance learning in civil engineering courses?

5. How can virtual reality and augmented reality be applied to civil engineering education?

6. What collaborative tools can facilitate project-based learning in civil engineering?

8.11 CONCLUSION

This chapter presents the common tools civil engineers use in the planning, design, and construction of civil engineering structures. FBDs are helpful in analyzing structures and applicable in all civil engineering branches. Therefore, sketching out an FBD is the first step in any civil structural analysis.

Drawings are a helpful tool used to communicate and visualize engineering information whether in planning, design, construction, operation, and maintenance of structures. The types of drawings a civil engineer must know are outlined in Section 8.6.

Civil engineer currently can learn through various pedagogical tools including software packages, visual aids, scale models, drawings, AI, generative AI, ML technologies, FBDs, the gym equipment and activities, graphs and charts, and pictures.

We have shed light on concepts such as roller, pin, and fixed support relating them to real-world applications and emphasizing the fact that these are ideals. In real-world applications, professional judgment is necessary to visualize the type of support most suitable for a structure.

REFERENCES

1. EngineeringSkills. (n.d.). Free body diagrams. Retrieved from: https://www.engineeringskills.com/posts/free-body-diagrams. Accessed 11/December/2024.
2. Meriam, J. L., & Kraige, L. G. (1998). *Engineering Mechanics: Statics* (4th ed.). John Wiley & Sons, New York.
3. Costanzo, F., Plesha, M. E., & Gray, G. L. (2010). *Engineering Mechanics: Statics & Dynamics*. McGraw-Hill, New York.
4. Civil engineering software development. Retrieved from: https://civilpracticalknowledge.com/civil-engineering-software-development/

FURTHER READING

Costanzo, F., Plesha, M. E., & Gray, G. L. (2010). *Engineering Mechanics: Statics & Dynamics*. McGraw-Hill, New York.
Meriam, J. L., & Kraige, L. G. (1998). *Engineering Mechanics: Statics* (4th ed.). John Wiley & Sons, New York.

Think like a Civil Engineer

9

9.1 INTRODUCTION

Tasking you to think like a civil engineer in project design and development primarily means thinking about safety—ensuring that, for example, the structural integrity, stability, and the overall public safety are well considered in project design and development. It means adhering to the ethical code of conduct and following applicable codes and standards in the design and development of structures. It means thinking durability, that is, design for longevity, resistance to wear and tear, and minimal maintenance. Is the project maintainable or will it be maintainable? That is one of the questions to answer.

Tasking you to think like a civil engineer means understanding the right construction technique to adopt for a task. Construction techniques provide a skillful or efficient way of constructing civil engineering structures.

Tasking you to think like a civil engineer means thinking reliability of the structures—the project should meet the intended purpose and fulfill requirements as expected, that is, it must be functional and operable throughout the lifecycle phases. It means considering the latest technologies—applying existing or new technologies to ensure the project is delivered to the highest standard required, cost-effectively and timely. How much time do you need to deliver a project?

It means considering sustainable development; the project is envisaged to be environmentally conscious and considers energy efficiency and long-term viability. It means considering cost-effectiveness, that is, balancing costs with benefits and considering appropriate materials while minimizing waste, labor, and maintenance. It means applying design intuition and logic to refine intuition. It means applying independent judgment to make informed decisions considering the safety and health of the people involved in the civil engineering business and the public and logically juggling 5 Ms of management for the benefit of the project stakeholders.

Thinking like a civil engineer means managing risk at all costs—management of technical, social, economic, and environmental risks. Civil engineers are continuously trained to prioritize safety and manage risk in their project design and development and decision-making processes. This involves identifying potential hazards and threats, assessing the likelihood and impact of risks, and then developing mitigation strategies and implementing safety protocols.

Effective risk management is crucial in civil engineering to protect human life and well-being, prevent environmental damage, minimize economic losses, and ensure compliance with regulations and standards. Through prioritizing risk management, civil engineers can create safer, more resilient, and sustainable infrastructure and systems. All these parameters, when well considered, the output is the engineer as a 'logical thinker.'

Thinking like a civil engineer means having the ability to exercise holistic independent judgment on civil engineering projects. You make decisions that consider the broader context, interconnectedness, and long-term consequences while maintaining autonomy, objectivity, and freedom from the external influence in the decision-making process.

Leading competencies for a civil engineer are management and leadership, design abilities, and commercial awareness/strength.

9.2 CIVIL ENGINEERS PIECE TOGETHER

A civil engineer is a logical thinker!

Civil engineers design, develop, operate, and maintain buildings, towers, transportation infrastructure, specialized structures (such as stadiums and arenas), industrial structures, water and wastewater structures, and many more.

Approach and strategy are crucial in engineering design, just as they are in building a romantic relationship—that is, seducing a cute lady for a romantic relationship—just a bit of hint: the first impression is key. A successful approach in engineering design requires a practical, hands-on, and iterative mindset, rather than a purely theoretical or academic one.

Do not try to apply a rigid, formulaic, or overly theoretical approach to engineering design. Instead, be flexible, adaptable, and open to experimentation and learning. Good engineering design requires a deep understanding of the problem, creativity, and a willingness to try new approaches. Engineering design is more like a dynamic, interactive process (like building a romantic relationship) than a static, theoretical exercise (such as writing a literature review). Design is not literature reviewing. No, not at all!

Civil engineers view their work holistically, and their attitude defines someone who is environmentally conscious, considers safety the top-most priority, and innovates reliable and durable structures that stand the test of time.

Did you know that depending on the project requirements and constraints, you may be able to start with various construction stages or features in the design? Say, you want a building structure that can allow natural light to reach almost everywhere in the building? Or you need a certain type of roof—such as transparent roof, roof tiles, or iron sheets? Or you need a structure that uses local materials such as bamboo for a house or logs for the bridge—to bridge two points!

DOI: 10.1201/9781003561620-9

Imagine starting with site constraints and regulations—say, who owns the land to situate the structure? How much land do we have for the project? You check all legalities in that case—regulatory (e.g., laws, standards, and policies). What is the project budget? All these might be constraints to influence the nature of design in the first place and consequently influence project development, thereafter. The world is full of constraints with make-or-break potential to do or achieve something you set out to accomplish—they greatly influence project outcomes.

> A constraint is a limitation, restriction, or boundary that affects the way something can be done, designed, or achieved. In engineering, constraints are limitations or restrictions that must be adhered to during the design and construction process, such as available resources, time, or safety standards, which impact the feasibility and final outcome of a project.

In various contexts, constraints can be physical (e.g., size, weight, and material), financial (e.g., budget and cost), time-related (e.g., deadline and schedule), regulatory (e.g., laws, standards, and policies), and environmental (e.g., climate and geography). Constraints can be challenging, but they can also drive creativity and innovation!

In modern design and construction, it is common to use iterative and flexible approaches, such as agile design methodologies, building information modeling (BIM), and modular design. These approaches allow designers and planners to start with different stages or features and then adapt and refine their designs as appropriate. Gone are the days of rigid design approaches. Advancing technologies have added flexibility in design.

Flexibility in design means you are at liberty to iterate the options much easily. You change the variables to fit the constructed models or equations and assumptions as much as you can, and as much as applicable. The starting point has never been a problem while solving problems that require divergent thinking, for many cases. That is the primary reason civil engineers check for the failure criterion for structures through an iterative process. Say, check whether the column fails in shear, bending, tension, or fire resistance, etc. Note that a school requires you to apply convergent thinking most of the time because you have to arrive at some unique answer, basically the same correct answer, generalized for the whole class.

The flexible design approach allows designers and planners to address specific challenges or priorities and make changes or adjustments as needed. They can collaborate with stakeholders or team members more effectively. It is a dynamic and adaptable way to approach construction design and planning!

However, it is important to note also that some construction projects may require a more linear or sequential approach, depending on factors such as regulatory requirements, site constraints, and technical complexities. In these cases, starting with a specific stage or feature might not be possible or practical.

In a more generalized sense, the client may specify materials to use in the project, and it becomes a fixed variable—almost a slave one. As a civil engineer, you can advise or do little to nothing, depending on the requirements of the client. For example, a client may propose to use bamboo to build an ecofriendly and sustainable building project. In case they propose using bamboo for columns, they are gameplaying with load-bearing elements of the building system. In that case, they specify materials for columns, which is key in the building frame, directly impacting the role of a civil/structural engineer.

9.2.1 Understand the Context in Design

When a client asks a question and you focus on **'why'** they asked the question, it's called 'understanding the context' or 'addressing the underlying concern.' This approach involves identifying the motivation behind the question; recognizing the questioner's goals, needs, or pain points; and addressing the root issue, rather than just providing a surface-level answer. Thinking like a civil engineer requires deep thought, not surface-level thinking. You must help the client by understanding the context to refine the problem. The right answers are obtainable only when the questions were right!

As an example, the client may dictate that the building must float on water—this requires you to diverge and think about several scenarios, from materials, to safety, to reliability, to cost, to sustainability, to the time to erect the structure, etc., exercising professional judgment right at the onset. In doing so, you might want it to look different, to save the cost and energy usage, and to be unique. This is where creativity and innovativeness come into play. However, understanding why they want it to float on water may help you understand the actual pain points and address the root issue. At the end of the day, any mistakes will rest on you, the designer, and you will take responsibility for errors or omissions. Things like you ought to have advised here and there come into play!

Iterative and flexible design approaches teach us that clients always define what they want, and we must make sense out of the way they define problems. You can advise the client what is possible or impossible based on regulatory requirements, that is, laws, polices, and regulations.

A client may want a building with a unique air circulation corridor with complex amenities. The functionality and operability of the structure will influence several decisions. And when professional judgment is taken, you ensure that everything is incorporated as the client desires, fitting within the necessary legal framework.

The client may define a problem that is complex to interpret. However, with engineering principles, fitting within the solution proposed, the engineer spends time coming up with conceptual designs, loading cases or configurations, and cost analysis based on the available financial envelope. Imagine, a client proposing to build a hospital with a ramp, founded on sandy soils, of a 2-acre piece of land, fitting within a $ 30 million financial envelope, etc.

Exercising independent judgment when designing is important. We must look at the cost of transporting materials, construction time, landscape complexities, excavations, machinery, etc. first and later check the underpinning theoretical literature. The theory supports in answering those questions. Site reconnaissance visits are very important.

When designing, you are solving societal problems, and everything must align with applicable legal frameworks in the region. The society may be facing a problem, and you need to

FIGURE 9.1   Tough terrain, Kisoro UG.

solve it using engineering principles lest assured that the solution provided must be compliant with social, economic, and environmental laws and policies, thus stimulating growth and sustainable development. In that regard, it must be practical and not theoretical. Take an example of constructing a community water supply scheme in a mountainous area as shown in Figure 9.1.

The immediate problem is that the community is terribly water-stricken. The community is facing a severe water shortage, with many residents struggling to access water. The road network is also equally terrible, posing additional challenges. The area is highly mountainous, with weak soils, and prone to landslides in some areas. Here, when defining the engineering problem, we do not focus only on the lack of water, which is the immediate problem. We also look at extraneous factors that may call for temporary works as well, which are secondary problems. We also look at a number of factors and constraints to define an engineering problem correctly. Defining the problem rightly is already half-way to finding the right engineering solution/marvel. In the case you have spent all your time in a highly developed setting, enjoying life on the urban balcony, perhaps lacking the experience of being in a countryside, imagine you are called to solve such a problem in a war-tone, undeveloped area—as a civil engineering expert. You may have to think about the following:

- The cost of transporting materials. This, of course, influences the project cost—for example, when the roads are in a poor state or site is far from where materials are fetched, the transportation becomes complex and hectic.
- The need to meet codes, standards, and desired specifications influences the cost and desired output.
- The terrain will definitely influence the type of pipework (metallic or plastic)—that is, transmission and

distribution pipelines. It can also influence the pipe sizes and pressure ratings due to the considerably high pressures, especially downhill.
- What is the source of water? Open or underground?
- What is the energy source? Gravity, solar, HEP, or hybrid?
- What is the level of water treatment required?
- What construction technology is required?
- What is the balancing volume required for reservoir storage?
- What type of reservoir tank is most appropriate? Will it be sited on dwarf walls or tower? Overhead water reservoir tank?
- What is the area climate?
- What is the available budget including O&M?
- What are the socioeconomic activities done in the area?
- What are the essential women and children rights and other considerations?
- What is the required public stand post (tap stand) yields or pressure in service connections?
- Who owns the land to situate the reservoir, the intake, treatment plant, the pumping station, the sump, or the booster station, etc.? What is the land tenure system?
- What are the essential geotechnical considerations?
- What is the intended system **design life**?
- And much more!

It is crucial to learn that sometimes **clients fail to define problems correctly,** and engineers are therefore tasked to **understand the context** and share insights with the clients at various stages throughout the design stages. In this project case, all the listed factors/questions must be answered with **studies** to come up with a solution for the appropriate community water supply scheme. However, the most important element is to approach the solution from a practical point of view. Theoretical means only refine, support, and defend arguments for the proposed solutions.

9.2.2 Logic and Design Intuition

You have perhaps heard the saying that 'logic is the outcome of engineering education.' And you have perhaps heard the saying that 'engineers always have a way out.' It is because engineers employ logic completely in practice. They strive to make everything tie together, leaving no gaps! When they spot a gap, they strive to find a constituent part or appropriate solution that can fill it, logically. Thus, engineers employ critical thinking all the time.

Logic is the outcome of engineering education which guides the design, analysis, and evaluation of engineering projects. Logic relies on reason and rational thinking by applying rules, principles, and evidence-based information. It is a systematic and analytical approach to evaluating information and making decisions. It is a science that studies reasoning and how to draw valid conclusions from given information. The word 'logic' comes from the Greek word logos, which means 'reason,' 'discourse,' or 'language.' Logic is a formal science that studies how conclusions follow from premises. It's concerned with whether arguments are correct or whether their premises support the conclusion.

For example, logic is often used in logistics, such as when determining the best route for delivering supplies. See case study 9.2. Issues like carbon reduction may be tackled by answering the question: how much carbon emissions shall we escape pumping into the atmosphere when we use the shortest route? Although we solve carbon issues here, we are indirectly determining the amount of gas/fuel to use in the transportation process.

Engineers apply logical reasoning. Logical reasoning is a systematic way of thinking that involves using evidence and reasoning to arrive at conclusions or solve problems by identifying patterns, relationships, and connections.

Logical reasoning can be deductive or inductive. Deductive reasoning starts from general premises and derives specific conclusions that must be true if the premises are true. For example, if you know that all metals conduct electricity and that copper is a metal, you can deduce that copper conducts electricity. Inductive reasoning starts from specific observations and infers general patterns or principles that are likely to be true, but not necessarily certain.

To achieve success in each stage of engineering design, logical reasoning is essential. It helps define the engineering problem accurately by identifying relevant facts, assumptions, constraints, and criteria. Additionally, it enables you to generate and evaluate alternatives by applying relevant rules, principles, or models to each option and comparing their advantages and disadvantages. Furthermore, logical reasoning helps you select and implement the best solution by weighing the evidence and arguments for each alternative. Finally, it allows you to test and refine the outcome by checking the validity and reliability of the solution and making any necessary adjustments or improvements.

Engineers follow design principles such a cohesion, consistency, flexibility, scalability, and feedback mechanisms. Engineers apply logical principles and use specialized tools and techniques to conduct studies, research, and design of structures. Through best practices such as collaborative design, iterative testing, feedback loops, modular design, and continuous refinement of designs, engineers enhance product reliability, optimize performance, reduce development time, and create innovative solutions that better meet user needs and adapt to evolving challenges.

Design intuition, on the other hand, involves the ability to make decisions based on experience, instinct, and creative insight. A deep understanding of the design principles, patterns, and human behavior is needed. The capacity to quickly recognize relationships, patterns, and opportunities that may not be immediately apparent occurs to a prepared mind ('the experienced mind'). Intuition is the ability to understand something immediately without needing conscious reasoning. Although sometimes intuition allows you to skip the friction of logical reasoning, it's by no means an effortless phenomenon. Intuition can inform professional judgment.

9.2.2.1 Basic sample cases

Sections 9.2.2.1.1–9.2.2.1.7 present typical sample cases.

9.2.2.1.1 Ferrying materials on a construction site
The driver at a construction site ferrying materials provides accountability for an itinerary where they claim to have used fuel of 700 L for 1,000 km distance traveled. However, the fuel consumption of the vehicle is (30–33) L/100 km. **Is this logical?** How possibly could it be logical, anyway?

9.2.2.1.2 Technical site visit case study [9.1]
An engineer tasked to inspect a technical site visit appears without a single tool. **Is that logical?** A client wanted to remodel an apartment. He got a phone directory and called a few contractors around town. Three appeared at the same time. He found the one who carried a tape measure quite interested and a committed contractor. So, he won the job! **Lesson:** once invited for a project, look the job you do!

9.2.2.1.3 Design of a bridge to span a river
Suppose a civil engineer is tasked with designing a bridge to span a river. The engineer must ensure that the bridge can safely support the weight of vehicles and pedestrians. Through several logical principles, they design a safe bridge. Some logical steps, among others, are listed below:

- **Step 1:** The engineer must understand how different design elements affect the bridge's overall performance. Knowledge and understanding about the cause-and-effect scenarios to collect relevant data are essential.
- **Step 2:** The engineer then uses established mathematical formulas (as per applicable standards (evidence-based approach!)) to construct equations to ensure accurate calculations.
- **Step 3:** Evaluation—the engineer weighs the advantages and disadvantages of different design options.

Through the application of these logical principles and methods, the civil engineer can create a safe, functional, and efficient bridge design that meets the required standards and specifications.

9.2.2.1.4 Supervision of a bridge to span a river

Exercising logic in bridge supervision requires having project management tools at play, ensuring that quality, cost, and time are achieved as expected. Logical reasoning during construction may typically include the following:

- **Developing a construction plan**: sequence and schedule activities and allocate resources.
- **Monitoring progress**: for example, use of the MS project or AI tools to monitor progress. These can be used to regularly track progress against the construction schedule.
- **Ensuring compliance**: for example, standard checklists must be obtained or prepared. Checklists help comply with standards, regulatory requirements, and project specifications (quality control/assurance).
- **Make informed decisions**: using logical reasoning to adjust construction schedules or allocating additional resources.
- Use of appropriate tools for the task is a logical step. No compromising safety!

9.2.2.1.5 Auditing the safety of the bridge

This needs a logical approach. Auditing the safety of the bridge involves a systematic evaluation of the bridge's design, construction, and maintenance to ensure it meets safety standards and regulations. Some logical principles to follow may include methodical examination of all aspects of the bridge's safety through systematic evaluation. Second, relying on empirical data and documentation to inform safety assessments, that is, the evidence-based decision-making which is logical! Third, conduct risk assessment to identify and evaluate potential safety hazards to prioritize remedial actions.

Audit tools and techniques that can be used to audit the safety of a bridge include checklists, risk matrices, and nondestructive testing. Checklists present a systematic list of safety criteria to be evaluated (check for). Risk matrices are usually visual tools that assess the likelihood and impact of safety hazards. The nondestructive tests are the chief techniques used, such as the ultrasonic pulse velocity testing or rebound hammer testing, to evaluate the bridge's structural integrity.

All the above are logical methods/steps to evaluate the safety of the bridge. By applying logical principles and using specialized tools and techniques, auditors can comprehensively evaluate the safety of the bridge and provide recommendations to ensure the safety of users.

9.2.2.1.6 Technical audits

Technical audits inform other audits such as value for money (VFM), safety audits, and forensic audits. The same logic we apply to design of civil engineering systems and structures can be applied to audit structures. Engineers rely so much on logic when auditing projects. That is why design cannot be straightforward but rather iterative and requiring divergent thinking. VFM audits and technical audits all rely on systematic

evaluation of data to draw logical conclusions that tie everything together.

9.2.2.1.7 Case study [9.2]: Procurement and logistics management

CMA was contracted to transport SPIL's full 40 ft container of goods to destination A, a distance of 240 km. Route planning was entirely at the discretion of the CMA agent. However, the truck driver ended up at destination B, 180 km away from destination A. In their argument, the CMA agent blamed the SPIL representative/engineer at destination A who was meant to offload goods that he misdirected the driver to destination B. The driver gave different reasons for ending up at destination B. SPIL had contracted CMA to transport eight containers full of construction materials to different destinations in the same period of time. Whose fault, was it?

Solution

This problem presents a basic, yet, profound case in effective communication being a necessary attribute for engineers. It showcases how the use of intuition properly is necessary to draw logical conclusions—thinking critically—and not relying on hearsay.

A number of questions were paused to solve this debate. The primary documents in contract management are the contract documents. Thus, contract documents took the precedence—everything else was immaterial. Whether it is true the driver was misdirected or not, that was immaterial. What was truly important was the final destination specified in the contract documents. No driver of sound judgment could change the original route involving cost implications on unsubstantiated information. Route planning was entirely at the CMA agent's discretion.

'Does the driver have the competence to interpret the documents and study the route plan that was shared with him while setting off?' is the question that dominated the debates! Any impromptu deviation had cost implications and would be reported prior, not only to 'CMA agent' but also to 'SPIL' (the client) as well.

Destination B was over 180 km from destination A. How could a driver of sound judgment be diverted 180 km without notifying the office, things that are contrary to the original route planned and documents (he held) that specified the final destination? It was resolved that there was negligence on the CMA agent's part, and not SPIL.

Lesson: as an engineer, we must exercise professional judgment and not simply rely on hearsay; the key items to look at here, that take precedence—which are at the epitome of sound logical planning, because everything in project delivery process has a cost implication, are as follows: contract agreement/documents, competence of the transporter and driver, and the security and safety of the goods. Such scenarios are very common in construction projects. So, safeguard yourself as an engineer all the time by taking **evidence-based decisions.**

9.2.2.2 Conclusion

Engineers follow evidence and not hearsay to draw logical conclusions. They base their arguments on available evidence to draw meaningful conclusions, for example, the top to bottom analysis of a storied structure progress. But you can also start from down reasoning upward if you have the good competence knowledge that things must tie. In other words, typically, analysis starts from the top (loads and forces) and works its way down to the foundation, ensuring that loads are properly transferred and supported. However, experienced engineers might start from the foundation (bottom-up approach) if they have a deep understanding of the structure's behavior and can ensure that all components tie together correctly. The logic is that engineers work in a well-defined structured framework, governed by the goals set out to achieve.

So, when someone mentions a unique beam size constructed of given compressive strength, as an engineer, your mind quickly runs to what particular code or codes this fellow must have followed or something of sorts. And remember codes and standards are developed with rigorous engineering, scientific, mathematical, and technological principles underpinning them. Knowing that standard, the logic is now activated, to question lots of things!

Engineers strive to make things tie. Even in communication, truthful information ties together. A lie is unsustainable; it is only a matter of time! A good engineer must have the ability to communicate complex ideas effectively even to non-technical people. He or she must understand the difference between hiding the truth and telling a lie! These concepts are the hallmark of sound professional judgment while delivering professional expertise!

Engineers conduct technical audits primarily at an output level. This means that, if for example, it is a hot dip galvanized (HDG) steel-pressed panel, and the thickness must fall within the desired/recommended range. An engineer instructing an increase in the thickness of the panel plate or the thickness of the bolt basically implies that when the cross-sectional area is increased, the stress is reduced in the panel/bolt, that is,

$$\sigma = \frac{F}{A_0} \qquad (9.1)$$

where σ is the stress, F is the force in the panel/bolt, and A_0 is the cross-sectional area.

If it is testing concrete, once a certain concrete strength or quality is dictated, it communicates several other associative scientific data! If it is a bolt, it must be of a certain diameter and of a certain quality/strength. Therefore, engineers approach their audit at the output level by measuring the sizes, qualities, and other output parameters. These output parameters are a gateway to understanding the integrity of a civil engineering project.

In case you wish to audit the beam, then you may use non-destructive tests, collect data, and analyze the data to ascertain whether the beam met the specifications. Destructive tests, on the other hand, would show you the materials used. In case you don't have a right to destructive testing, then a photo gallery, if available, is also important to rely on to a reasonable extent.

The need for a constantly evolving civil engineer to move with advances in technology cannot be overemphasized. It is self-evident. Technology is a tool that simplifies the engineer's work! You can only design up-to-date systems when you know the latest standards and codes of practice, the latest releases, latest materials and technology, and you propose designs that are eye-catching and up to date.

The theoretical framework, though important, must be guided by the law—and law operates on logic. Everything done must be complaint with standards, regulations, rules, and codes of practice—fitting within the legal frameworks.

It is essential to always start designs by studying the feasibility of materials and standards and codes—these govern the practicability of the proposed solution. It is not good news to finish the design and only discover that it is not workable, it is not compatible with emerging or new technologies, not adaptable, and unsafe, not to standard and unlawful.

A continuously evolving civil engineer must adapt with not only civil engineering technologies but also other technologies shaping the world. You must incorporate these technologies into your thought pattern to derive a modern solution to a civil engineering problem. That is logical thinking!

9.2.3 Intuition Informs Logic

Experienced engineers often rely on intuition to identify potential solutions or approaches. However, this intuition is typically grounded in their prior knowledge, experience, and logical understanding of the underlying principles.

Note: Intuition cannot happen to an unprepared mind! You must have the experience. Designers must apply their intuition to move forward with a design vision or initiative, even in the most data-rich environments. Almost everything worth building requires some level of blind faith that helps you move closer to a place of conviction. In situations of uncertainty, intuition-driven decisions are often the only source of meaningful creative breakthroughs.

Design intuition is an amalgamation of all your past experiences, knowledge, and passion for specific ideas. It emerges timidly in areas where your subconscious can see resemblances to what you already know about previously observed design problems. It shows strongly in areas where the issue at hand maps tightly to your most fundamental interpretations of the world.

9.2.4 Logic Refines Intuition

As engineers develop their ideas, they use logical reasoning to evaluate the feasibility of their intuitive solutions, analyze the potential consequences of their design choices, and refine their ideas based on logical analysis and evidence.

Intuition can spark initial ideas, but logic is essential for validating, refining, and optimizing those ideas. So, while design by intuition might seem unrelated to logic at first glance, it's actually an integral part of the engineering design process, where intuition informs and logic refines.

It's crucial to understand that intuition, in general, is a delicate tool. If you use it carelessly and don't calibrate it through practical reasoning, it will likely drive you into a dead-end. In the case of design, intuition needs to be applied with surgical precision, allowing data, facts, and anecdotes to enhance the quality and objectivity of the information that emerges instinctively.

Design intuition is the most disruptive skill any designer can develop. Still, it only works and effectively develops when built from a place of strict knowledge and deep reflection about our surrounding world. The difference between biased thinking and intuition is that the latter is a timid exercise of confidence, while the former is a negligent outburst of knowing everything.

9.2.5 Failure: An Option in Design, not in Construction

In design, failure is an option—you simply cannot say it is not an option—which is typical of biased thinking. Saying that you know the design is workable from the word '**go**' is illogical most of the time. This only happens in books for para-practitioners and gods! It is okay to fail. You can have a perfect draft! Fail faster and improve. Don't be shy! It is okay to have poor designs at the start, so long as you embrace teamwork and learn to listen and respect views and opinions of interest groups and stakeholders. Consult with colleagues and work with others. It is good practice to share concept designs, preliminary, and or draft designs with appropriate stakeholders for feedback—going through constant evaluation loops. You spot your weaknesses or design shortfalls or flaws and improve from the feedback you gather.

In some cases, displaying designs or conceptual sketches where stakeholders or colleagues can see is not a bad idea. All this is intended to gather much feedback to improve the designs. Engineers learn a lot from end-users and laypeople. They take their comments seriously to make lasting improvements.

What sets competent engineers apart from the rest of the pack is their understanding that failure is part of the process to innovate practical solutions to the problem. And before they decide the optimal solution, they try several options. That's how innovation works. Steve Jobs made this very clear—trust the process. See Section 2.5.1.2.

Note that even the most experienced engineers make mistakes, there is no doubt about that, and these can be seen in their design, constructed structures, or systems and in reports. However, logic (Section 9.2.2) is visibly seen from what they produce either on paper or practically. You can mess up as a professional, but a bystander can easily spot the difference between an error/mistake and negligence or not knowing what you are doing (ignorance)!

In government agencies, project designs are displayed on noticeboards or shared with stakeholders or colleagues, at different stages through the design process, fostering accountability and transparency. Planning with all stakeholders and incorporating views is aimed at making the project holistically feasible, inclusive, and sustainable. Grasp: doing things **with** stakeholders not **to** stakeholders, as a civil engineer, presents the best opportunity to make fantastic designs—it is the new normal for the modern world.

It is equally essential to note that in construction, failure is generally not an option. Civil engineering is very costly, and reworking things after they fail is not something so pleasant. However, failure could be an option for constituent parts that form the civil structures, say concrete units—but this is not so common, neither—it can only be pardoned when conducting applied research for the structure.

Materials and structural units are tested for use in construction. Failure, for example, of a substructure of a building is not tolerable, but failure of workability of a concrete mix can be pardoned because this can still be corrected. This must be noted. That is why **approval of critical building stages** or any other civil project is crucial to build reliability at all stages of a construction project.

In construction, failure is not an option. This is what sets civil engineers from other engineers because the reliability of civil engineering structures is built in the project right from the start through inception, design, construction, operation, and maintenance. It would be very costly to remodel or rework a casted slab foundation, for example! That means to demolish and cast afresh. So, as a civil engineer, note that your decisions during construction have a holistic impact both legally and technically.

9.2.6 Convergent Thinking versus Divergent Thinking

Engineers always innovate new solutions to civil engineering problems using existing or new technologies. In that case, they go through an iterative process. Unlike convergent thinking where a single solution is generated for a problem, divergent thinking allows enough room for failure! This does not mean constructing flawed infrastructure or structures. No, not at all! It means that design is iterative. Divergent thinking involves generating many ideas and exploring possibilities. You try many combinations before you decide the optimal solution for the problem. We think, 'This works, that can't work.' Things like, '...because it is not safe or the cost to ensure safety is exorbitant—I can't take it on!' Or the cheapest option may be of poor quality. Or, '...I lack the materials needed to achieve the strength required, or it is not sustainable.' All these are juggled to reach the optimal solution—looking at safety, cost, reliability, and sustainability.

Do not be stuck in the rut thinking that because you are not so intelligent or good at analytical techniques primarily tested in classroom that you cannot make a good engineer or designer. No, not all! Design is creativity. It is a creative process. Analytical methods are sometimes encountered or not, with numerical methods preferred!

As a piece of advice, you may normalize drawing, if you can, as part of your routine to arouse your creativity in civil engineering. Keep drawing, you will figure out many things or ideas.

It is essential to note that real design for engineers starts with sketch work, notes, and scale models. Do not attempt to design as if you were writing academic reports or some other kind of reports. It may work, though, but the perfect way is to envision the design through conceptual sketches, to scheme drawings, and preliminary drawings—and scale model development.

When you produce sketches, you visibly present the ideas that you question one by one. You tend to remember everything that could impact the design, from the loading configuration and type, the environmental conditions, constraints, challenges, and terrain, to complexity.

From whatever way you propose the design, the idea of developing sketches cannot be overemphasized. Engineers work with sketches and conceptual drawings to communicate ideas and question the design philosophy. The tank problem presented in Section 9.9 presents a perfect example of how a designer approaches an engineering problem.

The biggest mistake some engineers do is to think that convergent thinking creates enough room to foster innovation during design. It does not. Sketch development of the project you envision leads to divergent thinking, creates enough room for mind-mapping games and logic, and fosters innovation. You pose a sketch to be able to generate divergent questions that map the entire project, raising a cocktail of questions from which you assess their validity and utility to influence the design.

In design of structures, you quickly have the answer in several cases if the problem is well-defined—because suppose the client provides you with the purpose of the building structure, you work around that to know the loading to inform the type and nature of the structural members considering site conditions and constraints.

It is essential to note that engineers employ an integrated approach, where convergent thinking is used in combination with divergent thinking to create a comprehensive problem-solving approach. Therefore, engineers, as logicians, masterfully play the art of convergent solutions reciprocating with divergent ideas.

The question, for example, 'What will change in the next 10 years?' stimulates critical thinking—quite a mind-mapping one. This question, though open-ended, is straightforward, because many things will definitely change. No doubt about that. Engineers always deal with mind-mapping questions—questions that encourage divergent thinking. However, this question can be more valid for a constantly evolving civil engineer when phrased this way, 'What will not change in the next 10 years?' This question is also open-ended, a thought-provoking inquiry that encourages brainstorming, encourages critical thinking, idea generation, and notes and exploration of concepts. The difference is that the second question narrows down! The question does not only apply in innovation and engineering space but also in our daily lives. You can now quickly tell which question converges or diverges!

9.2.7 The Power of Playing with Sketches

Playing with sketches presents a window for idea generation. Playing with sketches is a powerful technique that helps visualize ideas. It is also a great tool that helps in proper decision-making. Sketching can be a powerful tool for brainstorming and coming up with new ideas, especially when approached with a playful and open-minded attitude.

The choices and decisions you make while playing with sketches, such as design strategy, resource management, puzzle-solving, and logical deductions, can produce better outcomes. Playing with conceptual sketches enables the use of 3D, 2D, and 1D visualizations. Through idea generation, you can come up with several constructs, including mathematical constructs.

Sketch development is so powerful that you can visibly see the unknown—a gateway to innovation. Start with sketches, and think of materials to use in the development of the project. Remember that design is a game of ideas and logic. It is not literature review but rather a game of ideas and placing logical thought at the center! Leave literature review to the academic world.

Innovation is that you trust the dots will connect in future. However, you don't know exactly whether they will connect, but you remain faithful that things will work out. Engineering,

though, requires that you must know what you are doing—you must be able to see ahead. For some reason, you must know where you are going. Sketch development arouses curiosity to innovate practical solutions weighing between cost, safety, durability, and sustainability.

Leonardo da Vinci masterfully applied art and sketches to be able to study and understand several disciplines, from engineering, to mechanics, and to science and biology. So, never underestimate the power of sketches, and art in general.

9.2.8 The Power of Developing Scale Models

Scale models are invaluable tools in engineering design and testing, serving as physical representations that accurately mimic the geometric proportions of their corresponding prototypes. These models can be smaller replicas of large structures like vehicles or buildings or larger representations of tiny objects like anatomical structures or subatomic particles, allowing engineers to test, analyze, and optimize designs before full-scale implementation.

9.2.9 Physical Prototypes

A prototype is an original or initial model of something, often used to test and refine a design, concept, or idea. It can be a physical object, a system, or even a digital model. The purpose of prototypes is to test assumptions, validate designs, gather feedback, identify and fix issues, and improve the final product.

The power of developing physical prototypes is immeasurable!

9.2.10 The Jigsaw

As stated in Section 2.5.1.2, according to Steve Jobs, when moving forward in life, you can't always see how your experiences, decisions, and actions will connect and lead to future outcomes or successes. You connect the dots moving backward.

Civil engineers piece together.

As shown in a jigsaw (Figure 9.2), when doing things, engineers ensure they fit into one another for a logical outcome. The interdependence of elements that govern an engineering function cannot be escaped for better outcomes. Engineers always

FIGURE 9.2 Jigsaw puzzle.

find a piece which adds to the picture (Figure 9.2). Doing things this way is one of the characteristics of an engineer. In career advancement, particularly, becoming an engineer, your work experiences must be put together to demonstrate what you have become—whether you have actually achieved the level of competence an engineer is required of! [1]

As an example, consider a building structure: the foundation designed must be commensurate with the type of soil, the size and materials of which load-bearing elements are made must be commensurate with the loading type or configuration, site orientation commensurate with site constraints, leveling and earthworks inform the structural design, esthetics are guided by the building functionality, and many more. These factors fit in one another, and you cannot find piles used where a strip foundation would be ideal, considering cost, safety, durability, reliability, and sustainability.

In engineering, constituent parts do not work in isolation. No, not at all! Although isolation techniques are deployed in design in some cases, still the outcomes communicate with and inform each isolated component and there should not be any discrepancy. If you are constructing a structure, or designing a system, things must be logical—they must tie. The substructure communicates with the superstructure, the foundation communicates with the underlying soil, the roof communicates with the superstructure, and the load-bearing elements communicate with each other!

That is why the concept of holism is essential in civil engineering design. When designing a civil engineering project, considering the interactions and relationships among these components is crucial. For example:

- How will the structural elements interact with the foundation and the surrounding soil?
- How will the various systems integrate and function together?
- How will the materials behave under different environmental conditions?

Elements must consult with each other and not only inform one another. Integration and coherence in design and construction is key. Components must fit together rationally and logically, following sound engineering principles and rules. Elements must influence and inform each other, ensuring that harmony and must communicate effectively. Engineers consider a holistic approach, viewing the system as a whole—components relying on each other and developing feedback loops for continuous interaction and adjustment or improvement.

Integrated design and construction ensure that components work together seamlessly, ensuring efficiency, effectiveness, and sustainability, thus mimicking the natural world to the fullest. See Chapter 6.

A sound design must be an integral of components parts but not add-ons. You cannot claim to be adding safety to already produced design or facility. It must be integrated in the process to ensure safe conditions are achieved throughout. You cannot add on environmental management systems later when the project is already designed or built.

Environmental sustainability is non-negotiable for today's civil engineer. Everything integrates in the whole system: safety, reliability, durability, sustainability, seamless functionality, marketability, and many others as far as goals set to achieve are concerned.

9.2.11 The Engineer's Thought Pattern

Engineers approach problems with a methodical, analytical, and often creative mindset, prioritizing logical reasoning, problem-solving, and finding practical and effective solutions.

Engineering is about thinking, not just knowing. Accreditation bodies evaluate your thought pattern, not just your knowledge to accredit you.

Engineers use a unique mode of thinking based on seeing everything as a system. Thinking in systems means that you can deconstruct (breaking down a larger system into its modules) and reconstruct (putting it back together). Engineers break down complex problems into manageable parts, analyze data, and evaluate evidence to develop solutions—analytical thinking.

They see structures that aren't apparent to the layperson (ability to see through walls), they know how to design under constraints, and they understand trade-offs. Adopting an engineering mindset can help you in any field by the way. When we think of the engineering mindset, we often think of a rational, methodical process. Both are important traits, but what about creativity? Engineering is often problem-solving. It requires an ability to think outside the box and around the box and visualize a number of alternative solutions and scenarios.

Engineers think creatively to design innovative solutions, considering multiple perspectives, and finding novel applications for existing or new technologies. Engineers balance idealism with practicality, recognizing that real-world constraints (e.g., budget, time, and resources) require adaptable and effective solutions.

The core of the engineering mindset is 'modular systems thinking.' It's not a singular talent, but a mélange of techniques and principles. System-level thinking is more than just being systematic; rather, it's about the understanding that in the ebb and flow of life, nothing is stationary and everything is linked.

The relationships among the modules of a system give rise to a whole that cannot be understood by analyzing its constituent parts—'holism'—the whole is more than the sum of its parts. Thus, the three essential properties of a mature engineering psyche/mindset are ability to see through walls, designing under constraints, and understanding trade-offs.

A structured systems-level thinking process would consider how the elements of the system are linked in logic, in time, in sequence, and in function—and under what conditions they work and don't work. A historian might apply this sort of structural logic decades after something has occurred, but an engineer needs to do this preemptively, whether with the finest details or top-level abstractions. This is one of the main reasons why engineers build models so that they can have structured conversations based in reality. Envisioning a structure involves having the wisdom to know when a structure is valuable and when it is not.

9.2.11.1 Ability to see through walls

The ability to see a structure where there's nothing apparent is a hallmark of great engineering judgment. Engineers have the ability to visualize and understand the underlying structures, systems, and mechanisms that are not immediately visible to a layperson. They can 'see' beyond the surface level, penetrating through the obvious to reveal the hidden patterns, relationships, and principles that govern how things work.

Great engineers employ their intuition properly and to the maximum. An engineer visualizes and produces structures through a combination of rules, models, and instincts. It's not only about what one sees; it's also about the unseen. The engineering mind gravitates to the piece of the iceberg underneath the water rather than its surface. It's not only about what one sees; it's also about the unseen.

9.2.11.2 Designing under constraints

Adeptness at designing under constraints is what makes engineers standout. The real world is full of constraints with make-or-break potential for project success. Engineers are expected to produce the best possible results under the given conditions. Even if there are no constraints, good engineers know how to apply constraints to help achieve their goals. That is where we derive the saying that 'engineers create their own problems is there is none!' You can't be an engineer, and you claim to fail to get what to do in the literary sense!

One of the common examples is how time constraints that engineers face fuel creativity and resourcefulness. Financial constraints and the physical constraints showcase how engineers exercise their skills to serve their purpose. These constraints, hinging on the laws of nature, for example, are common in many projects that engineers develop, coupled with an unpredictable constraint such as human behavior, and make engineers not only smart but also gifted—because they need to manage and lead under unprecedented constraints.

9.2.11.3 Understanding trade-offs

Engineers possess the ability to propose alternative ideas and make professional judgments. Engineers make design priorities and allocate resources by ferreting out the weak goals among stronger ones. For example, for a road pavement design and development, a typical trade-off could be to balance the demands of cost, quality, and time within the constraints of the given performance specifications. This type of selection pressure even trickles down to the question of whether all the categories of road users will be comfortable using the road. If constraints are like tightrope walking, then trade-offs are inescapable tugs-of-war among what's available, what's possible, what's desirable, and what the limits are [2].

9.3 TRANSITIONING FROM ACADEMIA TO INDUSTRY

Transitioning from academia to a professional engineer needs effort. Abiding by the ethical code of conduct is key, and professionalism and safety considerations are key issues in engineering practice. The process of transitioning from academia to other fields requires a strategic approach, including assessing your skills, researching potential career paths, networking, and tailoring your resume and cover letter to industry standards. This helps you gather relevant experience beyond steel and concrete.

Thinking like a civil engineer is distant from learning civil engineering. Learning civil engineering provides the foundation, but thinking like a civil engineer requires applying that knowledge in a real-world context, considering multiple factors, and making informed decisions. You cannot get all this in lecture rooms! Never!

This distinction emphasizes the importance of developing a professional mindset and problem-solving skills in addition to acquiring technical knowledge.

Practicing civil engineering is governed by the ethical standards that govern the profession. The American Society of Civil Engineers (ASCE) Code of Ethics outlines fundamental principles and guidelines for ethical and professional conduct, emphasizing integrity, honesty, and competence in civil engineering practice. It is not so different from the Institution of Civil Engineers (ICE).

The ICE code of conduct sets out how members should behave. The ICE takes the conduct of members very seriously. Being a member of the ICE means that a civil engineer has proved that they can work to the highest possible standards. So, thinking as a civil engineer means you are enveloped in the ethical code of conduct. This code of conduct typically includes the following:

- **Ethics:** upholding honesty, integrity, and transparency. Discharge professional duties with integrity and behave with integrity in relation to all conduct bearing upon the standing, reputation, and dignity of the institution and of the profession of civil engineering—undertaking work that you are competent to do [3]. Maintaining honesty and fairness in all professional dealings.
- **Objectivity:** striving for impartiality and objectivity when dealing with others.
- **Confidentiality:** protecting sensitive information [4].
- **Safety:** prioritizing public safety and well-being. Having full regard for the public interest, particularly in relation to matters of health and safety, and in relation to the well-being of future generations.
- **Responsibility and due care:** taking ownership of one's work and decisions, performing duties with diligence and care.
- **Sustainability:** considering environmental and social impacts, that is, showing due regard for the environment and for the sustainable management of natural resources.
- **Professionalism:** maintaining competence, objectivity, and respect for colleagues and clients. Maintaining and improving technical knowledge and skills. Keep developing professional knowledge, skills, and competence on a continuing basis and give all reasonable assistance to further the education, training, and continuing professional development of others. Benchmarking on the knowledge of others: Designers could utilize the knowledge and experience of contractors to assist in the process.
- **Professional behavior:** acting in a manner that reflects positively on the profession.

This code of conduct supports civil engineers to demonstrate their commitment to protecting the public, preserving the environment, and upholding the integrity of their profession.

9.4 THE DESIGN PROCESS

The design process is iterative. Design, analysis, and evaluation (DAE) is a fundamental framework in civil engineering design, which governs the iterative process. Usually, the design constantly communicates with the analysis and evaluation. DAE are interconnected processes.

Note: There is a preliminary evaluation before design. This evaluation hinges on the fact that engineers, to some degree, know where they are heading. Despite the fact that innovation is trusting the process, engineers are governed by codes and standards, and for that reason, they anticipate required outcomes to a significant level. Of course, where a project is unique, it must involve some degree of creativity and innovativeness, and that means you cannot know exactly what the final outcome will be.

Usually, a solution termed 'optimal' is always arrived at through a cyclic process. However, the preliminary evaluation involves assessing and measuring the constraints, complexity, goals, and environmental factors, among others items. Setting goals is essential here, alongside problem definition and evaluation. Appropriate research is necessary at this stage.

9.4.1 The DAE Cycle

The DAE cycle refers to design, analysis, and evaluation, as depicted in Figure 9.3. DAE are interconnected processes,

where design focuses on creating something, analysis examines its elements, and evaluation assesses its effectiveness against initial goals. The DAE are a cyclical process where each step builds upon the previous one, leading to continuous improvement and better outcomes. The interconnection is as follows:

- **Design informs analysis and evaluation:** The initial design sets the stage for what needs to be analyzed and evaluated.
- **Analysis informs evaluation:** By analyzing the design, areas that need to be evaluated are identified.
- **Evaluation informs design:** The results of evaluation can be used to improve the design in future iterations.
- In some cases, and depending on the nature and magnitude of the project, deployment typically occurs after the design and analysis phase where the plan is put into action, and before the evaluation phase, where project outcomes are assessed. It is essential to note that this sequence allows for testing and refinement of the deployed system or product based on the performance.

9.4.1.1 Design

This refers to the creation of a preliminary/design concepts, sketch or model development, draft designs, and final designs, considering requirements, constraints, and goals. The design involves creating plans and specifications for civil engineering projects, considering factors like functionality, safety, cost, and esthetics. Design is essentially planning and creating something, whether it's a product, a system, a process, or an idea. It is intended to develop a solution or a plan that

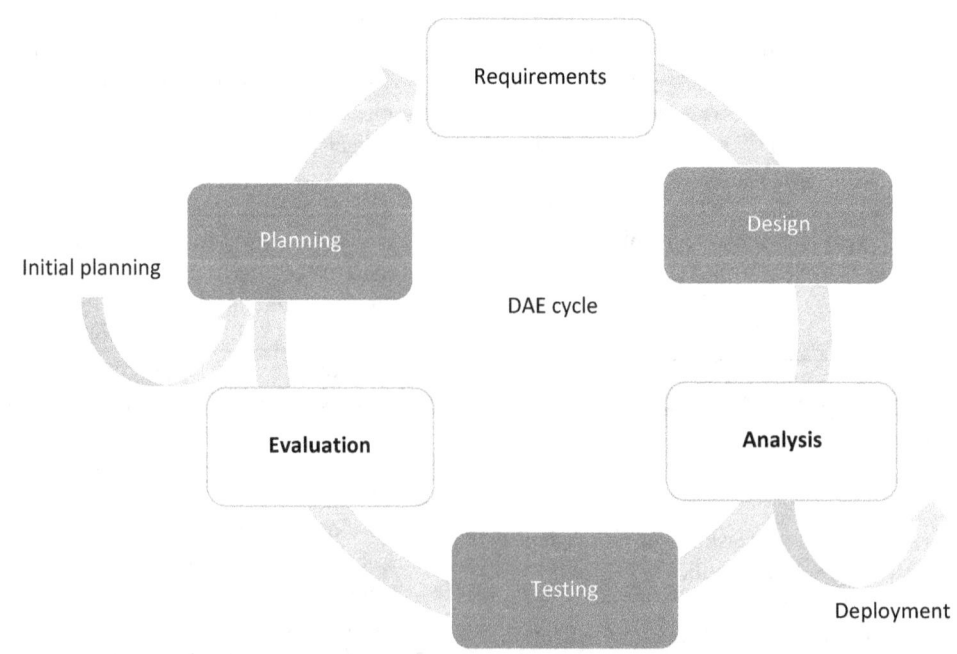

FIGURE 9.3 DAE cycle.

meets specific needs or requirements. Civil engineering design examples include the following:

- Structural designs such as beams, columns, and foundations,
- Hydraulic design (pipelines and dams), and
- Transportation design (roads, highways, and bridges).

These are communicated through drawings, plans, prototypes, and specifications.

9.4.1.2 Analysis

This is the examination and simulation of the design using various methods, for example, mathematical models/techniques or computer simulation techniques to predict its behavior and performance of the design. Examples include the following:

- Hydraulic design of pipelines requires energy and momentum techniques.
- Finite-element analysis (FEA) for structures.
- Stress and strain techniques for structures.

Analytical and numerical techniques are employed to analyze civil engineering systems. These are mathematical techniques to predict its behavior and performance of the design. The analysis examines the performance, reliability, durability, and stability of civil engineering designs. The analysis helps predict how the civil engineering system designed would behave, for example, under various loading configurations; under temperature variations, environmental differences, different materials; and of course, considering cost implications, safety, and sustainability issues.

Analysis involves examining the components, structure, or function of something to understand its characteristics and how it works in order to gain insights, identify problems, and make informed decisions.

9.4.1.3 Evaluation

This encompasses the assessment and measurement of the design's performance, comparing it to the requirements, constraints, and initial goals, and identifying areas for improvement. Evaluation assesses the project's effectiveness and success. Project design evaluation is the process of assessing the value, effectiveness, or quality of design based on specific criteria or goals in order to determine if a design or solution proposed is successful and to identify areas for improvement.

Evaluating the effectiveness of the proposed solution by comparing the actual performance of the project with the design specifications and objectives. It also includes assessing the project's impact on the environment and surrounding communities. Some of the examples of evaluation include assessing the safety and durability of a bridge (see Section 9.2.2.1.5), the efficiency of a water treatment plant, or the impact of a new road on traffic flow.

Some of the ways to evaluate a civil engineering design include the use of a self-review checklist, gathering feedback from stakeholders, simulation and testing the design, benchmarking and metrics, and reflection and learning.

9.4.1.3.1 Self-review checklist
One of the simplest and most effective ways to evaluate the design work is to use a self-review checklist. A checklist is a set of questions or criteria that you can use to check the quality, accuracy, and completeness of your design. One can create their own checklist based on the project specifications, codes, and standards or use existing ones from reputable sources. For example, professional societies such as the ASCE provide several checklists for different types of civil engineering projects, such as bridges, buildings, dams, and roads. A checklist can help you avoid common errors, identify gaps, and ensure consistency in the design. The intended goals and objectives can be answered by assigning potential indicators in the checklist.

Checklists are also used in construction—for example, concrete production—evaluating the quality of concrete and ensuring concrete quality is achieved, before mixing, during casting, and curing. Design can be continuous, for example, sectional designs or temporary works such as designing formwork, and checklists can be used. Shuttering formwork: this is done at the construction stage; the foreman or site/section engineer designs the formwork and uses checklists to evaluate the designs, including the workforce, and before casting, they check whether everything is in order, and then concrete is poured into the mold, the mold is vibrated, and then curing is initiated.

A checklist is vital during formwork design because it helps one to check against important criteria and minimum/maximum acceptable critical dimensions for different designs. Including the workforce in the design of temporary works solutions is essential, and it may add on the effectiveness of the proposed solution. A designer must take all reasonable steps to provide sufficient information about the design, construction, or maintenance of the structure and to adequately assist the client, other designers, and contractors to comply with their duties.

9.4.1.3.2 Simulation and testing
Simulation and testing tools can be used to evaluate the design work. Simulation and testing tools are software applications that can help analyze, visualize, and verify the performance and behavior of the design under various conditions and scenarios. For example, computational fluid dynamics (CFD) can be used to simulate the flow, heat transfer, and turbulence of the design. FEA can be used to simulate the stress, strain, and deformation of the design—particularly structural designs, for example, steel towers. Simulation and testing tools support design optimization. They help identify potential problems and ensure reliability and safety.

9.4.1.3.3 Feedback from colleagues or stakeholders
Feedback is an invaluable source of information and insight that can improve the design. The designer can improve their skills and learn from others/colleagues and stakeholders. Peers/colleagues may help review the design work and provide constructive comments and suggestions. Peer review platforms, such as **DesignSafe**[1] or **Civilax**, are available where you can share the design work with other civil engineers and get feedback from experts and peers. Feedback from colleagues can support the engineer to gain different perspectives, discover new ideas, and validate the design decisions.

9.4.1.3.4 Benchmarking and metrics

Benchmarking and metrics are methods that can help compare the design with the best practices, standards, and examples in that specific field. Benchmarking and metrics can help assess the design strengths and weaknesses and set realistic and achievable goals. For example, you can use benchmarking to compare the design work with similar projects that have achieved high performance, quality, and efficiency. You can also use metrics to measure and quantify the project outcomes and impacts of your design work, such as cost, time, energy, and emissions.

9.4.1.3.5 Reflection and learning

A fifth approach to evaluate your design work is to use reflection and learning. Reflection and learning are processes that can help review the design experience and learn from the successes and failures. For example, you can use reflection to examine your design process, methods, and choices and evaluate how they affect your design results. You can also use learning to identify your knowledge and skill gaps and seek opportunities to improve them. Reflection and learning can help you enhance your design competence, confidence, and creativity.

9.4.2 Iterative Process

The **DAE cycle** is iterative, meaning that the process is repeated until the design meets the desired requirements and constraints. Iteration basically means repetition of a mathematical or computational procedure applied to the result of a previous application, typically as a means of obtaining successively closer approximations to the solution of a problem.

Repetitive processes rely on **algorithms**. Algorithms break complex problems into manageable parts, making it easier to find optimal solutions. Algorithms can automate repetitive tasks, freeing up designers to focus on creative work. They can also process large amounts of data quickly, accelerating the design process. Algorithms can perform calculations and simulations with high accuracy, reducing the risk of human error and ensuring that designs meet specifications and regulations. Algorithms can be used to optimize designs for various criteria, such as cost, energy efficiency, or structural integrity.

Algorithms are extremely useful in civil engineering; especially the back-and-forth algorithms, also known as iterative algorithms or iterative methods, are a type of algorithm that use a repetitive process to converge on a solution.

In mathematics and computer science, an algorithm is a finite sequence of mathematically rigorous instructions, typically used to solve a class of specific problems or to perform a computation. Algorithms are used as specifications for performing calculations and data processing.

During civil engineering design process, algorithms are/may be built. An algorithm is a procedure used for solving a problem or performing a computation.

An algorithm is a step-by-step procedure or a set of rules designed to solve a problem or perform a task. It can be described in various ways, including natural language, pseudocode, or programming languages. AI and machine learning (ML) are integrated into civil engineering software to automate repetitive tasks, optimize designs, and predict project outcomes based on historical data.

Iterative and recursive algorithms are fundamental concepts in programming that a civil engineer should know. Popular high-level languages include **C++**, **Auto LISP**, **VBA**, **C**, **Java**, **Python**, and **C#**. Iterative and recursive algorithms support civil engineering software development and customization.

Both iterative and recursive algorithms can solve the same problems, but they do so in different ways. Iterative algorithms use loops, while recursive algorithms use function calls. Understanding an algorithm's time complexity is crucial for evaluating its efficiency, especially for large inputs. In natural languages, an algorithm can be described and then represented visually using a flowchart.

By applying the **DAE framework**, civil engineers can develop effective and efficient designs, ensure that designs meet requirements and constraints, identify potential problems and mitigate risks, and properly optimize designs for performance, cost, and sustainability.

9.4.2.1 Flowcharts

A flowchart is a diagram that visually represents the steps of an algorithm, using different symbols to represent different types of instructions or actions. It helps understand and communicate the logic of an algorithm in a clear and organized way. It is usually designed in the natural languages.

We can represent repetitive processes and conditional logic using flowchart symbols. If you compare a flowchart to a movie, then an algorithm is the story of that movie. In other words, an algorithm is the core of a flowchart. Actually, in the field of computer programming, there are many differences between algorithm and flowchart regarding various aspects, such as the accuracy, the way they display, and the way people feel about them. The following table illustrates the differences between them in detail.

ALGORITHM	FLOWCHART
• It is a procedure for solving problems.	• It is a graphic representation of a process.
• The process is shown in step-by-step instruction.	• The process is shown in block-by-block information diagram.
• It is complex and difficult to understand.	• It is intuitive and easy to understand.
• It is convenient to debug errors.	• It is hard to debug errors.
• The solution is showcased in natural language.	• The solution is showcased in pictorial format.
• It is somewhat easier to solve complex problem.	• It is hard to solve complex problem.
• It costs more time to create an algorithm.	• It costs less time to create a flowchart.

Flowchart example for illustrative purposes

The flowchart in Figure 9.4 provides a framework for high-level decision-making on whether a prefabricated bridge should be designed for a project. Therefore, during the design process, or decision-making process, the flowchart may be used in the process to decide whether to adopt prefabricated or conventional bridges. This kind of flowchart may be used in the design process at different stages, as circumstances may dictate.

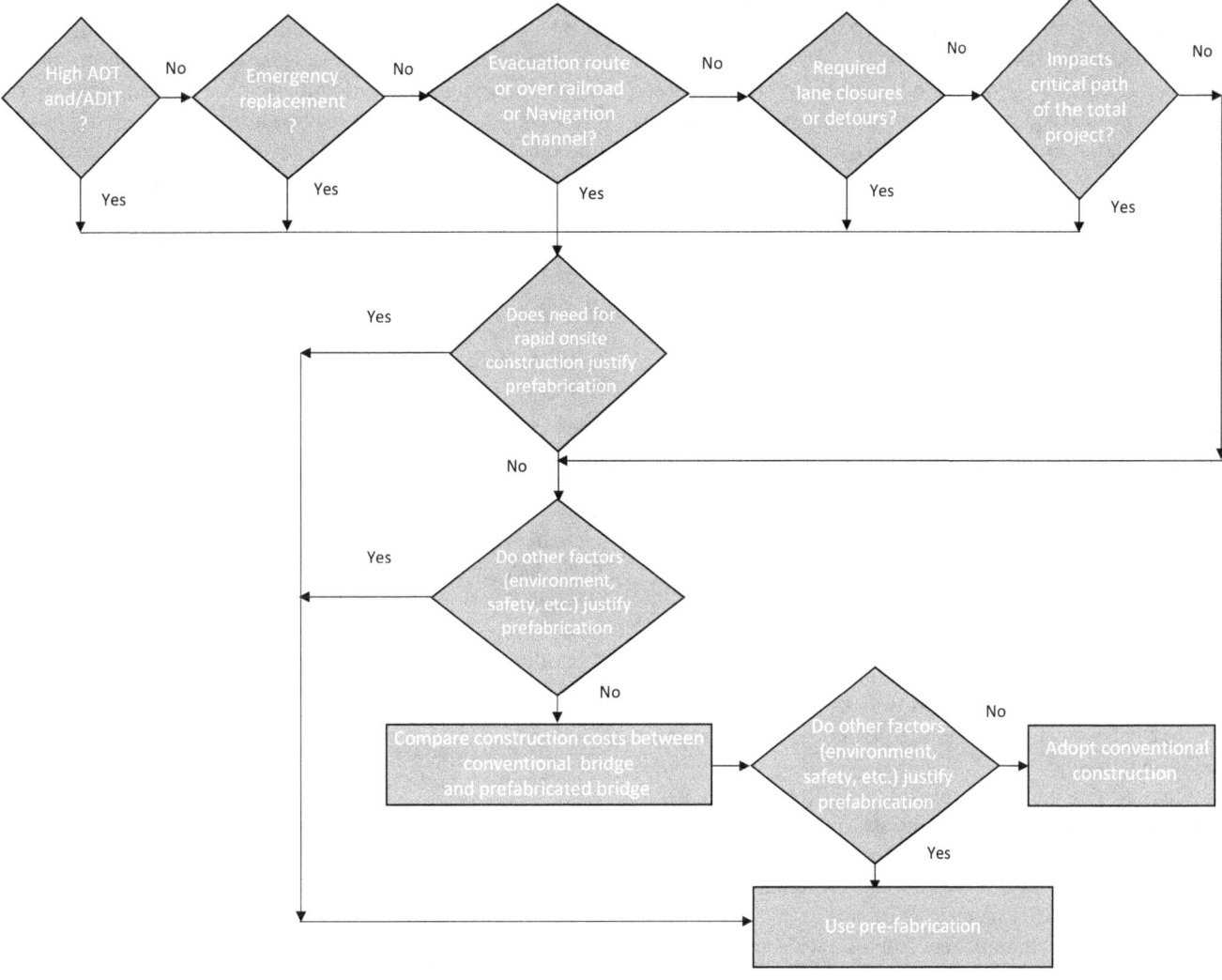

FIGURE 9.4 Flowchart.

9.4.2.2 *Iterative and recursive algorithms*

In programming, an algorithm is a set of well-defined instructions to solve a problem or perform a task. Or simply put, it is a step-by-step procedure or a set of instructions designed to solve a specific problem or perform a task. It's like a recipe that a computer follows to achieve a particular outcome. This iterative approach can be applied in program development when designing civil engineering structures. **Iteration** is a process of repeating a set of instructions or steps multiple times, until a condition is met, often to refine or improve a result. **Recursion** is a technique where a function calls itself to solve a problem. In essence,

- An algorithm is the overall plan for solving a problem.
- Iteration is a way to execute parts of an algorithm repeatedly.
- Recursion is a specific technique within algorithms where a function calls itself.

Thus, iterative and recursive algorithms are fundamental concepts in programming, and languages like C++, AutoLISP, VBA, and C# support them in various ways. Iterative algorithms use loops to repeat a set of instructions until a condition

is met. High-level languages that civil engineers use to develop systems support iterative algorithms. See comparisons below:

- C++: Uses 'for, while, and do-while loops' to implement iterative algorithms.
- AutoLISP: Uses while and repeat loops to implement iterative algorithms.
- VBA: Uses For, Do While, and Do Until loops to implement iterative algorithms.
- C#: Uses for, while, and do-while loops to implement iterative algorithms.

Recursive algorithms solve problems by breaking them down into smaller instances of the same problem. High-level languages that civil engineers use to develop systems/platforms support recursive algorithms. See the comparison below:

- C++: Supports recursive functions, which can call themselves to solve a problem.
- AutoLISP: Supports recursive functions, which can call themselves to solve a problem.
- VBA: Supports recursive functions, but with some limitations due to the language's design.
- C#: Supports recursive functions, which can call themselves to solve a problem.

These languages support both iterative and recursive algorithms essential for solving complex problems in computer science. It encompasses evaluating the effectiveness of the proposed solution. Each iteration refines the design, incorporating lessons learned from previous analyses and evaluations, new information or data, and the changes in requirements or constraints.

9.5 CONSTRUCTION TECHNIQUES

Construction techniques are a set of established methods, procedures, or tools that are used during the construction process. They are generally employed in the creation of buildings, infrastructure, and other physical structures.

Without knowing the right construction technique for a project task, typically outlined in codes and standards, one cannot claim to understand and select cost-effective or reliable construction technologies. In the same way, not knowing the right construction technique may compromise reliability and performance, safety risks and structural failures may also occur, and regulatory non-compliance may lead to legal issues.

Construction techniques provide a skillful or efficient way of constructing civil engineering structures. Understanding and implementing appropriate construction techniques is essential for ensuring the successful completion of a project within budget and on schedule.

Construction techniques encompass a wide array of methods ranging from traditional approaches such as timber framing and masonry to modern approaches such as 3D printing and modular construction. These methods are shaped by a variety of factors, including the materials available, the geographical location of the project, and the specific requirements outlined for the construction project.

Construction techniques have evolved significantly over time, with each technique presenting its own set of advantages and challenges. The selection of a construction technique depends on project requirements, budget considerations, and sustainability goals. Embracing new technologies and sustainable practices has proposed a potential drive for the future of construction toward increased efficiency, effective, resilience, and environmental sustainability.

There, several thousands of construction techniques and civil engineers need to know the appropriate construction techniques either during design or construction. In order to achieve durable structures and safe working environments, the selected techniques must be the right ones for the job.

9.5.1 Traditional Construction Techniques

Traditional construction techniques refer to the methods and practices used in building construction before the advent of modern materials and technologies. The term 'traditional' refers to things that are traditions, that is, they have been handed down from generation to generation. However, it may also be used to refer to things which are not 'modern.'

Traditional construction techniques have been passed down through generations and are often characterized by use of natural materials such as wood, timber, stone, clay, earth, and other natural materials. They often involve manual labor and handcraftsmanship. Traditional construction techniques are often sustainable, ecofriendly, and culturally significant and are still used today in various parts of the world.

9.5.1.1 Framing

Creating a skeletal structure using wood or steel. It includes timber framing and heavy steel framing. In timber framing construction, industry experts employ premium, heavy-duty lumber to craft robust building frames. A method of building frames using wooden beams and joints are often used for houses and other structures. This traditional technique involves meticulously joining lengthy timber beams to form the structural skeleton of a building.

Similar to timber framing, heavy steel framing involves assembling steel beams to form the structural framework of a building—structure's skeleton. Fabricated in factories, these steel beams are then transported to the construction site for assembly. Due to its exceptional load-bearing capacity, this construction method is particularly well suited for buildings requiring robust structural support, such as high-rise skyscrapers.

9.5.1.2 Masonry

Masonry is a broader term that refers to the art and craft of building structures using stone, brick, concrete block, and other masonry materials, bound together with mortar. Masonry construction involves the utilization of stone-based materials, where individual bricks or stones are interconnected with mortar to form a structural framework. This framework is then reinforced with concrete to enhance its stability and durability.

Masonry techniques are employed for both esthetic and functional purposes, providing a unique visual appeal, while also offering robust structural support. This construction method remains a popular choice globally, particularly in regions where durability and low maintenance are paramount. Masonry can include bricklaying, stonework, blockwork, and concrete masonry.

Bricklaying is the process of constructing structures with bricks, often using mortar and traditional techniques. Traditional bricklaying involves hand-crafting bricks from clay, using simple tools and techniques to build structures, often with a focus on durable, labor-intensive methods. Bricklaying is a specific type of masonry that focuses on laying bricks, blocks, or other masonry units in mortar to create structures.

9.5.1.3 Cofferdams

A cofferdam is a temporary enclosure built within a body of water to allow the enclosed area to be pumped out or drained, creating a dry working environment for construction or repair projects. This technique has been used for centuries, dating back to ancient civilizations such as the Egyptians, Greeks, and Romans.

Cofferdams are used to create a dry working space in areas where water is present, such as for building dams, oil platforms, bridge piers, or repairing underwater structures. They are typically constructed by driving piles into the ground to form a wall, which is then pumped out to remove the water within the enclosure. Cofferdams were used in the construction of many historic structures, including the following:

- The ancient Egyptian temples along the Nile River
- The Roman harbor of Caesarea Maritima
- The medieval castles and bridges in Europe (verify theses)

9.5.1.4 Formwork

Formwork, also known as shuttering, refers to the temporary or permanent molds used to contain and shape concrete until they are self-supporting. This refers to creating temporary molds to hold concrete in place during pouring.

Mold used to form concrete into structural shapes (beams, columns, slabs, and shells) for building. Formwork can be of timber, steel, plastic, or fiberglass. The inside surface is coated with a bond breaker (plastic or oil) to keep the concrete from sticking to the mold. It is used when using concrete materials to create foundations and structural frames.

Formwork refers to the mold construction workers create before pouring concrete. These molds can be temporary or permanent, depending on the building's needs. There are many types of formworks, including timber, engineered, reusable plastic, flexible, and permanent insulated. Shuttering is the special formwork for concrete works.

9.5.1.5 Scaffolds

Scaffolds represent an ancient construction technique. The use of scaffolding dates back to ancient civilizations, such as the Egyptians, Greeks, and Romans, who used wooden scaffolds to construct monumental buildings, temples, and other structures.

During building construction, scaffolds provide access to building exteriors, allowing workers to construct, repair, or maintain walls, windows, roofs, and facades. Scaffolds enable workers to perform routine maintenance, repairs, and inspections on buildings, bridges, and other structures. Scaffolding has been a crucial component of construction projects throughout history, providing a temporary platform for workers to access and work on buildings, bridges, and other structures.

While modern materials and technologies have improved the design, safety, and functionality of scaffolding, the fundamental principle remains the same. Scaffolding continues to be an essential tool in the construction industry, used in various projects, from small-scale renovations to large-scale infrastructure developments.

9.5.1.6 Earthbag construction

In this technique, bags are filled with soil or gravel are stacked to create walls. Earthbag construction is a building technique where bags, typically filled with soil, sand, gravel, or other natural materials, are stacked and tamped to create walls and foundations, offering a cost-effective and sustainable alternative to traditional methods.

9.5.1.7 Wattle and daub

This technique is used for building walls, where a lattice of wooden strips (wattle) is covered with a mixture of mud, clay, and straw (daub). It is an ancient building method where a woven lattice of wooden strips (the 'wattle') is covered with a sticky material, typically mud, clay, sand, and straw (the 'daub'), to create walls and structures.

9.5.1.8 Adobe construction

This involves using sun-dried clay bricks or mud to build walls, offering natural insulation. Adobe construction refers to building with sun-dried mud bricks, traditionally made from compacted sand, clay, and straw (or grass), mixed with water, and baked in the sun.

9.5.1.9 Concrete work

Concrete work is actually a combination of traditional and modern techniques. It involves pouring and shaping concrete for foundations, floors, and walls. The use of concrete dates back to ancient times, with the Romans being notable users of a primitive form of concrete, known as Opus caementicium, around 2000 years ago. However, modern concrete, with its precise mixture of cement, water, aggregates, and admixtures, is a relatively recent development, dating back to the mid-19th century.

In 1824, Joseph Aspdin, an English bricklayer, discovered modern concrete as distinct from the Roman variant. He invented Portland cement, a key ingredient in modern concrete, by heating chalk and clay and then grinding the resulting material into a fine powder. This process created a cement that resembled the high-quality building stones found in Portland, England, hence its name.

9.5.1.10 Bamboo construction

This is a method of building using bamboo, often used for houses, bridges, and other structures. Bamboo construction utilizes the strong, flexible, and rapidly renewable bamboo plant for building structures like houses, bridges, and scaffolding, offering a sustainable and cost-effective alternative to traditional materials.

9.5.2 Modern Construction Techniques

The construction industry faces persistent criticism for its inefficiencies and sluggish adoption of innovation. Despite some advancements, the fundamental construction methods, techniques, and technologies have remained largely unchanged since ancient times.

However, implementing innovation in the construction sector is fraught with challenges. Each project is unique, with distinct site conditions, and construction activities involve a complex interplay of personnel, equipment, and environmental factors. Moreover, unpredictable weather conditions and other variables can hinder the effective application of prior experience and knowledge.

The term modern construction techniques cover a wide range of 'advanced construction technology' and practices that encompass the latest developments in materials technology, design procedures, quantity surveying, facilities management, services, structural analysis and design, and management studies.

9.5.2.1 3D printing

3D printing, or additive manufacturing, is the construction of a three-dimensional object from a computer-aided design (CAD) model or a digital 3D model. It can be done in a variety of processes in which the material is deposited, joined, or solidified under computer control, with the material being added together, typically layer by layer. It involves layering materials such as concrete or polymers to create 3D structures. It offers design flexibility, faster construction, and reduced material wastage.

9.5.2.2 Tunnel formwork

Tunnel formwork is a construction method that uses a formwork system to cast walls and slabs simultaneously, particularly effective for repetitive cellular structures like residential blocks, hotels, and student accommodation.

This is a construction method employed to build large-scale structures comprising multiple units, such as hotels, apartment complexes, and office buildings. This technique utilizes reinforced concrete and enables professionals to achieve high-quality results while expediting the construction process on-site. The efficiency of tunnel formwork allows for rapid completion of buildings, making it an ideal choice for projects with tight deadlines.

Tunnel formwork emerged in the mid-20th century, with the introduction of new materials and technologies that enabled the creation of large, complex structures using reinforced concrete. This modern technique has since become popular worldwide, particularly for building large-scale structures like hotels, apartments, and office buildings, where speed, quality, and efficiency are crucial.

9.5.2.3 Slab decks

Slab decks, also known as slab-on-grade or slab-on-deck, represent a modern construction technique. This method involves pouring concrete over a steel deck or a pre-manufactured concrete slab, creating a solid, flat surface. Slab decks have become increasingly popular in modern construction due to their speed of installation, high strength-to-weight ratio, ability to support heavy loads, and resistance to weathering and corrosion. Slab decks are commonly used in various construction projects, including commercial buildings, industrial facilities, parking garages, and high-rise buildings. Their modernity lies in the use of advanced materials, such as steel decking and specialized concrete mixes, which have improved the efficiency, durability, and sustainability of this construction technique. See case study 5.1.

9.5.2.4 Cladding for buildings

Cladding refers to components that are attached to the primary structure of a building to form nonstructural, external surfaces. It is basically the application of one material over another to provide a skin or layer. This is as opposed to buildings in which the external surfaces are formed by structural elements, such as masonry walls, or applied surfaces such as render. In construction, cladding is used to provide a degree of thermal insulation and weather resistance and to improve the appearance of buildings.

9.5.2.5 Computer numerical control technology

Computer numerical control (CNC), also known as numerical control or computational numerical control, is the digital manipulation of machines such as drills, lathes, and other machine tools by computers and circuitry. It's a technology used to control machine tools and manufacturing equipment using computer programs. The process comprises a series of numerical values generated by a computer, each of which is assigned to a desired tool or control position to enable the machining of a blank piece of material to precise specifications without requiring a manual operator. Some applications include CNC milling, turning, and grinding.

9.5.2.6 IoT (Internet of Things) and smart construction

Using sensors and data analytics to monitor construction progress, optimize resource usage, and improve safety.

9.5.2.7 Augmented reality-assisted building

Augmented reality technology helps guide construction workers and improve the accuracy by visualizing designs through overlaying digital models onto real-world sites. It also assists in guiding workers through step-by-step instructions and real-time feedback.

9.5.2.8 Modular construction

Modular construction entails constructing entire building modules off-site and transporting them for assembly [5]. These modules are typically self-contained units that can be stacked or connected to form larger structures. Modular construction offers advantages such as rapid construction, cost savings, and quality control. It is suitable for various building types, including residential, hospitality, and healthcare facilities. It involves constructing entire building modules off-site and transporting them for assembly; is suitable for residential, commercial, and institutional buildings; and offers rapid construction, cost savings, and quality control.

9.5.2.9 Prefabrication

Prefabrication involves manufacturing building components off-site and assembling them on-site and constructing components off-site and assembling them on-site. These components can include walls, floors, and roofs, which are produced in controlled factory conditions. Prefabrication offers benefits such as reduced construction time, lower labor costs, and minimized material waste. It is commonly used in residential, commercial, and industrial construction projects.

9.5.2.10 Sustainable construction

Sustainable construction focuses on minimizing environmental impacts and conserving resources throughout the building lifecycle. It involves incorporating renewable materials, energy-efficient systems, and green technologies into the design and construction process. Sustainable construction practices aim to reduce carbon emissions, conserve water, and enhance indoor environmental quality. Examples include passive solar design, green roofs, and rainwater harvesting systems.

9.5.2.11 Cross-laminated timber

Cross-laminated timber (CLT) is an engineered wood product prepared by gluing together layers of solid-sawn lumber in alternating directions. CLT panels are lightweight, strong, and fire-resistant, making them suitable for structural applications in building construction. CLT offers advantages such as design flexibility, rapid installation, and sustainability. It is used for walls, floors, and roofs in residential, commercial, and institutional buildings.

9.5.2.12 Robotics and automation

Robots are used for tasks such as welding, bricklaying, and concrete pouring. Robotics and automation are used in construction to automate repetitive tasks and improve productivity. This technology includes robotic bricklaying, automated rebar bending, and drone-assisted site monitoring. Robotics and automation enhance construction efficiency, accuracy, and safety by reducing manual labor and human error. They are employed in various construction activities, such as site preparation, assembly, and quality control.

9.6 NEED FOR APPROPRIATE TOOLS TO PIECE A PROJECT THROUGH DESIGN AND CONSTRUCTION

As seen in Section 9.5, appropriate construction techniques are necessary for each project, and these come along with associated tools. Whether design or construction, the project would require appropriate tools, not simply going by whatever tool is available.

For example, imagine using the wrong power tool for a task during construction. It can pose hazards, creating mechanical and electrical stress that causes the equipment to become unsafe. The power tool you use should be suitable for the job.

Appropriate tools and technologies are necessary to successfully integrate and manage the various stages of a civil engineering project, from design to construction. Aerial photography, mapping, inspection, monitoring, and data collection all need appropriate tools. Some of the tools used in construction include the following:

- **Hand tools:** hammers, tape measures, levels, pliers, screwdrivers, wrenches, utility knives, drills, and many more.
- **Power tools:** circular saws, reciprocating saws, drill presses, impact drivers, rotary hammers, sanders, grinders, concrete mixers, and many more.
- **Heavy equipment:** cranes, excavators, bulldozers, backhoes, dump trucks, concrete pumps, asphalt pavers, road rollers, and many more.
- **Concrete and masonry tools:** concrete mixers, batch box, trowels, floats, formwork, edgers, masonry saws, and concrete finishing tools, among others.
- **Safety equipment:** hard hats, safety glasses, earplugs, respirators, steel-toed boots, and fall protection gear, among others.
- **Access equipment:** scaffolds, ladders, access towers, telescopic handlers, and scissor lifts, among others.
- **Surveying tools:** theodolites, levels, transit levels, GPS devices, and laser levels, among others.

Some of the tools used in civil engineering design include the following:

- **Surveying tools:** theodolites, levels, total station, drones, cameras, transit levels, GPS devices (such as Real-Time Kinematic (RTK)), and laser levels, among others.
- **Software:** Autodesk Civil 3D, AutoCAD, Revit, Bentley MicroStation, STAAD.Pro, ETABS, Safe, Navisworks, ArcGIS, QGIS, EPANET, and many more.
  - **CAD tools:** CAD drafting software, 3D modeling software, Building Information Modeling (BIM) software, and many more.
  - **Other tools:** Microsoft Office (Excel, Word, and PowerPoint), Project management software (Asana, Trello, and MS Project), Collaboration tools (Slack, Teams, and SharePoint), and many more.

9.6.1 Quality Control

Have you ever imagined the quality of concrete you would get mixing concrete constituents without a batch box? Or the quality of concrete finish you would get without proper formwork? The quality of concrete finish you obtain using steel molds or boards is definitely not the same as that using well-trimmed timber. There are always variations. Depending on the type of concrete specification you need, you must select the right tools to use through the mixing, casting, and curing stages, as specified in the applicable standard.

What about design and drafting complex civil engineering structures without software technologies such as AutoCAD, Civil 3D, and/or ArchiCAD?

Checklists are powerful tools used in construction to control the quality of output at different stages in the construction process. Supervisors heavily rely on checklists to monitor and supervise the works! For example, concrete production requires use of checklists, right from mixing, casting, and curing. Steelwork, formwork, and scaffolding also require checklists. Ensure that everything is properly checked before progressing to the next stage. This also enhances safety protocols.

To build reliability in civil engineering structures, engineers develop checklists to pre-check that the preceding stage is properly done to the satisfaction of the engineer prior to commencing the next phase/stage in the construction process.

9.6.2 Smooth Project Workflow

Appropriate tools support the seamless coordination, effective communication, and efficient problem-solving that are essential for successful project management.

Appropriate tools such as software applications ensure that all team members, stakeholders, and processes work together smoothly, without obstacles or miscommunication. Effective communication enables concise and timely exchange of information among team members, stakeholders, and clients.

Through interworked computer networks, an efficient problem-solving atmosphere is created, providing the means to identify, analyze, and resolve problems quickly, minimizing delays, and improving overall productivity.

9.6.3 Timely Completion

The right construction or design tools minimize delays and thus optimize the costs. Overall, timely completion is essential for delivering projects efficiently, effectively, and to the satisfaction of all stakeholders. The right construction tools help achieve timely completion in several ways:

- **Increased efficiency:** Specialized tools enable workers to perform tasks faster and more accurately. Efficiency refers to the ability to achieve a desired outcome or result with a minimum amount of effort, time, or resources.
- **Improved productivity:** The right tools reduce manual labor, allowing workers to focus on higher-value tasks. **Talk about production rates.** For example, a dozer production rates, or the amount of material moved per hour, can vary significantly based on factors like dozer size, terrain, material type, and operator skill, but generally range from 477 to 642 m³/h. This cannot in anyway be compared with manual labor.
- **Reduced errors:** Accurate tools minimize mistakes, rework, and delays, and this leads to timely completion.
- **Enhanced safety:** Proper tools ensure a safer working environment, reducing accidents and downtime.
- **Streamlined processes**: The right tools facilitate smooth workflows, eliminating bottlenecks and delays.
- **Better time management:** With the right tools, workers can manage their time more effectively, prioritizing tasks and meeting deadlines.

- **Meets client expectations**: Clients expect projects to be completed on schedule, and delays can lead to dissatisfaction.

9.6.4 Minimizing Errors

Having the right tools for the task minimizes errors. Minimizing errors in construction is crucial because errors can lead to costly rework, delays, and budget overruns. They can compromise the safety and quality of the building or infrastructure. Errors can damage the reputation of construction companies and professionals. They can also lead to legal disputes and financial losses. So, having the right tools is essential to achieve the desired quality at a reasonable cost and avoid all these messes.

9.7 THE IMPORTANCE OF MANAGING PROCEDURES IN CONSTRUCTION

A procedure is a set of instructions or guidelines that outline the actions to be taken, the methods to be used, and the resources required to achieve a particular goal. It is usually an established or official way of doing something. It includes a series of steps or actions taken in a specific order to achieve a particular goal or outcome.

Standards contain the specific rules (MUST DOs) and emphasized recommendations (SHOULD DOs). Guidelines and procedures are based on the associated standards and provide a context as to how to implement a given standard.

Codes and standards provide a framework for designing, operating, and maintaining systems, equipment, and processes safely. They are not step-by-step procedures, but rather guidelines and requirements that procedures should follow. However, some standards provide procedural guidance. In construction and design of civil engineering structures, standards and codes define the guidelines, and sometimes provide procedural guidance to use for any given task.

For example:

- **BS 8500** sets out five standard ways of specifying concrete. These five methods provide flexibility and clarity in specifying concrete for various construction projects. BS 8500-1: Concrete— Complementary British Standard to BS EN 2061 is the method of specifying concrete in the United Kingdom. It was first published in 2003 to align the specification, production, and conformity of fresh concrete with **BS EN 2062**. This guide has been revised to coincide with the latest amendment of BS 8500, published in 2023. The most recent amendments are predominately concerned with increasing the range of cementitious materials, in particular the use of multicomponent cements, which have increased the range of lower carbon concretes now available [6].

- It's worth noting that BS 3666 provided guidelines for mixing concrete (production) to achieve a desired type or quality but has since been withdrawn and replaced by BS EN 206 and BS 8500.

In the example given, the right procedure for mixing concrete (production) is properly outlined in BS 8500 for all five types of concrete specification, that is, designated mix, standardized prescribed mix, designed mix, prescribed mix, and proprietary mix. If you are a designer/specifier of concrete type, you use the standard, and at the same time, the constructor must adhere to the specification and complies with the specifics of the concrete type required. The standard sets out the 'how to' achieve the desired quality/type of concrete specification.

So, while BS 8500 is a standard, it does contain procedural guidance for specifying and producing concrete. BS 8500 is a standard, but it provides a procedure for specifying and producing concrete. In this case, BS 8500 is a standard that outlines a step-by-step procedure for specifying concrete, producing concrete, and testing concrete.

Note:

- A standard defines the tools, means, and methods for how to do something, and a procedure defines how to do these things.
- Standards can define procedures.
- Procedures can be outlined in standards.
- Not all standards are procedures, but some standards do contain procedural guidance.

Imagine assembling a cuboidal reservoir tank made out of steel panels. You must follow a procedure, from dwarf wall or tower construction, to tank assembling, itself. Sequence the activities chronologically as per the applicable standard procedure. Imagine a set of materials for a cuboid reservoir tank accessory, that includes rods, tank panels, bolts and nuts, and gaskets from the manufacturer, and you are told to assemble the tank. The first thing to ask for is the installation manual from the manufacturer, which recommends a standard method to do the work, including the safety protocols. In other activities, you may have to

- Prepare an inventory of materials you have.
- Check the standard installation procedure as recommended by the supervising engineer.
- Identify missing parts, if any.
- Check or draw the work plan.
- Mobilize the workers, tradesmen, and engineers/technicians.
- Organize construction drawings.

Proper procedural management reduces costs, minimizes waste, reduces rework, and optimizes resource allocation, leading to cost savings. You need to understand that correcting reliability problems is very expensive in civil engineering structures. So, procedural management is crucial. Efficient procedures make working environment safe, build reliability in the civil construction works, and support meeting the stakeholder goals.

Correcting reliability problems is very costly in civil engineering structures unlike simple electrical and mechanical devices. This is why procedural management is essential in constructing civil engineering structures. Don't risk, plan all the way to the end, and take the series of necessary steps or actions in a specific and right order to achieve a particular goal or outcome as required.

In systems that require installation, follow the installation guide, appropriate codes, and standards that outline the procedures. These include lifts installation, HAVC, tanks, and mechanical and electrical devices that enhance utilities and amenities for the civil engineering structures. The need to follow codes and standards cannot be overemphasized in this regard. These outline the guidelines necessary from which procedures follow.

Managing procedures in construction is important because it ensures a structured approach to projects, leading to improved efficiency, quality control, safety, and overall project success by clearly defining tasks, responsibilities, timelines, and quality standards, minimizing risks and confusion, waste, and potential delays while maximizing productivity and adherence to regulations across all stages of the construction process.

9.7.1 Improved Safety

We often solve health and safety problems by requiring workers to follow rules of procedure, for example, a construction worker wearing a hard hat. Managing and controlling hazards in construction activities is essential. Employers are tasked to prepare safe systems of work. Well-defined safety procedures minimize risks on construction sites by outlining the proper use of equipment/tools, hazard identification, and emergency response protocols.

A **safe system of work (SSoW)** is a formal procedure (usually written) resulting from an examination of any workplace activity in order to identify its hazards and assess its risks. This exercise culminates in the development and implementation of safe methods of work as a control measure to ensure that the residual risks—those which cannot be eliminated—are minimized. SSoW is a legal requirement for employers and must be put in place whenever a work task involves hazards that cannot be eliminated, thereby posing a risk to the safety of the employee. Examples of a formal SSoW include standard operating procedure (SOP), method statement, and permit-to-work. Examples of an informal SSoW are verbal instructions, list of do's and don'ts, and accepted customs and practices:

- A method statement is a written document that sets out the step-by-step process for completing a task safely, as determined by an SSoW.
- SOP is a detailed guide outlining the safe steps for starting, operating, and shutting down a piece of equipment.
- A permit-to-work is a type of SSoW that allows only certain individuals to carry out a specific task within a defined time frame.

An SSoW is a formal procedure based on a systematic examination of work in order to identify the hazards. It defines safe methods of working that eliminate those hazards or minimize the risks associated with them. We can identify three key elements from this definition of an SSoW:

- The SSoW is formal—documented or recorded in some way.
- It results from a systematic examination of work in order to identify the hazards—it is the result of risk assessment.
- It defines safe methods—it is the safe procedure or work instruction.

So simply put, the employer should carry out a systematic risk assessment, identify the hazards and precautions necessary, and then formally record the safe way to carry out the task taking this all into account.

The specific requirements for designing an SSoW will depend on an organization's needs and industry regulations. The higher risk the activity being performed, the more detailed an SSoW should be. For very low-risk activities, it may be sufficient to distribute an SSoW orally, though it is best practice to have all SSoWs in writing. If a task is particularly high risk, a permit-to-work may be required. A permit-to-work is a type of SSoW that allows only certain individuals to carry out a specific task within a defined time frame. The following work tasks are considered to be high risk and would therefore require a permit-to-work:

- Hot work
- Cold work
- Chemical work

- Working at height
- Working in confined spaces.

SSoW reduces workplace accidents and injuries. By implementing SSoW, employers can significantly reduce the risk of accidents and injuries in the workplace.

SSoWs are, therefore, derived from the risk assessment process. Whether an SSoW is an appropriate control measure depends on the outcome of the risk assessment process and consideration of the "principles of protection" as detailed in schedule 1 of the Management of Health and Safety at Work Regulations 1999.

There should be recognized procedures for the safe conduct of all work activities for an organization. These procedures should cover all foreseeable possibilities, for example, the operation of drilling equipment in different types of weather, rather than just a set of rules which ensure safety when the weather is good. Procedures should cover the routine day-to-day activities of the organization and the non-routine, occasional or one-off activities, as well as any foreseeable emergencies that might arise.

Generally, SSoWs are procedures designed to minimize workplace risks by outlining safe methods for specific tasks, including hazard identification, control measures, and emergency protocols. Examples are operating instructions for machinery, safety checklists for confined spaces, and guides for handling hazardous substances. See Table 9.1.

TABLE 9.1   Examples of safe systems of work

Operating instructions for machinery (e.g., standard operating procedures (SOPs) for an electric hand drill, Figure 9.5)	A detailed guide outlining the safe steps for starting, operating, and shutting down a piece of equipment.
A permit-to-work	A permit-to-work is a type of SSoW that allows only certain individuals to carry out a specific task within a defined time frame. The following work tasks are considered to be high risk and would therefore require a permit-to-work: • Hot work. • Cold work. • Chemical work. • Working at height. • Working in confined spaces.
Safety checklist for confined spaces	A checklist that workers must follow before entering and while working in a confined space, covering tasks like gas testing and ventilation.
Method statement	A method statement is a written document that sets out the step-by-step process for completing a task safely, as determined by an SSoW. It includes risks identified through the risk assessment and the measures that will be taken to manage those risks. A method statement is intended for use by the person carrying out the task, meaning it should clearly and concisely explain how that task is to be carried out.[a]
Guide to handling hazardous substances	A step-by-step guide detailing the safe handling, storage, and disposal of hazardous materials.
Safe work procedures for working at heights	Procedures that outline how to safely access and work at heights, including the use of fall protection equipment.
Lockout/tagout procedures	Procedures to ensure that machinery is safely shut down and locked out before maintenance or repair work can begin.
Emergency evacuation plan	A plan that outlines how to safely evacuate a workplace in the event of an emergency, including designated routes and assembly points.

Note
- All SSoWs are SOPs, but not all SOPs are necessarily SSWs.
- A SOP can be an SSoW if it includes safety considerations and controls.

[a] https://www.highspeedtraining.co.uk/hub/safe-systems-of-work/

Advantages of written SSoW

- SSoW may be recorded in the form of short notes, or perhaps manuals **detailing exactly** what steps need to be taken when carrying out more complex and lengthy procedures, such as calibrating and setting up grinding wheels.
- Written SSoW establishes a standard that can be audited and provide the employer with a written record, which may be required for legal reasons, such as in incident investigations or during enforcement action.
- Documenting SSoW provides a precise reference for all workers and ensures consistency of method, especially as the procedure may be complex or detailed—passing information via "word of mouth" is an unreliable method of communication and prone to errors.
- SSoW documentation can be accompanied by checklists (tool) for employees to use as aids to ensure that all the correct steps are taken and to tick off details before continuing with the next step or starting operations.
- Written SSoW provides a reference for use in training and instruction in safe procedures and, as most procedures will need to be consulted on more than one occasion, the creation of a definitive document is a way of ensuring consistency whenever the process is carried out.

9.7.1.1 SOP for an electric hand drill

Figure 9.5 is an electric hand drill, a portable power tool used for drilling holes in various materials such as wood, metal, plastic, or walls. It drives screws and other fasteners. Typical features include an electric motor, adjustable speed, keyless chuck, ergonomic handle, and trigger switch.

A sample SOP for an electric hand drill is provided in Appendix 9.1.

FIGURE 9.5 Electric hand drill.

9.7.2 Minimizes Reworks

Efficient procedures minimize waste, reduce rework, and optimize resource allocation, leading to cost savings. Without proper procedures, reworking tasks or double works are inevitable.

Note that rework

- Can generate waste, including materials, energy, and resources. Minimizing rework helps reduce waste and promote sustainability.
- Can cause delays, impacting project timelines and schedules. Minimizing rework helps ensure that projects are completed on time.
- Can be expensive as it requires additional resources, labor, and materials. Minimizing rework helps reduce these costs.

9.7.3 Quality Control

Rules of procedure do not only apply in courts or legislative assemblies. No, not really; they apply in engineering and construction, too. Procedures or protocols help achieve consistency and efficiency and compliance with regulations and standards. This is because procedures ensure that tasks are performed in a standardized way, reducing variability and errors, thus leading to improved/better quality.

Quality control (QC) is a process through which a business seeks to ensure that product quality is maintained or improved. Quality control requires the company to create an environment where management and employees strive for perfection. Achieving consistency and efficiency leads to both improved quality and increased productivity. See how concrete quality is achieved in the example provided above (Section 9.5)—BS 8500. In addition:

- **ISO 19011:2018:** This standard provides guidelines for auditing management systems, including quality management systems. It outlines procedures for conducting audits, including planning, preparation, and reporting. ISO 19011 is an international standard that provides guidelines for auditing management systems, including quality management systems (ISO 9001) and environmental management systems (ISO 14001). It outlines the principles of auditing, managing audit programs, and conducting management system audits.
- **ISO/IEC 20000-1:2018:** This standard provides requirements for IT service management, including procedures for service delivery, control, and improvement.

Note: ISO 9001:2015 Quality management systems—Requirements is a globally recognized standard for quality management. It helps organizations of all sizes and sectors to improve their performance, meet customer expectations, and demonstrate their commitment to quality. Its requirements define how to establish, implement, maintain, and continually improve a quality management system (QMS) [7].

Implementing ISO 9001 means your organization has put in place effective processes and trained staff to deliver flawless products or services time after time. Quality control procedures for inspecting and testing materials and built structures are essentially vital, meeting the expectations of stakeholders on the civil engineering project. They help verify that materials conform to project requirements, industry standards, and regulatory requirements and that structures are built correctly—meeting design specifications, safety standards, and quality expectations.

Procedures help detect defects or non-conformities early and take corrective action or remedial actions to prevent further issues. Procedures clearly outline roles, responsibilities, and expectations, eliminating confusion and miscommunication. Procedures enable repeatable processes, ensuring that tasks are performed consistently, even by different personnel. This improves the quality of output.

9.7.4 Cost Control

This is essentially obvious. If you do things haphazardly, without well-defined protocols, the cost escalates! Effective procedures minimize errors by providing clear instructions and checks, reducing rework and corrective actions. Time saving: effective procedures save time by providing a clear, step-by-step approach, enabling personnel to complete tasks quickly and accurately. If you are not chronologically following activities as they are supposed to be aligned (as per written procedures), you may not get it cost-effectively. Never. Thus, you need the following:

- **Streamline processes:** effective procedures eliminate unnecessary steps, reduce waste, and increase productivity.
- **Improve decision-making:** effective procedures provide clear guidelines, reducing the need for repetitive decision-making and minimizing delays. With effective procedures, Fayol's principle of centralization/decentralization can be applied effectively. Henri Fayol recommended an appropriate balance of the two (centralization/decentralization) depending on the size, nature of work, situation, and weight of decision—see Section 2.5.4.1.8.
- **Improve communication**: Effective procedures facilitate clear communication among team members, stakeholders, and clients, ensuring everyone is on the same page.

9.7.5 Builds Reliability in the Project

A procedure being a series of steps or actions taken in a specific order to achieve a particular goal or outcome helps build reliability in construction projects—through failure reduction—by reducing failures in process, reliability is built.

9.7.6 Minimizes Risks

Managing procedures identifies potential risks and mitigates them, minimizing the likelihood of delays, cost overruns, and disputes. By proactively identifying potential issues through standardized procedures, project teams can implement preventative measures to manage risks.

9.8 1D, 2D, AND 3D MODELS IN CIVIL ENGINEERING

Civil engineering DAE of structures can be done by either developing or adopting 1D, 2D, or 3D models. As outlined in several sections earlier, developing sketches and models, as discussed in Sections 9.2.7 and 9.2.8, respectively, is essential—these sketches and/or models can be 1D, 2D, or 3D. This entirely depends on the level of simplification required and the complexity of the problem. Simple linear systems, planar or axisymmetric systems, and non-uniform systems, all call for different approaches to achieve more accurate design and analysis.

- **1D represents a single dimension:** length, typically used for simple, linear systems.
- **2D represents two dimensions:** length and width, or length and depth, typically used for planar or axisymmetric systems.
- **3D represents all three dimensions:** length, width, and depth, typically used for complex, non-uniform systems.

Therefore, civil engineers can decide to work their problems in different modes, either 1D, 2D, or 3D. It depends on the expertise, skill, knowledge, available resources, and time to model the civil engineering problem.

9.8.1 1D Analysis

The examples are beam analysis, column design, and pipe flow simulations. For 1D water pipeline design, St. Venant equations, energy equations, and Bernoulli's equation are considered.

For example, designing a drainage channel and applying energy and momentum equations are usually done in 1D analysis; see the conceptual diagram in Figure 9.6. Advantages of adopting a 1D model are that it is simple, fast, and easy to understand. However, it has some limitations such as oversimplifying complex systems and neglecting interactions between dimensions.

x

A

B

FIGURE 9.6   1D analysis.

9.8.1.1 Example 9.1, 1D

Compute the head loss due to pipe friction and the power required to maintain flow in a circular pipe of 40 mm diameter and 750 m laid horizontal when water flows at a rate 30 L/min. Take dynamic viscosity of water equal to $1.14 \times 4 \times 10^{-3}$ N s/m^2. Assume that for the pipe absolute roughness, k is 0.00008 m.

Answer:

- Head loss = 4.89 m of water
- Power = 24 W

This is a 1D problem because the key characteristics being analyzed are primarily dependent on a single spatial dimension, meaning the system can be effectively represented and analyzed along a single line or axis—in that case the length of the pipeline.

9.8.2 2D Analysis

These 2D models are applied in beam design (statically determinate structures). A 2D planar structure is a type of structural system that lies in a single plane (x-y plane), has no significant thickness in the third dimension (z-axis), and resists the loads applied in the plane (in-plane loads). See Figure 9.7. The advantages of adopting a 2D model include the following: being more accurate than 1D models and can capture some interactions between dimensions. However, it poses some limitations due to oversimplifying complex systems and neglecting some interactions between dimensions.

9.8.2.1 Example 9.2, 2D

See Example 5.6 and an example in Section 8.5.4.2. In Section 8.5.4.2, the reaction forces and corresponding shear force diagram (SFD) are categorized as a 2D problem because the key characteristics being analyzed are primarily dependent on two spatial dimensions, meaning the system can be effectively represented and analyzed along the x and y axes—in that case the length of the beam in the horizontal direction and forces acting on it in the vertical direction.

9.8.3 3D Analysis

3D analysis represents all three dimensions, that is, length, width, and depth, typically used for complex, non-uniform systems. See Figure 9.8. It captures complex interactions between dimensions, materials, and systems. Some of the examples which adopt the 3D models include building simulations, bridge analysis, steel towers, and soil–structure interactions. The advantages of adopting 3D models include being the most

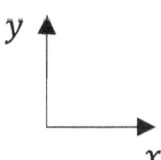

FIGURE 9.7   Single plane (x-y plane).

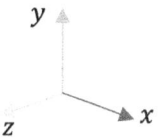

FIGURE 9.8   3D analysis represents all three dimensions in x, y, and z.

accurate and can capture complex behaviors and interactions between dimensions. However, it has some limitations—it is computationally intensive and requires significant resources and expertise.

9.8.3.1 Example 9.3, 3D

See Example 8.1. This example analyzes the 3D bar in 3D, that is, x, y, and z spaces.

- **Note:** In BIM, 4D adds the time dimension to the 3D model, while 5D integrates cost data.

9.9 CASE STUDY [9.3]: WATER RESERVOIR TANK PLANNING AND DESIGN, KISORO UG

This community solar-powered water supply scheme developed in Kisoro UG, constructed in a terrain shown in Figure 9.1, formed part of a broader US$ 35 million water supply project, and had several small-to-medium reservoir tanks. Besides being solar-powered, the project employed local subcontractors and suppliers, contributing to broader sustainability goals, particularly in alignment with UNSDG 8. The small-sized reservoir tank problem presented here is intended to provide an illustrative example bringing out several civil engineering branches in a single unit to foster understanding—and again to illustrate 'how to think like a civil engineer' taking a number of factors or constraints into context to design or develop a cuboidal reservoir water tank seated on dwarf walls or steel tower, whatsoever. It has elements of water resource management and hydraulics, structural system analysis, geotechnical analysis, engineering software applications, and project management.

In addition, this overhead water reservoir tank problem is particularly selected because it has a component of hydraulics (particularly, water pipeline design), the structural design of structural members, that is, the beams and columns; geotechnical investigations, that is, soil and rock testing. Therefore, it is an ideal case to learn about the application of 1D, 2D, and 3D models in civil engineering.

It shares the software you may need to model the tank and corresponding structural system to carry the tank, that is, dwarf walls or tower. It is quite insightful for a civil engineer in the making or training to appreciate how to approach a civil engineering problem.

This 90 m^3 reservoir tank problem described here for illustrative purposes, indicating holistic independent judgment at play, is concerned with a rectangular (cuboidal) water tank—a three-dimensional solid tank—with six rectangular faces (Figures 9.9 and 9.10) developed in Kisoro UG—in its developmental phase.

FIGURE 9.9   Isometric view of the cuboidal water tank.

FIGURE 9.10   Front view of the cuboidal water tank.

In Section 9.2.1, understanding the context, we shared a picture of a mountainous terrain that poses several constraints for a civil engineering project. This portable water supply scheme developed in the area (Figure 9.1) needed stage pumping depending on the types and sizes of the pump being influenced by the financial envelope. Staged pumping comes with booster stations depending on pump sizes and types specified and feasibility studies conducted alongside volume of water present for abstraction or extraction and to be pumped to the final reservoir or point of use.

The socioeconomic feasibilities informed the technology to adopt. The project must be feasible and viable, which means the specified materials and technologies must be commensurate with the project outputs and outcomes, avoiding unnecessary redundancy or high-tech materials/systems that may increase costs unreasonably.

Kisoro, UG, experiences a temperate climate with cool temperatures due to its high altitude. Daytime temperatures typically range from 20°C to 22°C (68°F to 72°F), while nighttime temperatures drop to around 10°C–12°C (50°F–54°F).

Climatological data collection was inevitable considering that the effect of temperature and environmental conditions on concrete works and other materials generally must not be forgotten in the planning phase! See Chapter 5.

Kisoro is a mountainous region with elevations ranging from 1,200 to 4,127 m above sea level. It is part of the Virunga Mountains, a chain of volcanoes that extend into Rwanda and the Democratic Republic of Congo (DRC). The designers bore in mind the idea to develop cuboidal tanks of pressed steel panels due to their simplicity, practicality, and cost-effectiveness.

This overhead water reservoir tank is provided as the focal structure for analysis because the reliability of the scheme centers around it. The storage capacity must be correctly determined, and the location must be feasible to distribute water to the intended community. All the survey work conducted ensured that the most appropriate tank location was selected, enabling all intended beneficiaries to receive water. This careful planning avoided a situation where some individuals would be left out due to inaccurate surveys! Both the transmission and distribution networks must perform reliably, and therefore, the topographical survey is an essential component in the planning of a reliable scheme. The hydraulic models always use the reservoir tank as a crucial node that greatly influences the output of the model. The elevation at which it is located provides the necessary heads (static and dynamic) for both the distribution and transmission networks, and the subsequent power required to have water entering and exiting the tank.

The primary factor that influenced the choice of structure to lift the reservoir tank, either dwarf walls or tower, was the required elevation/head. This is obtained particularly from topographical surveys. The Kisoro UG location has enough elevation. Other factors such as security required, wind loads, seismic loads, space constraints, cost, tank size, and material availability in the region became secondary. Wind loads, though, when it comes to tower design, are paramount—this needs not to be forgotten and must be able to allow wind to pass through the diagonal bracings effectively but not deflecting it—thus the design must be robust. However, the primary factor remains elevation, and all other factors must be customized to achieve a workable design that can distribute water to people because that's the primary goal of the water supply scheme.

It is essential to note that water is pumped from the intake/source or treatment plant or (may be a gravity system, and there is no need for pumping) to the reservoir and then distributed by gravity to end-users. However, in this particular case, water was pumped—with two-stage pumping (one as a booster station)—due to the mountainous area, and feasibility issues dictated two-stage pumping. Energy equations were applied to find the static and dynamic heads along the transmission pipelines. Elsewhere, water flew by gravity in the distribution networks:

- The pipeline design adopted a 1D model, which means the design and analysis are represented in a single dimension or axis, in this case length of the pipeline (x-axis).
- The slab bed/beam and dwarf wall design adopt a 2D model, which means that the design and analysis are represented in two dimensions (x-y plane or elevation)—designed in 2D planar geometry. In case

it was a tower, the frame analysis (structural analysis) would adopt a 3D model, as appropriate, which means the design and analysis are represented in three dimensions (x-y-z space).
- Tank modeling adopted a 3D approach. **Note:** if required, the entire geometry may be analyzed in 3D. Dwarf walls, rectangular sections for the bed, and the reservoir tank.

Dwarf walls are suitable where you have enough elevation, and water would be able to flow at a required pressure in the pipeline. Particularly, for gravity-fed systems (gravity flow water schemes) where the water flows downhill to the point of use, dwarf walls are used—because the primary source of energy is gravity and the water intake is located upstream, typically in the hills. Towers, on the other hand, are used when you need more elevation/head to distribute water and water cannot flow in the pipelines by gravity at a pressure required (including service/intensification lines) when the enough head is not achieved. This is the first step to propose that you are now thinking like a civil engineer on this reservoir tank problem! In this particular case, we have enough elevation. This area is mountainous! So, dwarf walls were used.

After this preliminary decision to adopt a tower or dwarf walls, sketch development was commenced to determine how the system would look like, following site constraints and complexity! See sketches in Figures 9.11 and 9.12. Figure 9.11 shows an opened reservoir tank side. This helped not to forget any essential components and accessories on the system, for example, pipes, fittings, washouts, inlet, outlet, breather, overflow, water level indicator, internal and external ladders, tank cover, and inspection chambers. The orientations of these on the tank were assessed during site visits and during surveys. A sectional sketch or a cut-through view of the tank may be developed as well to show internal details. Site visits also evaluated the challenges, constraints, and complexity, thus coming up with proper site orientation, leveling requirements, and excavation depths. All these must be captured in the sketches, especially the terrain.

- HDG water tank, 7.32 (L)×3.66 (W)×3.66m(H)
- D: Bottom panel
- A: 1st panel
- B: 2nd panel
- C: 3rd panel
- G: Roof panel

Note: The panel thicknesses vary by the position of the HDG panel. Bottom panels are thicker than the rest, and the roof panels are the thinnest. It is quite logical because the amount of hydrostatic pressure exerted on each panel varies by position. These panels come in sizes of 1.22 m by 1.2 m, as per BS 1564:1975 (specification for pressed steel sectional rectangular tanks), which influences the design of the whole reservoir tank system right from the foundation.

The sketches helped the engineer visualize the forces and ensure that safety considerations are taken into account to prevent accidents and ensure the structure's integrity. In Chapter 4, we saw forces in civil engineering. The engineer can identify the forces acting on each element and isolate a specific component, such as a side or floor panel, to draw its

			C1	C2	C2	C2	C2	C1							
			B1	B2	B2	B2	B2	B1							
			A1	A2	A2	A2	A2	A1							
C1	B1	A1	D1	D2	D2	D2	D2	D1	A1	B1	C1	G	G	G	G	G	G
C2	B2	A2	D2	D3	D3	D3	D3	D2	A2	B2	C2	G	G	G	G	G	G
C1	B1	A1	D1	D2	D2	D2	D2	D1	A1	B1	C1	G	G	G	G	G	G
			A1	A2	A2	A2	A2	A1							
			B1	B2	B2	B2	B2	B1							
			C1	C2	C2	C2	C2	C1							

FIGURE 9.11   Opened reservoir tank sides.

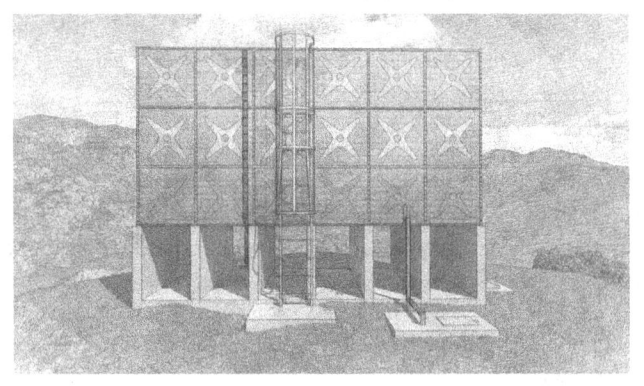

FIGURE 9.12   (a–e) Sketch of a cuboidal reservoir tank on dwarf walls.

free-body diagram (FBD), if required. The designer considers several forces that may be involved in the system. These include seismic forces, wind loads, compression forces, hydrostatic forces, tension, gravitational force, and geotechnical forces. In Chapter 3, we saw how the codes and standards of practice influence engineering choices. In other words, they tie the engineer in a specific room to do things lawfully and up to the required standard. In this particular case, the codes and standards for geotechnical investigations, steel reservoir tanks, seismic analysis, and concrete designs, among others, were interplayed. In particular, energy equations couldn't be escaped in pipeline design, and empirical equations were very helpful in structural and geotechnical analyses.

In terms of safety considerations, the engineers evaluated the structural safety, choosing materials that could withstand water pressure and external loads, such as steel and concrete—of course, these needed no vote considering several constraints surrounding cost and material availability. Second, by designing walls and floors that could resist bending moments, shear forces, and earth pressure, the engineers were sure that safety considerations were taken into account. However, they did not stop at that; pipes that could handle the maximum flow rates and prevent overfilling were considered alongside pressure relief devices such as safety valves, thus preventing overpressure and tank rupture. And survey results were correctly used in deciding pipeline routes and sizes—and also identifying where we need air-release valves, washouts, and minimize negative pressures as much as possible. Through the sketches, engineers were able to remind themselves of the need to provide adequate ventilation to prevent vacuum formation during water withdrawal. This component is very important. Providing an effective drainage system around the tank was another safety consideration so that flooding is never spotted around the tank. Perimeter fencing, though not included in the sketch, was another observation to enhance security at the station. Firefighting devices and fire-resistant materials were equally considered, among other systems/factors.

The height of dwarf walls from the base was proposed = 1 m. Is 1 meter feasible? That was the question to answer—from both the hydraulic models and structural analysis. To provide a glimpse of what would happen in case of a tower, say 12 m high, arriving at a 12 m high tower or 1 m high dwarf walls, you must have done hydraulics and determined the required pressures to distribute water including service connections (intensification lines), ensuring that negative pressures are minimized or eliminated completely from the system. The topographical survey conducted for the scheme provided sufficient details about what the terrain warranted and of course provided the priority communities that needed to receive the water. Their elevations were recorded and modeled in the hydraulics modeling platform with the tank sited on 1 m high dwarf walls—of course taking into consideration the required public tap stands (PSPs) and required flow rates—this must not be forgotten. When all this was done, the preliminary evaluation process was well executed, with due consideration given to the system's design life typically 25 years for solar-powered water supply schemes, as specified in the UG manuals and standards.

Selecting the HDG water tank 7.32×3.66×3.66 m (H) due to its corrosion-resistant properties enhanced the reliability of the system. We notice that the manufacturer manufactures tank panels of a certain standard, 1.22 by 1.22 m. Each HDG pressed steel panel measures 1.22 m (height) by 1.22 m (width) according to BS 1564:1975 (specification for pressed steel sectional rectangular tanks). Therefore, a tank of 98.06 m^3 would measure (7.32×3.66×3.66 m) with the panel thickness ranging between 4 and 6 mm, that is, side, roof, bottom panels, and FOS = 1.09. The required yield strength achieved from the calculations must be rounded off to the nearest standard thickness for each respective face of the tank and then dimensions that fall within the 1.22 factor computed. However, appropriate standards of the required steel grade were complied with to ensure that the output rhymes well. We can now imagine 1.22, a coefficient of sorts in finding the optimal tank dimensions. See the role of coefficients (Section 7.6). For a panel size of 1.22 m × 1.22 m, the recommended thickness for a tank face depends on the specific design requirements, including the tank's height, material, and intended use.

The DAE process for the reservoir tank design

Step 1: Data collection and preliminary design

The first step is the determination of the required storage capacity. This is determined following standard manual recommendations, feasibility studies, and available funds—taking into consideration the balancing volume/storage required and special water needs. Therefore, appropriate data are collected. Then, decide on the height of the tank based on water pressure requirements. This a key consideration; in some manuals, a height of more than 5 m is not recommended in some jurisdictions. As far as the location and soil investigations are concerned, the bearing capacity of the soil for foundation design is assessed. The foundation type is key—that is, adopting pad, raft, or strip foundation. Other data considerations include the wind and seismic data considerations—checking local codes for earthquake and wind load requirements.

Step 2: Selection of tank type

Geometry of the tank matters, whether circular or rectangular tank (depends on space constraints and cost). A cuboidal tank was proposed because of being easy to construct, requiring only straight cuts and welds, while circular/cylindrical tanks demand curved cuts and extremely complex welding. Ease of transportation, construction, and installation: cuboidal tanks can be simpler to transport because of their lightweight panels and being simpler to build and fit into specific spaces, especially in buildings or areas with limited dimensions. They simplify the manufacturing process for panels and better resistance to external loads such as wind and seismic forces. On the other hand, cylindrical (or round) tanks are actually the most common due to their structural advantages and cost-effectiveness. However, cuboidal tanks may be more cost-effective in some situations. For example, for smaller, less demanding applications, the cost of a cuboidal tank might be lower. In some cases, cuboidal tanks can fit better in corners or against walls, optimizing available space.

Step 3: Load considerations

The loads considered were self-weight of tank, water load (1,000 kg/m^3), live load on roof (as per code requirements), imposed/wind load (as per BS EN 1991-1-1), seismic load (as per US 319: 2003), hydrostatic pressure (due to water), and the soil pressure on foundation.

Step 4: Structural design

Design of tank components.

Standards

- **Reservoir tank design:** In accordance with BS 1564:1975 for rectangular and sectional steel tanks.
  - BS 729: 1971 (for HDG).
- **RC wall design:** In accordance with EN1992-1-1:2004 incorporating corrigendum January 2008 and the UK national annex.
  - **BS EN 1992-1-1+A1** of **Eurocode 2** gives a general basis for the design of structures in plain, reinforced, and prestressed concrete made with normal and lightweight aggregates together with specific rules for buildings.
  - BS EN 1992-1-1:2004+A1:2014 is the UK National Annex to Eurocode 2, which provides the rules for the design of concrete structures.
  - Reinforced Concrete: BS EN 1992-1-1: (2014).
- **Foundation analysis:** In accordance with EN1997-1:2004+A1:2013 incorporating corrigendum February 2009 and the UK National Annex incorporating corrigendum No. 1.
  - Foundations: EN1997-1:2004+A1:2013.
- **Geotechnical investigations:** In accordance with BS 1377 (Methods of test for soils for civil engineering purposes) and BS 5930 (Code of Practice for Site Investigations).
  - BS 5930:2015+A1:2020 is a British Standard that provides a code of practice for ground investigations.
- **Seismic design considerations:** In accordance with the seismic code of practice for structural designs, 2003 (US 319: 2003).

    These standards, among others, worked hand in hand to engineer a reliable/cost-effective design.

    Tools
    - STAAD Pro
    - GIS and mapping tools
    - EPANET

Step 5: Analysis

The cuboidal overhead tank was proposed seated on dwarf walls 1 m above the ground, with HDG steel-pressed panels (thickness ranging from 4 to 6 mm), bolted together with rubber gaskets, connecting rods, and several other accessories that help tighten the tank firmly in position restraining several forces. The combination of panel thicknesses is determined by the depth of the tank. See Table 9.2. At that elevation, the reservoir tank can ably distribute water to the intended community.

The site was predominantly underlain by inorganic clayey gravel and well-graded gravel with clay

TABLE 9.2 Maximum loads per water storage

MAXIMUM LOADS PER WATER STORAGE	
DEPTH OF TANK	LOADS ON WALLS PER FOOT RUN
1.22 m (4 ft)	0.5 tons
2.44 m (8 ft)	1.0 tons
3.66 m (12 ft)	1.5 tons
4.88 m (16 ft)	2.0 tons

of intermediate plasticity. The dynamic cone penetrometer (DCP) test, conducted at depths between 1.0 and 2.2 m, indicated that the allowable bearing capacity of the soil across the site ranged from 127 to 748.7 kPa. The natural moisture content of the soils collected ranges from 13.2% to 17.0% for both trial pits. No water table was encountered in both trial pits and recorded on the test pit logs.

Furthermore, structural analysis and design of the proposed 90 m^3 reservoir tank was conducted. Mathematical techniques were used to compute the necessary parameters—analyzing system components. The 7.32×3.66×3.66 m steel tank is supported on base plates connected on to the dwarf RC walls. The base plates transfer the loads on to the reinforced dwarf walls. The reinforced dwarf walls later transfer the loads through the strip foundation, which in turn dissipates the loading safely into the founding soils.

By selecting the appropriate dimensionality for the model, ensuring that the analysis was accurate, efficient, and effective, 2D analysis was the chief dimension adopted.

Step 6: Evaluation

Evaluation was basically conducted to ascertain the effectiveness of the design. The design team took all reasonable steps to provide sufficient information about the design and construction of the structures, to assist the client and contractor to comply with their duties. In so doing, the output designs were assessed following a checklist to check the quality, accuracy, and completeness of the design. The project's impact on the environment and surrounding communities, safety, and durability of the system were also assessed. The checklist was designed to ensure that the objective or goal of the design is met and that nothing that would actually contribute to the reliability of the scheme was left out. A checklist helped avoid common errors, identify gaps, and ensure consistency in the design. Where gaps were identified, design reviews were conducted.

Simulation and testing were conducted. Simulation and testing tools helped analyze, visualize, and verify the performance and behavior of the design under various conditions and scenarios. For example, by using STAAD Pro FEA features to simulate the stress, strain, and deformation of the tank design, the design output was improved drastically. The stress and strain analysis under various loads (e.g., water pressure, wind, and seismic) was conducted, enabling the evaluation of the structural stability and potential failure modes for the reservoir tank. EPANET software was used to simulate the behavior of water distribution systems, focusing on pipe flow, node pressures, and water quality. These simulation and testing tools supported design optimization and identified potential problems to ensure reliability and safety.

Step 5: Output information to the manufacturer

Finally, the output information supplied to the manufacturer included the following:

- Capacity of the tank with dimensions in multiples of 4ft (1.22 m).
- Location of the proposed installation, site conditions, and accessibility.
- Whether water is provided for testing after the erection of the tank.
- Whether the tank is to be galvanized, that is, whether to use HDG panels or cold-pressed panels.
- Accessories, that is, number of partitions, type of cover and position of manholes, level indicator and height, and internal/external ladder height.

9.10 CONCLUSION

Thinking like a civil engineer is a task. The essential responsibility engineers hold is ensuring that humans and other living beings find the systems they develop safe and comfortable to use. In that regard, exercising professional judgment is a logical process, which requires an engineer to place safety, environmental stewardship, and quality at the forefront.

NOTE

1 NSF NHERI DesignSafe | DesignSafe-CI, https://www.design-safe-ci.org/

REFERENCES

1. Waterhouse, P., & MacDonald Steels, H. (2015). *Successful Professional Reviews for Civil Engineers* (4th ed.). ICE Publishing, London, UK. https://www.emerald.com/books/book/17579/Successful-Professional-Reviews-for-Civil
2. Farnam Street. (n.d.). The engineering mind-set. Retrieved from: https://fs.blog/the-engineering-mind-set/. Accessed 7/July/2025.
3. ICE code of professional conduct. (2017).
4. Ethics. (2008). *Guidelines for Professional Conduct for Civil Engineers*. American Society of Civil Engineers, Virginia, USA.
5. Methods of building construction. Comprehensive fundamental architectural techniques course on methods of construction. Retrieved from: https://www.architecturecourses.org/learn/methods-building-construction. Accessed 7/July/2025.
6. How to design concrete structures using Eurocode 2. BS 8500 for building and civil structures. (2024). The Concrete Centre. 1st Floor, 297 Euston Road. London NW1 3AD. Retrieved from: https://www.concretecentre.com/Resources/Publications/How-to-design-concrete-structures-using-Euroco-(3).aspx
7. ISO 9001:2015 Quality management systems — Requirements. Retrieved from: https://www.iso.org/standard/62085.html

APPENDIX 9.1

DO NOT use this machine unless you have been trained in its safe use and operation

Description of Work:	Using an electric hand drill
	**Potential Hazards:** Exposed moving spindle and cutter. Electrical hazard with the potential to cause harm through entanglement, cutting, impact, abrasion, exposure to noise, projectiles, sharp objects, and friction. **Important Notice:** This SOP is intended as a reference guide to complement M&N Co. Ltd.'s Safety Training. This document by itself does not constitute shop/equipment training.

Personal Protective Equipment (PPE) Required *(Check the box for required PPE):*

Respirator /Dust Mask/ Extraction	Eye Protection	Employ Entanglement Precautions	Cutting Hazard	Hearing Protection	Protective Clothing
☒	☒	☒	☒	☒	☒

Risk and Hazards Prevention:
- Entanglement Hazard—Ensure that the tool is gripped firmly and the operator is prepared to overpower torque. No loose clothing, no jewelry or accessories, and long hair must be contained
- Cutting—Be cautious of moving bits/blades
- Use a dust mask to avoid inhalation of fumes and dusts
- Pinching from work piece shifting use proper work holding strategies during operation

Safe Work Procedure Checklist:

1. **Pre-operation:**
   - Always begin by setting up your work area safely to avoid the following: trip hazard, uncomfortable working conditions, electrical hazards, crowded work spaces (other students), and other dangerous situations
     - Use only in designated area. Use screens or curtains to contain hazards
     - Ensure that the appropriate bit is correctly installed and that it is appropriate for the task and material
     - Work piece—ensure that the workpiece is securely clamped
     - Be prepared to firmly grip the tool when in use
     - Plan your cut to ensure that your blade will not cut anything unintentionally: Work table
     - Identify ON/OFF switch
     - Inspect electrical cord if so equipped

2. **Operation:**
   - Leave the machine unplugged/unpowered until ready for use
   - Ensure all other students are clear of the immediate work area
   - Keep fingers, hands, and cables free of cutters and the surrounding area
   - Check that the bit runs 'true' and does not wobble. Excess vibration must be identified and addressed immediately
   - Do not make adjustments while the machine is plugged in—un-plug to make adjustments
   - Increase working loads gradually. Never abruptly. Use smooth fluid motions. Use caution when drilling though holes
   - Be aware of flying debris, dust, and particulate matter. Aim at debris away from self and other people
   - Turn the drill off when not in use
   - Ensure machines have been isolated from power sources before being cleaned, adjusted, maintained, or repaired
   - Un-plug when not in use

3. **Post-operation:**
   - Remove cutters from the chuck before putting the tool away.
   - Store the cord (if equipped) in such a way that it is not in contact with the blade. Never carry any power tool by its cord.
   - Remove (if equipped) battery and charge if less than 50% power.

Competent Persons (The following persons are authorized to operate, supervise, and test trainees on the equipment/process).

Name:	Title:	Contact Details:

This SOP does not necessarily cover all possible hazards associated with the machine and should be used in conjunction with other references It is designed to be used as an adjunct to teaching Safety Procedures and to act as a reminder to users prior to machine use

Date of last review: _____ **Signature:** _____

Must-Know Terminologies in Civil Engineering

10

10.1 INTRODUCTION

The field of civil engineering is broad. For someone new to the field, they need to familiarize themselves with the terminologies used throughout civil engineering.

Civil engineering deals with everything built around us by applying physical and scientific principles to planning, designing, constructing, operating, and maintaining artificially created and natural environments, the things we take for granted but would find life very hard to live without.

Civil engineers are responsible for designing and constructing roads, power plants, railroads, bridges, canals, aqueducts, tunnels, buildings, airports, dams, water supply and distribution systems, sewage systems, airfields, schools, offices, hospitals, and much more. They are responsible for ensuring the safety, durability, and sustainability of the infrastructure.

The field of civil engineering has a list of must-know terminologies you cannot escape to grasp, especially as a passionate civil engineer.

For those who wish to master the profession, particularly, this glossary is highly valuable. It is a valuable resource to support you build your knowledge and expertise.

These terms cover a wide range of topics and concepts throughout the branches of civil engineering. The glossary runs from A to Z.

10.2 A

10.2.1 Abutment

An abutment is a structure that supports the ends of a bridge or viaduct. It is typically built on a solid foundation, such as rock or a concrete footing. The abutment resists the horizontal forces from the bridge deck and transfers them to the ground below. Abutments are often made of reinforced concrete or masonry.

10.2.2 Analogizing Nature

Analogizing nature means making comparisons or drawing parallels between natural phenomena, processes, or structures and human-made systems, designs, or technologies. It involves identifying similarities or patterns in nature and applying them to solve problems, design solutions, or understand complex systems.

10.2.3 Active Forces

These are also known as applied forces and are forces that are exerted by an external agent or object on another object.

10.2.4 Aquifer

An aquifer is a body of porous rock or sediment saturated with groundwater. This is a layer of permeable rock or soil that stores and transmits water. Groundwater enters an aquifer as precipitation seeps through the soil. It can move through the aquifer and resurface through springs and wells.

10.2.5 Asphalt

Asphalt is a mixture of aggregates, binder, and filler, used for constructing and maintaining roads, parking areas, railway tracks, ports, airport runways, bicycle lanes, sidewalks, and also play- and sport areas. Asphalt is a petroleum-based material used for paving roads, parking lots, and other surfaces. It is made from a mixture of aggregates, such as sand and gravel, and a binder, such as asphalt cement. Asphalt is durable, flexible, and waterproof, making it ideal for harsh climates and heavy traffic. It can be laid hot or cold, depending on the application.

10.2.6 Analytical Model

This is a mathematical model that uses a closed-form solution to represent a system and answer specific questions or design decisions.

10.2.7 ASTM

This is the acronym for the American Society of Testing and Materials. The American Society for Testing and Materials (ASTM) is an organization that develops standards for materials.

DOI: 10.1201/9781003561620-10

10.2.8 ASCE

Acronym for American Society of Civil Engineers.

10.2.9 Ambient Temperature

The temperature of the surrounding environment.

10.2.10 Anchor

A device used to secure a structure or object or component to the ground, thereby enhancing the stability and preventing movement or displacement.

10.2.11 Angle of Repose

The angle of repose is the steepest angle at which a pile of loose, granular material, like sand or soil, can stand without slumping or sliding—the angle at which a material will naturally rest. It's a measure of how stable a material is when piled and is influenced by factors like particle size, shape, and moisture content. The angle of repose is the angle between the sloping side of a heap of particles and the horizontal plane.

10.2.12 Aggregate

An aggregate is a mixture of sand, gravel, crushed stone, or other materials used in concrete, asphalt, and road construction. It provides bulk and strength to the final product. The size and shape of the aggregate can affect the workability, durability, and appearance of the finished product. The aggregate is typically sourced locally to minimize transportation costs.

10.2.13 Artificial Intelligence

Artificial Intelligence (AI) refers to the ability of a digital computer or computer-controlled robot (computational systems) to perform tasks typically or commonly associated with intelligent beings, such as learning, reasoning, problem-solving, perception, and decision-making.

10.2.14 Architect

An architect is a licensed design expert who is trained in the art and science of designing buildings or other spaces. An architect plans, designs, and oversees the construction of buildings and structures and other spaces. Architects play a significant role in shaping the built environment, from residential homes and commercial buildings to public spaces and urban landscapes.

10.3 B

10.3.1 BSI

Acronym for British Standards Institution.

10.3.2 BS

Acronym for British Standard.

10.3.3 Brittleness

The property of breaking of a material with little permanent distortion. Brittleness of the material has opposite characteristics to ductility. Brittle materials withstand compression load, and when subjected to tensile loads, they snap off without giving any sensible elongation. Cast iron is the brittle material.

10.3.4 Blend

In the context of materials science, a blend is a mixture of two or more components often with the goal of achieving a specific property or performance. It is a mixture of two or more substances, but not necessarily at the molecular level.

10.3.5 Beam

A beam is a horizontal structural element that supports loads by bending. It is typically made of wood, steel, or reinforced concrete. Beams can be designed to resist different types of loads, such as bending, shear, or torsion. They are often used in building construction to support floors, roofs, and walls.

10.3.6 Bearing Capacity

The maximum load that a soil can support without failing. Bearing capacity is the maximum load that a soil can support without failure. It depends on the type of soil, its density, and its moisture content. Engineers use various methods to measure the bearing capacity of a site, such as the Standard Penetration Test (SPT) or the Cone Penetration Test (CPT). The bearing capacity of a site is a critical factor in the design of foundations and structures.

10.3.7 Bending Moment

Bending moment is the force that causes a beam to bend. It is calculated by multiplying the load on the beam by its distance from the support. Bending moment can cause stress and deflection in the beam, which must be accounted for in its design. Engineers use various methods, such as the moment distribution method or the slope-deflection method, to analyze the bending moment in a beam.

10.3.8 Bitumen

Bitumen is a viscous liquid material derived from crude oil that serves as the primary binding medium in asphalt pavements due to its superior viscoelastic and water-proofing properties. It consists mainly of carbon and hydrogen, with small amounts of heteroatoms like nitrogen, oxygen, sulfur, and traces of metals.

10.3.9 Bridge

A structure that spans a physical obstacle, like a river or valley. A bridge is a spanning structure that creates a passage over an obstacle such as a river, gorge, valley, road, and railway. A bridge is a structure which is built over some physical obstacle such as a body of water, valley, or road, and its purpose is to provide a means of crossing over that obstacle. It is built to be strong enough to safely support its own weight as well as the weight of anything that should pass over it.

10.3.10 Building Code

Regulations governing building design, construction, and safety. A set of ordinances or regulations and associated standards intended to regulate the aspects of the design, construction, materials, alteration, and occupancy of structures which are necessary to ensure human safety and welfare, including resistance to collapse and damage.

10.4 C

10.4.1 CAD

Acronym for Computer-Aided Design.

10.4.2 Closed-Form Solution

The solution to the equations that describe changes in a system can be expressed as a mathematical analytic function.

10.4.3 Caisson

A caisson is a watertight structure used to work on the foundation of a bridge or other structure. It is typically made of concrete or steel and sunk into the ground to the desired depth.

Once in place, workers can enter the caisson and excavate the soil beneath it to create a foundation. Caissons can be open or closed, depending on the conditions at the site.

10.4.4 Creep

When a material is subjected to constant stress at high temperature for a long period of time, it will undergo a slow and permanent deformation called *creep*. This property plays a major role in the design of internal combustion engines, boilers, and turbines.

10.4.5 Colloid

A mixture where small particles (solid, liquid, or gas) are dispersed in another substance, but not dissolved. Not an alloy.

10.4.6 Camber

The slight convex curvature of a road surface for drainage. Camber is the curvature built into a beam or other structural element to counteract deflection due to load. It is measured as the difference in height between the center of the beam and its supports. Camber can improve the appearance and durability of a structure by reducing sagging and cracking.

10.4.7 Cement

A binding agent used in concrete. A binding material used in construction that, when mixed with water, sets and hardens, forming a strong mass. It's a crucial ingredient in concrete and mortar, used to bind sand, gravel, and other materials together.

10.4.8 Column

Vertical structural elements that act as supports and primarily support axial compressive loads. They are slender members designed as a support to hold the ceiling and roof and the weight acting on them.

10.4.9 Concrete

Concrete is a composite material made of cement, water, and aggregates, such as sand and gravel. It is one of the most widely used construction materials in the world and is valued for its strength, durability, and versatility. Concrete can be molded into different shapes and sizes, and can be reinforced with steel or other materials to resist different types of loads.

10.4.10 Construction Management

The process of planning, coordinating, and controlling construction projects. Construction management involves planning, budgeting, coordinating, and supervising construction projects from start to finish. As a construction manager, you may work on various construction projects, including buildings, roads, bridges, and other structures.

10.4.11 Contract

A contract is a legally binding agreement between two or more parties that creates mutual obligations and is enforceable by law. It involves an exchange of something of value (consideration) and requires mutual assent (offer and acceptance), legal capacity, and a legal purpose.

10.4.12 Culvert

A culvert is a structure designed to allow the passage of water, such as a stream or drainage, under a road, railway, or similar obstruction. A structure that allows water to flow under a road or railroad.

10.4.13 Cantilever

A structural element that extends horizontally and is supported at one end. A rigid structural element, like a beam or slab, that is supported only at one end and extends horizontally, often over a void. This allows for overhanging structures without additional support beneath the free end.

10.4.14 Cofferdam

A cofferdam is a temporary enclosure built around a construction site, typically in water or wet environments, to allow for dry excavation and construction.

10.4.15 Compaction

The process of densifying soil or aggregate using rollers or other equipment. It is the mechanical process of increasing soil density and reducing voids by applying force or vibration. This process is used to improve the soil's strength, stiffness, and load-bearing capacity, making it more suitable for construction purposes. Compaction reduces settlement under loads and is crucial for projects like roads, embankments, and foundations.

10.4.16 Compression

Compression is a type of stress that occurs when a material is squeezed or pressed together. On the contrary to tension, it occurs when a material is stretched. Compression can cause a material to deform or fail, depending on its strength and stiffness. Engineers must account for compression in the design of structures and materials, especially in load-bearing elements.

10.4.17 Concrete Mix Design

The process of selecting the proportions of cement, water, and aggregate to achieve the desired properties. It involves determining the optimal proportions of cement, water, and aggregates (coarse and fine) to achieve desired concrete properties like strength, durability, and workability.

10.4.18 Cracking

The formation of cracks in a material due to stress or shrinkage.

10.4.19 Cantilever

A cantilever is a structural element that extends horizontally from a supporting point and is unsupported at its other end. It is commonly used in bridge construction, where one end of the cantilever is attached to the abutment and the other end extends over the gap to support the deck. Cantilevers can also be used in buildings to support balconies or overhangs.

10.4.20 Compression Member

A compression member is a structural element that primarily resists compressive loads. It is typically a column or post that supports the weight of a building or other structure. Compression members can be made of wood, steel, or reinforced concrete and must be designed to resist buckling and other forms of failure.

10.4.21 Consolidation

Consolidation is the process by which a soil becomes more compact and denser due to the weight of overlying materials. It can occur naturally over time or can be accelerated through the use of dewatering or compaction techniques. Consolidation can affect the stability and bearing capacity of a soil and must be taken into account in the design of foundations and structures.

10.4.22 Construction Survey

Construction survey is the process of determining the position, elevation, and alignment of structures and elements during the construction process. It typically involves the use of surveying instruments, such as total stations and GPS, to measure and record data. A construction survey is critical to ensuring that structures are built according to the design plans and specifications.

10.4.23 Curb

A curb is a raised edge or barrier along the edge of a road or sidewalk. It is typically made of concrete or stone and helps prevent vehicles and pedestrians from crossing over into adjacent areas. Curbs can also be used to control drainage and provide a decorative element to the streetscape.

10.5 D

10.5.1 DCP Test

The Dynamic Cone Penetration Test provides a measure of a material's *in situ* resistance to penetration. The test is performed by driving a metal cone into the ground by repeated striking it with an 8 kg weight dropped from a distance of 575 mm. The penetration of the cone is measured after each blow and is recorded to provide a continuous measure of shearing resistance up to 1.5 m below the ground surface. Test results can be correlated to California Bearing Ratios, *in situ* density, resilient modulus, and bearing capacity.

10.5.2 Degrees of Freedom

In structural analysis, the term 'degrees of freedom' refers to the following:

1. The six possible movements that can occur at a point: three linear (horizontal, vertical, and rotational) and three rotational (around x, y, and z axes).
2. The minimum number of independent variables required to define the position of a rigid body in space.
3. The number of independent coordinates needed to describe the position of a body in space.

10.5.3 Detailing a Drawing

Means to add precise and thorough information to a drawing, focusing on the specific details of a design, including dimensions, materials, connections, and assembly methods, essentially 'zooming in' on a specific part of a larger design to provide a clear and complete picture for construction or manufacturing purposes; it's like providing a magnified view of a component to ensure accurate execution.

10.5.4 Ductility

The ductility property of a material enables it to be drawn into the wire with the application of a tensile force. A ductile material must be strong and plastic. The ductility is usually measured by the terms, percentage elongation, and percentage reduction in area. The ductile materials commonly used in engineering practice are mild steel, copper, aluminum, nickel, zinc, tin, and lead.

10.5.5 Dam

A dam is a barrier constructed to hold back water and raise its level, creating a reservoir, typically across a river or stream to control water flow. Dams are built for various purposes, including storing water for irrigation, hydroelectric power generation, flood control, and municipal water supply.

10.5.6 Dead Load

Dead load is the weight of a structure and its permanent components, such as walls, floors, and roofs. It is a constant load that is always present and must be accounted for in the design of structures. Dead load can be calculated based on the material properties and dimensions of the components.

10.5.7 Density

The mass per unit volume of a substance.

10.5.8 Drainage System

A network of artificial structures that collect, convey, and dispose of surplus ground or surface water. In other words, it is a system of pipes, channels, and structures used to manage and remove surface water and groundwater from a site or structure. Proper drainage is critical to prevent flooding, erosion, and other forms of damage. Engineers must consider the topography, soil conditions, and climate of a site when designing drainage systems.

10.5.9 Drawdown

Drawdown is a process whereby the water level on a lake or pond controlled by a dam or other structure is lowered and held at a reduced level for some period of time, typically for several months to 2 years depending on the goals of the project. It is generally the drop in water level due to pumping or other factors.

10.5.10 Deflection

Deflection is the deformation or displacement of a structural element under load. It is a critical factor in the design of structures as excessive deflection can cause damage or failure of the structure. Engineers use various methods, such as finite-element analysis or the virtual work method, to predict and analyze deflection in structures.

10.5.11 Dewatering

Dewatering is the process of removing groundwater or surface water from a construction site or foundation. It is typically done using pumps or other drainage systems to lower the water table and allow for excavation or construction. Dewatering can be a critical aspect of construction in wet or flood-prone areas.

10.5.12 Dowel

A dowel is a cylindrical rod or pin used to reinforce a joint between two structural elements. It is typically made of steel or wood and can be used in concrete or masonry construction. Dowels can help resist lateral loads and improve the stiffness and strength of a structure.

10.6 E

10.6.1 Energy

Energy is the ability to do work. It comes in various forms, such as kinetic energy (the energy of motion), potential energy (stored energy), and thermal energy (the energy of heat).

10.6.2 Elasticity

The property of a material to regain its original shape after deformation, when the external forces are removed. This property has desirable for materials used in tools and machines.

It may be noted that steel is more elastic than rubber.

10.6.3 Earthquake

An earthquake is a sudden shaking or movement of the Earth's crust caused by the release of energy from tectonic plates. Earthquakes can cause significant damage to structures and infrastructure, especially in areas prone to seismic activity. Engineers must design structures and foundations to resist the forces generated by earthquakes and to minimize damage and loss of life.

10.6.4 Elasticity

Elasticity is the ability of a material to deform under stress and return to its original shape when the stress is removed. It is a critical property in the design of structures and materials as it affects their strength, stiffness, and durability. Engineers must consider the elasticity of materials when designing load-bearing elements and structures.

10.6.5 Elevation

Elevation is the height of a point or object above a reference point, such as sea level. It is an important factor in surveying, construction, and planning as it affects the visibility, drainage, and accessibility of a site. Engineers must consider the elevation of a site when designing foundations, roads, and other structures.

10.6.6 Embankment

An embankment is a raised structure or mound of earth, rock, or other materials used to support a road, railway, or other structure. It can be built using cut and fill techniques, where soil is excavated from one area and placed in another, or using retaining walls and other structures. Embankments must be designed to resist the forces of gravity, erosion, and settlement.

10.6.7 Erosion

Erosion is the process by which soil, rock, or other materials are removed from a site by wind, water, or other natural forces. It can be accelerated by human activity, such as deforestation or construction. Erosion can cause damage to structures, reduce soil fertility, and increase sedimentation in waterways.

10.6.8 Excavation

Excavation is the process of removing soil, rock, or other materials from a site to create a foundation or to prepare for construction. It can be done using manual labor, heavy machinery, or a combination of both. Excavation can be a hazardous activity and must be done carefully to avoid damage to underground utilities, structures, or adjacent properties.

10.6.9 Expansion Joint

An expansion joint is a gap or joint in a structure that allows for expansion and contraction due to temperature changes or other factors. It can be found in concrete, steel, and other materials and is designed to prevent cracking, buckling, or other forms of damage. Expansion joints must be carefully designed and located to allow for movement without compromising the integrity of the structure.

10.6.10 Emulsion

A mixture of two or more liquids that don't normally mix, like oil and water. Not an alloy.

10.6.11 Earthquake Engineering

It is a specialized field of structural engineering focused on designing and constructing buildings and infrastructure that can withstand earthquakes.

10.6.12 Embankment

A raised bank of earth/soil or rock or other materials used to support roads, railways, or other infrastructure, often to elevate them above the surrounding terrain, particularly in low-lying or flood-prone areas.

10.6.13 Excavation

Excavation in construction refers to the process of digging and removing earth, rock, or other materials to create a cavity or depression, often to prepare a site for building or infrastructure. It is the process of digging or removing earth or rock.

10.6.14 Earthwork

Earthworks are engineering works carried out in and with granular soils, that is, the movement of earth by means of excavation and filling. It is the process of moving or excavating soil or rock. Earthworks are one of the cornerstones of civil engineering and form the basis for other fields of construction.

10.6.15 Embankment Dam

An embankment dam, also known as an earthfill or rockfill dam, is a large artificial dam constructed from compacted natural materials like soil, rock, and other locally available materials. These dams are characterized by their earthen or rockfill structure and are designed to impound water, creating reservoirs for various purposes such as irrigation, power generation, and flood control.

10.6.16 Excavator

Also known as a digger, it is a piece of heavy construction equipment designed for digging, excavating soil or rock, and other material handling tasks.

10.7 F

10.7.1 Factor of Safety

Factor of safety is the ratio of the maximum load a structure can support to the actual load applied. It is a measure of the safety margin or redundancy built into a structure and is typically expressed as a number greater than 1. A higher factor of safety indicates a more robust and reliable structure.

10.7.2 Force

This is either a pull or a push. According to Sir Isaac Newton's second law of motion:

$$\text{Force} = \text{mass}\,(m) \times \text{acceleration}\,(a)$$

10.7.3 Fatigue

Fatigue is the progressive weakening of a material or structure due to repeated loading and unloading cycles. It can occur in metal structures, such as bridges and aircraft, and can lead to cracks, fractures, and failure. Engineers must consider fatigue in the design of structures and use techniques such as fatigue testing and analysis to ensure their safety and reliability.

10.7.4 Flexure

Flexure is the ability of a material or structure to bend or undergo deformation under load. It is a critical property in the design of beams, columns, and other load-bearing elements. Engineers must consider the flexure of materials when designing structures to ensure their safety and reliability.

10.7.5 Footing

A footing is a structural element that supports the weight of a building or other structure and transfers it to the ground below. It is typically a shallow, spread foundation made of concrete or masonry. Footings must be designed to resist vertical and horizontal loads, and to prevent settlement and displacement of the structure.

10.7.6 Foundation

A foundation is the part of a structure that supports the weight of the building or other structure and transfers it to the ground below. It typically consists of a footing, a column, and a beam or slab. Foundations can be shallow or deep and can be made of concrete, steel, or other materials. The design of a foundation depends on the soil conditions, the load requirements, and other factors.

10.7.7 FIDIC

FIDIC stands for Fédération Internationale des Ingénieurs-Conseils, which is French for 'The International Federation of Consulting Engineers.' FIDIC is an international organization that represents the consulting engineering industry.

10.8 G

10.8.1 Gabions

Gabions are rectangular, interconnected, stone-filled wire baskets formed from welded or woven mesh, used to form an aesthetic retaining wall.

10.8.2 Geosynthetics

Geosynthetics are polymeric (plastic) manmade products used to solve geotechnical problems in construction projects [1]. They may be planar, strips or three-dimensional structures. The different types of geosynthetics are uniquely designed to deliver varying functions such as stabilization, reinforcement, separation, filtration, drainage, erosion control, or containment of liquids. Some products may be designed to deliver a combination of functions. The purpose of geosynthetics can be derived easily from the word itself. 'Geo' means earth, implying ground or land, and 'synthetic' means manmade, implying an artificial substance. There are nine different types of geosynthetics, each differing in the way that they're designed and created, as well as what they're used for. The categories of geosynthetics are as follows:

- Geogrids
- Geotextiles
- Geocells
- Geonets
- Geopipes
- Geofoam
- Geocomposites
- Geomembranes
- Geosynthetic clay liners

10.8.3 Granular Materials

A granular material is a multiphase material made up of a large collection of closely packed solid particles surrounded by a gas or a liquid. These include sand, gravel, or crushed stone used in construction. Because the ratio of the volume of solid to fluid phases is very high, the particles are in very close contact with each other.

10.8.4 Grading

Grading is the process of shaping and leveling the surface of a site to prepare it for construction. It can involve cutting and filling soil, creating drainage systems, and removing obstacles. Grading is critical to ensure the stability, safety, and functionality of a construction site.

10.8.5 Geotechnical Engineering

Geotechnical engineering is the branch of civil engineering that deals with the behavior of soil and rock and their interactions with structures and foundations. It includes the study of soil mechanics, foundation design, slope stability, and other geotechnical issues. Geotechnical engineers use various methods, such as soil testing and analysis, to design safe and reliable structures.

10.8.6 Girder

A girder is a horizontal structural element that supports the weight of a bridge or other structure. It is typically made of steel or concrete and can be designed to resist different types of loads, such as bending or shear. Girders are often used in bridge construction to support the deck and transfer the load to the abutments or piers.

10.8.7 Grade

Grade is the slope or incline of a road, railway, or other surface, expressed as a percentage or ratio. It affects the speed and safety of vehicles and the drainage and erosion of the surface. Engineers must consider the grade of a site when designing roads, railways, and other infrastructure.

10.8.8 Grade Beam

A grade beam is a reinforced concrete beam that spans between two foundation elements, such as piles or footings. It is typically used to support a load-bearing wall or other structural element. Grade beams must be designed to resist bending and shear and to transfer the load to the foundation elements.

10.8.9 Groundwater

Groundwater is the water that exists beneath the Earth's surface in saturated soil or rock. It can be a critical factor in the design of foundations and other structures as it can affect the stability and bearing capacity of the soil. Engineers must consider the groundwater conditions of a site when designing construction projects.

10.8.10 Grout

Grout is a fluid mixture of cement, water, and sand or other fine aggregates used to fill gaps or voids in concrete or masonry structures. It can be used to reinforce joints, repair cracks, and improve the durability of a structure. Grout must be carefully formulated and applied to ensure its strength and performance.

10.9 H

10.9.1 Hardness

It is a very important property of metal and has a wide variety of meanings. It embraces many different properties such as resistance to wear, scratching, deformation, and machinability. Hardness, the ability of materials to resist being permanent deformed when a load applied (i.e., bent, broken, or shape change). The greater the hardness, the greater the resistance to deformation. It also means the ability of a metal to cut another metal. The hardness usually expressed in numbers which are dependent on the method of making the test.

10.9.2 Health

The absence of disease or ill health. For example, asbestos creates a health risk because if you inhale asbestos dust, you may contract lung cancer (a disease) at some stage later in life (perhaps 10 or 20 years after you inhaled the dust). Health relates not only to physical ill health but also to psychological ill health (e.g., exposure to extreme stress can lead to a nervous breakdown).

10.9.3 Hydrostatic Pressure

The pressure exerted by a fluid at equilibrium at any point of time due to the force of gravity. Hydrostatic pressure is proportional to the depth measured from the surface as the weight of the fluid increases when a downward force is applied.

10.9.4 Hydrology

Hydrology is the study of the movement, quality, and distribution of water on the Earth's surface and in the soil and rock beneath it. It includes the study of precipitation, evaporation, runoff, and other aspects of the water cycle. Hydrology is critical to the design of drainage systems, water supply systems, and other infrastructure.

10.10 I

10.10.1 ICE

Acronym for the Institution of Civil Engineers. The ICE is a professional organization that supports civil engineers in the United Kingdom and around the world. The ICE's work includes promoting the civil engineering profession, providing qualifications, advising on infrastructure development, sharing knowledge, promoting ethical standards, and liaising with industry, academia, and government.

10.10.2 I-beam

An I-beam is a steel or aluminum beam shaped like the letter 'I.' It is commonly used in construction and engineering to support heavy loads over long spans. I-beams can be designed to resist different types of loads, such as bending, shear, or torsion. See Figure 10.1 for I-beam nomenclature.

10.10.3 Infill

Infill is the material used to fill the space between two structural elements, such as a wall or a frame. It can be made of various materials, such as concrete, masonry, or timber, and can provide insulation, soundproofing, or fire resistance. Infill

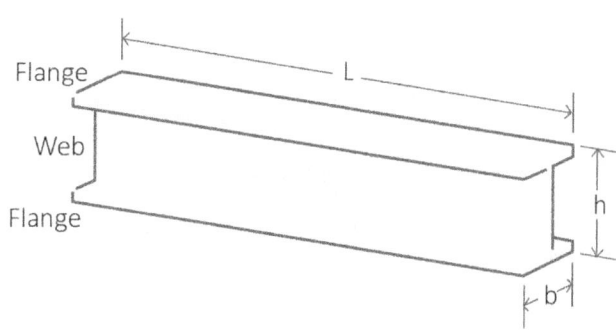

FIGURE 10.1 I-beam nomenclature.

must be carefully designed and installed to ensure its structural integrity.

10.11 J

10.11.1 JSCE

Japanese Society of Civil Engineers.

10.11.2 Joist

A joist is a horizontal structural element that supports the weight of a floor or ceiling. It is typically made of wood, steel, or concrete and can be designed to resist different types of loads, such as bending or shear. Joists are often spaced at regular intervals to support decking or other materials.

10.12 K

10.12.1 K—Kerb

A kerb (or curb) is a stone or concrete edging that forms the boundary between a road or sidewalk and an adjacent area, such as a garden or a parking lot.

10.13 L

10.13.1 Landfill

It is a well-engineered and managed facility for the disposal of solid waste by burial.

10.13.2 Landslide

The movement of rock or soil down a slope.

10.13.3 Lateral Load

A lateral load is a horizontal load applied to a structure or element, such as wind or seismic forces. It can cause bending, shear, or torsion in the structure and must be accounted for in the design of load-bearing elements. Engineers use various methods, such as wind tunnel testing or seismic analysis, to predict and analyze lateral loads.

10.13.4 Lateral Force

A lateral force is a force acting horizontally, or sideways, on an object or structure. It's a force that pushes or pulls in a direction perpendicular to the object's length or the direction of its primary movement.

10.13.5 Live Load

A live load is a variable load applied to a structure or element, such as the weight of people, vehicles, or equipment. It can cause bending, shear, or deflection in the structure and must be accounted for in the design of load-bearing elements. Engineers use various standards and codes to determine the appropriate live load for a given structure or application.

10.13.6 Load-Bearing Wall

A load-bearing wall supports the weight of a building or other structure. It is typically made of concrete, masonry, or other materials and can be designed to resist different types of loads, such as bending or shear. Load-bearing walls must be carefully designed and located to ensure the stability and safety of a structure.

10.13.7 Load Factor

Load factor is the ratio of the maximum load a structure can support to the design load. It is a measure of the safety margin or redundancy built into a structure and is typically expressed as a number greater than 1. A higher load factor indicates a more robust and reliable structure.

10.14 M

10.14.1 Mass of Material

The amount of matter in an object. It is a measure of the material's quantity and is determined by the number and type of atoms in the object.

10.14.2 Moment of Inertia

The moment of inertia of an object is a measure of its tendency to maintain its rotational motion or its resistance to changes in its rotation. It depends on mass distribution and distance from an axis: How mass is distributed within the object. Distance from the axis: How far mass elements are from the axis of rotation.

10.14.3 Monolithic Connections

A monolithic connection is a type of structural connection where two or more elements are joined together to form a single, integral unit, without any visible gaps or interfaces. The connection is rigid and continuous, with no relative movement between the connected elements.

10.14.4 Malleability

It's a special case of ductility which permits materials to rolled or hammered into thin sheets, making wire. A malleable material should be plastic, but it's not essential to be so strong. The malleable materials commonly used in engineering practice are lead, soft steel, wrought iron, copper, and aluminum.

10.14.5 Machinability

It's the property of a material which refers to the relative ease with which a material can be cut. The machinability of a material is measured as follows:

- Comparing the tool life for cutting different materials
- Thrust required to remove the material at some given rate
- The energy required to remove a unit volume of the material.

For example, brass can be easily machined than steel, which means the machinability property of brass was higher than that of steel.

10.14.6 Masonry Construction

Building method using brick, stone, or concrete blocks. Masonry construction is a building method that utilizes individual units such as bricks, stones, or concrete blocks, bound together with mortar to create walls, structures, and other elements. It's a durable and time-tested technique known for its strength, stability, and aesthetic appeal.

10.14.7 Micro-Pile

A small-diameter pile used for foundation support.

10.14.8 Moisture Barrier

A material used to prevent moisture from entering a building.

10.14.9 Masonry

Masonry is the construction of structures using individual units, such as bricks, concrete blocks, or stones, held together by mortar. It can provide strength, durability, and fire resistance to a structure. Masonry must be carefully designed and installed to ensure its structural integrity and performance.

10.14.10 Modulus of Elasticity

Modulus of elasticity is a measure of the stiffness of a material, defined as the ratio of stress to strain under linear conditions. It is a critical property in the design of structures and materials as it affects their strength, stiffness, and durability. Engineers must consider the modulus of elasticity of materials when designing load-bearing elements and structures.

10.14.11 Moment of Inertia

Moment of inertia is a measure of an object's resistance to rotational motion, defined as the sum of the products of the mass and the square of the distance from the axis of rotation. It is a critical property in the design of beams, columns, and other load-bearing elements. Engineers must consider the moment of inertia of materials when designing structures to ensure their safety and reliability.

10.15 N

10.15.1 NEC

Acronym for New Engineering Contract.

10.15.2 Numerical Models

These models are mathematical representations of physical systems that use numbers and equations to describe real-world scenarios. They are often used to simulate complex systems and are a common tool in engineering, geology, and groundwater management. A numerical model is a combination of a large number of mathematical equations that depends upon computers to find an approximate solution to the underlying physical problem.

10.15.3 Negative Pressure

Negative pressure, also known as suction or partial vacuum, occurs when the pressure in a given area is lower than the surrounding pressure, and where pressure is smaller in one place relative to another place. It doesn't refer to the actual or static pressure being negative since that cannot go lower than 0. You will often hear about negative room pressure. That means the air pressure inside the room is lower than the pressure outside the room and air will flow into the room from outside.

10.16 O

10.16.1 Offset

An offset, also called construction offset, refers to a deviation or a shift in the alignment of a road, pipeline, or other infrastructure from its original or intended path. It is a distance or a difference in position, measured from the original or intended alignment, used to locate a point or feature relative to that alignment.

10.17 P

10.17.1 Power

Power is the rate at which work is done or energy is transferred. It's measured in watts (W). Power is calculated as follows:

$$\text{Power } (P) = \frac{\text{Work } (W)}{\text{Time } (t)}$$

or

$$\text{Power } (P) = \text{Force } (F) \times \text{Velocity} (v)$$

10.17.2 Poisson Ratio, v

Poisson ratio is the ratio of transverse contraction (or expansion) strain to longitudinal extension strain in the direction of stretching force. Tensile deformation is considered positive, and compressive deformation is considered negative. Poisson's ratio is a measure of the Poisson effect, the phenomenon in which a material tends to expand in directions perpendicular to the direction of compression.

10.17.3 PUNDIT

Portable Ultrasonic Nondestructive Digital Indicating Tester is the equipment used to conduct UPV tests.

10.17.4 Plasticity

The plastic property of a material refers to the deformation material under load permanently. This property of the material is necessary for forgings, in stamping images on coins and in ornamental work.

10.17.5 Pavement

A hard, flat surface for roads or walkways. A highway pavement is a structure consisting of superimposed layers of processed materials above the natural soil subgrade, whose primary function is to distribute the applied vehicle loads to the subgrade.

10.17.6 Pile

A deep foundation element used to transfer loads to a deeper soil layer.

10.17.7 Pipe

A hollow tube used for conveying fluids.

10.17.8 Pipeline

A system of pipes used for conveying fluids. Such as the distribution pipeline, transmission pipeline, crude oil pipelines, etc.

10.17.9 Prestressed Concrete

Concrete that has been compressed using steel cables or wires.

10.17.10 Pavement

Pavement is the surface layer of a road, runway, or other structure designed to support vehicular or foot traffic. It can be made of various materials, such as concrete, asphalt, or pavers, and can be designed to resist different types of loads and weather conditions. Pavement must be carefully designed and installed to ensure its durability and performance.

10.17.11 Permeability

Permeability is the measure of the ability of a material or soil to allow water or other fluids to pass through it. It is a critical property in the design of drainage systems, foundations, and other structures. Engineers must consider the permeability of soils and materials when designing construction projects.

10.17.12 Pile Foundation

A pile foundation is a type of deep foundation that uses piles to support a structure. It is typically used in areas with weak or compressible soils, where a shallow foundation would not provide sufficient support. Pile foundations can be designed using various types of piles and techniques, depending on the soil conditions and load requirements.

10.17.13 Plate Load Test

A plate load test is a field test used to determine the bearing capacity of soil. It involves placing a large steel plate on the ground and measuring the deflection or settlement of the plate under a known load. Engineers use plate load tests to evaluate the strength and stiffness of the soil and to design foundations and other structures.

10.17.14 Point Load

A point load is a concentrated load applied to a single point or area on a structure or element, such as the weight of a column or a machine. It can cause bending, shear, or deflection in the structure and must be accounted for in the design of load-bearing elements. Engineers use various methods, such as finite-element analysis, to predict and analyze point loads.

10.17.15 Post-tensioning

Post-tensioning is a method of reinforcing concrete or other materials using high-strength steel strands or bars. It involves tensioning the strands or bars after the concrete has hardened to create compression and increase the strength and stiffness of the structure. Post-tensioning is commonly used in bridges, buildings, and other structures that require high strength and durability.

10.17.16 Purlin

A purlin is a horizontal structural element that supports the weight of a roof or other element. It is typically made of wood, steel, or concrete and can be designed to resist different types of loads, such as bending or shear. Purlins are often spaced at regular intervals to support roofing materials or other elements.

10.18 Q

10.18.1 Quantity Surveyor

A quantity surveyor (QS) is *a construction industry professional* with expert knowledge on construction costs and contracts [2]. A quantity surveyor is the financial linchpin of any construction job. You'll become an expert at precisely costing buildings from conception through completion, while ensuring each project meets all legal and quality requirements.

10.19 R

10.19.1 Reactive Forces

These are also known as reaction forces that arise in response to an active force. They are the forces exerted by an object on the external agent or object that is applying the active force.

10.19.2 River Training

River training refers to the practice of modifying a river's flow and channel configuration by constructing manmade structures like dikes, groins, and revetments, aiming to stabilize the riverbed, prevent erosion of the banks, and control the flow of water, especially during high flood events, to protect populated areas along the riverbank.

10.19.3 Resilience of Material

It is the property of a material to absorb energy and to resist shock and impact loads. Resilience measures the amount of energy absorbed per unit volume within the elastic limit. This property plays a major role in the design of spring materials.

10.19.4 RCC

10.19.4.1 Reinforced Cement Concrete

RCC is a type of concrete that contains reinforcement, typically in the form of steel bars (rebars) or fibers, to improve its strength, durability, and resistance to various types of loads.

10.19.5 Riprap

A layer of large stones or rocks used to prevent erosion. Riprap describes a range of rocky material placed along shorelines, bridge foundations, steep slopes, and other shoreline structures to protect from scour and erosion. Rocks used range from 4 inches (101.6 mm) to over 2 feet (609.6 mm).

10.19.6 Rockfill

Rockfill refers to a type of embankment construction that utilizes a mixture of constituent materials with specific gradation to influence stress–strain characteristics and permeability, aiming to minimize settlement and maintain structural integrity. It is composed of large rocks or boulders. The use of rockfill allows a steeper angle in forming road embankment when compared with earthfill so that it results in a small amount of fill. As such, it helps save the cost of construction. Rockfill, typically, has a higher angle of repose than earthfill.

10.19.7 R-Value

R-value is a measure of the resistance of a material to heat flow, typically used to evaluate the insulation properties of a building or other structure. A higher R-value indicates a more effective insulation material. Engineers must consider the R-value of materials when designing buildings and other structures to ensure their energy efficiency.

10.19.8 Raft Foundation

A raft foundation is a type of shallow foundation that covers the entire area of a building or other structure. It is typically used in areas with weak or compressible soils, where a traditional foundation would not provide sufficient support. Raft foundations can be designed to resist vertical and horizontal loads and to prevent settlement and displacement of the structure.

10.19.9 Rebar

Rebar, or reinforcing steel, is a steel bar or mesh used to reinforce concrete or masonry structures. It can provide strength and durability to a structure and resist tensile and bending stresses. Rebar must be carefully designed and placed to ensure its effectiveness in reinforcing a structure.

10.19.10 Reinforced Concrete

Reinforced concrete is a type of concrete that is reinforced with steel bars or mesh to provide additional strength and durability. It can resist compressive, tensile, and bending stresses and can be used in various applications, such as beams, columns, walls, and foundations. Reinforced concrete must be carefully designed and placed to ensure its structural integrity and performance.

10.19.11 Retaining Wall

A retaining wall is a structure designed to hold back soil or other materials and prevent them from sliding or eroding. It can be made of various materials, such as concrete, masonry, or timber, and can be designed to resist different types of loads and soil conditions. Retaining walls must be carefully designed and located to ensure their stability and safety.

10.19.12 Rigid Frame

A rigid frame is a structural system consisting of rigidly connected beams and columns, typically made of steel or concrete. It can resist bending, shear, and torsion and can provide stability and stiffness to a structure. Rigid frames can be used in various applications, such as buildings, bridges, and industrial structures.

10.20 S

10.20.1 Slenderness Ratio

Slenderness ratio is a geometrical parameter, defined for a compression member (column). It is the ratio of effective length and lateral dimension of the compression member. It is also a measure of the structural vulnerability to the failure of the structure.

Slenderness ratio can also be defined as the ratio of the effective length of the column to the minimum radius of gyration. Generally, we design the columns to resist the axial compression load. Sometimes, a combination of bi-axial/uni-axial moments acting on it.

10.20.2 Statically Determinate Structure

A stable structure that can be analyzed using only the equations of equilibrium to determine all of its unknown reactive forces. The equations of equilibrium state that the sum of forces and moments acting on a structure must be 0.

10.20.3 Statically Indeterminate Structure

A structure where the internal forces and support reactions cannot be determined solely by applying the equations of static equilibrium (force and moment equilibrium). That means there are more unknowns than equations, and additional information, like compatibility equations or material properties, is needed for a complete analysis.

10.20.4 Strength of a Material

The ability of a material to resist the externally applied forces without breaking or yielding. The internal resistance offered by a part to an externally applied force is called stress.

10.20.5 Stiffness

Stiffness refers to the ability of a material to resist deformation under stress. Modulus of elasticity measures the stiffness of the material. *Stiffness* is the ability of a *structure* to withstand bending, twisting, compression, or stretching without losing its original shape.

10.20.6 Suspension

A mixture where solid particles are dispersed in a liquid or gas. Not an alloy.

10.20.7 Safety

The absence of risk of serious personal injury. For example, walking under a load suspended from a crane during a lifting operation is not safe because if the load falls, serious personal injury or death could result. Staying out of the danger area results in safety.

10.20.8 Smart Systems

In civil engineering, **smart systems** encompass the integration of advanced technologies like sensors, data analytics, and AI to enhance infrastructure performance, safety, and sustainability. This includes areas like smart structures, smart buildings, and intelligent transportation systems.

10.20.9 Serviceability Limit State

The Serviceability Limit State (SLS) in structural engineering refers to a condition where a structure or its components no longer perform their intended function, even if they are still structurally safe against collapse. Therefore, during design, the SLS ensures comfort and functionality. The SLS focuses on the structure's performance during normal, everyday conditions. SLS design ensures that

- **Deformations**: Deflections or movements within the structure remain within acceptable limits so that they don't affect the building's appearance or function.
- **Vibrations**: Movements caused by occupants, machinery, or environmental factors don't cause discomfort or alarm to users.
- **Cracking**: Limiting crack widths in concrete or other materials to prevent water ingress and corrosion.
- **Operational efficiency**: Doors, windows, and other components function correctly without sticking or jamming due to structural movements. SLS considerations are crucial for the comfort, usability, and longevity of the structure. They ensure that even if the building is structurally safe (as per ULS), it also meets the occupants' expectations for a comfortable and functional space.

10.20.10 Schedule

A timeline for completing a construction project. A detailed plan that outlines the sequence, timing, and resource allocation for all project activities. It serves as a roadmap for the entire construction process, ensuring efficient execution and on-time delivery

10.20.11 Settlement

The downward movement of a structure's foundation or the ground itself due to applied loads or changes in soil conditions. This downward movement of a structure or soil is due to various factors, such as the weight of the structure, the compressibility of the soil, or changes in moisture content. Settlement can cause structural damage, cracks, and other problems and must be carefully monitored and managed during construction and operation.

- **Differential settlement:** Different parts of the foundation settle at different rates. This can cause cracking in walls, floors, and other structural elements.
- **Uniform settlement:** The entire foundation settles at a relatively even rate. This is less likely to cause significant damage.

10.20.12 Shear Force [3]

A force acting in a direction parallel to a surface or to a planar cross-section of a body. Shear is a force that causes one material layer to slide past another layer along a parallel plane. Shear forces often result in shear strain. **Note:** resistance to such forces in a fluid is linked to its viscosity.

10.20.13 Soil Mechanics

The study of soil behavior under various loads.

10.20.14 Soil Stabilization

The process of improving soil strength and stability.

10.20.15 Steel Reinforcement

The use of steel to add strength to concrete.

10.20.16 Structural Analysis

Structural analysis is the process of analyzing the behavior and performance of structures under various loads and conditions. It includes determining the effects of loads on a structure. It involves using mathematical models, simulations, and experiments to predict and evaluate the strength, stiffness, and stability of structures. Structural analysis is a critical discipline in civil engineering, and engineers must have a deep understanding of it to ensure the safety and reliability of their designs.

10.20.17 Substructure

The substructure refers to the portion of the building that sits below the ground. The substructure distributes the weight of the building to the ground beneath the building. The superstructure is thus built right against the soil. The substructure can make or break a structure.

10.20.18 Superstructure

The part of a structure above the ground surface. The superstructure is the component constructed above ground level. The superstructure may seem superior to the substructure as it is typically much longer than the substructure. However, both parts are critical to the stability of any building type.

10.20.19 Sedimentation

The process of settling or depositing sediment. In water treatment, sedimentation is the process where suspended particles settle out of water or wastewater under the influence of gravity, effectively removing them from the liquid. This process is crucial for clarifying water by removing larger, heavier particles like sand, silt, and debris before further treatment

10.20.20 Shear Wall

A structural element that resists shear forces.

10.20.21 Shoring

Temporary support systems used to hold up excavations or structures.

10.20.22 Site Investigation

The process of gathering information about a site's geology, hydrology, and other conditions.

10.20.23 Slurry Wall

A type of retaining wall constructed using a mixture of cement, water, and soil.

10.20.24 Soil Nailing

Soil nailing is a remedial construction measure used to treat unstable natural soil slopes or unstable manmade (fill) slopes as a construction technique that allows safe over-steepening of new or existing soil slopes. In other words, it is a ground stabilization technique used to reinforce existing slopes or excavations. It involves installing closely spaced, slender steel elements (nails) into the soil mass to enhance its stability. This method is often employed for both temporary and permanent support, particularly in situations where other methods like driven piles might be less suitable.

10.20.25 Soil Stabilization

The process of improving soil strength and stability.

10.20.26 Spillway

A structure that allows excess water to flow out of a dam or reservoir.

10.20.27 Spread Footing

A type of foundation that consists of a shallow footing that spreads the load.

10.20.28 Stability Analysis

The process of evaluating the stability of a structure or slope.

10.20.29 Steel Reinforcement

The use of steel to add strength to concrete.

10.20.30 Stormwater Management

The process of controlling and managing stormwater runoff.

10.20.31 Stress Concentration Point

A stress concentration is a **localized area** in a structure at which the stress is significantly higher than in the surrounding material. Stress concentration occurs at joints, holes, bends, sharp corners, and even within the same material if it changes the cross-sectional area or shape.

10.20.32 Structural Integrity

The ability of a structure to withstand loads and stresses.

10.20.33 Structural System

A system of structural elements that work together to support loads.

10.20.34 Subsurface Drainage

Subsurface drainage refers to the removal of excess water from below the ground surface, typically to lower a high-water table or manage waterlogged soils [4]. This is achieved through various systems, including buried perforated pipes, open drains, or wells, that collect and redirect water away from the root zone of plants or infrastructure.

10.20.35 Subsidence

The sinking or settling of the ground surface.

10.20.36 Superimposed Load

A load that is placed on top of an existing load.

10.20.37 Survey Monument

A physical marker used to reference a survey point.

10.20.38 Suspended Floor

A floor that is suspended between walls or beams.

10.20.39 Surveying

The process of measuring and mapping the Earth's surface.

10.20.40 Shear Force

Shear force is a force that acts parallel to a surface or element, causing it to slide or deform. It can occur in various structural elements, such as beams, columns, and walls, and must be accounted for in the design of load-bearing elements. Engineers use various methods, such as shear diagrams and equations, to predict and analyze shear forces.

10.20.41 Shear Strength

Shear strength is a measure of a material's ability to resist shear forces, such as those caused by bending or sliding. It is the maximum shear force that a material can withstand. It is a critical property in the design of load-bearing elements and structures as it affects their safety and reliability. Engineers must consider the shear strength of materials when designing structural elements.

10.20.42 Slab

A slab is a flat, horizontal structural element used to support the weight of a floor or roof. It can be made of various materials, such as concrete, steel, or timber, and can be designed to resist different types of loads, such as bending or shear. Slabs are often reinforced with steel bars or mesh to provide additional strength and durability.

- Two-way slab
- One-way slab

10.20.43 Slope Stability

Slope stability is the ability of a slope or embankment to resist sliding or collapsing due to gravity or other forces. It is a critical factor in the design of transportation infrastructure, such as roads, railways, and airports. Engineers must consider the slope stability of a site when designing and constructing infrastructure.

10.20.44 Soil Mechanics

Soil mechanics is the study of the behavior of soils and their interaction with structures and the environment. It involves analyzing the physical, chemical, and mechanical properties of soils and using this knowledge to design and construct structures and infrastructure. Soil mechanics is a critical discipline in civil engineering, and engineers must have a deep understanding of soil behavior and properties to ensure the safety and reliability of their designs.

10.20.45 Surveying

Surveying is the process of measuring and mapping the physical features of a site or area. It involves using various instruments and techniques, such as total stations, GPS, and aerial surveys, to determine the location, elevation, and characteristics of the site. Surveying is a critical discipline in civil engineering, and engineers use survey data to design and construct infrastructure and structures.

10.20.46 Span

A span is the distance between two supports, such as the length of a beam or bridge. It is a critical factor in the design of load-bearing elements and structures as it affects their strength and stiffness. Engineers must consider the span of structural elements when designing them to ensure their safety and reliability.

10.20.47 Span-to-Depth Ratio

The span-to-depth ratio is the ratio of the length of a structural element to its depth or height. It is a critical factor in the design of beams, columns, and other load-bearing elements as it affects their strength and stiffness. Engineers must consider the span-to-depth ratio of structural elements when designing them to ensure their safety and reliability.

10.20.48 Specific Gravity

Specific gravity is the ratio of the density of a material to the density of a reference material, typically water. It is a critical property in the design of structures and materials as it affects their weight, strength, and other properties. Engineers must consider the specific gravity of materials when designing load-bearing elements and structures.

10.20.49 Splice

A splice is a joint or connection between two structural elements, such as two steel beams or two concrete columns. It is typically designed to provide continuity and strength to the structure and must be carefully designed and located to ensure its effectiveness. Engineers use various methods, such as welding, bolting, or grouting, to create splices in structural elements.

10.20.50 Spread Footing

A spread footing is a type of shallow foundation that spreads the weight of a structure over a wider area of soil or rock. It is typically used in areas with strong and stable soils and can be designed to resist vertical and horizontal loads. Spread footings must be carefully designed and located to ensure the stability and safety of a structure.

10.20.51 Strain

Strain is a measure of the deformation or elongation of a material or structure under loading, expressed as the change in length or dimension divided by the original length or dimension. It is a critical property in the design of load-bearing elements and structures as it affects their strength and durability. Engineers must consider the strain of materials and structures when designing them to ensure their safety and reliability.

10.20.52 Stress

Stress is a measure of the internal force or load in a material or structure, expressed as the force or load per unit area. It is a critical property in the design of load-bearing elements and structures as it affects their strength and durability. Engineers must consider the stress of materials and structures when designing them to ensure their safety and reliability.

10.20.53 Structural Engineering

Structural engineering is a branch of civil engineering that deals with the design, analysis, and construction of load-bearing structures and infrastructure. It is a field of engineering that deals with the design and analysis of structures. It involves considering various factors, such as loads, materials, and environmental conditions, to ensure the safety, reliability, and sustainability of structures. Structural engineering is a critical discipline in civil engineering, and structural engineers play a vital role in ensuring the safety and functionality of the built environment.

10.20.54 Subgrade

The subgrade is the natural or prepared surface on which a pavement or other structure is built. It is the layer of soil or material beneath a pavement or foundation. It can be made of various materials, such as soil, rock, or concrete, and must be carefully designed and prepared to ensure the stability and durability of the structure. Engineers must consider the subgrade conditions when designing and constructing pavements and other structures.

10.21 T

10.21.1 Toughness

It's a property of a material to resist fracture due to high impact. The amount of energy that a unit volume of the material has absorbed after being stressed up to the point of fracture. This property is desirable in parts subjected to shock and impact loads. Normally, the toughness of the material decreases when subjected to heat. This property plays a major role in the design of the hammer and press machine.

10.21.2 Tunnel

A tunnel is a manmade underground or undersea passageway, dug through soil, a mountain or under a body of water, earth, or rock, or laid under water, and is usually completely enclosed, except for the portals at each end, typically used for transportation, utilities, or other purposes.

10.21.3 Tensile Strength

The maximum tensile force that a material can withstand.

10.21.4 Tension

A force that stretches or pulls apart a material.

10.21.5 Timber

A type of construction material made from wood.

10.21.6 Topography

The study of the shape and features of the Earth's surface.

10.21.7 Topographical Survey

A topographical survey, also known as a topo survey, is a detailed survey of a landscape that maps the terrain, including natural and manmade features, and their elevations.

10.21.8 Traffic Engineering

The field of engineering that deals with the planning, design, and operation of transportation systems.

10.21.9 Truss

A triangulated structural system used for roofs and bridges.

10.21.10 Tailings

Waste rock or mineral processing waste.

10.21.11 Tensile Strength

The maximum tensile force that a material can withstand.

10.21.12 Tieback

A type of anchor used to stabilize a retaining wall or slope.

10.21.13 Topographic Survey

A survey that maps the shape and features of the Earth's surface.

10.21.14 Torsion

A twisting force that causes rotation. It is a force that causes an object to twist along its longitudinal axis. It occurs when forces act in opposite directions, causing a twisting motion. Torsional forces are significant in shafts, propellers, and other rotating components.

10.21.15 Traffic Count

A count of the number of vehicles passing a point on a road.

10.21.16 Transverse Slope

The slope of a road or surface perpendicular to the direction of travel.

10.21.17 Trenching

The process of digging a narrow excavation or trench.

10.21.18 Tunnel Boring Machine

A machine used to dig tunnels.

10.21.19 Turnkey Project

A turnkey project is a type of project where a single entity is responsible for all aspects of a project, from design and construction to completion and handover, ready for immediate operation. In other words, it is typically a project where the contractor is responsible for designing, building, and delivering the project.

10.21.20 Tension

Tension is a type of stress that occurs in a material or structure when it is subjected to a pulling or stretching force. It can cause elongation, deformation, or failure in the material or structure and must be accounted for in the design of load-bearing elements. Engineers use various methods, such as stress analysis, to predict and analyze tension in structural elements.

10.21.21 Topographic Map

A topographic map is a detailed, accurate representation of the physical features and terrain of a site or area. It can be used to determine the location, elevation, and characteristics of the site and to plan and design infrastructure and structures. Topographic maps are a critical tool in civil engineering, and engineers use them to ensure the accuracy and reliability of their designs.

10.21.22 Torsion

Torsion is a type of stress that occurs in a material or structure when it is subjected to a twisting or rotational force. It can cause deformation, buckling, or failure in the material or structure and must be accounted for in the design of load-bearing elements. Engineers use various methods, such as stress analysis, to predict and analyze torsion in structural elements.

10.21.23 Truss

A truss is a structural system consisting of interconnected triangular elements, typically made of steel or timber. It can resist compressive and tensile stresses and provide stiffness and stability to a structure. Trusses can be used in various applications, such as roofs, bridges, and towers.

10.22 U

10.22.1 UPV Test

An ultrasonic pulse velocity test is an *in situ*, nondestructive test to check the quality of concrete and natural rocks. In this test, the strength and quality of concrete or rock are assessed by measuring the velocity of an ultrasonic pulse passing through a concrete structure or natural rock formation. This test is conducted by passing a pulse of ultrasonic wave through concrete to be tested and measuring the time taken by the pulse to get through the structure. Higher velocities indicate good quality and continuity of the material, while slower velocities may indicate concrete with many cracks or voids.

10.22.2 Underpinning

Underpinning is the process of strengthening the foundation of a structure by extending it deeper or wider. It can be used to repair or prevent foundation settlement, instability, or other problems and must be carefully designed and executed to ensure the safety and stability of the structure. Underpinning is a critical technique in civil engineering, and engineers use it to ensure the durability and safety of buildings and other structures.

10.22.3 Uplift

Uplift is a type of force or load that acts in an upward direction, typically caused by soil or water pressure. It can cause uplift or flotation of a structure or foundation and must be accounted for in the design of load-bearing elements. Engineers use various methods, such as soil analysis and foundation design, to prevent uplift in structures and ensure their stability and safety.

10.22.4 UIPE

Acronym for the Uganda Institution of Professional Engineers.

10.22.5 Ultimate Limit State

Ultimate Limit State (ULS) is about safeguarding the structure against structural failure. The ULS pertains to the maximum load-carrying capacity of a structure before it reaches a point of collapse or failure. ULS design focuses on ensuring that structures can endure extreme conditions without experiencing catastrophic failure. This involves complex mathematical models that account for the following: Material Strength: Assessing the maximum stress materials can withstand before failing. Extreme Load Types: Including rare but severe events like earthquakes, hurricanes, floods, and accidental overloads. Safety Factors: Applying margins of safety to account for uncertainties in material properties and load estimations. Environmental Conditions:

Considering factors like temperature extremes that could affect material performance. By designing for the ULS, engineers make sure that even under the most demanding conditions, the structure will remain intact, thus protecting lives and property.

10.22.6 User-Centered Design (UCD)

User-centered design (UCD) is a design philosophy that prioritizes the needs and preferences of the end-user throughout the design process. Key principles include understanding users and their context, involving them early and often, using an iterative design process, and focusing on usability, efficiency, and user satisfaction.

10.23 V

10.23.1 Void Ratio

The void ratio is the ratio of the volume of voids or empty spaces in a soil or material to the volume of the solid particles. It is a critical property in the design of foundations and soil structures as it affects their strength, stiffness, and deformation behavior. Engineers must consider the void ratio of soils and materials when designing load-bearing elements and structures.

10.24 W

10.24.1 Work

Work is done when a force is applied to an object, causing it to move. The force applied can be a push or a pull. The amount of work done is calculated as follows:

$$\text{Work } (W) = \text{Force } (F) \times \text{distance } (d)$$

10.24.2 Welfare

Access to basic facilities such as toilet facilities, hand wash stations, changing rooms, rest rooms, places where food can be prepared and eaten in relatively hygienic conditions, drinking water and basic first-aid provision.

10.24.3 Wall

A vertical structural element that encloses or divides a space.

10.24.4 Water Supply System

A system that provides water for drinking, sanitation, and other uses.

10.24.5 Water Table

The upper surface of the saturated zone in soil.

10.24.6 Weld

A weld is a joint or connection between two metal parts or elements, typically made by heating the parts and adding a filler material. Welds can provide strength and continuity to a structure and must be carefully designed and executed to ensure their effectiveness. Engineers use various types of welds, such as butt welds, fillet welds, or plug welds, in the design and construction of metal structures.

10.24.7 Wind Load

Wind load is the force or pressure exerted by the wind on a structure or element. It can cause bending, twisting, or collapse in the structure and must be accounted for in the design of load-bearing elements. Engineers use various methods, such as wind tunnel testing and computational fluid dynamics, to predict and analyze wind loads on structures.

10.24.8 Welding

A process that joins two metal parts together using heat and pressure.

10.24.9 Work Breakdown Structure

A hierarchical decomposition of a project into smaller tasks and activities.

10.24.10 Wall Footing

A wall footing, also known as a strip footing, is a continuous strip of concrete that forms part of a shallow foundation. It's designed to spread the weight of a load-bearing wall over a wider area of soil, preventing the wall from settling or shifting.

10.24.11 Water Main

A pipe that distributes water to a community or building. Usually the main distribution line.

10.24.12 Weathering

The process of breaking down rocks or materials due to exposure to the elements.

10.24.13 Weep Hole

A small opening, typically found in masonry walls (like brick walls), retaining walls, and other structures, designed to allow water and moisture to drain out.

10.24.14 Working Stress

The stress that a material is expected to withstand during normal use.

10.25 X

10.25.1 X-bracing

X-bracing, also known as cross bracing, is a type of structural bracing system used in civil engineering to provide lateral support and stability to buildings, bridges, or other structures. It consists of diagonal members that intersect in an 'X' shape to resist external loads and stresses.

10.26 Y

10.26.1 Young's Modulus, *E*

Young's modulus is a measure of the **stiffness of a material—** it describes the relationship between stress and strain.

10.26.2 Yield Strength

The stress at which a material begins to deform plastically. Yield strength is the point at which a material begins to deform plastically under loading, typically expressed as the stress at which a material starts to yield or change shape permanently. It is a critical property in the design of load-bearing elements and structures as it affects their strength and durability. Engineers must consider the yield strength of materials when designing load-bearing elements and structures.

10.27 Z

10.27.1 Zero Lot Line

A property line that coincides with the edge of a building or structure.

10.27.2 Zoning

Zoning is the process of dividing land into different zones or areas for specific uses or purposes, such as residential, commercial, or industrial. Zoning regulations can affect the design and construction of buildings and infrastructure, and engineers must comply with zoning regulations when designing and constructing structures. Zoning regulations can include restrictions on the height, size, and use of buildings; setbacks from property lines; and other requirements. Engineers must carefully consider zoning regulations and obtain the necessary permits and approvals to ensure the compliance and safety of their designs.

10.27.3 Zoning Ordinance

A zoning ordinance is law or regulation that governs land use in a specific area. It **outlines permitted uses for various sections of land**. Thus, some lands may be dedicated to industrial use, while others may be set aside for residential or commercial use only. A zoning ordinance may also regulate building height, restrict sound, or require open space.

10.28 CONCLUSION

Whether you are a seasoned professional or just starting out in your career, understanding these must-know terminologies is essential for designing and building safe, durable, and sustainable infrastructure. This list is of course not exhaustive but provides much information of the mostly used terms across the civil engineering discipline in a clear and concise way that is accessible even by those just starting out to learn about civil engineering.

REFERENCES

1. Tensar International. (n.d.). Introduction to Geosynthetics. Retrieved from: https://www.tensarinternational.com/resources/guides/introduction-to-geosynthetics. Accessed 7/July/2025.
2. RICS. (n.d.). What is a Quantity Surveyor? Retrieved from: https://www.rics.org/surveyor-careers/surveying/what-surveyors-do/what-is-a-quantity-surveyor. Accessed 7/July/2025.
3. Costanzo, F., Plesha, M. E., & Gray, G. L. (2010). *Engineering Mechanics: Statics & Dynamics*. McGraw-Hill, New York.
4. Ssempeera, P. (2023). *Integrated Drainage Systems Planning and Design for Municipal Engineers* (1st ed.). CRC Press. https://doi.org/10.1201/9781003255550

Index

Note: **Bold** page numbers refer to tables; *italic* page numbers refer to figures and page numbers followed by "n" denote footnotes.

aerobics 151, 155, 157, 158
analysis 1, 7, 11, 18, 27, 29–31, 42, 46–47, **54**, 67, 71–72, 74–76, 82, 86, 88, 99–100, 109, 111–12, 118, 121–22, 127, 129, 149, 152, 157–58, 160–61, 163–64, 166–67, 169, 172, 174, 178–79, 183–**84**, 188, 193, 195–96, 200, 205, 212, 215–16, 218–22, 224–25, 227, 229–33, 243–51, 253, 255–57, 262–63, 268, 274–75, 277, 279–80, 287, 289–90, 292, 295–98, 300–02
analytical method 160, 164–66, 178, 258
anatomical systems 132, 136, 148, 151, 160
anatomy 36–38, 40, 42, 128, 130–32, 135, 143, 145, 151, 160, 224
arteries 129–30, 138–39
artificial intelligence 6, 28, 59, 88, 135, 160, 162, 245, 284
apprenticeship 4, 14, 16, 21
attributes 7, 23, 26, 36, 44–46, 51, 99–101, 127, 177
    civil engineer's 23

Bazalgette, J. W. 16–17, 21, 50
beam 26, 69, 70, 89, 92–94, 102, 107, 112, 114–18, 125–26, 151, 155–58, 160, 183–**84**, **194**, 212–13, 215, 226–35, 239, 246, 257, 263, 266–67, 274–75, 277, 284–86, 289–91, 293, 296, 299–300
bending moment diagram 155, 224, 229, 234, 236
BIM 6–7, 59, 86, 248, 253, 269
biomimicry 128–32, 135, 160
blood vessel 129–31, 138–39
British standard 55, 79, 87–88, 270, 280, 284
Brunel, I. K. 16, 21

calculus 68, 71–72, 162, 164–65, 169, 181–82, 190–92, 202, 204–06, 208, 211–13, 215, 219–20, 222–23
circulatory system 129, 132, 138–39
civil engineering surveying 13
climate change 5, 7, 11, 25, 50, 59, 131, 222
code 5–7, 14, 26, 29–32, 38, 43, 51, 55, 57–**58**, 59–61, 63–67, 83, 86–88, 119, 122, 149, 193, 201, 220, 234, 239, 246, 249–50, 252, 254, 257, 261–64, 266, 270–71, 279–81, 285, 292
coefficients 60, 74, 102–07, 127, 166, 171, 179, 188–89, 190–93, 202, 216–17, 222, 250, 279
    role of 60, 127, 179, 188, 279
column 26–27, 70–80, 89, 93–94, 96–97, 112, 114, 118, 124–25, 129–31, 144, 151, 158, 170, 183, 192, **194**, 215, 221, 227, 234, 239, 253, 263, 267, 274–75, 285–86, 289, 293, 295–96, 299–300
communication 3–5, 7, 23, 26–28, 31, 36, 40, 45–46, 82–83, 132, 135, 137, 138, 150, 163–64, 178, 229, 235, 243, 256–57, 270, 273–74
concrete 2, 14, 18, 26, **30**–31, 50, 60–61, 66–67, 77–81, 87–88, 99–**103**, 108–10, 112, 115, 121–26, 143, 145, 158, 160, 171, 190, 193–94, 216, 220, 224–25, 227,

234, 243, 257–58, 261, 263, 266–67, 268–71, 273, 277, 279–81, 283–300, 302–03
constraints 47, 164, 174, 207, 219, 232, 252–55, 258–64, 266, 275–77, 279
construction engineering 13, 54, 88
    management 231
construction management 6, 13, 44, 86, 222, 285
construction techniques 18, 54, 59, 101, 252, 266–67, 268–69
convergent thinking 253, 258–59
cost control 46, 274
creativity 1, 3, 23–24, 26, 28, 37–40, 131, 224, 229, 246, 252–53, 258, 260–62, 264

DAE 262, 264, 274, 279
Dammond, W. H. 19
data 6–7, 10, 13, 29–31, 47, 59, 66–67, 110, 135–37, 138, 163–65, 166, 169–71, 173–75, 178, 188, 192–96, 198–200, 205, 207, 215–17, 219–20, 222, 224, 245–49, 256–57, 260, 264, 266, 268–69, 277, 279, 286, 297, 299
design optimization 67, 158, 164, 193, 235, 263, 281
differential equations 82, 164–65, 167, 169, 176, 180, 183–84, 192, 196–97, 205–06, 215, 219
digestive system 132, 139
distribution 1, 9, 44, 111, 114, 128–29, 139, 160, 188–89, 191–92, 195, 199, 204–05, 215–17, 219, 229, 234, 248, 254, 277, 280, 283, 292, 294, 303
divergent thinking 253, 256, 258, 259
drawings 40–41, **53**, 220, 224, 235–36, 239, 242–44, 246, 248, 251, 258, 263, 271
durability 29, 60–61, 65–66, 70, 89, 91, 95, 99, 101–02, 108, 110–11, 119, 121, 124, 126–27, 136–37, 145, 160, 252, 259, 260, 263, 266, 268, 280, 283–86, 288, 291, 293–96, 299–300, 302, 304

Eaves, E. 18
economic 6, 8, 10, 23, 25, 31, 40, 47, 49, 50–51, 59, 64, 91, 121, 166, 193, 194, 195, 229, 252, 254
eigenvalues 180, 219
eigenvectors 180, 219
empirical method 166, 222
endocrine system 132, 139–40
energy
    conservation of 67, 70, 72, 82, 87, 134
    dissipation 76
    dissipators 72, 77
engineering ethics 31–32
environment 1, 3–7, 9–11, 13–14, 21, 24–25, 27, 29, 31, 46, 49, 50–51, 65, 91, 99–100, *121*, 125, 128–29, 131–33, 136–37, 144–46, 148, 151, **153**, *158*, 160, 220, 231, 245, 249–**50**, 257, 262, *265*–66, 270–71, 273, 280, 283–84, 286, 299, 300
environmental engineering 3–4, 6, 9–10, 21, 29, 222, 231
environmental impact assessment 47, 50
equation formulation 70, 162, 180

equipment 11, 13, 24, 29–31, 46–47, 49, 57, 67, 79–80, 82–83, 138, 140, 145–51, 206, 217, 224, 247, 251, 268–72, 282, 286, 289, 292, 294
estimating 86, 173
ethical code 5, 7, 21, 26, 32, 51, 58, 64, 252, 261
evaluation 33, 47, 49, 91, 100, 150, 152, 164, 167, 174, 191, 195, 221–22, 248, 250, 255–56, 258, 262–63, 266, 279–80

Fayol, H. 43, 45, 46, 274
financial 24, 44–45, 47, *48*, 49, 51, 59, 177, 193, 253, 261, 270, 276, 295
fixed support 107, 192, 225, 227, 235, 246, 251
force 11, 18–19, 23, 33, 41, 50, 53, 57, 58na, 62–64, 66, 68–75, 77, 82, 84–**85**, 87, 89, 91–101, 107–15, 117, 120–21, 124–25, 133–34, 136, 138, 140, 146, 149, 151, 154–57, 163–64, 169, 171, 175–78, 181–83, **184–85**, 189–93, 208, 212, 215, 224–35, 243–44, 246, 257, 275, 277, 279–80, 283–84, 286–89, 291–92, 294–303
formative research 63, 194–96
formula 21, 68, 70–71, 87–88, 95–96, 112, 135, 152–54, 161, 166–71, 173–77, 180–83, 185–93, 196–97, 203–05, 210–11, 219–20, 223, 255
FOS 60–61, 192–93, **194**, 222, 279
free-body diagram 73, 118–19, 224, 229, 251, 279

Gantt, H. 42
geotechnical engineering 6, 10–11, 17, 34, 58nd, 77, 122, 166, 176, 184, 193–94, 204, 215, 222, 224, 231, 244, 247, 290
    software 248
geotechnics 167, 222
GIS 6, 7, 13, 248
    arc 269
    Q 269
gravity 2, **30**, 35–36, 60, 72, 75, 84, 91–92, 95, 97, 129–30, 133, 175–76, **194**, 206–08, 226, 230–31, 254, 277, 288, 291, 298, 299–300
gym 7, 70, 145–52, 154–58, 160, 226–27, 247, 251

health 9–10, 23–25, 31–33, 38, 50, 64, 66–67, 122, 131–32, 137, 140, 145–48, 150, 160–61, 195, 224, 252, 261, 271–272, 291
heart 5, 20, 24, 35, 42, 50, 129, 138, 139, 147, 220
highway engineering 8, 76, 224
hinge 140–42, 227, 246, 262
Hooke, R. 69–70, 113
Hooke's law 67, 69–70, 77, 87, 113, 156, 169, 183, 185
horsepower 63
humanity 1, 5, 21, 23–25, 27–28, 51, 53, 129, 132, 136, 151, 174, 224, 246
hydraulic engineering 40, 52, 86, 171, **183–85**, 222, 224
hydraulics 10, 34, 60, 69, 75–76, 170, 174, 176, 178, 212, 222–23, 244, 275, 279
hydrology 6, 10, 20, 60, 86, 205, **216**, 245, 291, 298

immune system 132, 144
imperial units 84, 184
infiltration system 51, 144, 145
innovation 5, 7, 14, 18, 23–26, 28, 31, 36–40, 42, 49, 52–53, 59–60, 66, 71, 83, 87, 122, 128–29, 130–31, 162–64, 174, 196, 224, 246, 253, 258–59, 262, 267
Institution of Civil Engineers 1, 13, 21–22, 32, 51–52, 58–59, 170, 261, 291
institutions 2, 4–5, 14, 20, 32, 45, **58**
integration 29, **54**, 59, 68, 75, 129, 136, 137, 166, 169, 204–05, 208–10, 213–14, 222, 227, 248–50, 260, 297
integumentary system 132, 140
interpersonal networking 23, 27
intuition 36, 39, 91, 163, 176, 220, 222, 224, 246, 252, 255–57, 261
Isambard, K. B. 16, 21

jigsaw 220, 259

knowledge 1, 5–6, 8, 10–11, 14–15, 17, 21, 23–33, 35–36, 38, 40, 42–46, 51, 53–54, 59–60, 65, 71, 83, 86, 101, 112, 132, 145, 162–63, 173, 182, 191, 222, 245–46, 250–51, 255, 257, 260–61, 264, 268, 274, 283, 291, 295, 299

Leonardo, da Vinci 20, 36–38, 40–42, 51–**53**, 128, 155, 160, 259
lifelong interdisciplinary learning 5
loads 11, 17, 33, 50, 62–63, 68, 70, 76, 81, 89, 92, 95–101, 111–14, 119–21, 124, 130, 140, 143, 145–46, 149, 151, 155, 158, 160, 171, 183, 190, 192–93, 204–05, 213, 215, 217, 224–35, 248, 257, 268, 275, 277, 279–80, 282, 284–86, 288–92, 294–300, 302–03
logarithms 61, 162, 174, 197–98, 200–02, 222
logic 3, 27–28, 51, 169–70, 174, 176, 185, 220–21, 252, 255–56, 257–59, 260, 264
log–log plot 175, 199–200
lymphatic system 132, 144

machine 6, 11, 15, 28, 31, 36, 40–42, 54, 59, 66–67, 71, 78–81, 128, 132, 135–37, 140, 142, 145–47, 151–55, 162, 189, 219, 230, 243, 245, 253, 264, 268, 282, 288, 295, 300–01
machine learning 6, 59, 67, 135–37, 162, 219, 245, 264
management 5–9, 13, 23–26, 31–33, 36, 42–51, 54, **58**, 66–67, 75, 86–88, 130–31, 137–39, 144, 146–48, **183**, 188, 191–92, 205–06, 216–17, 222, 231, 245–50, 252, 256, 259–61, 268–73, 275, 281, 285, 293, 298
manpower 49
Mao Y. 19, 20
mass **53**, 71–72, 77, 82, **84–85**, 89, 94, 115–**16**, 122, 146–47, 155, 169, 171, 176–78, 181–82, 188–89, 210, 225, 230–32, 285, 287, 289, 292–93, 298
materials, science of 99, 101, 111–12, 127
mathematics 3–5, 9–11, 21, 27, 36–38, 53, 60, 85–86, 162–64, 166–69, 180, 187, 195, 200–02, 204–05, 215, 219–20, 222–23, 264
McCorkindale, A. A. 17
membership 5, 14, 19, 51, 147
metric system 62, 64, 72, 83, 84, 170–71, 177, 183, 200, 274
military engineer 1, 61
mind mapping 51, 86, 97, 126, 160, 222, 250

modeling 6, 11, 41, **54**, 59, 86, 97, 162–64, **185**, 188–89, 197, 205–06, **216**, 219–20, 248, 253, 269, 277, 279
Mokshagundam, V. 19, 22
moment 40, 69, **85**, 91, 95, 112, 114–15, 117, 119, 149, 151, 154–58, 169, 176, 178, 182–84, 211–15, 223–25, 227–30, 233–36, 274, 279, 284–85, 292–93, 296
moment of inertia **85**, 112, 114–15, 156, 183, 212–14, 292–93
money 49, 177, 256
moral dilemma 32–34
municipal engineering 11, 13
muscular system 132–33, 140

nature-based solutions 129
nature-inspired art 131
nervous system 131–38
Newton, I. 67–68, 87, 181–82, 205, 228, 289
Newton's laws 63, 68, 70, 77, 87, 169, 181
numerical method 162–67, 176, 198, 205, 219, 258

ordinance 285, 304

Pascal 62, 83–85, 96, 113–14, 133–34, 171
pedagogical illustration 151, 155
philosopher 26–28, 35, 51
philosophy 23, 25, 27–28, 32, 35, 38–39, 45, 51, 64, 128, 195, 258, 302
physiology 38, 128, 131–32, 145
pollution 9, 23, 50
power 1, 5, 7, 9, 14–15, 17, 21, 23, 28, 38, 40, 42–43, 49, 51, 53, 63–65, 69, 71–72, 75, 82–83, 87, 128, 132, 154, 160, 164, 170, 179, 188, 192, 199–200, 202, 206, 208, 211, 219, 246, 259, 269, 273, 275, 277, **282**–83, 287, 289, 293–94
pressure 32, 40–41, 60–62, 69, 75, 77, 83–85, 90–91, 95, 97, 99–101, 109–11, 116, 121, 124, 126–27, 129–30, 133–34, 136, 138, 145–46, 148, 171–72, 174–75, 179, **183–85**, 190, 200, 204–07, 212, 215, 231–32, 254, 261, 277, 279–81, 291, 293, 302–03
pressure nominal (PN) 61
professional 4–8, 13–14, 18, 20–21, 23, 25–34, 39, 51, 54, 58, 64–65, 81–82, 87, 128, 146, 148, 150, 160, 163, 216, 224, 246, 250–51, 253, 255–58, 261, 263, 268, 270, 281, 291, 295, 302, 304
project management 8, 13, 33, 42–45, 51, 54, 66–67, 246–49, 256, 269–70, 275

quality control 13, 29, 42, 46–47, 61, 67, 101, 136, 188, 191, 192, 205, 216–17, 256, 268–69, 271, 273–74

railway systems engineering 8
rebound hammer 66, 69, 77–78, 87, 256
reliability 23, 30–31, 44, 60–61, 65–66, 91, 97, 110, 119, 121–22, 127, 129, 131, 164, 171, 178, 188, 192–93, 205, 216–20, 228, 249, 252–53, 255, 258–60, 263, 266, 270–71, 274, 277, 279–81, 289, 293, 297, 299–301
reliability-based design 217
Renard's preferred numbers 61–62
reproductive system 132, 143–44
respiratory system 132–34, 138, 160
risk 23–25, 29, 33, 35–36, 39, 42–43, 45–47, 50–51, 66–67, 95, 100, 109, 124, 131, 147, 149–50, 158, 178, 193, 195, 205, 215–16, 229, 245, 248, 250, 252, 256, 264, 266, 271–72, 274, 282, 291, 296

roller 110, 137, 151–52, 225–28, 231, 246, 251, 269, 286
Rugumayo, A. I. 20

safety, factor of 60, 112, 149, 192–93, 289
science 1, 3–6, 9–11, 14–15, 19–21, 27–30, 32, 36–38, 40, 42, 52, 53–**54**, 57, 60–62, 67–68, 70, 72, 79, 83–84, 86–87, 98–99, 101–02, 111–13, 122–23, 127–28, 132, 145, 148, 154, 162–64, 166–67, 169, 173, 181–82, 188, 192, 197–98, 200–02, 204, 219, 224, 244, 249, 255, 259, 264, 266, 284
scientific laws 67–68, 77, 87
shear force 93–96, 108–09, 112, 120, 155–57, 224, 229–30, 233–35, 275, 279, 297–99
  diagram 155, 224, 229, 233, 275
simple support 225–26, **228**
skeletal system 132, 140–41, 143, 149
skills 1, 2–8, 14, 20, 25–32, 36–37, 40, 43–46, 48, 51, 63, 83, 148–50, 164, 245, 247, 250, 261, 263
slab 77, 112, 121, 125–26, 158, 227, 239, 258, 267–68, 277, 286, 290, 299
Smeaton, J. 1–2, 14–15
social 3, 5–7, 9, 23–25, 27, 31–32, 42, 45, 47, 50, 52, **54**, 121–22, 131, 143, 147–48, 160, 166, 195–96, 246, 252, 254, 261
software 6–7, 20, 30, 45, 54, 67, 86, 132, 137–38, 172, 176, 186, 195, 224, 231, 236, 239, 243, 245–51, 263–64, 269–70, 275, 280
soil mechanics 10, 17, 95, 184, 193–94, 205, 215, 222–23, 231, 290, 297, 299
standards 6–7, 14, 18, 24–25, 27, 29–32, 34, 36, 44, 46–48, 54–55, 57–67, 80, 83, 86–87, 112, 122, 124, 126, 149–50, 164, 177, 193, 207, 232, 234, 246, 249–250, 252–57, 261–64, 266, 270–71, 273–74, 279–80, 283–85, 291–92
steel 14, 24, 26, 33, 51, 55, 60, 62, 72–73, 87, 89, 100–02, **106**–08, 110–27, 140, 145, 152, 158, 160, 183, 190, 193, 207, 220, 225, 227, 243–44, 257, 261, 263, 266–71, 275, 277, 279–81, 284–88, 290–300, 302
STEM 3–5
stewardship 23, 31, 44, 132, 281
strain 70, 72, 76, 89, 93, 100, 108, 110–11, 113–15, 120, 149, 157, 160, 178, 182, 185, 188, 224, 244, 263, 280, 293, 294, 297, 300, 304
  energy 72, 76, 108
strength 11, 18, **30**, 42, 44, 55, 60, 62, 65–66, 70, 77–79, 81, 88–89, 92, 99–102, 108–12, 114–15, 119–21, 124–27, 130, 135, 140, 145–47, 149, 151, 157–58, 160, 171–72, 187–88, 192–94, 196, 213, 216, 246, 248, 252, 257–58, 264, 268, 279, 284–86, 288, 291, 293–304
stress-strain curve 113–14
structural engineering 6, 11, 41, 70–72, 75–76, 81, 91, 126, 140, 143, 151, 176, 178, 181–83, 192, 224, 229, 231, 289, 297, 300
SUDs 122, 131–32, 144–45
surveying 13, 17, 34, 86, 222, 268–69, 286, 288, 299, 304
sustainability 3, 6–7, 9, 14, 23–27, 29, 31–32, 44–47, 59, 66–67, 86–87, 91, 99, 119, 121–22, 125, 127–29, 131–32, 136–37, 145, 149, 158, 164, 247, 253, 258–61, 263–64, 266, 268–69, 273, 283, 297, 300

sustainable development 13–14, 25, 47, 121–22, 132, 136, 160, 252, 254
    pillars 25, 121
sustainable drainage system 58, 122, 128, 131

technician 4–5, 24–25, 27–31, 45, 65, 271
technologist 25, 27–30
technology 3–4, 6, 8, 13, 18–19, 20, 23, 25–26, 28–30, 32, 39, 42, 49, 53–55, 57, 59–61, 63, 65–68, 70–72, 80, 82–84, 87, 122, 134, 137–39, 154, 160, 163–64, 173–74, 219, 223–24, 229, 246–47, 250, 254, 257, 268–69, 276
Telford, T. 15–16, 21
temperature 72, 84, 85
tertiary education 4
Terzaghi 18, 20, 21, 42, 51
Terzaghi bearing capacity equation 171
Terzaghi, K. 17, 171–72, 184, 193, 223
Terzaghi's equating 171–72
Terzaghi's principle: of effective stress 171

thermal expansion 99–100, 102–**03**, 107–08, 110, 112, 115, 124, 190, 226, 289
thought pattern 257, 260
tools 6–8, 24, 26, 45, 49, 53–54, 59, 66–67, 72, 86, 148–49, 151, 161–64, 174–75, 176, 178, 181, 195, 200, 208, 215, 220, 222, 224, 229, 244–48, 250–51, 255–56, 259, 263, 266, 268–69, 270–71, 280–81, 288
traffic engineering 8, 205, 301
transportation engineering 8, 11, 86, 222, 224, 231
    software, 248
Tredgold, T. 1, 14–15, 21, 63, 132
trolley problem 35–36
troubleshooting 30–31, 150

Uganda Institution of Professional Engineers 20, 302
ultrasonic pulse velocity 66, 79, 256, 302
units 50, 62, 64, 72, 74, 83–85, 89, 135, 143, 145, 153, 162, 166, 171, 176–79, 183–**84**, 187, 189–91, 193, 213, 250, 258, 266, 268, 293

urinary system 132, 143–44

values 5, 27–28, 31–32, 35–36, 61–62, 112, 114–**15**, 160, 166, **167–68**, 171–72, 175–76, 179–80, 186, 190–91, 193–94, 198–99, 200, 205, 208, 215, 219, 234–35, 245, 268
vein 129–30, 138–39
visual aids 247, 251
Vitruvian Man 37–38, 128, 149, 155

water resources engineering 9, 11, 75, 224
Watt, J. 41, 53, 63–64, 72, 83, 154, 293
welfare 23, 25, 32, 65, 150, 285, 303
work–energy theorem (WET) 71
Wright, B. 17, 21

Yoshida, T. 18
Young's modulus 70, 100, 108, 110, 112, 113, 114, 115, 116, 182, 183, **185**, 304

zone 41, 134, 226, 298, 303–04

For Product Safety Concerns and Information please contact our EU
representative GPSR@taylorandfrancis.com
Taylor & Francis Verlag GmbH, Kaufingerstraße 24, 80331 München, Germany